LANCASTER PRESS, INC., LANCASTER, PA.

PREFACE

This report upon the fossil mammals of the San Juan basin of New Mexico is based in part upon material in the Cope Collection of the American Museum of Natural History and in part upon the collections made in that region by several American Museum expeditions, culminating in that of 1913, which added to the available material many fine skulls, jaws, and even a few more or less complete skeletons representing various archaic and extinct families of placental mammals. The fauna belongs to the two earlier phases of the Paleocene epoch, a transitional period following that of the great reptiles of the Mesozoic era and preceding the Lower Eocene phase, when the direct ancestors of the horses, tapirs, rhinoceroses, ruminants and other progressive placental families first became established in western North America.

The author of this report was at the time of his death in 1930 the leading authority on American Paleocene and Eocene faunas and his contributions to this subject were summed up in his 1927 address before the Zoological Society of London.

The greater part of the present work was written in 1916 and 1917 but the author later revised and extended it, especially in 1928, 1929 and 1930; in fact, his last scientific labors virtually completed the text of this monograph. Active students of the subject may well have been somewhat impatient over the long delay in publishing this important report but a great deal of work had to be done in filling in references, in preparing many additional illustrations called for in the text, and in editing, while the "depression" greatly added to the congestion of other works awaiting publication at the American Museum of Natural History.

However, all minor difficulties have at last been overcome and, thanks in large part to the generous action of the American Philosophical Society, publication of this work is now assured. The text is presented as written by the author and without material emendation, except the addition of references.

Special acknowledgments are due to the artists who prepared the illustrations: the late Mrs. L. M. Sterling, Mrs. E. M. Fulda, Mr. and Mrs. John Germann, Mr. Sydney Prentice, Mrs. Mildred Clemens and Mrs. Margaret Colbert.

Dr. George Gaylord Simpson has very kindly aided the editors by critically reading the manuscript and by suggesting changes that have helped to unify this work. The arduous task of preparing the manuscript for publication and of editing the galley and page proofs has been most ably carried out by Miss Florence Milligan, under the supervision of one of the editors (E. H. C.).

WALTER GRANGER,
WILLIAM K. GREGORY,
EDWIN H. COLBERT.

December 23, 1935.

Published by permission of the
Trustees of the American Museum
of Natural History

PALEOCENE FAUNAS OF THE SAN JUAN BASIN, NEW MEXICO

By William Diller Matthew

CONTENTS

I. INTRODUCTION

1. HISTORICAL REVIEW OF DISCOVERIES AND RESEARCHES

DISCOVERY OF THE PUERCO FAUNA

This fauna was first discovered in 1880 by David Baldwin, a collector employed by Professor E. D. Cope. Brief notices of the first fossils found were published by Cope in the American Naturalist for April, August and October, 1881, and in the Proceedings of the American Philosophical Society for 1881, Vol. XIX, pp. 484–91, printed October 21, 1881, he described thirteen species and tentatively referred the fauna to the Puerco formation described by him in 1874 as underlying the Wasatch in New Mexico on the southeast margin of the San Juan basin.

From 1880 to 1888 Baldwin kept up an intermittent search for fossils in these beds, and the new material was described by Cope in various short papers and notices, chiefly in the Proceedings of the American Philosophical Society and the American Naturalist. In his great volume in the Hayden Survey reports commonly known as "Tertiary Vertebrata" [1] Cope included descriptions and illustrations of fifty-two species. In 1888 he published in the Transactions of the American Philosophical Society a "Synopsis of the Vertebrate Fauna of the Puerco Series," in which he listed one hundred and six species of vertebrata, mostly mammals. This was Cope's last systematic contribution to this fauna.

In 1892 Professor Osborn sent an expedition for the American Museum to the San Juan basin, in charge of Dr. J. L. Wortman. The collections secured were described by Osborn and Earle in 1895. A few new species, and additional and better preserved remains of others, were described. In the same year Professor Cope's collections of North American fossil mammals were purchased for the American Museum, and the cataloguing of the collection and the revising of the Puerco fauna were intrusted to the present author, the revision being published in 1897. In 1896, Wortman again explored parts of the San Juan basin and secured an important collection, and Matthew's revision included this material. The number of species was considerably reduced, but seventy-five species of Mammalia were admitted. The total number of specimens catalogued, however, was nearly fifteen hundred. Wortman in 1896 and 1897 published two papers revising and redescribing the Taeniodonta, or Ganodonta, which he maintained were Edentates and to which he referred four genera from the Puerco and Torrejon.

Osborn in 1898 described the skeleton of *Pantolambda* and added some further notes on the skeleton of *Tetraclaenodon*. In 1901 Matthew redescribed two creodont genera from these horizons, *Triisodon* and *Claenodon*. In 1907 Mr. J. H. Gardner of the U. S. Geological Survey made a reconnaissance of part of the Puerco-Torrejon in connection with stratigraphic studies in the San Juan Cretaceous, and collected a few teeth and fragments which

---

[1] Cope, E. D., 1884, The *Vertebrata* of the Tertiary formations of the West. Book I. Report U. S. Geological Survey of the Territories . . . Washington, iii, pp. i–xxxv, 1–1009; pls. i–lxxva. [This work was not actually issued until 1885 and throughout the present volume Dr. Matthew generally uses that date and refers to the work as "Tertiary Vertebrata."—THE EDITORS.]

were identified by Mr. J. W. Gidley.   In 1913 and 1916 the Survey parties in charge of Bauer and Reeside added additional stratigraphic notes on the Paleocene in connection with more detailed field work upon the Cretaceous formations, and collected a number of specimens, principally Chelonia, which were described by Gilmore in 1919.

No further mammal collections of importance were made in the San Juan basin after 1896 until 1913, when an American Museum party in charge of Walter Granger, assisted by W. J. Sinclair and George Olsen, explored the area systematically and secured a large and valuable collection, which forms the basis of the present revision.   The stratigraphic results of this expedition were briefly outlined by Sinclair and Granger in 1914. In 1916 Granger explored a fossiliferous horizon on the northern margin of the basin, intermediate between the Torrejon and Wasatch, which he has named Tiffany beds.   Descriptions of this fauna appeared in the American Museum Bulletin for 1917.

None of the descriptions or revisions of the Paleocene fauna heretofore published has been complete, or adequately illustrated.   No photographs of the numerous finely preserved dentitions of the Puerco and Torrejon mammals have been published.   The material now available is so much more complete than any known previously that the systematic relations of the fauna can be far more certainly determined.   A number of new species will be described in the present revision, which is chiefly devoted to describing and illustrating the dentition, skulls and skeletons of the better-known types.   The original descriptions are cited or summarized for each species.

PRESERVATION OF SPECIMENS

By far the greater number of the specimens from the Puerco and Torrejon consist of parts of jaws and other fragments found on the surface.   A considerable number, however, were found still buried partly or wholly in the shale, and some of these are skulls or skeletons more or less complete.   The weathering breaks up the specimens, but does not often destroy the surface, especially of the teeth, which are usually perfect if not buried in concretionary crusts or broken into fragments.

The specimens in shale are more or less crushed, distorted or broken, owing to the slipping and compression of the material in which they are enclosed during its conversion into shale.   As a rule the surfaces of bone are clean and part readily from the matrix, but not uncommonly they are covered with a hard calcareous, silicious or hematite crust.   The hematite crusts are very difficult to remove, but the teeth or bones still remain beneath them.   The harder stony crusts, on the other hand, are often associated with the partial or entire absorption and disappearance of the substance of the fossil bone or teeth underneath them, so that even if they are chiselled or ground away, little or nothing is left of the bone which one expects to expose.   This is a feature which limits the possibilities of preparation of many specimens which appear very promising.

The majority of jaws and teeth maintain their form and proportions without serious distortion by crushing.   Many specimens, however, especially in the Puerco, are very much flattened by crushing, losing half or more of their thickness.   Usually this crushing, if examined carefully, can be seen to consist in the cracking up of the surface of bone or tooth into larger or smaller fragments, which are then squeezed down upon each other and tipped this way or that.   In such cases it is possible to reconstruct a mental image of the original proportions of the bone or tooth or to compare the fragmented bits in detail with the corresponding parts of uncrushed teeth or bones. Sometimes, however, there is no evidence at all of fragmentation; the bones, even the crowns of the teeth, seem to have become actually plastic and to have reduced their thickness uniformly and regularly without any breaking up of the surface. The results of this process are very deceptive in the comparative study of the teeth of various species. Certain genera and species which we regard as founded upon crushed types have been described by Cope and Scott as distinct, and the most careful study of the specimens fails to show conclusive evidence of crushing. They correspond in structure point for point with other well-known species, except for the reduced transverse diameters of the teeth. This, with the fact that other specimens show intermediate grades of crushing, that no additional specimens are found with the narrowed teeth on clearly uncrushed jaws, and similar details of evidence, convinces us that certain species must be reduced to synonyms.

2. STRATIGRAPHIC RELATIONS OF THE CRETACEOUS AND PALEOCENE FORMATIONS IN THE SAN JUAN BASIN

GENERAL REMARKS

The stratigraphy of the Puerco, Torrejon and Tiffany was described by Sinclair and Granger in 1914 and by Granger in 1917, and their relations with the underlying Cretaceous and overlying Wasatch are discussed in various papers by the staff of the U. S. Geological Survey dealing with stratigraphic problems of the San Juan basin. I include here a very brief summary based upon these researches.

The San Juan basin is a structural basin in northwestern New Mexico lying in part in the valley of the San Juan River and partly across the continental divide in the headwaters of Rio Puerco and other tributaries to the Rio Grande. The Paleocene formations overlie the Upper Cretaceous and are overlain in turn by the Eocene, as follows:

Geologic Age	Formations		Vertebrate Faunas
Lower Eocene........	Wasatch	1000–2000 ft.	*Coryphodon*
Upper Paleocene......	Tiffany beds	?300 ft.	*Plesiadapis*
Unconformity of overlap			
Middle Paleocene.....	Torrejon	100–600 ft.	*Pantolambda*
Lower Paleocene......	Puerco	100–300 ft.	*Taeniolabis*
Erosional unconformity			
Upper Cretaceous ...	Ojo Alamo,[2] *s.s.*	60–400 ft.	
	McDermott	50–300 ft.	*Pentaceratops*
	Kirtland	200–1200 ft.	
	Fruitland	200–500 ft.	

[2] The Ojo Alamo of Brown, 1910, was quite obviously intended to include the greater part, if not all, of the formations subsequently distinguished by Bauer as Kirtland and Fruitland. Brown estimated the thickness of the beds he examined at 200 feet but placed no lower limit on the formation, the upper limit being the unconformity beneath the Puerco. Sinclair and Granger gave some further details on the succession of strata in the Ojo Alamo, but likewise failed to specify any lower limit, doubtless because they regarded the name, as Brown evidently did, as being proposed for the beds referred to the Laramie group in the earlier stratigraphic work. i.e. for the entire series between the marine Cretaceous and the Puerco.

THE CRETACEOUS FORMATIONS

The marine Upper Cretaceous, of Pierre age, underlies this sequence of fresh-water (epicontinental) formations, all of which have yielded very considerable vertebrate faunas and can be correlated with a corresponding degree of accuracy with vertebrate faunas elsewhere.   It will be observed that the stratigraphic breaks do not correspond with the faunal breaks.

The Upper Cretaceous "formations" as distinguished by Bauer in 1916 contain an identical fauna as far as known.   On the other hand the Torrejon and Puerco, with very distinct faunas, as fully shown in a later chapter of this memoir, are indistinguishable by any stratigraphic break or lithologic change, so that Gardner in 1910 united them into a single "Nacimiento group," and Bauer in 1916 and Reeside in 1924 appeared somewhat skeptical of the distinction.[3]

There is an unconformity of erosion and overlap at the base of the Wasatch, which over most of the area is a thick, massive non-fossiliferous sandstone, capped by fossil-bearing clays with the *Coryphodon* fauna of Suessonian age.   North of the San Juan, where it crosses the Colorado line, the base of the Wasatch is fossiliferous and the fauna is the *Plesiadapis* fauna of Thanetian age, while above these Tiffany beds the main body of the Wasatch contains the *Coryphodon* fauna of Suessonian age.   Stratigraphically the Tiffany beds belong to the Wasatch, but palaeontologically their fauna is Upper Paleocene, equivalent to the Cernaysian fauna of France,[4] while the Wasatch in the San Juan basin has two levels corresponding approximately to the Sparnacian and Cuisian phases of the Suessonian, but not so clearly separated as the European faunas.   It is a matter of interest that the relations of the Cernaysian of France to the overlying Suessonian correspond to those of the Tiffany and Wasatch, the Suessonian being "à peine séparée, stratigraphiquement de la faune thanétienne, mais très différente d'elle zoologiquement."

The advanced character of the Ceratopsia at the base of this sequence would indicate that none of it is older than Edmonton.   The character of the dinosaurs in its uppermost member would indicate that none of it is as recent as Lance.   The remaining vertebrates have no value for exact correlation, as they all range through the Upper Cretaceous and Paleocene without progressive change.

The important collections made by Sternberg in the Fruitland beds (*auct.* Mook and Osborn) are not yet prepared or studied except for the fine skull of *Pentaceratops* described by

---

Bauer in 1916 limited Brown's Ojo Alamo to the upper sandstones of his Kirtland formation, 63–110 feet.   Reeside in 1924 gave the name of McDermott shales to beds underlying these sandstones and from which Brown had obtained his typical Ojo Alamo fauna.   The rest of the "Laramie" is now included under the Kirtland and Fruitland formations of Bauer.   In any event, whatever stratigraphic names be applied, the entire series is a palaeontologic unit, as far as the vertebrates are concerned, from the marine Cretaceous to the base of the Puerco, and the endlessly varying character and thickness of the strata do not afford a very reliable basis for formational divisions.   The stratigraphic relations will be further discussed on a later page of this memoir.—W.D.M.

[3] "Owing to the close similarity in lithology such separation of the Puerco and Torrejon as has been made in the past has depended entirely on fossil collections, and although the fossils seem to indicate an unconformity between the two formations the writer has yet found no stratigraphic reason for such division" (Bauer 1916, p. 277).   What the fossils actually indicate is a wide gap in time, "hiatus," diastem or disconformity, *not* an unconformity.   The field work of Granger's party in 1912 and 1913 failed to detect any stratigraphic evidence of this time gap.—W.D.M.

[4] Teilhard, 1922, Annales de Paléontologie, XI, pp. 47–9, 100.

OJO ALAMO GROUP

Revised List of Vertebrate Fauna

	Fruit-land	Kirt-land	McDer-mott	Ojo Alamo s.s.
Dinosauria:				
Sauropoda:				
Alamosaurus sanjuanensis scapula, etc.	. . . .	. . . .	. . . .	× ·
Ceratopsia:				
Pentaceratops skull, part skeleton	×	. . . .	. . . .	. . . .
?Anchiceratops skull	×	. . . .	. . . .	. . . .
Indet., ?above genera, fragments; includes material referred				
to *Monoclonius* and *Ceratops*	×	×	×	×
Hadrosauridae:				
Kritosaurus navajovius skull and jaws	. . . .	?	×	?
Kritosaurus, or related genus, skeleton	×	. . . .		
Kritosaurus, or related genus, fragments	. . . .	×	×	×
Deinodontidae:				
Deinodonts, indet., teeth and fragments	×	×	×	×
Nodosauridae:				
Genus indet. fragments	. . . .	×	×	×
Crocodilia:				
Skeleton, not studied	×	. . . .		
?Crocodilus and *?Brachychampsa* fragments	. . . .	×		×
Chelonia:				
Adocus bossi, A. kirtlandius	. . . .	×		
Adocus vigoratus	. . . .	. . . .		×
Adocus sp. indet.	×	. . . .		
Baëna nodosa	×	×	×	
Baëna sp. indet.	. . . .	×		
Neurankylus baueri	. . . .	×		
Plastomenus robustus	×	×		
Plastomenus sp. indet.	. . . .	×		
Basilemys nobilis	. . . .	. . . .		×
Aspideretes vorax, fontanus, austerus	. . . .	. . . .		×
Aspideretes	×	×		
Thescelus rapiens	. . . .	. . . .		×
Compsemys sp.	. . . .	. . . .		×

Osborn. The occurrence of so progressive a Ceratopsian at the base of the sequence makes it improbable that the fragmentary specimens from the Fruitland, Kirtland and Ojo Alamo, provisionally identified by Gilmore as *Monoclonius* and *Ceratops*, really belong to these more primitive genera; more probably they are congeneric with more complete skulls from the Fruitland which resemble these genera in size and in some proportions of the skull but are in reality more progressive, comparable to *Anchiceratops*. The evidence as far as available at present would tend to correlate the fauna with the Edmonton of Alberta, or perhaps later; not so late probably as the Lance, although further studies may indicate that while not closely related or identical it is not far from equivalent. Knowlton[5] con-

[5] U. S. G. S. Professional Paper 98-S, 1916, p. 330.

cludes from comparison of a considerable flora from these formations that they are of Montana age, which would make them more nearly equivalent to the Judith and Belly River than to the Lance although not excluding their correlation with Edmonton. He expressed a suspicion, however, that a small collection of plant remains from the Ojo Alamo "may ultimately be shown to be of Tertiary age," while the McDermott flora is a mixture of Cretaceous and Tertiary types.

Stanton [6] from the study of a considerable fauna of freshwater invertebrates chiefly from the Fruitland beds concluded that they are "considerably later than Judith River and possibly somewhat earlier than Lance . . . and the sequence from the base of the Fruitland up to the top of the Ojo Alamo . . . may include the equivalents of everything from the Fox Hills to the Lance inclusive."

The vertebrate evidence is very strongly in favor of the entire sequence being a unit faunally, and correlation with the Edmonton appears to be most in harmony with the evidence as a whole.

Relations of the Puerco and Torrejon Faunas

In Cope's early papers the fauna was regarded as a unit and referred to the "Puerco." In his later papers he recognizes a partial distinction between Upper and Lower Puerco, giving a list in 1888 of twenty species peculiar to the lower beds. In 1895 Osborn and Earle further emphasized this distinction; but it was not until 1897, as a result of a critical revision of all Cope's collections, and comparison with the collections made by Wortman in 1892 and 1896 in which the localities were recorded and the stratigraphic relations better understood, that it was possible to distinguish the two faunas completely and to show that they had no species and very few genera in common.

Granger in 1913 made a careful stratigraphic study of the area, aided by the survey made in 1907 by J. H. Gardner for the U. S. Geological Survey, and his observations have been closely coördinated with the further stratigraphic work by U.S.G.S. parties.

The Puerco and Torrejon are in fact very distinct faunas. No species passes through from one to the other unchanged; most of the genera are different, and in those genera that pass through (*Periptychus, Anisonchus, Ellipsodon*), the representative species in each horizon are very distinct, those of the Torrejon in all cases much more specialized. The genera of certain families, closely allied and primitive in the Puerco, are decidedly more specialized and divergent in the Torrejon.

The Puerco fossils all come from two layers, one about fifteen to twenty feet above the base, the other thirty to thirty-five feet higher. The lower level is characterized by abundant periptychids and is called the *Ectoconus* level; the upper one is called the *Taeniolabis* level, as this genus has been found only at that level. The Torrejon fossils are more scattered through the formation, but it may also be divided into upper and lower levels, with certain differences in the occurrence or abundance of the species found. The study of the collections has failed to show any true zonal distinctions in these upper and lower collecting levels. Such species as pass through from one to the other pass through without change, nor is there anything to indicate that the upper collecting level is later faunally than the other. The differences may indicate slight or considerable changes in facies, or they may be due merely to the accidents in collecting. They are therefore not properly to be called "zones."

[6] U. S. G. S. Professional Paper 98-R, 1916, p. 310.

CORRELATION OF THE WASATCH AND TIFFANY

The Wasatch south of the San Juan River yielded the large fauna described by Cope in 1874–7, and the more extensive later collections made by the American Museum expeditions of 1911–13 were described in part by Matthew and Granger in 1915–18. The thickness of the formation in this area is estimated by Granger (1914) at 1000 feet, the greater part being fossiliferous, but the basal sandstone barren. Two faunal horizons were distinguished, the lower two-thirds of the formation here called Almagre, the upper third Largo. The Largo fauna is correlated with the Lost Cabin, the Almagre regarded as near to the Lysite, but both somewhat older, or at all events more archaic, than the correlated zones in Wyoming. The lower part of the Gray Bull, the Sand Coulée and Clark Fork horizons of the Bighorn basin do not appear to be represented faunally in the Wasatch south of the San Juan River; they may of course be partly or wholly represented in the barren basal sandstone, but that is mere guessing.

North of the San Juan the Wasatch is of much greater thickness—2275 feet according to Gardner—but in general very barren so that there is no evidence as to its exact correlation, save for the basal beds east of Ignacio from which Gidley secured a few fossils in 1909 and Granger in 1916 obtained a considerable fauna. This fauna, the Tiffany, is of Paleocene type in the absence of the principal modern orders of mammals (Ferae excepted), the fauna being quite nearly related to the Torrejon, but more progressive, and the phyla that pass through into the Eocene are represented by species rather closely related to the species of the Sparnacian or Lower Wasatch fauna. It was correlated by Granger in 1914 with the Clark Fork horizon of the Bighorn basin and by Teilhard in 1922 with the Cernaysian fauna of France. Rather wide differences in facies prevent these correlations from being other than provisional.

The correlation of the Wasatch with the Suessonian is a well-documented and undisputed conclusion. The correlation of the Tiffany with the Cernaysian and Upper Fort Union is rather provisional. The Torrejon and Puerco cannot be directly correlated as there is no identity and little, if any, close correspondence between their characteristic mammals and those of any other formation. The upper and lower limits between which they must lie are the Cernaysian-Tiffany, Upper Paleocene or Thanetian above, and the Ojo Alamo-Edmonton of the Upper Cretaceous succession, probably equivalent to Maestrichtian or Lower Danian of the European succession. The correlation of the Torrejon and Puerco must therefore be with the Montian and Upper Danian of Europe.

3. VALIDITY OF THE PALEOCENE AS A PRIMARY DIVISION OF THE TERTIARY (OR CRETACEOUS) COÖRDINATE WITH EOCENE OR OLIGOCENE

A comparison of the vertebrate faunas will show how those of the Puerco, Torrejon and Tiffany are related to the faunas of other Tertiary formations. In making these comparisons it is well to keep in mind that in so far as they rest upon fragmentary material, rare or little-known genera and doubtful or disputed points in taxonomic arrangements, they are uncertain and subject to revision. The real weight of evidence is the whole of the complex detailed record of evolutionary change and adaptation, which is illustrated best in the more abundant and well-known types of each fauna, sketchily and often doubtfully in the rare, little-known or apparently absent types. In all such comparisons it is necessary to assess the importance of each item of the evidence in the light of several lines of probabilities: the accidents of collecting, the interpretation of the stratigraphy and conditions and probabilities as to preservation of different animals, the facies of the local fauna represented by the fossils and its probable limitations, the probabilities as to geographic limitation of species, genera and larger groups, the greater or lesser certainty in systematic reference of more or less complete material, and the extent to which accepted taxonomy represents the real phyletic relations of the forms or groups compared. All these lines of evidence cannot be summarized, condensed or in any way exactly represented in a faunal list. But it is from this evidence as a whole that our conclusions as to relationship are really drawn, and one can only say that the statistical results of such a summary comparison accord in substance with our actual conclusions as to relationship. It is not the basis of them; it is probably misleading or inaccurate in many details; but the errors cancel out so that the resultant, as far as we are able to judge, is not far wrong.

4. FAUNAL LISTS

PUERCO

Pisces:
  Lepisosteus sp.
Reptilia:
  Rhynchocephalia:
    Champsosaurus saponensis Cope, 1882
  Loricata:
    Crocodilus sp.
  Chelonia:
    Pleurosternidae:
      Compsemys vafer Hay, 1910
      Compsemys parva Hay, 1910
      Compsemys puercensis Gilmore, 1919
    Baënidae:
      Baëna sp.
    Dermatemydidae:
      Adocus hesperius Gilmore, 1919
      Hoplochelys crassa (Cope, 1888)
      Hoplochelys bicarinata Hay, 1910
      Hoplochelys laqueata Gilmore, 1919
    Plastomenidae:
      Plastomenus sp.
    Trionychidae:
      Aspideretes sagatus Hay, 1908
      Aspideretes puercensis Hay, 1908
      Aspideretes reesidei Gilmore, 1919
      Aspideretes vegetus Gilmore, 1919
      Aspideretes quadratus Gilmore, 1919
      Aspideretes perplexus Gilmore, 1919
      Conchochelys admirabilis Hay, 1905

Mammalia:
  Marsupialia:
    Didelphiidae:
      Thylacodon pusillus Matthew and Granger, 1921
  Multituberculata:
    Taeniolabididae:
      Catopsalis foliatus Cope, 1882
      Taeniolabis sulcatus Cope, 1882
      Taeniolabis taöensis (Cope, 1882)
      Taeniolabis attenuatus (Cope, 1885)
      Taeniolabis triserialis Granger and Simpson, 1929
    Ptilodontidae:
      Eucosmodon americanus (Cope, 1885)
  Creodonta:
    Arctocyonidae:
      Carcinodon filholianus (Cope, 1888)
      Loxolophus interruptus (Cope, 1882)
      Loxolophus priscus (Cope, 1888)
      Loxolophus attenuatus (Osborn and Earle, 1895)
      Loxolophus hyattianus (Cope, 1885)
      Oxyclaenus cuspidatus Cope, 1884
      Oxyclaenus simplex (Cope, 1884)
      Protogonodon pentacus (Cope, 1888)
      Protogonodon stenognathus Matthew, 1897
      Protogonodon protogonioides (Cope, 1882)

Protogonodon kimbetovius, new species
Paradoxodon rütimeyeranus (Cope, 1888) [7]
Eoconodon gaudrianus (Cope, 1888)
Eoconodon heilprinianus (Cope, 1882)
Taligrada:
  Periptychidae:
    Ectoconus ditrigonus (Cope, 1882)
    Ectoconus majusculus, new species
    Periptychus (Plagioptychus) coarctatus (Cope, 1883)
    Conacodon entoconus (Cope, 1882)
    Conacodon cophater (Cope, 1884)
    Anisonchus gillianus Cope, 1882
    Hemithlaeus kowalevskianus Cope, 1882
Condylarthra:
  Hyopsodontidae:
    Ellipsodon priscus, new species
    Oxyacodon turgidunculus (Cope, 1888)
    Oxyacodon apiculatus Osborn and Earle, 1895
    Oxyacodon agapetillus (Cope, 1884)
    Oxyacodon priscilla, new species
Taeniodonta:
  Stylinodontidae:
    Wortmania otariidens (Cope, 1885)
    Onychodectes tisonensis Cope, 1888
    Onychodectes rarus Osborn and Earle, 1895

TORREJON

Pisces:
  Lepisosteus sp.
Reptilia:
  Rhynchocephalia:
    Champsosauridae:
      Champsosaurus australis Cope, 1881
      Champsosaurus puercensis Cope, 1882
      Champsosaurus saponensis Cope, 1882
  Loricata:
    Crocodilus sp.
  Serpentes:
    Helagras prisciformis Cope, 1883

[7] See Addendum, p. 361.

Chelonia:
  Pleurosternidae: ·
    Compsemys torrejonensis Gilmore, 1919
    Compsemys parva Hay, 1910
  Baënidae:
    Baëna escavada Hay, 1908
    Baëna sp.
  Dermatemydidae:
    Adocus substrictus (Hay, 1908)
    Adocus onerosus Gilmore, 1919
    Adocus annexus (Hay, 1910)
    Hoplochelys saliens Hay, 1908
    Hoplochelys paludosa Hay, 1908
    Hoplochelys elongata Gilmore, 1919
  Plastomenidae:
    Plastomenus acupictus Hay, 1907
    Plastomenus torrejonensis Gilmore, 1919
    Plastomenus sp. indet.
  Trionychidae:
    Aspideretes singularis Hay, 1907
    Aspideretes sp.
    Platypeltis antiqua Hay, 1907
    Amyda eloisae Gilmore, 1919
Mammalia:
  Multituberculata:
    Taeniolabididae:
      Catopsalis fissidens Cope, 1884
    Ptilodontidae:
      Ptilodus mediaevus Cope, 1881
      Ptilodus trovessartianus Cope, 1882
      Eucosmodon molestus (Cope, 1886)
      Eucosmodon teilhardi Granger and Simpson, 1929
  Creodonta:
    Arctocyonidae:
      Chriacus pelvidens (Cope, 1881)
      Chriacus baldwini (Cope, 1882)
      Chriacus truncatus Cope, 1884
      Chriacus schlosserianus Cope, 1888
      Tricentes subtrigonus (Cope, 1881)
      Tricentes crassicollidens Cope, 1884
      Mixoclaenus encinensis Matthew and Granger, 1921

Deltatherium fundaminis Cope, 1881
Triisodon·quivirensis Cope, 1881
Triisodon antiquus (Cope, 1882)
Triisodon crassicuspis (Cope, 1882)
Goniacodon levisanus (Cope, 1883)
Claenodon ferox (Cope, 1883)
Claenodon corrugatus (Cope, 1883)
Neoclaenodon procyonoides, new species
Mesonychidae:
Dissacus navajovius (Cope, 1881)
Dissacus saurognathus Wortman, 1897
Microclaenodon assurgens (Cope, 1884)
Miacidae:
Didymictis (Protictis) haydenianus Cope, 1882
Condylarthra:
Phenacodontidae:
Tetraclaenodon minor[8] (Matthew, 1897)
Tetraclaenodon puercensis (Cope, 1881)
Tetraclaenodon subquadratus (Cope, 1881)
Tetraclaenodon pliciferus (Cope, 1882)
Hyopsodontidae:
Mioclaenus turgidus Cope, 1881
Mioclaenus lydekkerianus Cope, 1888
Ellipsodon lemuroides (Matthew, 1897)
Ellipsodon acolytus (Cope, 1882)
Ellipsodon inaequidens (Cope, 1884)
Ellipsodon aequidens, new species
Protoselene opisthacus (Cope, 1882)
Taligrada:
Periptychidae:
Anisonchus sectorius Cope, 1881
Haploconus angustus (Cope, 1881)
Haploconus corniculatus Cope, 1888
Periptychus rhabdodon (Cope, 1881)
Periptychus carinidens Cope, 1881

Pantolambdidae:
Pantolambda bathmodon Cope, 1882
Pantolambda cavirictus Cope, 1883
Taeniodonta:
Stylinodontidae:
Psittacotherium multifragum Cope, 1882
Psittacotherium aspasiae Cope, 1882
Conoryctes comma Cope, 1881
Insectivora:
Leptictidae:
Prodiacodon puercensis (Matthew, 1918)
Acmeodon secans Matthew and Granger, 1921
Centetidae:[9]
Palaeoryctes puercensis Matthew, 1913
Mixodectidae:
Mixodectes pungens Cope, 1883
Mixodectes crassiusculus Cope, 1883
Mixodectes malaris (Cope, 1883)
Pantolestidae:
Pentacodon inversus (Cope, 1888)
Pentacodon occultus, new species

[8] See p. 188 and Addendum, p. 362.

[9] See Addendum, p. 362.

5. COMPARATIVE LIST OF PUERCO AND TORREJON MAMMALS

	Puerco	Number of Species	Torrejon	Number of Species
Plagiaulacidae.........	? *Catopsalis*...........	1	*Catopsalis*...........	1
	Taeniolabis...........	2	*Ptilodus*...............	2
	Eucosmodon...........	1	*Eucosmodon*...........	1
Didelphiidae..........	*Thylacodon*...........	1	....	
Miacidae..............			*Didymictis*..........	1
Arctocyonidae.........			*Claenodon*...........	3
	Eoconodon...........	3	*Triisodon*...........	1
			Goniacodon..........	1
			Microclaenodon.......	1
	Oxyclaenus..........	2	....	
	Loxolophus..........	2	*Chriacus*............	4
	Carcinodon..........	1	*Tricentes*............	2
	Protogonodon.........	3	*Deltatherium*.........	1
Mesonychidae.........	....		*Dissacus*............	2
Phenacodontidae.......	....		*Tetraclaenodon*.......	2
Mioclaenidae..........	*Ellipsodon*...........	1	*Ellipsodon*...........	2
			Protoselene..........	1
			Mioclaenus..........	2
	Oxyacodon...........	3	....	
Periptychidae.........	*Hemithlaeus*..........	1	....	
	Anisonchus..........	1	*Anisonchus*..........	1
	Conacodon...........	2	....	
	Periptychus..........	1	*Periptychus*..........	2
	Ectoconus...........	2	....	
Pantolambdidae.......	....		*Pantolambda*..........	2
Stylinodontidae........	*Wortmania*...........	1	*Psittacotherium*.......	3
Conoryctidae..........	*Onychodectes*..........	2	*Conoryctes*...........	1
Leptictidae...........			*Palaeolestes*..........	1
Centetidae............			*Palaeoryctes*.........	1
Pantolestidae.........			*Pentacodon*..........	1
Mixodectidae.........			*Mixodectes*..........	2

From the above list it appears that there are in the Puerco 17 genera representing 7 families and 6 orders and in the Torrejon 26 genera, representing 14 families and 7 orders. No new *orders* except Insectivora appear in the Torrejon, but 8 new families (half of them Insectivora), and all but 5 (?4) of the genera are new. The absence of the Torrejon Insectivora from the Puerco and of one of the Puerco families (Didelphiidae) from the Torrejon may reasonably be attributed to the accidents of collecting, as they are rare forms. Omitting these, we have 5 new families appearing in the Torrejon, of which two, Miacidae and Pantolambdidae, find no ancestral types in the more generalized Puerco mammals, while the Mesonychidae, Phenacodontidae and Mixodectidae may fairly be regarded as evolved from more primitive ancestors approximately represented by known genera of the Puerco.

Four to five genera pass through .... 15–19 per cent of Torrejon fauna
Nineteen Torrejon families come through .... 73 per cent of Torrejon fauna
The orders are unchanged..................... 100 per cent

6. COMPARATIVE LIST OF WASATCH AND BRIDGER MAMMALS

	Wasatch	Bridger
Plesiadapidae......................	*Trogolemur*	*Trogolemur*
		Apatemys
		Uintasorex
Adapidae..........................	*Pelycodus*	*Notharctus*
Tarsiidae..........................	*Tetonius*	
	Omomys	*Omomys*
	Absarokius	*Washakius*
	Cynodontomys	*Hemiacodon*
		Anaptomorphus
		Microsyops
Miacidae..........................	*Didymictis*	
	Miacis	*Miacis*
	Viverravus	*Viverravus*
	Uintacyon	*Uintacyon*
	Vassacyon	*Oödectes*
	Vulpavus	*Vulpavus*
		Palaearctonyx
Arctocyonidae.....................	*Anacodon*	
	Thryptacodon	
Oxyclaenidae......................	*Chriacus*	
Oxyaenidae........................	*Palaeonictis*	*Patriofelis*
	Ambloctonus	*Machaeroides*
	Oxyaena	*Thinocyon*
	Prolimnocyon	*Limnocyon*
Hyaenodontidae...................	*Sinopa*	*Sinopa*
		Tritemnodon
Mesonychidae.....................	*Pachyaena*	*Mesonyx*
		Synoplotherium
	Dissacus	*Harpagolestes*
	Hapalodectes	
Pantolestidae......................	*Palaeosinopa*	*Pantolestes*
Talpidae...........................	*Entomacodon*	*Entomacodon*
	Nyctitherium	*Nyctitherium*
Leptictidae........................	*Diacodon*	
	Didelphodus	*Phenacops*
	Creotarsus	
Tillotheriidae	*Esthonyx*	*Tillotherium*
		Anchippodus
Paramyidae........................	*Paramys*	*Paramys*
		Sciuravus
		Mysops
Stylinodontidae....................	*Calamodon*	*Stylinodon*
Hyopsodidae.......................	*Haplomylus*	
	Hyopsodus	*Hyopsodus*
Phenacodontidae...................	*Phenacodus*	
	Ectocion	
Meniscotheriidae...................	*Meniscotherium*	
Coryphodontidae...................	*Coryphodon*	

6. COMPARATIVE LIST OF WASATCH AND BRIDGER MAMMALS (*Continued*)

	Wasatch	Bridger
Uintatheriidae		*Uintatherium*
Helaletidae.	*Heptodon*	*Hyrachyus*
		Isectolophus
Tapiridae.	*Systemodon*	*Eomoropus*
		Helaletes
Equidae.	*Eohippus*	*Orohippus*
Titanotheriidae.	*Eotitanops*	*Palaeosyops*
		Limnotherium
		Telmatherium
Homacodontidae.	*Diacodexis*	*Homacodon*
		Microsus
	Wasatchia	*Sarcolemur*
		Helohyus
	Bunophorus	*Lophiohyus*
Metacheiromyidae.	*Palaeanodon*	*Metacheiromys*
Notostylopidae.	*Arctostylops*	

In the Wasatch 46 genera are listed, representing 24 families and 12 orders. In the Bridger 51 genera are listed, representing 24 families and 10 orders.

Nine genera pass through. . . . . 20 per cent
Twenty-two families pass through. . . . . 92 per cent
The orders are unchanged. . . . . 100 per cent

Comparison of Torrejon with Wasatch:

Three genera pass through. . . . . 7 per cent (of Wasatch fauna)
Eight families come through. . . . . 33 per cent (of Wasatch fauna)
Five of the orders come through. . . . . 42 per cent (of Wasatch fauna)

The White River fauna, although larger and better known, is in certain respects more limited than the Bridger. It includes but six orders, five of the Eocene orders—the Edentata, Tillodontia, Condylarthra, Amblypoda and Primates—having disappeared at the end of the Eocene. The Notoungulata are known only from the Lower Eocene.

Of sixty-three genera, all are new . . . . 0 per cent pass through
Of thirty-three families, twenty-two are new. . . . . 33 per cent pass through
No new orders, but five are extinct. . . . . 40 per cent pass through

Comparison of the above will indicate that:

1. The Puerco and Torrejon faunas are more distinct from each other than is the Wasatch from the Bridger.

2. The Torrejon is nearly as distinct from the Wasatch as is the Bridger from the White River.

3. The Paleocene faunas are distinguished from the Eocene and the Eocene from the Oligocene by a marked change in appearance and by the extinction of orders and families. Between the successive Eocene or Paleocene faunas there is little change in the families, practically none in the orders.

II. SYSTEMATIC AND DESCRIPTIVE REVISION OF THE PALEOCENE MAMMALS OF NEW MEXICO

EUTHERIA

1. ORDER CARNIVORA

Suborder *CREODONTA* or *Primitive Carnivora*

GENERAL DISCUSSION

Alone among the dominant orders of the Tertiary and modern faunas, the Carnivora is well represented in all the Paleocene horizons. But an examination of the Tertiary history of the order and the relations of the several groups shows that the anomaly is rather apparent than real, and conditioned upon taxonomic convenience rather than upon any fundamental difference in its history and evolution from that of other mammalian groups.

All the Paleocene Carnivora belong to the more primitive division of the order, the Creodonta, and they represent the earlier portion of its adaptive radiation. One of the groups of this adaptive radiation, the Miacidae, appearing first in the Torrejon, underwent a secondary adaptive radiation in the Oligocene and later Tertiary, culminating in the present diversity of modern Carnivora. The remainder of the creodonts continued their adaptive radiation through the earlier Tertiary, but became extinct one after another, mostly before the end of the Eocene. In the Puerco fauna we see the early stages of the first radiation.

The diagnostic characters of the Carnivora are:

Placental tooth formula, molars frequently, front teeth rarely reduced in number; incisors small, set in a transverse row; canines enlarged, prehensile or laniary; premolars usually simple, trenchant, molars typically tuberculo-sectorial, varying into sectorial shearing adaptations and tubercular crushing types. Orbit open posteriorly; lachrymal foramen within orbit. Cranium moderately long, median portion of skull short, face long to short.

Dorso-lumbar formula twenty (with rare exceptions).

Limbs flexible, feet five- to four-toed, plantigrade to digitigrade, phalanges never keeled, rarely hinge-jointed, unguals clawed (except Mesonychidae). Ulnar shaft complete, separate, carpals alternating, more or less interlocking, centrale separate or united with scaphoid and lunar, pollex imperfectly or not opposable. Tarsals nearly serial, astragalus with convex head and distinct neck, fibula separate, slender-shafted, articulating laterally with astragalus and sometimes distally with calcaneum.

In the family classification of the Creodonta the customary division distinguishes Oxyclaenidae, Triisodontidae and Arctocyonidae as families. The distinctions, however, are not so great as those which separate different genera of Mustelidae, Viverridae or Procyonidae, and it is more logical to unite them under the single family name of Arctocyonidae. Scott in defining Oxyclaenidae distinguished them by the tritubercular teeth with raised trigonid as against the low and quadritubercular teeth of Arctocyonidae and the massive tritubercular teeth of Triisodontidae. The differences are not so great as those between *Bassariscus*, *Procyon* and *Potos;* between *Mustela*, *Gulo* and *Taxidea;* between

14

Herpestes, Viverra and *Paradoxurus;* or as those between various members of the same families in other orders.   They are further reduced by the better knowledge of intermediate genera and species such as *Tricentes, Neoclaenodon, Thryptacodon* and *Eoconodon.*   Nor are there any more fundamental differences in skull or skeleton as far as known.   It has been argued that these differences are the beginnings of phyla which would later become more diverse; but the same argument might equally well be applied to any of the modern genera as an excuse for splitting up the modern families.   The fact is that on present evidence the degree of relationship between Arctocyonidae, Triisodontidae and Oxyclaenidae is closer than between the subfamilies of modern Procyonidae, Mustelidae or Viverridae, and they should logically be reduced to or below the rank of subfamilies.

The suborder Creodonta may be defined as follows:

Teeth primitive or specialized into various predaceous or omnivorous adaptations. Brain-case small to medium, feet mostly pentadactyl, tympanic bulla usually absent, carpals usually separate.   Not satisfactorily definable in any definite structural way. Includes a variety of archaic specializations as well as the primitive ancestors of these and the modern fissipeds and pinnipeds.

Three families are represented in the Paleocene, the primitive Arctocyonidae by eighteen genera, while the more specialized families of Miacidae and Mesonychidae have each a single precursor in middle and upper Paleocene.

ARCTOCYONIDAE Giebel, 1855

Arctocyoninae Giebel, 1855, Die Säugethiere, p. 755.   Arctocyonidae Murray,
    1866, Geographical Distribution of Mammals, pp. 117, 329; Gervais, 1877,
    Journal de Zoologie, VI, p. 75.

DEFINITION AND DISCUSSION

Gill in 1871 (under Synoptical Tables, p. 59) gives the following definition of the family: "Teeth 44 (M_3^3, PM_4^4, C_1^1, $I_3^3 \times 2$) ? last pre-molar of *upper* jaw trituberculate; true molars tuberculate."

Under this family name we include the Oxyclaenidae as well as the more typical group of Arctocyoninae to which the term has hitherto been applied.   This is in substance the view taken by Schlosser, who placed *Mioclaenus* in the family, that genus thus including most of the genera which Scott in 1892 distinguished as new and grouped under the new family Oxyclaenidae.   Scott's family distinction in essence amounts to nothing more than that the teeth are tritubercular with raised trigonids, as against the more flattened and quadrate teeth ascribed to Arctocyonidae.   The differences, not very great in the first instance, have been largely broken down by the discovery of genera, such as *Loxolophus, Tricentes, Neoclaenodon* and *Arctocyonides,* intermediate in the tooth characters that were the sole basis of distinction, and also by the evidence that the characters of skull, feet and other characteristic features were very much alike in all genera in which they are known (*Arctocyon, Claenodon, Neoclaenodon, Thryptacodon, Loxolophus, Chriacus, Deltatherium*).

Revised Diagnosis: Teeth primitive, tritubercular, varying towards tuberculo-sectorial or bunodont.   No carnassials or specialized shearing teeth.   Premolars mostly simple, acute, an inner cusp on p^4, sometimes on p_4^3, canines large, acute, incisors small.   Skull moderately long, brain-case small, sagittal and occipital crests strong, occiput narrow and high, tympanic bulla not ossified.

Limbs of moderate length, ulnar and fibular shafts relatively stout, toes 5–5, the hallux and pollex a little reduced, phalanges narrow and long, the unguals compressed and claw-like, astragalus with little or no trochlear groove, the head wedged between tibia and fibula, astragalar foramen large, fibulo-calcanear facet distinct and fibulo-astragalar facet facing almost as much proximad as ectad. Tibiale present in one, probably in all the genera. Centrale united to scaphoid in *Claenodon*, unknown in others.

The largest and best known genus is *Claenodon*, with which *Arctocyon* of Europe is practically identical. Skulls of *Arctocyon* and some of the skeleton bones have been known for many years, but the American genera were known practically only from jaws and a few fragments of the skeleton until the expedition of 1913 secured finely preserved skulls, limbs and feet of *Claenodon*, skulls of *Deltatherium* and *Eoconodon*, and skull with parts of skeleton of *Loxolophus*. Some parts of the skeleton of *Chriacus* and *Thryptacodon* had also been secured, and a skull and part of a skeleton of *Neoclaenodon* were secured in the Fort Union for the National Museum.

The skull, humerus and some other fragments of skeleton figured by Blainville in 1841 under the name of *Arctocyon primaevus* were referred by him to the subursid group of Carnivora, essentially corresponding to the Procyonidae. Laurillard and some other authorities argued for reference of this skull, as well as *Hyaenodon* and other creodonts, to the Marsupialia. Gervais in 1859 combatted this view and placed the genus provisionally in the Canidae.[1] In 1877, however, he changed his opinion as a result of the study of the brain-case, and placed it among the marsupials. Cope regarded it as one of the types of the Creodonta or Primitive Carnivora in 1877 and 1885 (see his discussion of their affinities in *Tertiary Vertebrata*). Subsequent opinion has quite generally accepted this view. Lemoine in 1878 described certain new species which were raised to the rank of genera by Cope. Teilhard, however, in his recent revision states that all the remains of *Arctocyon* from the French Paleocene belong to a single species, somewhat variable but well defined. He shows furthermore that *Claenodon* is doubtfully separable from *Arctocyon*.

The remainder of the genera and species of this family were of much later discovery. The American genera were described by Cope in 1881–8 and the Cernaysian forms by Lemoine about the same period. Most of them were at first referred by Cope to *Mioclaenus* or related to it, and regarded as a primitive group ancestral to both Creodonta and Condylarthra. Scott in 1892 undertook the task of separating this group into several component elements, related some to creodonts, others to Condylarthra, Artiodactyla and Tillodontia, and split up the creodont components into three families of primitive tritubercular Creodonta, the Oxyclaenidae, Arctocyonidae and Triisodontidae.

The family Oxyclaenidae was proposed by Scott [2] to include the genera *Oxyclaenus*, *Chriacus*, *Protochriacus*, *Epichriacus*, *Pentacodon*, *Loxolophus*, *Tricentes* and *Ellipsodon*, with the following diagnosis:

"Superior molars tritubercular, not trenchant; cusps erect and acute; inferior molars tuberculo-sectorial with trigonid moderately elevated above the talon, but not forming a shearing blade; premolars simple and trenchant, p^4 with a deuterocone and $p_{\overline{4}}$ sometimes with a deuteroconid." He remarks further: "The genera associated to form this family

[1] Zoologie et Paléontologie Françaises, 1859, pp. 220–1, figs. 22, 23.
[2] Scott, W. B., 1892, Proc. Acad. Nat. Sci. Phila., XLIV, p. 294.

are known almost entirely from the dentition, and their relationship with'one another, even their ordinal position, is very obscure, the teeth being of that generalized and primitive character to which all mammalian types of dentition converge, as we trace them back in time."

Osborn and Earle in 1895 [3] included all the above genera except *Oxyclaenus* and *Pentacodon* in a new family Chriacidae, but placed the two excluded genera as "incertae sedis"! The diagnosis is: "This family includes forms more primitive than the *Adapidae* but with a similar dental formula."   They observe: "It is exceedingly difficult in the present state of our knowledge to decide with certainty as to the ordinal affinities of the genera which Scott has included under the Oxyclaenidae.   We think it probable, however, that *Chriacus* and its allies are more closely related to the Primates than to any of the Creodonta to which Cope has referred them."   The evidence for Primate relationships consisted chiefly in the lemuroid form of the upper molars.

Matthew in 1897 included in the Oxyclaenidae *Oxyclaenus, Chriacus, Protochriacus* and *Tricentes*.   The genus *Epichriacus* was regarded as a synonym of *Chriacus, Loxolophus* of *Protochriacus, Ellipsodon* of *Mioclaenus*, and *Pentacodon* as *incertae sedis*.

Deltatherium was placed by Cope and Scott in the Proviverridae.   In 1899 Matthew [4] transferred it to the Oxyclaenidae.   In 1909 [5] *Carcinodon* and *Paradoxodon*, placed by Scott as *incertae sedis*, were also included in this family.

Although it appears unnecessary to separate the arctocyonine phylum as a distinct family from the rest of the group, the additional evidence confirms the view expressed by Scott in 1892 and cited above.   The Arctocyonidae, broadly speaking, represent the central primitive stock from which not only other creodont families but probably the Condylarthra are derivable.   *Oxyclaenus* is known only from the teeth, but these suggest affinities to the triisodonts and through them to the Mesonychidae.   *Protogonodon* at the other extreme suggests affinity to *Tetraclaenodon* and through that genus to the phenacodonts.   *Loxolophus* and *Mixoclaenus* suggest relationship to the earliest Mioclaenidae in the teeth, and the former resembles them still more in the skeleton.   *Tricentes* has distinct features of approach to the *Arctocyon* series, and in *Thryptacodon, Neoclaenodon, Claenodon, Arctocyon* and *Anacodon* this specialization is progressively advanced.   *Chriacus* and *Deltatherium*, although peculiarly specialized, show in some features a distinct approach to the oxyaenid-hyaenodont group.   *Eoconodon, Goniacodon* and *Triisodon* are placed in a separate subfamily Triisodontinae, which shows certain points of approach to the Mesonychidae in the structure of the molars, but still retains many of the primitive characters, while the short deep jaw and robust teeth indicate a peculiar line of adaptation.   The skeleton in the larger genera is insufficiently known and in *Goniacodon* is very like that of *Loxolophus* and the oxyclaenine group.

Probably in none of these cases is the relationship that of direct descent, as far as the known species are concerned.   In some it clearly cannot be, as they are contemporary with the more specialized stages.   They may nevertheless be regarded as either approximate or structural ancestors.   None of them, however, has acquired the specialized characters distinctive of the more specialized groups to which they are severally related, and only

[3] Osborn, H. F., and Earle, Charles, 1895, Bull. Amer. Mus. Nat. Hist., VII, p. 20.
[4] Matthew, W. D., 1899, Bull. Amer. Mus. Nat. Hist., XII, p. 29.
[5] Osborn and Matthew, 1909, U. S. G. S. Bull. No. 361, p. 91.

upon an extreme view of phyletic classification could they be distributed into the several families to which they appear to be related. Such distribution would ignore the fact that they are all very much alike in structure and undoubtedly closely related to each other, and would make it impossible to define the several families by their distinctive specialized characters.

Synopsis of Described Genera of Arctocyonidae

Subfamily Arctocyoninae Giebel, 1855.

Molars low-crowned, bunodont, rounded quadrate with hypocones distinct and paraconids vestigial to absent except on m_1. Premolars simple, compressed, reduced.

1. *Arctocyon* Blainville, 1841; type *A. primaevus.*

   Syn. *Hyodectes* Cope, 1880, type *A. gervaisii* Lemoine; *Heteroborus* Cope, 1880, type *A. deuili* Lemoine; and *Arctotherium* Lemoine, 1896, type *A. cloezi* Lemoine.

   Cernaysian formation, *A. primaevus* (including the above species as synonyms), known from skulls, jaws and some skeleton bones.

2. *Claenodon* Scott, 1892; type *Mioclaenus ferox* Cope.

   Torrejon formation, *C. ferox*, *C. corrugatus*, known from skulls, jaws and most of skeleton. Also found in Tiffany and Fort Union formations (*Claenodon* sp. indet.).

3. *Neoclaenodon* Gidley, 1919; type *N. montanensis.*

   Fort Union (*N. montanensis*, *N. silberlingi*, *N. latidens*) and Torrejon (*N. procyonoides*) formations, known from most of the skull and parts of the skeleton.

Subfamily Oxyclaeninae Scott, 1892 (as a family).

Molars tritubercular with hypocone rudimentary or absent and paraconid distinct. Trigonid moderate to low, premolars simple.

4. *Oxyclaenus* Scott, 1892; type *Mioclaenus cuspidatus* Cope.

   Puerco formation, *O. cuspidatus*, *O. simplex*, known from upper and lower jaws. Transitional to Chriacinae.

5. *Loxolophus* Cope, 1885; type *L. hyattianus* Cope.

   Syn. *Protochriacus* Scott, 1892; type *Chriacus priscus* Cope = *L. hyattianus.* Puerco formation, *L. hyattianus*, *L. attenuatus*, known from skull, lower jaws and parts of skeleton.

6. *Protogonodon* Scott, 1892; type *Mioclaenus pentacus* Cope.

   Puerco formation, *P. pentacus*, *P. stenognathus*, known from upper and lower jaws. May be transitional to Phenacodontidae.

7. *Carcinodon* Scott, 1892; type *Mioclaenus filholianus* Cope.

   Puerco formation, *C. filholianus*, known from lower jaws and a doubtfully referred upper jaw.

8. *Tricentes* Cope, 1883; type *T. crassicollidens* Cope.

   Torrejon formation, *T. crassicollidens*, *T. subtrigonus*, known from upper and lower jaws, etc. Transitional to Arctocyoninae.

9. *Mixoclaenus* Matthew and Granger, 1921; type *M. encinensis.*
   Torrejon formation, *M. encinensis*, known from upper and lower jaws. Transitional to Mioclaenidae.
10. *Thryptacodon* Matthew, 1915; type *T. antiquus.*
   Wasatch (*T. antiquus*, *T. olseni*) and Tiffany (*T. australis*) formations. Known from skull, jaws and parts of skeleton. Transitional to Arctocyoninae.
11. *Arctocyonides* Lemoine, 1891; type unnamed.
   Syn. *Creoadapis* Lemoine, 1893; type *C. douvillei.*
   Cernaysian formation, *A.* sp., *A. trouessarti*, *A. douvillei.* Known from part of lower jaw and isolated teeth. Transitional to Arctocyoninae.
12. *Paroxyclaenus* Teilhard, 1922; type *P. lemuroides.*
   Phosphorite formation, *P. lemuroides*, known from upper and lower jaws. Transitional to Miacidae?
13. ?*Coriphagus* Douglass, 1908; type *C. montanus.*
   Fort Union formation, *C. montanus*, known from lower jaw. Transitional to Mioclaenidae?

Subfamily Chriacinae Osborn and Earle, 1895 (as a family).

Molars approaching tuberculo-sectorial, higher trigonids, paraconids strong, fourth premolar more complex.

14. *Chriacus* Cope, 1883; type *Pelycodus pelvidens* Cope.
   Syn. *Epichriacus* Scott, 1892; type *Chriacus schlosserianus* Cope.
   Torrejon (*C. pelvidens*, *C. baldwini*, *C. truncatus*) and Wasatch (*C. gallinae*) formations, known from upper and lower jaws and parts of skeleton.
15. *Deltatherium* Cope, 1881; type *D. fundaminis.*
   Syn. *Lipodectes* Cope, 1881; type *L. penetrans* (*D. fundaminis*).
   Torrejon formation, *D. fundaminis*, known from skulls, jaws and a few bones of the skeleton.

Subfamily Triisodontinae Scott, 1892 (as a family).

Molars tritubercular with round conic cusps. Hypocone rudimentary, paraconid reduced, paracone + metacone and paraconid + metaconid progressively connate. Premolars becoming small, massive, with deep short jaw and heavy symphysis. Canines large, round-oval.

16. *Triisodon* Cope, 1881; type *T. quivirensis.*
   Torrejon formation, *T. quivirensis*, *T. antiquus*, *T. crassicuspis*, known from upper and lower jaws.
17. *Eoconodon* Matthew and Granger, 1921; type *Triisodon heilprinianus* Cope.
   Puerco formation, *E. heilprinianus*, known from skulls and jaws.
18. *Goniacodon* Cope, 1888; type *Triisodon levisanus* Cope.
   Torrejon formation, *G. levisanus*, known from upper and lower jaws and parts of skeleton.

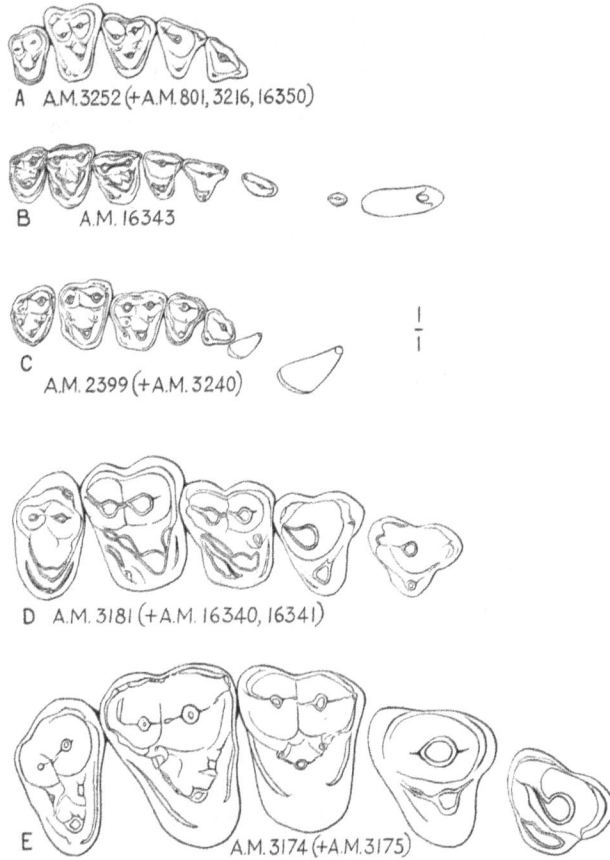

FIG. 1.   Upper teeth of Arctocyonidae, crown views, all natural size.   *A, Oxyclaenus cuspidatus,* type, A. M. No. 3252, supplemented by details from A. M. Nos. 801, 3216, 16350.   *B, Loxolophus hyattianus,* from skull, A. M. No. 16343.   *C, Tricentes subtrigonus,* A. M. No. 2399, supplemented by details from A. M. No. 3240.   *D, Eoconodon heilprinianus,* type, A. M. No. 3181, supplemented by details from A. M. Nos. 16340, 16341.   *E, Triisodon antiquus,* A. M. No. 3174, type of *Sarcothraustes conidens,* supplemented from A. M. No. 3175.

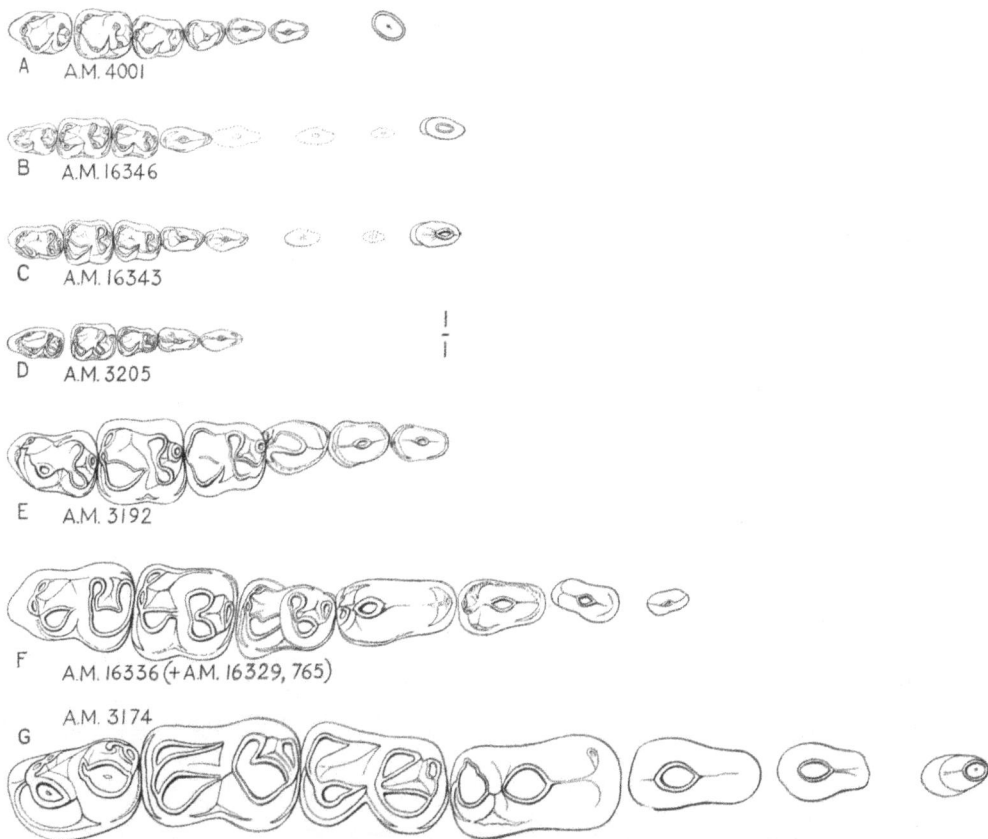

Fig. 2.   Lower teeth of Arctocyonidae, crown views, all natural size.   A, *Tricentes subtrigonus*, A. M. No. 4001. B, *Oxyclaenus cuspidatus*, A. M. No. 16346.   C, *Loxolophus hyattianus*, A. M. No. 16343.   D, *Carcinodon filholianus*, type, A. M. No. 3205.   E, *Protogonodon pentacus*, type, A. M. No. 3192.   F, *Eoconodon heilprinianus*, A. M. No. 16336, supplemented from Nos. 16329 and 765.   G, *Triisodon antiquus*, A. M. No. 3174, type of *Sarcothraustes conidens*.

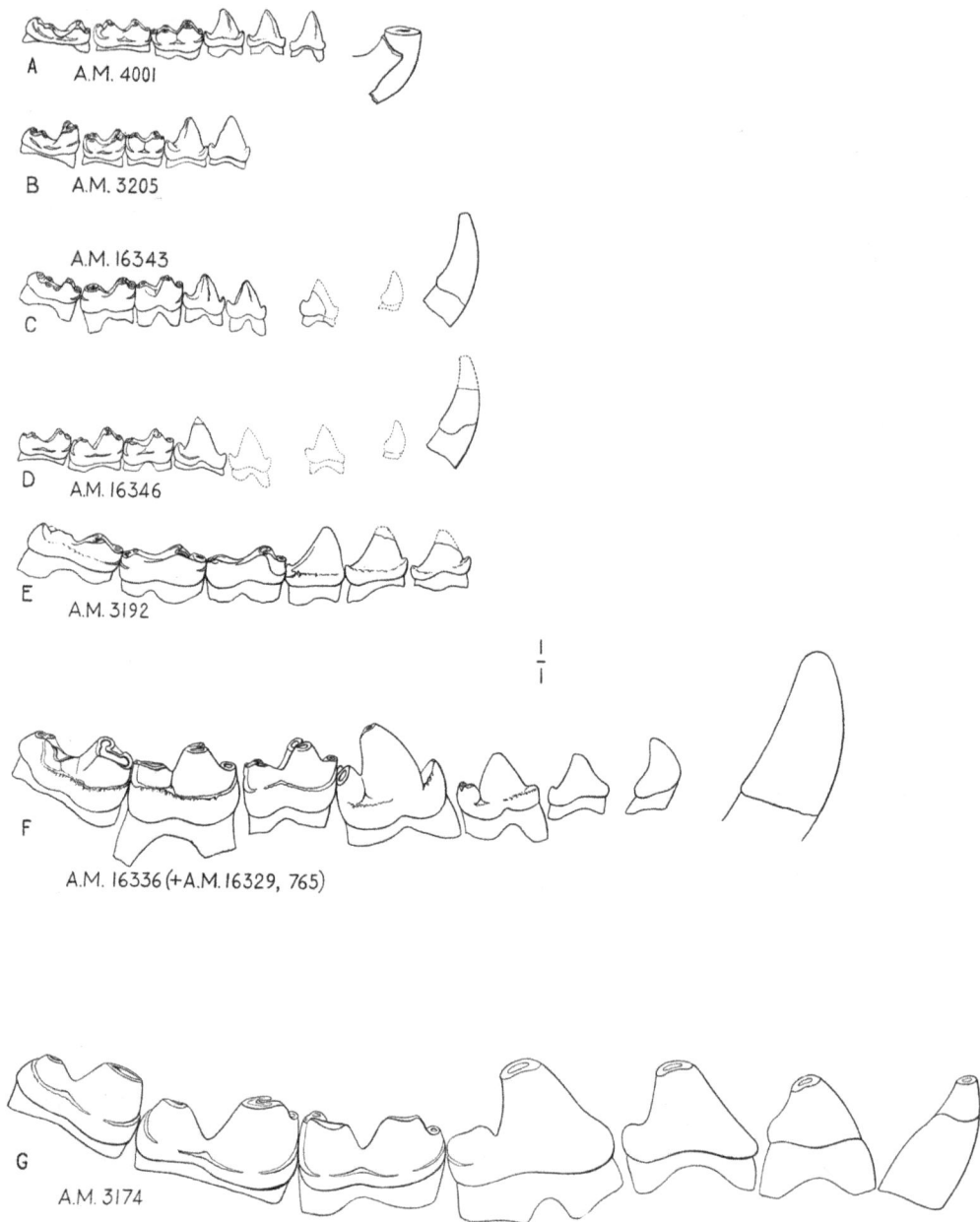

Fig. 3. Lower teeth of Arctocyonidae, external views, all natural size. *A, Tricentes subtrigonus,* A. M. No. 4001. *B, Carcinodon filholianus,* type, A. M. No. 3205. *C, Loxolophus hyattianus,* A. M. No. 16343. *D, Oxyclaenus cuspidatus,* A. M. No. 16346. *E, Protogonodon pentacus,* type, A. M. No. 3192. *F, Eoconodon heilprinianus,* A. M. No. 16336, supplemented from Nos. 16329 and 765. *G, Triisodon antiquus,* A. M. No. 3174, type of *Sarcothraustes conidens.*

ARCTOCYONINAE Giebel, 1855

Giebel, C. G., 1855, Die Säugethiere, p. 755

Claenodon Scott, 1892

Scott, W. B., 1892, Proc. Acad. Nat. Sci. Phila., XLIV, p. 298

Type: Mioclaenus ferox Cope, 1883,[6] from the Upper Paleocene of New Mexico.

Author's Diagnosis: A not unimportant difference from *Arctocyon* is the less completely quadritubercular character of the upper molars. From *Hyodectes* it differs in the greater simplicity of the molars and lack of secondary tubercles, as well as in the less extreme reduction of $m_{\overline{3}}$, while it may be distinguished from *Heteroborus* by the presence of $p_{\overline{1}}$, which is absent in that genus. In *Claenodon* the upper molars are subquadrate in outline, with fairly well developed hypocone. M^2 is the largest of the series, $m^{\underline{3}}$ the smallest. The anterior lower premolars are small and feeble; $p_{\overline{1}}$ is implanted by a single fang, and is separated by a diastema from $p_{\overline{2}}$, which has two roots. $P^{\underline{4}}$ is much the largest of the series and consists of a high, acute and trenchant cone with a strong cingulum, which forms minute anterior and posterior basal cusps. The lower molars are longer and narrower than the upper; the talon is larger than the trigonid and the paraconid is much reduced or absent. $M^{\underline{3}}$ has a distinct hypoconulid. The mandible is long and stout, with regularly curved inferior border and large, deeply-marked masseteric fossa. The zygapophyses of the lumbar vertebræ display the involuted and interlocking shape characteristic of the creodonts. The manus is pentadactyl, plantigrade, and remarkable for the very slight degree of interlocking of the metacarpals. The fibula is very stout and forms an exceedingly massive external malleolus. The astragalus is much like that of *Arctocyon*, but has a longer, narrower and somewhat flatter trochlea and, as in that genus, is perforated by a foramen.

Claenodon ferox (Cope)

Mioclaenus ferox Cope, 1883, Proc. Amer. Phil. Soc., XX, Feb. 14, p. 547, and Paleont. Bull. No. 36.

Type: A. M. Cope Coll. No. 3268, teeth and parts of skeleton.

Horizon and Locality: Middle Paleocene, Torrejon formation,[7] San Juan basin, New Mexico.

Cope's original description of this species is detailed and quite exceptionally long, because the fragmentary skeleton supplied data that enabled him to fix the systematic position of the genus *Mioclaenus*. He observes: "Although we do not possess the corresponding parts of the *Mioclaenus turgidus*, the type of the genus, it is probable, if not certain, that they agree in generic characters."

It is rather difficult to see why Cope referred this species to *Mioclaenus*, with which even the molars do not agree closely, instead of to *Arctocyon*, with which he had evidently compared it. But in this and later papers he used the genus as a sort of catch-all for unspecialized "Bunotheria" which have since been distributed into various families and orders. Perhaps this distribution has been in a way unfortunate, for it has obscured what was very prominent in Cope's concepts and discussions, namely, the rather close cousinship and common primitive structure of most or all these early Tertiary placentals.

However this may be, the skeleton structure of *M. turgidus* is still unknown. Probably it was much like that of *M. lemuroides* and *Hyopsodus walcottianus*, i.e. primitive condylarthran, and not so far removed from the primitive creodont construction which in the

---

[6] Type fixed by Hay, 1902, U. S. G. S. Bull. No. 179, p. 747. The originally included species are *M. ferox*, *M. corrugatus* and provisionally *M. protogonioides*.

[7] Cope, E. D., 1883, Proc. Amer. Phil. Soc., XX, p. 555. [Upper Puerco of Cope, 1883 = Torrejon.]

main is exemplified by "*M.*" *ferox.* The exigencies of classification demand that they be placed in different families and even orders, but Cope's view was not so far from the truth as the present classification might seem to imply.

Author's Description: Char. specif. The canines are well developed, and have a robust root. The crown is rather slender and is very acute. It is rounded in front, but has an acute angle posteriorly. It is not grooved, and the enamel is smooth. The single-rooted first superior premolar is situated close to the canine, and behind it is a short diastema. I have the probable first true molar or fourth premolar. The external cusps are rather small, and are well separated from each other. The inner outline of the crown is rather broadly rounded. The internal tubercle is connected on wearing, with an anterior transverse crest which terminates near the inner base of the anterior external cusp in an intermediate tubercle. There is a posterior intermediate tubercle. There is a cingulum all round the crown excepting at the posterior intermediate tubercle. The second (? first) true molar is like the one just described, but has relatively greater antero-posterior width. In this tooth the cingulum extends all the way round the crown.

There are but two inferior molars of this individual preserved, the second and third true. The former of these has a parallelogrammic outline with rounded angles. There are two posterior and two anterior rather large tubercles; an anterior transverse ledge; and a narrow external and posterior cingulum, the latter running into the internal posterior tubercle. The latter has a circular section, and is much smaller than the external posterior, which has a wide crescentic section. Of the anterior tubercles the anterior is much the larger, judging from its worn base. The third true molar is triangular in outline. Its crown includes two anterior and an external median tubercle. The inner and posterior parts of the crown form a wide shelf, with the internal edge denticulate. A weak external cingulum. [Measurements follow.]

The second individual includes part of the superior walls of the skull. The fragment displays a high sagittal crest, which is fissured in front so as to keep the temporal ridges apart to near its anterior apex. The brain surfaces show small, smooth, flat hemispheres, separated by a constriction from the wide and large olfactory lobes. The navicular bone shows three well defined distal facets, indicating probably five digits in the pes. The teeth of this specimen include a posterior superior molar, and an inferior third or fourth premolar, with other teeth. The premolar is like that of a creodont. Its principal cusp is a simple cone. To this is added a short wide heel, whose superior surface is in two parts, a higher and a lower, divided by a median ridge. A low anterior basal lobe, and a weak external cingulum.

The third specimen belonged to an individual a little smaller than the other two. It includes the first inferior true molar, a tooth lost from the others. Its form is somewhat narrowed anteriorly, where it has two low, but well separated anterior inner tubercles, which form a V with the external anterior.

Specimen No. 1 is accompanied by fragments of vertebræ and limbs. The former are principally from the lumbar region, but fragments of the atlas remain. This vertebra is of moderate length, and the cotylus is somewhat oblique. The vertebrarterial canal is rather elongate, and its anterior groove-like continuation in front of the diapophysis is not deeply excavated. The lumbar vertebræ are remarkable in the characters of their zygapophyses. These display subcylindric surfaces of the posterior pair, which indicates that the anterior ones are involuted, as in the specialized Artiodactyles and Perissodactyles of the later geological ages. Such a structure does not exist among carnivora, nor to my knowledge among creodonta, nor in any mammals of the Lower Eocene.[8] I do not find it in *Didelphys* nor *Phascolarctos*, but it exists in a moderately developed degree in *Sarcophilus*. The articular surface forms more than half of a cylinder, and its superior portion is bounded within by an anteroposterior open groove. The surface within this is not revolute, as in *Bos* and *Sus*, but the articular surface disappears, as in *Cervus*. Eight such postzygapophyses are preserved, all disconnected from their centra. Two of them are united together. There are two other separated zygapophyses of smaller size, which have but slightly convex surfaces. One is probably a prezygapophysis of a dorsal vertebra. No centrum is preserved.

Of the anterior limb there is a probable distal half of a radius.[9] It is of peculiar form, and resembles that of *Sarcophilus ursinus* more than any other species accessible to me. One peculiarity consists in the

---

[8] Compare this statement with Scott's selection of the character (1892, p. 299) as characteristically creodont. The facts lie between these two extremes of inference.—w.d.m.

[9] This was really the distal end of the tibia.—w.d.m.

outward look of its carpal surface, which makes an angle of about 45° with the long axis of the shaft. The obliquity in *S. ursinus* is less. The external border of the shaft in *M. ferox* is, however, straight, and terminates in a depressed tuberosity. Beyond this, the border extends obliquely outward to the carpal face, which it reaches at a right angle. The internal border of the shaft is gradually curved outwards to the external border of the carpal face. Its edge is obtuse, while the external one is more acute for a short distance, and rises to the anterior (superior) plane of the shaft. The carpal face is a spherically subtriangular with rounded angles. It displays two slightly distinguished facets, one of which is superior, and the other is larger and surrounds it, except on the superior side. The internal marginal projection, or "styloid process," is not so prominent as in *S. ursinus*, and is a roughened raised margin. Joining it on the inferior edge of the carpal face is another rough projection of the margin. Immediately opposite this, on the superior edge of the carpal face, is a rough tuberosity, which encloses a small rough fossa, between itself and the styloid process. Internal to it is a shallow groove for an extensor tendon of the manus; then a low short ridge, and internal to that a wide shallow depression for other extensors. The carpal face differs greatly from those of *Sarcophilus* and *Didelphys* in having the inner portion wider than the outer, instead of the reverse, and in having no distinct styloid process. It indicates that the manus was turned outwards much more decidedly than in those genera.

Of carpal bones the only recognizable one is the unciform.[10] Its proximal articular surface rises with a strong convexity entad, and descends to an edge ectad. The metacarpal surface is concave in anteroposterior section, forming a wide shallow groove, extending in the direction of the width of the foot. Its two metacarpal areas are not distinguished. The entire first and second metacarpals,[11] with the heads of the third and fourth are preserved. They considerably resemble those of *Sarcophilus ursinus*. The distal articulations are injured in both, but both display a sharp trochlear keel posteriorly, which on the second extends nearly to the superior face of the articulation. The condyle is subround, and is constricted laterally, and at the base above. The second metacarpal is short and robust, shorter than in *Sarcophilus ursinus*. The first is also robust, but is relatively longer, as it is three-quarters the length of the second. Its head is expanded, especially posteriorly, and the large trapezial face is subtriangular, with round apex directed inward as well as forward. The posterior face of the head is notched ectad to the middle. On the external side of the head there is a vertical facet with convex distal outline, for contact with the second metacarpal.[12] The head of the latter is narrow, and is concave between the sides. The concavity is bounded posteriorly by a raised edge. The anterior part of the proximal facet is decurved. The shaft is deep proximally, but on the distal half is wider than deep. The lateral distal fossæ are remarkably deep and narrow, the condyle very much contracted. The head of the supposed third metacarpal is as wide as the second anteriorly, but narrows to the posterior third, and then contracts abruptly to a narrow apex. The supposed external side of the head is perfectly straight, and is continuous with the side of the shaft without interruption. The entad side displays no facet, but has a depression below the head which adapts itself very well to the head of the first metacarpal. In fact, if the metacarpals just named second and third, exchange places, so that second is placed third and third second, the metacarpal series fits far better. The fourth fits the so-called second much better than the so-called third. This may therefore be the true order, although that first used agrees better with the carpus of *Sarcophilus*. The head of the so-called third is slightly convex anteroposteriorly, and is oblique laterally, descending a little to the inner side. The fourth metacarpal is wider anteriorly than either the second or third. The inner edge is straight, while the outer is concave, the head being narrower before than behind. It has a lateral facet on each side; the inner plane, the external concave in the vertical as well as in the anteroposterior direction. It thus approaches the form of a metatarsal, but is not so strongly excavated, nor is the head notched on either side. The unciform face is convex anteroposteriorly and plane transversely.

The *femur* is broken up so that I cannot restore it. The head of the *tibia* is gone, but a considerable part of the astragalar face is preserved. This is transverse to the long axis of the tibia. It is narrowed anteroposteriorly next the fibular facet. Malleolus lost. The shaft is robust, and does not expand distally for articulation with the astragalus. Three centimeters proximal to the distal end, the external side throws

---

[10] The scapho-centrale and trapezium were also present, but were not recognized; they are described below as unrecognized.—W.D.M.

[11] These are not metacarpals but metatarsals.—W.D.M.

[12] Error. No facet here? But the bone is a metatarsal I, not metacarpal I.—W.D.M.

out a low, rough, ridge-like tuberosity. Above the middle the crest turns outwards, leaving the internal face convex. There is a broken patella, which has one facet much wider than the other.

The *astragalus* has the trochlear portion a little oblique. That is, the internal crest is a little lower than the external, and the inner face is a little sloping. The latter is impressed by a fossa above the posterior part of the sustentacular facet, which runs out on the neck. The trochlea has a shallow groove which is nearer the external than the internal crest, and which passes entirely round the posterior aspect to the plane of the inferior face of the astragalus. The groove for the flexor tendon is thus entirely enclosed, and issues on the inferior face at the posterior extremity of the groove which separates the sustentacular from the condylar facets. The external crest of the trochlea is less prominent posteriorly than the internal, thus reversing the relations of the superior part. The internal ridge becomes quite robust but does not flatten out and project sub-horizontally as in *Oxyæna forcipata*. The fibular face is vertical; neither its anterior nor posterior angles are produced. The neck is somewhat contracted (the internal side is injured). The head is a transverse oval, strongly convex vertically, moderately so horizontally, and without flattening. A *mesocuneïform* (or possibly *ectocuneïform*) bone is wedge-shaped in horizontal section, without posterior tuberosity, and its anterior face is a slightly oblique square. The narrower facet is oblique in the transverse sense.

The *metatarsals* are represented, excepting the first and second.[13] The only complete one is the fifth. The heads of the third and fourth are much like those of *Oxyæna forcipata* and of about the same size. Their anterior width is equal, and in both the external side is more oblique than the internal. Both have a notch at the middle of the internal side, but they differ in that the third has an open notch on the external side which is wanting to the fourth. The lateral excavations of the external sides are deep and rather large and thin out the anterior external edge. The lateral facets are correspondingly large on the fourth and fifth; on the third metatarsal it is small, and a mere decurvature of the proximal surface. That of the fourth is longer proximo-distally than transversely. That of the fifth is about as long as wide, and presents more anteriorly; or, to express it more accurately, the shaft and head present more outwardly than those of the fourth. The proximal, or cuboid facet is narrow anteroposteriorly, and is curved, the external side being concave. On the external side just distal to this facet, the head of the bone expands into a large outward-looking tuberosity, which is separated from the posterior tuberosity by a strong notch. Between it and the head proper, on the anterior face, is a large fossa. The entire form is something like that of the proximal extremity of a femur with head, neck, great trochanter and trochanteric fossa. A somewhat similar form is seen in the corresponding bone of *Oxyæna forcipata*. The shaft of the fifth metatarsal is one-fifth longer than that of the second metacarpal (?3rd) above described. Its direction is straight, but it is somewhat curved anteroposteriorly. Its section is subtriangular, the apex external. The condyle is narrow and sub-globular above, and spreads laterally behind, the external expansion being wide and more oblique. The keel is prominent, and is only visible from above (in front) as an angle. The distal extremities of some other metatarsals differ in being flatter at the epicondyles, and concave between them on the posterior face. The condyles are more symmetrical, and are bounded above on the anterior face by a profound transverse groove. Several *phalanges* are preserved, including part of an unguis. They are all depressed, and with well-marked articular surfaces, of which the distal are well grooved, and the proximal notched below. The lateral areas of insertion of the tendons of the flexors are well marked on the edges of the posterior faces. An ungual phalange is much compressed at the base. The basal table is well marked and has a free lateral edge. The nutritive foramen enters above the posterior extremity of this edge. No trace of basal sheath. . . .

The specimen which has been partially described in the preceding pages as No. 2, has many pieces which are identical with those preserved in specimen No. 1. Among these may be mentioned the glenoid cavities of the squamosal bone. These display, besides the large postglenoid process, a well developed preglenoid ridge, as in *Arctocyonidæ*, *Oxyænidæ* and *Mesonychidæ*. A large distal caudal vertebra of elongate form, indicates a long tail. An articular extremity of a flat bone is intermediate in form between the proximal end of the marsupial bone of *Didelphys* and that of *Sarcophilus*. Its principal and transverse articular surface is transversely convex, as in the latter (*S. ursinus*), but the lesser articular face is separated from it by an even shorter concave interspace than in the opossum. It has almost exactly the form of that of the latter animal. It is a short, flat cone, with two faces presenting on the same side, the one part of

[13] The first and second metatarsals are described above as metacarpals.—w.d.m.

the concavity mentioned, the other flat and presenting away from it. This piece has a slight resemblance to the very peculiar head of the fibula in the opossum, but is not like that of *Sarcophilus ursinus*. I, however, think it much more probably the proximal extremity of a marsupial bone.

A supposed *cuneïform* is subtransverse in position, and resembles in general those of *Oxyæna* and *Esthonyx*. It has the two large transverse proximal facets, the anterior one-quarter wider than the posterior. The distal facet (trapeziotrapezoidal) is simple. The *navicular* is much like that of *Oxyæna forcipata*, but is more robust. Its external tuberosity is flattened anteroposteriorly and is produced proximally. The three distal facets are well marked, the median a little wider than the external, while the internal is subround, convex, and sublateral in position. The *entocuneïform* is a flat bone, with cup-shaped facet for the navicular, and narrow facet for the first metatarsus. This facet is transverse transversely, and concave anteroposteriorly. It shows (1), that there is a pollex [hallux]; (2), that it is probably small; and (3), that it was not opposable to the other digits as is the case in the opossum. (4) It does not show whether the pollex [hallux] has an unguis or not. . . .

Two other bones of specimen No. 2 I cannot positively determine. The first resembles somewhat the trapezium of *Sarcophilus ursinus*, and still more that of *Didelphys* [this is the trapezium]. I will figure it, as a description without identification would be incomprehensible. The next bone is of very anomalous form. It may be the magnum, which is the only unrecognized bone of importance remaining, or it may be a large intermedium [this is the scaphocentrale]. It has no resemblance to the magnum of any mammal known to me. It was evidently wedged between several bones, as it has eight articular facets. Two are on one side; the largest (convex and oval) is on one edge; three are on one end, and two, the least marked, are on the other flat side, opposite to the first.

Restoration. We can now read the nature of the primitive mammal *Mioclaenus ferox*, in so far as the materials above discussed permit. It was a powerful flesh eater, and probably an eater of other things than flesh. It had a long tail and well-developed limbs. It had five toes all around, and the great or first toe was not opposable to the others, and may have been rudimental. The feet were plantigrade and the claws prehensile. The fore feet were well turned outwards. There were in all probability marsupial bones, but whether there was a pouch or not cannot be determined. These points, in connection with the absence of inflection of the angle of the lower jaw, render it probable that the nearest living ally of the *Mioclœnus ferox* is the *Thylacynus cynocephalus* of Tasmania. The presence of a patella distinguishes it from Marsupials in general. Its dentition, glenoid cavity of the skull and other characters, place it near the *Arctocyonidæ*. Should the forms included in that family be found to possess marsupial bones, they must probably be removed from the *Creodonta* and placed in the *Marsupialia*.

Matthew in 1897 and 1901 [14] restudied the type material of *Claenodon ferox* together with some additional specimens referred to the genus, revised the identification of several skeletal elements misidentified by Cope, and discussed the affinities of the genus and family at some length, with the following conclusions: "Judging from Cope's descriptions of the skeletal material of *M. ferox* at his command, he was evidently thoroughly impressed with its resemblance to marsupials, an idea which led him astray in parts of his description. I do not think that this resemblance is wholly a case of parallelism, for all the Creodonta of the Basal Eocene were quite nearly allied, and certain of them show primitive marsupial characters that it is very difficult to explain without admitting a closer connection between marsupials and placentals than their modern differentiation would lead one to believe. *Claenodon* has, however, no marsupial characters except such as must be considered an inheritance from the common stock which gave rise to both marsupials and placentals. Its progressive characters [these are listed on pp. 16–17] are placental carnivore." The arguments for and against regarding the Arctocyonidae as ancestral to the Ursidae are presented on pages 17–19 of the same paper.

Wortman in 1901 [15] appears disposed to revert to Cope's earlier view that the "nearest

[14] Matthew, W. D., 1901, Bull. Amer. Mus. Nat. Hist., XIV, pp. 12–19.
[15] Wortman, J. L., 1901, Amer. Jour. Sci., XII, pp. 283–5.

living ally [of *Mioclaenus ferox*] is the *Thylacynus cynocephalus* of Tasmania," which he quotes with approval, observing: "In my judgment Cope's position rests upon very sound anatomical reasoning, the force of which is rendered all the more patent by these later discoveries." These statements, however, are part of a characteristic diatribe inspired apparently by the fact that Matthew had unwittingly anticipated him by a few months in publishing a revised classification of the Creodonta, and do not represent Wortman's real views, which are shown by his inclusion of the Arctocyonidae in the suborder Creodonta of the order Carnivora, and by a foot-note in regard to the Oxyclaenidae on a later page,[16] in which he says: "There is as yet no evidence sufficient to determine the position of this group satisfactorily. They have always been assumed to be Placentals, and have been placed among the Creodonts. They may quite as well be Implacentals, as far as any very good evidence to the contrary is concerned, or they may prove to be Insectivores, with numerous transitional or Implacental Metatherian characters." Wortman's skepticism was perhaps more or less justified by the scantiness of the evidence then available, but the much more complete material now at hand confirms the view generally accepted.

Matthew in 1915 [17] associated the Arctocyonidae and Oxyclaenidae as "Procreodi" and described some parts of the skeleton of a species of *Chriacus* from the Wasatch, the new genus *Thryptacodon*, typically from the Wasatch, since identified in the Upper Paleocene, and some additional specimens of *Anacodon*, the Wasatch genus. These specimens add to the weight of evidence for "a near affinity between the less specialized Arctocyonidae (*i.e.*, excluding *Anacodon*), the Cercoleptoid Miacidae and the Oxyclaenidae, although part of the resemblance is due to parallelism (*i.e.*, in the cercoleptoid Miacidae)."

Gidley in 1919 [18] redefined the genus *Claenodon* on the basis of new specimens secured in 1913 by Granger and Sinclair, figuring the skull and hind foot, and proposed the new genus *Neoclaenodon* upon a skull and parts of the skeleton from the Fort Union beds of Montana. In Gidley's opinion, although not published at this time, the Arctocyonidae were more or less nearly ancestral to the Ursidae, a view which had been advocated by Matthew in 1897 but subsequently withdrawn.

In 1922 Teilhard de Chardin [19] redescribed the Cernaysian *Arctocyon* and discussed its affinities. He pointed out the near relationship of *Claenodon*, the distinctions made by Matthew being inconstant, but notes the following as distinctive characters:

Clænodon: p^{3-4} à deutérocône faible; m^2 avec cingulum antérieur faible; fémur avec 3e trochanter rapproché du 2e; facette péronéale du calcanéum large.

Arctocyon: p^{3-4} à deutérocône assez fort; m^2 avec ectocône; fémur avec 3e trochanter éloigné du 2e; facette péronéale étroite.

Arctocyon est vraisemblablement plus évolué que *Clænodon;* mais, au moins à en juger par la seule dentition, les deux formes se suivent de tout près.

Teilhard's view of the affinities of *Arctocyon* is thus expressed: "Si l'on veut tenir compte de *tous* les éléments du problème, je pense qu'on doit considérer les Arctocyonidés (et avec eux les Oxyclénidés, dont il va être question ci-après), non pas comme des Condylarthrés, mais comme une branche (une 'radiation') carnivore issue du seul et même tronc

[16] Wortman, J. L., 1902, Amer. Jour. Sci., XIII, p. 434.

[17] Matthew, W. D. (and Granger, Walter), 1915, Bull. Amer. Mus. Nat. Hist., XXXIV, pp. 5, 9.

[18] Gidley, J. W., 1919, Bull. Amer. Mus. Nat. Hist., XLI, pp. 541–55.

[19] Teilhard, P., 1922, Annales de Paléontologie, XI, pp. 28–31.

qui a donné les Condylarthrés dans la direction Ongulés, les Chiromyidés dans la direction Rongeurs, les Lemuriens dans la direction Primates et peut-être les Pantolestidés dans l'axe Insectivores." The hypothesis of ursid affinities he regards as untenable.

Description of Skull and Skeleton

The most important specimens of *Claenodon* obtained in 1912 and 1913 are a skull, No. 16545, considerably crushed and lacking the zygomatic arches and premaxillae; and a fragmentary skeleton, No. 16543, including fore and hind limbs with the feet found partly articulated and nearly complete. In addition a series of upper and lower jaws and fragmentary skeletons supplement the characters of specimens previously described. They range in size from somewhat larger than the type of *C. ferox* down to the size of the type of *C. corrugatus*, and are of more or less intermediate character in the teeth as far as comparisons can be made.

SKULL. The skull is stout and massive with the usual creodont type of small braincase, high sagittal and occipital crests, moderately heavy zygomata, cranial region rather long and wide, with exposed auditory prominences, the tympanic not ossified into a bulla and the ring not preserved. Glenoid articulation transverse, postglenoid process not very high and no preglenoid crest, postglenoid foramen large, basicranial foramina not sufficiently preserved for description. The muzzle is rather short and heavy, the lower jaw deep and massive.

The incisors are unknown. The canines are long, sharp, almost equally recurved in upper and lower jaw, oval in cross-section, crested and serrate posteriorly when unworn. They vary greatly in size, the smaller individuals (*C. corrugatus*) having relatively small canines and smaller and more compressed premolars, the larger ones (*C. ferox*) relatively larger canines with more massive roots.

The premolars are somewhat reduced in size, the first one-rooted, quite small, the second, third and fourth progressively larger, with short diastemata separating $p_{\overline{1}}^{1}$ and p^2 from their neighbors. P^3 and p^4 are three-rooted, subtrigonal, with a small inner cusp on p^4, rudimentary or absent on p^3. P_{2-4} are moderately compressed and the principal cusp somewhat recurved, not very high, a well-developed heel on all and a small anterior basal cusp on p_4.

The molars are low-crowned, quadrate, the cusps very low, massive and with rugose enamel. Hypocone moderately developed on m^{1-2}, absent on m^3. Encircling cingula on all the upper molars, the protoconule small, metaconule large on m^{1-2}, absent on m^3. Paraconid distinct on m_1, vestigial and twinned with metaconid on m_{2-3}. Trigonids low, with protoconid and metaconid of equal size. Talonids forming shallow basins, the hypoconids flat and rugose internally, nearly as high as protoconids, entoconids lower, marginal, hardly more than crests, hypoconulids rudimentary and twinned with entoconids on m_{1-2}, large but low and not very distinct on m_3. The second molar is considerably larger than the first, and the talonids are broader than the trigonids. M_3 is about as long as m_2 but narrower, the talonid elongate and of nearly the same width as the trigonid.

The skull is not exceptionally large in proportion to the skeleton, as shown by comparative measurements of skull, humerus and pes in other Carnivora. It has much the same proportions as in some other omnivorous Carnivora, compares fairly well with the civets and is smaller than in the more predaceous creodonts. The statement made by

Cope and repeated by most subsequent writers that *all* creodonts are distinguished by relatively large skulls has been shown by Matthew [20] to be incorrect.

SKELETON. The vertebrae are imperfectly known. A few dorsal, lumbar and caudal vertebrae are preserved in No. 16543, and show that the dorsal centra are of moderate length, rather wide and flattened; the lumbars are not much larger than the dorsals and have strongly convex and slightly reflected zygapophyseal facets, approaching the characteristic artiodactyl construction. The caudals (Nos. 16543, 16541, 16010) are large and indicate a very long and heavy tail as in *Patriofelis* and most other Creodonta.

The ribs are slender and round-oval in cross-section, moderately convex, apparently not very long.

The fore and hind limbs are stout, the fore limb not much smaller than the hind, but a little shorter, the proportions not differing materially from the average of modern Carnivora. Of the scapula only a few fragments are preserved in No. 16542, indicating a wide, short blade, a stout and heavy acromial process. The coracoid process is broken off so that its form and size are unknown. The humerus (complete in No. 16542, parts of both humeri in No. 16543) is much like that of *Arctocyon*, with high, stout deltoid crest extending to about the middle of the shaft and ending abruptly, greater tuberosity massive and a little higher than the head; the lesser tuberosity is large but not prominent, the head comparatively broad and flat. On the inner border of the shaft almost opposite to the deltoid crest is a rather prominent elongate marginal process probably for the triceps, an unusual feature in Carnivora, whether creodonts or fissipeds, but apparently indicated in *Arctocyon*. It is perhaps to be connected with the length and massiveness of the olecranon of the ulna. The upper part of the shaft is obliquely expanded by this process, and the deltoid process is at an angle of 50° to 60° to the expansion of the lower part by the supinator crest and the entepicondylar process, so that very little is left of the straight cylindrical shaft. The supinator crest is high and relatively long, but the entepicondyle scarcely projects beyond the distal facets; the entepicondylar process, on the other hand, is remarkably heavy, prominent, flattened anteroposteriorly, and projects inward beyond the distal facets about two-thirds of their total width. The entepicondylar foramen is large, continued as a marked groove some distance above the bridge, which is stout and very oblique. There is no supratrochlear foramen. The radial facet is comparatively small and flattened towards its external border; the ulnar facet is wide and with very prominent internal flange; the anconeal fossa is quite shallow.

The ulna and radius (Nos. 16542, 16543) are rather stout, short bones, the radius only about three-fourths the length of the humerus. The ulna has an extremely long, heavy olecranon, shallow anconeal fossa and wide, flattened, heavy shaft, much wider than that of the radius, although not so thick. The distal end is as wide as that of the radius. The lesser sigmoid fossa for the head of the radius is wide, flat and shallow; the articulation for the radius at the distal end is strongly convex, extended anteriorly and posteriorly, and the styloid process is short and very wide. A prominent crest on the posterior face marks the border of the pronator quadratus attachment on the ulna, extending up obliquely from the middle of the posterior face toward the internal border. The head of the radius is considerably expanded laterally, its articulation with the ulna wide and comparatively flat, the inner border of the humeral facet extended and everted. The shaft is stout, nearly straight,

[20] Matthew, W. D., 1909, Mem. Amer. Mus. Nat. Hist., IX, p. 348.

cylindrical only in its upper part, the lower part becoming trihedral with prominent internal and postero-external ridges, and towards the lower end a strong antero-external ridge.   The distal end carries a single rather shallow facet for lunar and scaphoid, concave external facet for the ulna, and between the two apparently a lateral facet for the cuneiform.

The above characters would seem to indicate a stout, heavily muscled fore limb, very much flexed at the elbow (shallow anconeal fossa), with some reduction of the rotation of ulna and radius at their proximal end, approaching Condylarthra to some extent in this and other features.   It has some points of resemblance to the bears (heavy acromion and wide scapula, prominent entepicondyle, wide ulnar shaft), but much more to *Phenacodus*.   The form of the olecranon, head of radius, anconeal fossa, humeral shaft, etc., are rather in contrast with the bears.

The manus is nearly complete and is shown in Pl. V.   It is five-toed, rather short and stout, the digits all somewhat divergent and the pollex semi-opposable—a term which we continue to consider appropriate in spite of the objections raised by Gidley and considered more fully on a later page.   The manus and pes of *Claenodon* have already been described, but additional specimens show more clearly their characters and variability. The carpals are all separate in No. 16543, the centrale not having united with the scaphoid. In No. 3269 (paratype of *C. ferox*), No. 16541 (*C. corrugatus?*) and No. 2456 (figured, 1897, as *C. corrugatus*) they are united, although the teeth are unworn and the animal presumably not old.   The union or separation cannot therefore be associated with species or age; probably the difference is individual.   The carpus is broad and short, the scaphoid quite thin, with the radial facet rolled over the dorsal surface so as to nearly reach the distal facets.   Of these the facet for the trapezium is larger, the facet for the centrale about two thirds as wide; beneath, the scaphoid has a very short, blunt internal hook, and externally a well-fitting facet for the lunar.   The centrale is a wide, short, somewhat rectangular bone lying chiefly beneath the scaphoid but with a short extension beneath the lunar.   Dorsally it is thinned out to very slight thickness, and distally it rests chiefly on the trapezoid, but at the inner end on the magnum.   The trapezoid is also a short, wide bone, sub-rectangular but expanded distally at the dorsal surface which has an oval outline; the four faces of the rectangle are for the centrale, magnum, second metacarpal and trapezium.   The magnum has as usual a small dorsal surface but is expanded below with a proximal keel fitted into the inner side of the distal face of the lunar, with the inner process of the centrale intervening dorsally between magnum and lunar so that at the dorsal surface the contact is limited.   The external face of the magnum rests against the unciform, and the distal, at right angles to it, rests upon the third metacarpal.   The lunar is broad, with the usual facets, the distal end wedged between unciform and magnum plus centrale; the proximal facet conforms to that of the scaphoid, and, like it, is strongly rolled over the dorsal surface; the distal facets are deeply concave and there is a rather obscure external facet for the cuneiform.   The unciform is wide; proximally it is wedged in between cuneiform and lunar, the lunar facet being much narrower, especially toward the palmar surface; there is a short palmar hook but no external extension of the bone beyond the cuneiform facet, the facets for cuneiform and fifth metacarpal coming practically together at their external margins. Distally the facets for the fourth and fifth metacarpals are semi-distinct, the latter being much larger towards the palmar surface.   The cuneiform is a rather thick wide bone, with a short, stout palmar hook, large distal facet for unciform, small and not very distinct internal facets for lunar and radius, a large proximal ulnar facet and rather smaller palmar proximal facet for pisiform. The trapezium is not preserved in No. 16543. In No. 3269, paratype of *C. ferox*, it is a trihedral bone with a flat proximal facet for the scaphoid, three external facets for centrale, trapezoid and the head of the second metacarpal, and a broad distal-internal facet, strongly concave from side to side, concave and dorsally overlapping in a dorso-palmar direction, and deeply notched at the middle of the dorsal margin. This facet has a marked resemblance in form to the proximal facets of the proximal phalanges in Carnivora, and the movement of this particular joint must have been of similar type, allowing for the reversal of the form in both dorso-palmar and proximal-distal direction. That is to say, the first digit had the same degree, manner and limits of flexure upon the trapezoid that the distal ends of the metatarsals have upon the phalanges in Carnivora, only reversed as to dorso-palmar direction. The ordinary movement in walking is to lift the palm upward to a right angle upon the resting phalanges. In the present case it should mean flexing the metacarpal downward to a right angle upon the side of the wrist, the digit standing out from the side at an angle of about forty-five degrees. Some further in-turning of the claw of digit I would be accomplished through flexure of the phalanges. The trapezium itself has a more limited mobility upon the adjacent bones, the flat facets permitting limited sliding movement while the peg-like head of the bone, projecting beneath, facilitated a sliding and turning movement, which would serve to rotate the bone to a limited degree on the rest of the inner carpals. The movement of the first digit thus permitted was briefly described by Matthew in 1897 as semi-opposability. *Claenodon* is quite similar in the form and relations of the corresponding bones and facets of the first digit to the lemurs and to the Eocene *Notharctus*, except that the digit bears a heavy claw instead of a nail. In the higher Primates and man the form and character of the corresponding parts have been further modified, perfecting the mobility of the pollex in man, variously modifying it in others. Gidley [21] has objected to the term semi-opposable as applied to *Claenodon* and other primitive placentals, and declares that the first digit is merely "divergent" like the divergent digits in the manus of Reptilia. But there is nothing in the form or relations of the facets of the first or any digit in reptiles that is in the least comparable to the above described structures in *Claenodon*, which are highly characteristic and peculiar, clearly analogous in a reversed way to the structure of the proximal end of the phalanges, and indicate a corresponding reversed movement, which clearly involves a partial flexure of the first digit obliquely across the palm. They are moreover very much alike in all details of construction and relation to the corresponding parts of the Lemuroidea, for which the term of semi-opposability was coined and is in general use. It appears, therefore, quite proper to apply the term to the construction seen not only in *Claenodon* but to a greater or less extent in many other primitive Tertiary Mammalia, as will be seen in description of other well-preserved Paleocene skeletons. On the other hand, it is incorrect and misleading to compare it to the merely divergent digits of Reptilia which are not at all similar in details of structure and do not have this especial adaptation, although they may often accomplish in a crude way the grasping movement for which this particular adaptation in mammals appears to be designed. The question of the arboreal adaptation of *Claenodon* may be deferred for the present; it obviously does not turn upon this single feature.

The first metacarpal is the only one preserved complete in No. 16543. The proximal

[21] Gidley, J. W., 1919, Jour. Wash. Acad. Sci., IX, p. 278.

ends of Mc. II–V, and distal ends of Mc. II–IV, also the five proximal phalanges, the third, fourth and fifth intermediate, and first to fourth unguals were found.   The exact length of the metacarpals except the first therefore remains somewhat uncertain; they were evidently considerably longer but not quite so stout in the shaft as Mc. I, the proportion probably corresponding to No. 2456 (*C. corrugatus*) in which Mc. II and V are complete.   The foot as a whole is broad and the metacarpals somewhat divergent (in the proper sense of the term).   The overlap of the metacarpal heads is moderate, the second having a narrow but deep externally facing proximal facet for the magnum, the third having a similar facet for the unciform, wide dorsad but narrowing and disappearing palmad.   The fourth and fifth have the usual closely corresponding lateral facet, somewhat peg-like and permitting a limited rotation downwards of the first digit on the others.

The first row of phalanges is comparatively long, about equal in length to Mc. I, somewhat more flattened, the distal ends comparatively narrow.   The second row of phalanges shows little trace of the peculiar obliquity of the shaft seen in the modern Felidae and rather generally in primitive Fissipedia.   This has been noted by Scott in *Daphaenus*, is equally seen in *Cynodictis*, and some trace of it is seen in most of the Eocene Miacidae.   It is indicative when present of a slight degree of retractility of the claws, not carried so far as in the modern cats.   This apparently was not possible in *Claenodon*.   The ungual phalanges are long, high, sharp, compressed and curved, slightly fissured distally, with moderate sub-ungual processes and no trace of hoods.   The claws of *Claenodon* were much more bear-like than cat-like or dog-like in size and proportions, but primitive in the lack of the hoods developed by most modern Carnivora except Viverridae.

Parts of the pelvis are preserved in No. 16543, and a few fragments in other specimens, but not enough to give any complete idea of its characters.   The ilia are flattened and moderately wide, the chief expansion being on the superior border; the ischia are rather long, flattened, and considerably expanded distally, with a prominent ischial spine.

Parts of the femora are preserved in a number of specimens, the entire length of the bone in No. 16543 but with the upper part badly crushed.   Other specimens aid in showing the true form of this part.   The femur is long, with nearly straight axis of the shaft, head of moderate size with deep central fossa for the crucial ligament set near to the postero-internal border of the facet but completely surrounded by it.   The great trochanter is prominent, well-separated, nearly as high as the head; the lesser trochanter is more internal than posterior; the digital fossa is deep rather than spacious and the third trochanter is a distinct and prominent external process situated one-fourth from the proximal end of the bone.   The shaft is fairly round, the distal end moderately expanded, of somewhat unusual depth for Carnivora, the patellar trochlea shallow, long and not very wide, the condyles prominent and the fossa between them deep.   The patella is small and broadly oval.

The tibiae are complete in No. 16543 and the fibulae incomplete, portions of the shafts being missing.   The tibia is moderately long, deep proximally for a carnivore, the upper portion of the shaft being strongly compressed.   The cnemial crest is high proximally, fading out gradually towards the middle of the shaft; opposite it, on the posterior face of the shaft, a prominent ridge extends downward from the internal articular surface of the head, fading out about one-third of the way down the shaft.   The distal end has a nearly flat and very oblique facet for the proximal face of the astragalus, and a wide prominent internal malleolus extending around to a considerable extent behind the astragalus, the trochlear groove being conspicuously lacking and the astragalar body held in place by the internal malleolus on the inner side, the fibula on the outer side. The fibula has a comparatively stout shaft and heavy ends, the proximal end abutting from below against the nearly horizontal flat fibular facet of the tibia, the distal end scarcely touching the tibia and resting on the oblique fibular facets of the astragalus and calcaneum, which are some forty-five degrees from the vertical. Both distal facets of the fibula are somewhat concave, the astragalar facet larger, flatter and facing a little more inward, the calcanear facet smaller, more concave and facing a little more distad. The external malleolus of the fibula is a prominent process.

The right and left pedes are almost perfectly preserved in No. 16543, and the pes is less completely represented in a number of other specimens, including No. 3268 (type of *C. ferox*) and No. 16542. The pes is somewhat longer and slightly more robust than the manus, but the metapodials and phalanges are very much alike in the two. The astragalus has no trochlea, a very oblique fibular facet, short but distinct neck and wide flattened convex head, the external part of the facet articulating with the cuboid being obscurely distinguishable from the navicular facet, while the internal dorsal part of the facet is also obscurely distinct, is not covered by the navicular, but is articulated to a small compressed bone which is either a sesamoid or represents the tibiale of the primitive vertebrate foot. Its position and relations strongly suggest the latter interpretation; but if this be so, *the astragalus of mammals does not represent either the tibiale of primitive reptiles or the consolidated intermedium + tibiale, but is the intermedium alone.* For this interpretation there is much to be said, but the discussion is best postponed until after the description of other Paleocene mammals in which the pes is completely preserved. The peculiar relations of the astragalus wedged in proximally between tibia and fibula, with the facet for the fibula very oblique, is seen in other Paleocene creodonts, Condylarthra and Taligrada, not in the insectivore-taeniodont group. The astragalar foramen in *Claenodon* is large and limits the backward movement of the tibia on the pes, as also generally in the early members of the creodont-condylarth-taligrade group.

The calcaneum is massive, with stout and moderately long tuber calcis, short distal portion, astragalo-calcanear facet unusually transverse, flanked externally by a good-sized convex fibular facet facing outward and upward, a large and prominent peroneal tubercle, and the cuboid facet facing somewhat palmad as well as distad and entad. The navicular is rather shallow and not very wide; it bears the usual facets for the three cuneiforms and on the inner face a fourth facet for the supposed tibiale. The cuneiforms have the usual carnivore proportions, and the entocuneiform does not indicate anything more than the usual degree of relative mobility of the first digit. The ?tibiale is a small flat thin bone, looking a little like a much compressed entocuneiform. Proximally it articulates on its external face with the astragalar head and distal to that is a small facet for the navicular; the distal end of the bone is thin and extended dorso-plantar and ends in a rugose imperfectly ossified surface apparently having a cartilaginous plate continuing it in a distal and plantar direction.

The cuboid is large, with a considerable astragalar facet concave, facing proximad-internad, a large slightly convex calcanear facet facing proximad-externad and to a considerable degree dorsad, the calcaneum overlapping on the cuboid (as in many ungulates).

The groove for the peroneus longus is nearly vertical and close to the distal end; the facets for Mt. IV and V are not distinct, and face somewhat plantad in the usual manner.

The five metatarsals are almost equally stout, but the first is considerably shorter than the others, as in most primitive Carnivora. The line of median symmetry of the pes lies between the third and fourth digits, as in Miacidae and most fissiped Carnivora—also in Mesonychidae. The relations of the metapodials are much as in modern Carnivora of corresponding foot-proportions, and the characters of the phalanges are nearly the same as in those of the manus, except that the unguals appear to be slightly fissured at the tips.

The above detailed description of the skeleton is that of a rather large but very primitive creodont, and typical in most respects of the Paleocene Creodonta. With suitable allowance for diversity of size and massiveness it indicates a near relation to *Loxolophus* of the Lower Paleocene, and to what is known of *Chriacus*, *Thryptacodon* and *Deltatherium*. Nor is it far removed in skeleton characters from the Miacidae, the most primitive Oxyaenidae and Hyaenodontidae. The modern Carnivora are to a varying extent advanced and variously specialized in the construction of manus and pes, reduction of the tail, lengthening and straightening of limbs and feet and adaptation for digitigrade terrestrial locomotion. *Claenodon* represents in most respects the ancestral carnivore type, which may be reconstructed as follows:

The tail is extremely long and heavy, a character most nearly retained in some Viverridae or in such more primitive orders as Edentata. The lumbar region is relatively reduced and inflexible with strongly interlocked and somewhat revolute zygapophyses, a character which can hardly be primitive, and suggests relations with the ungulate orders. The limbs were rather short and stout, the fore limbs very much flexed and apparently everted at the elbow—quite in contrast to the bears and suggestive of relations to Taligrada and Condylarthra. The hind limbs are somewhat longer but also apparently with considerable flexure at the knees, unlike the bears and again suggesting primitive ungulates. The feet seem to have been plantigrade or nearly so, the articulations of carpus with fore limb bones and of tarsus with hind limb bones indicating a sharp angle at this point and not allowing any complete straightening of the joint. It does not appear that *Claenodon* could raise the metapodials to anything like a vertical position in walking, as many so-called plantigrade animals can do. It was at least much less capable of such movement than are the bear and the wolverene.

The adaptation of the teeth is evidently to an omnivorous diet, comparable with the bears, to a less extent with the smaller arboreal omnivora like Procyonidae, Paradoxures or Primates. It appears probable that *Claenodon* was a terrestrial animal, feeding largely on fruits, nuts, seeds, insects and similar food, but as capable as are the bears of attacking small animal prey. Its chief advantage lay in its size and powerful proportions and large claws, suited for digging but also efficient in fighting off attack. This might compensate for the lack of activity indicated by the rather awkward, clumsy proportions. The opposable thumb, while probably a primitive inheritance, would be useful in prehension of fruits, etc., search for insects or other food in decaying wood or in the earth, or other activities such as characterize its ursoid adaptation. That it was related to the Ursidae appears wholly improbable in view of the evidence for derivation of this family from primitive Canidae.

Claenodon corrugatus (Cope)

Mioclaenus corrugatus Cope, 1883, Proc. Amer. Phil. Soc., XX, March 16, p. 556, and Paleont. Bull. No. 36; 1884, Amer. Nat., XVIII, p. 349, fig. 16.

Type: A. M. Cope Coll. No. 3258, upper jaw with p^4–m^3, right.

Horizon and Locality: Middle Paleocene, Torrejon formation, San Juan basin, New Mexico.

Author's Description: This species is intermediate in size between the *M. protogonioides* and *M. ferox*, as the following measurements of the second superior true molar show:

	M. protogonioides	*M. corrugatus*	*M. ferox*
Diameter transverse......	.011[22]	.0118	.015
" anteroposterior .	.008	.010	.013

The superior molars are more nearly quadrate than in the other species of the genus, owing to the better development of the posterior internal tubercle, which is, however, as in the others, a mere thickening of the posterior cingulum. It is wanting from the last superior molar. The cusps on the true molars are as in the *M. ferox*, small, and not large and closely placed as in the *M. protogonioides*. The intermediate ones are nearly obsolete. The crowns are all entirely surrounded by a cingulum. The entire enamel surfaces wrinkled so as to be rugose, although the teeth are those of an adult and well used. The second superior molar is larger than the first, exceeding it in the transverse rather than the fore-and-aft diameter. The third is the smallest, and is of oval form with obliquely truncate external face. It is less reduced than in the *M. turgidus*.

The fourth premolar consists of a strong compressed-conic cusp with three basal cusps of small size, viz., an anterior, a posterior, and an internal. The last is the larger, though small, is formed like a heel, and is connected with the others by a cingulum. No external cingulum. [Measurements follow.]

Claenodon corrugatus is a species of doubtful validity, as there is little except size and the relative robustness of the front teeth to distinguish it from *C. ferox*. The measurements indicate a species about 5/7 to 3/4 as large as *ferox*, the canines being relatively small and the limbs and feet smaller and of less robust proportions. The series of jaws, more or less fragmentary, referred to the two species indicates a good deal of variability in the characters of the molar and premolar teeth. From *Neoclaenodon procyonoides* it is quite clearly distinct. To *corrugatus* may be referred, besides the type: No. 2456, upper and lower jaws with parts of the skeleton associated, including most of the fore foot; No. 16541, lower jaw and fragmentary skeleton including parts of fore and hind feet and some distal caudals; No. 16009, lower jaw and some skeleton fragments; and, doubtfully, Nos. 3260, 3266, 16548.

The characters of the teeth and skeleton agree throughout with those of the genotype, except for smaller size, less robust premolars, the inner cusps somewhat less developed on p^{3-4}, decidedly smaller and more slender canines, limb and foot bones smaller and of more slender proportions throughout.

[22] Cope's measurements were in meters and are so retained in his citations in this volume.—THE EDITORS.

Neoclaenodon Gidley, 1919

Gidley, J. W., 1919, Bull. Amer. Mus. Nat. Hist., XLI, p. 547

Type: Neoclaenodon montanensis, from the Fort Union of Montana.

Author's Diagnosis: Dental formula as in *Clænodon;* cranial portion of skull relatively long and deep; interorbital space apparently much narrower, and postorbital constriction longer and more slender than in *Clænodon:* anterior premolars, upper and lower, much reduced; in upper jaw distinct diastemæ behind p¹, and between p² and p³; the first premolar, above and below, lies closely appressed to the canine; hypocone in m¹ and m² rudimentary, wanting in m³; m³ much reduced, suboval in outline with relatively small metacone.   Carpus and tarsus much as in *Clænodon*, but differing in minor details as follows: lunar relatively small, radial facet being not more than two-thirds as wide as that of the scaphoid; fibular facet of the calcaneum much reduced or wanting; neck and head of astragalus relatively thin and broad; cuboid with facet for the astragalus, navicular and ectocuneiform arranged horizontally, nearly parallel and merging into each other.

The characters used by Gidley as diagnostic do not appear to be very constant in the material (*N. procyonoides*) which I refer to this genus.   The genus is distinguished from *Claenodon* by the smaller size, unreduced premolars and lack of the heavily rugose enamel; from *Thryptacodon*, with which Gidley makes no comparisons, it is distinguished by the more elongate skull and smaller brain-case, high sagittal and occipital crests, greater transverse diameters of the molars and the unreduced premolars.   I refer to this genus an undescribed species, *N. procyonoides*, represented by a number of well-preserved upper and lower jaws from the Torrejon, and by more fragmentary jaws referred hitherto to ?*Claenodon protogonioides* (Cope), but not the type of that genus which is from the Puerco and is a species of *Protogonodon*.   The skeleton is unknown in the Torrejon species, but Gidley has given a number of skull and skeleton characters in the genotype, among which some appear to be distinctive of the genus.   The scaphoid and centrale are fused, as in some but not all specimens of *Claenodon;* the lunar is relatively small and shallow; the fibulo-calcanear facet is stated to be much reduced or absent; the navicular and ectocuneiform facets of the cuboid bone are separate in *Claenodon*, confluent in *Neoclaenodon*.

Arctocyonides Lemoine, as figured by Teilhard,[23] appears to be rather nearly related to *Neoclaenodon*, but the premolar cusps are somewhat higher and more compressed.   The last molar appears to be somewhat reduced, as it sometimes is in this genus.   The type species is quite small, about the size of *Tricentes subtrigonus* (*q.v.*), but Teilhard refers to the genus other species of larger size and known only from scattered teeth.

Neoclaenodon procyonoides, new species

Mioclaenus protogonioides (in part) Cope, 1882, Amer. Nat., XVI, Sept. 28, p. 833; 1885, Tertiary Vertebrata, pp. 325, 340, pl. xxivg, fig. 9; (*Claenodon*) Scott, 1892, Proc. Acad. Nat. Sci. Phila., XLIV, p. 299; (reference doubted) Matthew, 1897, Bull. Amer. Mus. Nat. Hist., IX, p. 291.

Type: No. 16554, upper and lower jaws from the upper level of the Torrejon formation, east fork of Torrejon Arroyo, San Juan basin.

Diagnosis: Hypocones of m¹ and m² and metacone of m³ well developed; protoconids of m₂ and m₃ vestigial; premolars with prominent basal cusps and cingula.   Closely comparable to the generic type, *N. montanensis.*

[23] Teilhard, P., 1922, Annales de Paléontologie, XI, pl. i, figs. 18, 19.

This species includes specimens from the Torrejon referred by Cope and Scott to *Mioclaenus protogonioides.* The reference was questioned by Matthew in 1897, and better specimens obtained in 1913 show clearly that the two species are distinct, the Puerco form, including the type of *M. protogonioides*, being allied to *Protogonodon pentacus*, while the Torrejon form was rightly referred by Scott to the neighborhood of *Claenodon.* It agrees with this genus and differs from *Protogonodon* in the vestigial character of the protoconids on m_{2-3}, the more distinct hypocones on m^{1-2}, prominent basal cusps and cingula on the premolars; but the flattening and corrugation of the molar crowns is not so marked as in the larger species, *C. corrugatus* and *C. ferox.* It is more nearly related to the species described by Gidley under the name of *Neoclaenodon*, although it does not conform to Gidley's definition of that genus, which distinguishes it from *Claenodon* proper by the lack of hypocones on the first and second upper molars and the reduced size and oval form of m^3 with reduced metacone. The hypocones of m^1 and m^2 and metacone of m^3 are well developed in *N. procyonoides.* A second specimen however, No. 16553, consisting also of upper and lower jaws with p_3^3–m_3^3 well preserved, approaches Gidley's species to some extent in both these features, although it agrees so nearly with the type in the characters of all the teeth preserved that we hesitate to separate it specifically from No. 16554. Other specimens referred here — No. 3255, m^{1-2}; No. 2462, p_3–m_3; No. 16008, p_3–m_3; No. 3254 + No. 3320, p_2–m_3 — agree fairly well with each other and with the type. It appears probable that *N. procyonoides* is closely related to *N. montanensis* and possible that the species are identical, but in view of the stated difference between the types and of their having come from different formations wide apart geographically, it seems better to hold them distinct. If *Neoclaenodon* be regarded as a valid genus, which it may well be, although upon other grounds than those cited by Gidley, the present species will undoubtedly be referable to it rather than to *Claenodon* proper.

OXYCLAENINAE, new subfamily

(= Oxyclaenidae) Scott, W. B., 1892, Proc. Acad. Nat. Sci. Phila., XLIV, p. 294

Oxyclaenus Cope, 1884

Cope, E. D., 1884, Proc. Amer. Phil. Soc., XXI, p. 312

Type: Mioclaenus cuspidatus Cope, from the Lower Paleocene of New Mexico.

Author's Diagnosis: Distinguished from *Mioclaenus* by "third superior premolar without internal tubercle." Besides the type Cope included *Oxyclaenus corrugatus* and, very probably, *O. ferox.*

Cope did not maintain *Oxyclaenus* or refer to it in his revision of the Puerco fauna in 1888, perhaps because a more careful examination of the genotype convinced him that it did not actually show the character on which he had based the genus.

Scott's Diagnosis, 1892: The anterior premolars form simple, compressed and trenchant cones; on p^4 there is also a well developed deuterocone. The molars are simply tritubercular, with small, erect and acute cusps. M^2 is the largest of the series, especially in the transverse direction. The para- and metacones arise close to the outside of the crown, the latter somewhat nearer to the median line. The protocone is the largest of the elements. There is no distinct hypocone, merely a thickening of the cingulum at that point, which is most marked in m^2. Minute but very distinct proto- and metaconules are present. M^2 is very much reduced in size and more oval than triangular in shape, but preserves all the cusps.   One species: *O. (Mioclaenus) cuspidatus* Cope.

Matthew in 1897 cites the above diagnosis and adds (p. 276): Under this genus are included two or three species of doubtful family relationships, being intermediate between the Chriacidæ and Triisodontidæ.   They are small forms with a trigon like *Chriacus*, but with somewhat more rounded cusps.   The type is *O. cuspidatus* Cope, represented by one specimen whose horizon is not recorded.   A closely allied or identical species is represented by most of the specimens referred by Osborn and Earle to *Protochriacus simplex*.   A third species represented by the type specimen only of *C. simplex* Cope can be placed here only provisionally, the premolars being unknown.

Revised Diagnosis: Upper molars tritubercular with conic cusps of moderate height, hypocones rudimentary or absent, transverse diameter considerably exceeding the anteroposterior, strong encircling cingula except upon inner face of protocone.   External face of protocone flat, conules small but distinct.   Lower molars with trigonid higher than talonid, metaconid and protoconid subequal, well separated, paraconid smaller, nearly internal in position, talonid (except on m₃) as wide as trigonid, deeply basined, the basin more open internally than in *Loxolophus*.   Cusps of lower molars not inflated, tending to be conic. Hypoconulid small except on m₃, which has a narrow, more or less elongate talonid pinched up posteriorly into a large hypoconulid equalling the hypoconid in height.

Fourth premolar above and below with high, acute, conical protocone; p⁴ with large inner cusp and small cusps at outer angles; p₄ with protocone somewhat compressed and prominent sharp anterior basal cusp and heel.   Anterior teeth unknown.[24]

Oxyclaenus is distinguished from *Loxolophus* by the high acute premolars and by the higher and more conical cusps of the molars.   In both features it approaches *Eoconodon*. As the genus is known only from the teeth its affinities remain a little uncertain.   It may be more nearly related to *Eoconodon* than to *Loxolophus*, the molar pattern differing but little in the two genera.   The details of molar construction and proportion are so much alike, however, in *Loxolophus* and *Oxyclaenus* that it is not easy to distinguish the two, while *Eoconodon* may be structurally derived from *Oxyclaenus* although contemporary in time, but without evidence as to the feet of the latter genus it is hardly advisable to embody this possibility in the classification.   *Oxyclaenus* is therefore associated in the same family with the better-known *Loxolophus*.

Oxyclaenus cuspidatus Cope

Mioclaenus (Oxyclaenus) cuspidatus Cope, 1884, Proc. Amer. Phil. Soc., XXI, p. 312.

Type: A. M. Cope Coll. 3252, parts of upper jaws, with premolars and molars; lower jaws associated, buried in hard matrix.

Horizon and Locality: "Puerco of New Mexico," i.e. Nacimiento group of the San Juan basin, but whether true Puerco or Torrejon is not recorded.   Referred specimens are all from the true Puerco, to which the type undoubtedly belongs.

Author's Description: The *Mioclœnus cuspidatus* is distinguished among its congeners, by the transverse character of its superior molar teeth, that is, by the relatively smaller anteroposterior diameter as compared with the transverse; and by the prominence and acuteness of their principal cusps. They thus stand at the opposite extreme of the genus from the *M. turgidus*, where the teeth are characterized by the robustness and obtuseness of the cusps, although in the triangular basis of the second superior molar

---

[24] P³ is stated by Cope and Scott to have no internal root, but the type is insufficiently preserved to prove this and no referred specimen shows it.—W.D.M.

they agree. The external cusps are compressed cones, and in contact at the base; the intermediate tubercles are small and distinct. The internal cusp is large and prominent. The base of the fourth premolar is T-shaped, and is as long as wide. Its internal and external cusps are well developed. The cingulum of the true molars is complete all round on the last one, and on the two others except at the internal base, where it is interrupted. The second molar only displays a posterior inner tubercle of the cingulum, which is small, and does not give a truncate interior outline of the crown, characteristic of *M. opisthacus, M. ferox,* etc. On the ms. i and ii, the cingulum is expanded at the external angles of the crown, most so anteriorly. The anterior expansion rises in a low cusp on P-m. iv. The enamel is smooth.

. . . The fourth premolar [in *M. opisthacus*] . . . is narrower and more transverse, and with larger conical cusps, much as in *M. turgidus;* in the present species it has the trilobate outline seen in *M.* [*Tricentes*] *subtrigonus.* As to the latter species, the teeth are wide [i.e. broad, quadrate, flat-topped, not wide transversely, however] and the cusps smaller and separated at the base, and the cingulum is crenate and lobate, in a manner quite different from the smoothness and compactness of structure seen in the *M. cuspidatus.*

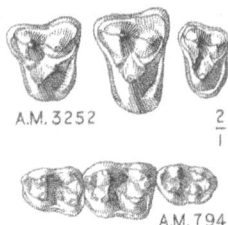

FIG. 4. *Oxyclaenus cuspidatus,* type upper molars, A. M. No. 3252; lower molars, A. M. No. 794, crown views. Twice natural size.

The above specific diagnosis consists almost wholly of the generic distinctions that separate *Oxyclaenus* from *Loxolophus* and *Chriacus,* and from *Claenodon, Mioclaenus* and *Protoselene* of other families. The type has not hitherto been figured. Additional material referred to *O. cuspidatus* by Matthew in 1897 consisted of various parts of jaws identified by Osborn and Earle in 1895 as *Chriacus simplex,* but which are too large for that species. To *O. cuspidatus* are now referred, besides the type, about twenty specimens of upper and lower jaws, more or less fragmentary. Three of them have upper and lower teeth associated. Most of them differ from the type in the character of m_3^3, which are less reduced in size, the upper molar being more triangular in outline and transversely extended. They are all from the *Taeniolabis* or upper level of the Puerco.

Oxyclaenus simplex (Cope)

Chriacus simplex Cope, 1884, Paleont. Bull. No. 37, Jan. 2, and Proc. Amer. Phil. Soc., XXI, p. 314.

Type: A. M. Cope Coll. No. 3107, fragments of upper and lower jaws with m^{1-2} and part of m^3, m_1, m_2 and p_4.

Horizon and Locality: Paleocene of San Juan basin, undoubtedly from Puerco formation, Lower Paleocene.

Author's Description: Posterior cingulum without tubercle; small species.

. . . The true molars are about the size of those of the *C. truncatus,* but of very different detailed structure, as already pointed out. The posterior cingulum is stronger than the anterior, but does not support a trace of a cusp, and they do not unite on the inner face of the crown. External cingulum present. External cusps rather small, separate. Intermediate cusps present; V large and distinct. Enamel smooth.

The inferior true molars support Vs; in the second the anterior is smaller and more elevated than the posterior.   The latter is continued as a raised posterior, and partly interior border of the heel, without prominent cusp.   The crown has a distinct external and a very faint internal cingulum.   In the supposed first true molar, the anterior V is more prolonged anteroposteriorly as in the corresponding tooth of *Mioclænus ferox*, etc., and the fourth premolar of *Phenacodus primævus*.   The anterior cusp is the lowest.   The heel supports three low cusps, of which the external has a crescentic section, and the posterior is the smallest.

It is probable but not certain that the fourth premolar has an internal cusp, as the tooth, presumably this one, is injured at that point.   Should the internal cusp be absent, this species cannot be referred to *Chriacus*.

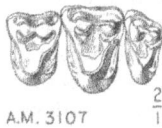

Fig. 5. *Oxyclaenus simplex*, type upper molars, A. M. No. 3107, crown view.   Twice natural size.

Loxolophus Cope, 1885

Cope, E. D., 1885, Amer. Nat., XIX, March 6, p. 386

Type: Loxolophus adapinus = Chriacus (Protochriacus) hyattianus, from the Puerco formation, New Mexico.

Synonyms: Chriacus (in part) Cope.

Protochriacus Scott, 1892; type: *Chriacus priscus* Cope, 1888.

Author's Diagnosis: Char. gen.—Known only from inferior molars.   Crowns with three cusps anteriorly and a basin posteriorly.   The internal and external anterior so connected as to form a transverse crest on a little wear; anterior or fifth cusp distinct.   Rim of basin elevated on the external side and extending as a crest to the base of the anterior cusps.   Internal rim acute, and so near the external as to resemble a large cingulum.   Third true molar with a small heel.   The position of this genus cannot be determined without further material.   The oblique direction of the crests resembles what is seen in the genus Adapis Cuv.

This group of species was included by Cope in his revision of 1888 under the genus *Chriacus*, of which the type is *Pelycodus pelvidens* Cope, 1881, of the Torrejon.   Scott in 1892 separated *Chriacus priscus* and *hyattianus* as types of distinct genera—*Protochriacus* and *Loxolophus*—with definitions as follows:

Protochriacus gen. nov. . . . is closely allied to *Chriacus*, but differs from it in a number of details. $P_{\overline{3}}$ has no distinct deuterocone; the upper molars are less extended transversely, the hypocone is smaller and the protostyle absent.   In the lower molars the trigonid and talon are of nearly equal height. . . .

Loxolophus Cope . . . The superior molars are tritubercular with very minute hypocone, and are remarkable for their anteroposterior as compared with their transverse extent.   The lower molars have a high trigonid with all three cusps well developed and basin-shaped talon with elevated hypoconid.[25]

Matthew in 1897 (p. 268) united *Loxolophus* with *Protochriacus*, defining the genus as follows:

Dentition: $I_{\frac{2}{3}}$, $C_{\frac{1}{1}}$, $P_{\frac{4}{4}}$, $M_{\frac{3}{3}}$.   Upper molars tritubercular with hypocone little developed and no protostyle.   Lower molars broad and low, approaching the *Protogonodon* type; p³ and p₄ with rudimentary dentocene [deuterocone].   Intermediates minute or absent on upper molars.

Cope's *Loxolophus adapinus* was founded on a crushed specimen of his *Chriacus hyattianus*.   The distinctions so far as made were based on error, and Scott's name, *Protochriacus*, is therefore preferred.   The type is *P. priscus* Cope; another species, *P. attenuatus*, was described by Osborn and Earle, and a third,

[25] Scott, W. B., 1892, Proc. Acad. Nat. Sci. Phila., XLIV, pp. 296, 297.

P. hyattianus Cope, is probably referable, although the premolars are unknown. Scott's second species, *P. simplex* Cope, is, as remarked by Osborn and Earle, widely different from *P. priscus*, and may perhaps be provisionally placed under *Oryclænus*. All the above species are from the lower beds or true Puerco.

With regard to the use of the name *Protochriacus* instead of *Loxolophus*, it was based upon the principle strongly insisted upon by Professor Cope and others that a generic name, to be valid, must be accompanied by a definition. The basis for this requirement was that, if not enforced, it left the way open for anyone with or without a knowledge of the subject to propose new names and thus secure an apparent credit for discovery of new genera the characters and distinctions of which were really discovered by others at a later date, a credit to which he was in no degree entitled. In practice it was intended to prevent among authors more ambitious than scrupulous a tendency to give new names to forms which they suspected might prove to be new, though they were either unable or unwilling to take the trouble to prove them new.

If this requirement were granted, it would obviously follow that no new genus could be held valid unless the definition specified some distinctive character which the type species really possessed. The author could hardly deserve credit for attributing to an animal characters which it did not possess, and while it would of course be admitted that errors in the description do not invalidate a genus,[26] yet it would not seem fair to admit them as validating it, and if *all* the distinctions alleged were errors of fact there would be no valid description left. Moreover, the enforcement of this ruling would be equally important as a means of preventing the same ambitious type of research worker from proposing innumerable new genera upon unproved or fictitious characters attributed to species which he did not trouble to examine or verify.

Whether or not it is wholly practical so to sift new definitions of genera or species is, however, an academic question now. The rules of procedure in nomenclature have passed beyond the stage of adjustment in this particular, as it is held by all recent authority that a new genus is sufficiently validated by reference to a published species as type, or a new species by reference to a published or illustrated specimen as type. Whether wise or not the rule must be followed. *Loxolophus*, accordingly, must be preferred to *Protochriacus* if the two genera are synonymized. The same procedure is followed in the similar case of *Tetraclænodon* and *Euprotogonia*.

Revised diagnosis: Dentition $\frac{3 \cdot 1 \cdot 4 \cdot 3}{3 \cdot 1 \cdot 4 \cdot 3}$. Incisors small; canines large, crowns sharp, recurved, not compressed; $p_{\overline{1}}^{1}$ one-rooted, small, close behind canines; p^2 two-rooted, compressed; p^{3-4} triangular, three-rooted; protocone minute on p^3, large on p^4; molars tritubercular with rudimentary hypocone and no protostyle; cusps low, conules small, m^2 largest, m^3 little reduced; lower premolars compressed, with small heels, no accessory cusps or cingula; lower molars with low trigonids and large, wide basin heels, paraconids well developed, protoconid and metaconid subequal, entoconid strong, hypoconulid very rudimentary on m_{1-2}, well developed on m_3; heel of m_3 narrower but longer than trigonid.

Skull elongate, brain-case very small, sagittal crest well developed, arches moderate, no preglenoid crest.

Lower jaw moderately long, slender anteriorly, compressed and not thickened, condyles not transversely extended, angle deep, coronoid process moderately high, somewhat recurved.

---

[26] Hay, O. P., 1899, Science, N.S., IX, p. 593.

Limbs of creodont type, humerus with high deltoid crest abruptly ending, distal end wide, with entepicondylar foramen; femur with well-developed third trochanter, second trochanter internal in position; tibia with moderate cnemial crest fading gradually towards the middle of the shaft; astragalus with head of primitive type, wedged between tibia and fibula, no trochlear groove, tibial facet limited backwardly, astragular foramen will developed, fibular facet facing obliquely outward, head of astragalus oval, neck distinct.

Metapodials slender, inner digit semi-opposable but not so large nor so long as the others; phalanges long and slender, unguals claw-like.

The above diagnosis is based chiefly upon *L. hyattianus* but is supported at various points by the other species.

Loxolophus is easily distinguished from *Oxyclaenus* by the character of the premolars, which are lower-crowned, more robust, with massive crowns much more like those of *Pelycodus* than of the high, sharp cusps of *Oxyclaenus*. The molars are much more alike in the two genera, but in *Loxolophus* the upper molars are more quadrate or rounded and less extended transversely; the trigonids of the lower series scarcely exceed the talonids in height, the heels are broader and more fully basined and the entoconids more fully developed. The lower molars have usually no external cingulum; the paraconids are somewhat more median in position.

Two species are placed in this genus:

L. hyattianus (Cope): $m^{1-3} = 15.5$ mm.; premolars moderately compressed, last molar somewhat reduced.

L. priscus (Cope): larger, $m^{1-3} = 19.5$ mm.; teeth more robust, last molar unreduced.

Loxolophus hyattianus (Cope)

Chriacus hyattianus Cope, 1885, Amer. Nat., XIX, p. 385.

Loxolophus adapinus Cope, 1885, *ibid.*, p. 386.

Protochriacus attenuatus Osborn and Earle, 1895, Bull. Amer. Mus. Nat. Hist., VII, p. 22; Matthew, 1897, *ibid.*, IX, p. 269.

Type: A. M. Cope Coll. No. 3121, an upper jaw fragment with m^{1-3} of the left side, considerably distorted by crushing. Type of *L. adapinus*: A. M. Cope Coll. No. 3124, a part of the lower jaw with m_{1-3} left, much crushed. Type of *P. attenuatus*: A. M. No. 790, a fragment of the lower jaw with m_{1-2}, uncrushed.

Horizon and Locality: All from the Puerco formation of the San Juan basin.

Author's Description of Chriacus hyattianus: Represented by two maxillary bones with molar teeth, one of which is accompanied by a broken mandibular ramus, which supports the second true molar and parts of other teeth. The superior molars are quite peculiar, and are especially characterized by their small transverse as compared with their anteroposterior diameter. The crowns are surrounded by a cingulum, except on the inner side, where distinct traces of it are visible. The external cusps are small and low and flattened on the external side, and are connected at their bases by a low ridge. They send inwards each an angular ridge which unites with its fellow in an angular internal cusp of little elevation, enclosing a triangular fossa. Small angular intermediate tubercles exist at the internal bases of the external cusps. The posterior cingulum is a little better developed than the anterior, and rises into a very small cusp or tubercle, which is not of sufficient size to truncate the internal outline of the crown. The crown of the second true inferior molar displays a contracted triangle of three well developed cusps anteriorly, and a wide basin posteriorly. The rim of this basin is elevated all round and develops into a cusp on the external side. An external, no internal cingulum. Enamel longitudinally wrinkled.

*Author's Description of Protochriacus attenuatus:*Paraconid well marked, on a line with meta-conid, trigonid not raised above talon, hypoconulid distinct . . . smaller than the *P. priscus;* the jaw is very narrow and slender. The crescents of the inferior true molars are very sharply marked, and the cusps are sharper than in the allied species. The paraconid is well marked on the first true molar, but is rudimentary on the second.

Revised Description, Matthew, 1897: This species is much smaller than *P. priscus*, which it other-wise closely resembles. The teeth are not so wide, the paraconid is more internal, the notch between proto-conid [?hypoconulid] and entoconid deeper.

The type of *C. hyattianus* is evidently crushed but the crushing is of a peculiar kind, involving not much shattering of the teeth but a uniform reduction in the transverse diameters of every cusp or valley. Save in this respect it agrees very closely in the structure proportions and form of the teeth with the specimens here referred to it, in particular with No. 16343 (described below).

The type of *L. adapinus* shows the crushing more clearly. The enamel is broken into pieces which are obliquely displaced and compressed. Examined in detail it is seen to agree closely with the lower molars of No. 16343 and other specimens referred.

The type of *P. attenuatus* also agrees closely with the lower teeth of No. 16343.

Cope in 1888 made *L. adapinus* a synonym of *C. hyattianus*. Scott in 1892 admitted the genus *Loxolophus* as valid, with a diagnosis based upon the type of *hyattianus* (which he did not recognize as crushed), and distinguished it from *Protochriacus*, to which he referred *C. priscus* and *simplex* Cope. Matthew in 1897 united *hyattianus* generically, but held it as distinct specifically from *attenuatus*, although recognizing the crushed con-dition of the type.

To this species are referred No. 16343, a skull, lower jaws and parts of the skeleton, all fairly well preserved, collected by W. J. Sinclair, of the American Museum Expedition of 1913, three miles east of Kimbetoh in the lower fossiliferous level of the Puerco forma-tion, and a considerable series of parts of upper and lower jaws, most of them obtained by the expedition of 1892.

Description of No. 16343

Skull. The general proportions of the skull are long and slender, with rather weak zygomatic arches, very small, low brain-case, moderate sagittal and occipital crests, long muzzle, basicranial region elongate. The proportions are suggestive of the genets and other long-skulled Viverridae, save for the longer muzzle, shorter basicranial region, the very much smaller brain-case, and the more backward position of the orbits, which are above and partly behind the last molar, m^3 in this case, while in the viverrids and most other Carnivora they are above p^4–m^3. The skull has been considerably distorted by crushing and broken into somewhat displaced fragments, the original relation of which is not always readily seen. Very little can be certainly recognized of the skull sutures.

Only a part of the premaxilla is preserved, containing the three small incisors; it is relatively small and weak. The muzzle behind the premaxilla is long and slender, the palate narrower and apparently more vaulted than in *Hemigale* or *Didelphys*, the first premolar close behind the canine, the second with diastemata before and behind it about equal to its own length. The width of the palate between the cheek teeth is not so exactly seen, but it was materially narrower and more vaulted than in *Hemigale*. The outer border of the row of cheek teeth is less convex than in *Hemigale*, more than in *Didelphys*. There are no indications of palatal vacuities; if present at all they must have been quite

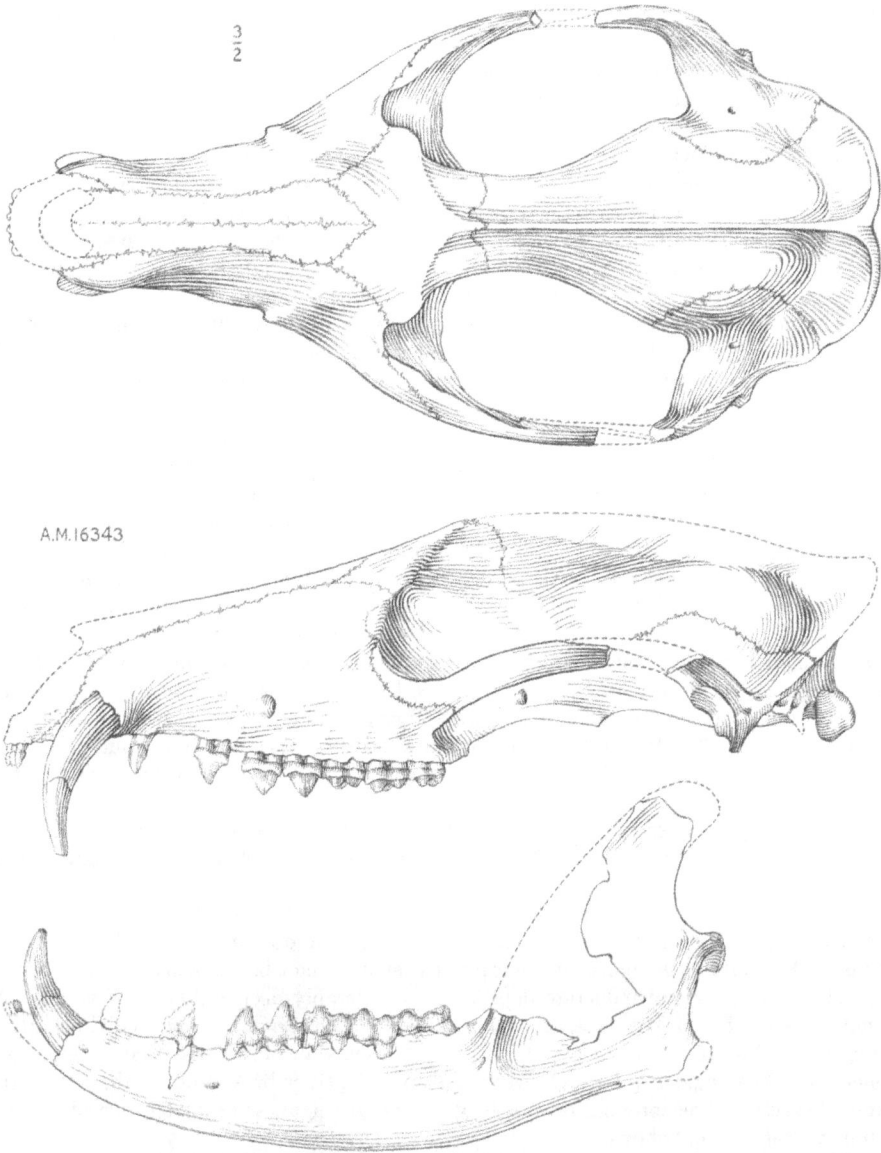

FIG. 6. *Lozolophus hyattianus*, skull and lower jaw, A. M. No. 16343, top and side views.   One and a half times natural size.

small; the palate lacks the raised posterior border of *Didelphys* (and *Sinopa*) and is not extended backward as in many placentals (a little in *Hemigale*), but its posterior border is opposite the posterior end of the tooth row.   Between the anterior end of the pterygoid crest and the posterior border of the tooth row is the usual notch of the Carnivora, but the maxilla behind the tooth row is extended into a rather prominent crest, supported by a buttress from the external side of the pterygoid crest.

The anterior base of the zygomatic arch springs from above m^{2-3}; the jugal appears to be small and does not extend very far backward, its posterior tip not approaching the glenoid articulation as nearly as it does in most modern Carnivora, and the anterior end, although its sutures are very poorly preserved, appearing to have no great expansion upon the face.   The anterior portion of the zygomatic arch has about the same depth and form as in *Hemigale;* the posterior squamosal process is heavier.

The postorbital region is long and narrow, owing to the small brain-case, resembling Insectivora and marsupials rather than any Carnivora, and most nearly approached among the creodonts by *Didymictis* and *Sinopa*.   The pterygoid region is badly disorganized, so that little can be determined save that the posterior nareal gutter is narrow, but not in any degree overarched by the pterygoid processes of the palatines.

The glenoid fossa is shallow, of little transverse width, with a moderate postglenoid process and no preglenoid crest.   The postglenoid foramen is very large.   The tympanic ring is not preserved; there is no indication of a tympanic bulla.   The petrosal is present, but displaced, so that its inner border appears inferior; in normal position it would apparently be oval or pear-shaped, with a flattened infero-external face and a large fenestra rotunda near its posterior end.   The mastoid has a considerable exposure, chiefly inferior, and a short but well-defined mastoid process buttressed anteriorly by a very slight post-tympanic process.   The paroccipital processes are broken off.   The condyle is of moderate size, longer anteroposteriorly than in *Hemigale*, but much less extended transversely and less convex.   It compares more nearly with *Didelphys*, but shows a deep notch on the external margin.   The foramen for the ?vertebral artery perforates the occiput just above the condyle as in Carnivora, but is external instead of internal to the margin of the foramen magnum.

The occiput is imperfectly preserved, but shows the usual deep fossa above the condyles characteristic of the Creodonta.   The occipital crest is broken off so that its height cannot be determined.

The sagittal crest is broken off, but was apparently of moderate height, continuous forward to the postorbital crests, which are very slight and obscure, with no fossa between them.   The limits of the bones of the top of the skull cannot be certainly distinguished.

The lower jaw is of moderate depth, but considerably deeper than in *Hemigale*.   It compares with that of Miacidae, especially *Miacis* and *Didymictis*.   The condyle is rather large, but of less transverse extent than in any miacids and much less than in fissipedes generally.   The angle is quite wide (vertically) but the tip is broken off, so that its length cannot be seen.   The coronoid process is of moderate height, somewhat recurved, but its complete length is not shown.

The teeth are well preserved, except that three of the anterior teeth are out of the alveoli.   The upper incisors are quite small, short, subspatulate teeth, the crowns not expanded and the tips of oval outline, the edges flattened.   A part of the premaxilla with

i^2 in place shows that they were all of subequal size, the lateral incisor but little larger than the others.   The upper canines are strongly recurved, the roots very long, flattened oval in cross section, and with no trace of either anterior or posterior ridge.   The enamel extends down less than a third the entire length of the canine (including the root) even upon the outer surface, and less than a quarter upon the inner face.   This peculiar hook-like character of the canines is not like any other creodonts with which I am acquainted; some of the hyaenodonts and oxyaenids approach it more nearly than the Miacidae.   The first upper premolar was a small one-rooted tooth, close to the canine; a small tooth which had dropped out of the alveolus is either the first upper or first lower premolar.   It has a flattened oval crown, moderately recurved and crested in front and behind, with round-oval root much longer than the crown.   The second premolar is two-rooted, with the usual compressed form, of moderate height, somewhat recurved and pointed, with a very slight posterior heel, and no trace of cingula.   The third premolar is triangular, somewhat wider than long, with a small internal cusp and more rudimentary cusps at the external angles, and an external cingulum.   The principal cusp (paracone) is a little less in height than in length, somewhat crested anteriorly and posteriorly, with no posterior accessory cusp.   The fourth premolar is like the third save for greater transverse width and much larger and subcrescentic internal cusp (protocone), with slight basal cingula on its anterior and posterior sides in addition to the external cingulum.   The protocone is set farther forward, so that it is more antero-internal than internal to the paracone.

The upper molars are tritubercular, with very rudimentary hypocones appearing on the postero-internal cingula of m^1 and m^2.   The cusps are low, the paracone and metacone well separated, conules small but distinct.   The external cingula are well developed and continuous, and basal cingula extend around the inner half of each tooth, interrupted only on the internal face of m^1.   They are between triangular and quadrate in outline, the second molar considerably larger than the first, the third somewhat smaller, with the metacone and adjoining portions of the tooth reduced so as to give it a more oval outline.

The lower incisors are unknown.   The lower canine is shorter and smaller than the upper, procumbent at the base, but considerably recurved, round-oval in cross section, with an obscure anterior crest.   The first lower premolar is small, one-rooted, close to the canine; the second two-rooted, apparently rather small and much compressed, with short diastemata before and behind it; but the crowns of these two teeth are not preserved. The third premolar is compressed ovate at the base, wider behind than in front, the principal cusp slightly recurved, crested anteriorly and with a short median crest on the posterior face.   It has a small heel and minute anterior basal cusp, but no cingula and no accessory cusps.   The fourth premolar is a little larger and more robust, with larger heel and anterior basal cusp, and a rudimentary postero-internal cusp on the flank of the protoconid, corresponding to the better developed cusp in this position in *Chriacus* and *Deltatherium*.   The lower molars have low trigonids and wide-basined heels, the heels of m_1 and m_2 being slightly wider and longer than the trigonid.   The metaconid and protoconid are of equal size and are well separated; the paraconid is lower and smaller, but well developed, and lies between the internal border and median line in position.   The hypoconid is subcrescentic, slightly, if at all, more externally placed than the protoconid; a median notch on the posterior margin separates it from the entoconid ledge of nearly equal height and size, with a very rudimentary hypoconulid median-internal in position but obscurely separated from the entoconid by a slight notch.   The second molar is a little larger than the first; the third is narrower but somewhat longer, with paraconid more median, heel much narrower but longer, the hypoconulid well developed, median in position, but still closely connate with the entoconid.

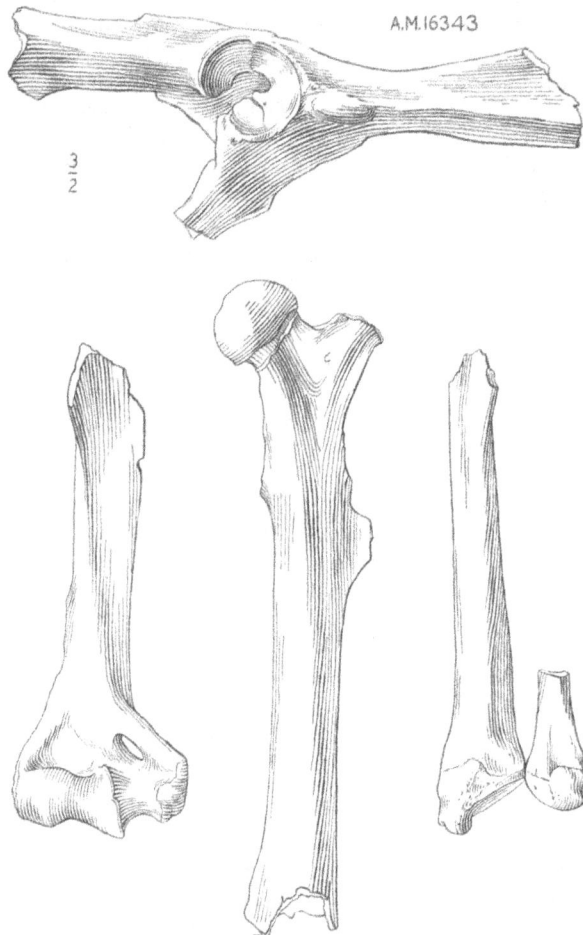

Fig. 7.   *Loxolophus hyattianus*, pelvis, right side, acetabular view; left humerus, anterior view; left femur and distal parts of tibia and fibula, anterior view, A. M. No. 16343.   One and a half times natural size.

SKELETON.   The portions of the skeleton associated with the skull and jaws described above consist of four incomplete limb bones and a large part of the pelvis, the astragali, calcanea and navicular, and several metapodials and phalanges.

The humerus is of about the size of that of the kinkajou (*Potos*), from which it differs in the prominent deltoid crest, ending somewhat abruptly a little above the middle of the shaft, supinator crest weaker than in *Potos*, and the distal end of the bone expanded much more abruptly with the entepicondylar bridge somewhat heavier and decidedly more oblique, the entepicondylar foramen larger and more roundly oval.   The entepicondylar process is heavier and somewhat more prominent.   Nothing is preserved of radius or ulna or of any of the carpal bones.

In the fore foot the second, third and fifth metacarpals and the proximal end of the fourth are recognizable.   They have about the same proportions as in the kinkajou but considerably larger size, in spite of the equal sizes of the humeri in the two.   The shafts are somewhat heavier and less regularly cylindrical, the distal facets are not so convex dorsally.   The phalanges of the first row are somewhat longer than in *Potos* and the shafts not so much flattened dorsally, with a distinct tendency to a median ridge on the dorsal face.   The phalanges of the second row are much longer than in *Potos* or any other modern Carnivora so far as we can find, rather narrow and compressed as in Carnivora, hence adapted to a compressed claw, but none of the ungual phalanges is preserved.

The pelvis is of the usual creodont type, with long, heavy ischium, the ilium somewhat expanded along its upper border, the ischial process distinct but not very prominent, neck of ilium long, inferior border thickened by a heavy external crest, superior half thinned out to a narrow border.

The femur is of moderate length, the shaft oval, head well offset inwardly, great trochanter broken so that its full outline cannot be determined, lesser trochanter well developed, more internal than posterior in position, and rather low down; third trochanter quite strong, situated about a third down the shaft.   The distal end is not preserved save for a small fragment.

The tibia lacks the proximal third, but was apparently nearly as long as the femur. The shaft is straight and not flattened or oval, the cnemial crest comes down nearly to the middle of the shaft, fading out gradually.   The distal end has a very oblique astragalar facet, nearly flat save for the internal malleolar portion; the internal malleolus is prominent and heavy, projecting well below the rest of the shaft, and ending in a large tendinal rugosity.   The fibula apparently barely touches the tibia distally, as there is no facet on either bone; it is comparatively large and heavy in the shaft and distal end, the distal facet nearly as large as that of the tibia would be with the malleolar portion left off; it is correspondingly oblique, somewhat convex from within outward, somewhat concave anteroposteriorly.   Not all of this facet is for the astragalus; a part of it rests upon the calcaneum, the facets for the two bones being obscurely separated.

The hind foot is represented by the right and left astragali and calcanea, right navicular, the first and fifth metatarsals right and left, and the proximal ends of the second and fourth metatarsals.

The astragalus is of very primitive creodont type, the body being short, wide and rather shallow, with trochlear groove very indistinct, the inner crest not developed, the fibular facet very oblique, facing almost as much proximad as ectad, the astragalar foramen large. The neck is wide, flat and rather short, the head wide and flattened oval, showing on the internal distal face a distinct facet which is probably for the tibiale (cf. *Claenodon*, in which this bone is preserved).   In general the proportions and construction of the astragalus compare best with those of *Claenodon, Miacis* and other primitive creodonts, but it retains more fully the general primitive structure that we ascribe to the pro-creodont type.

The calcaneum is likewise of uniformly primitive type, with wide stout tuber calcis, little compressed and of moderate length; the sustentacular and astragalo-calcanear facets are wide, the external border of the latter very oblique to the axis of the bone (about 40°, while in modern Carnivora it is usually nearly parallel), and external to this border is a distinct and considerable facet for the fibula; the cuboid facet has the usual quadrate form and dorsal overlap of the primitive creodonts, facing plantad and entad as well as distad.

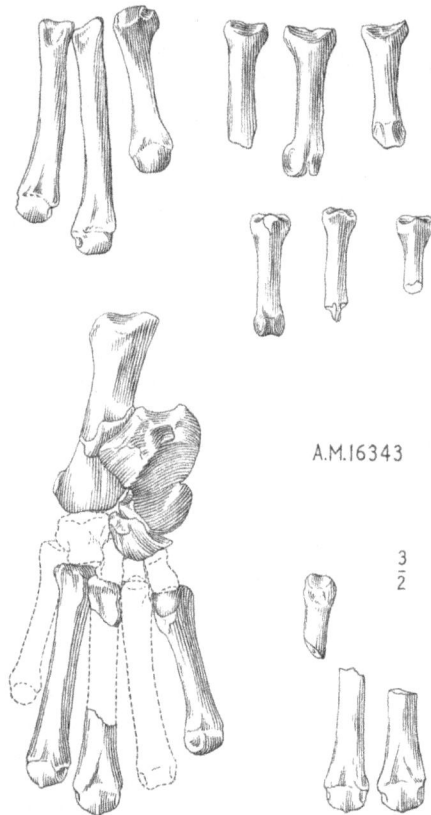

FIG. 8.   *Loxolophus hyattianus*, foot bones and right pes, A. M. No. 16343, anterior views.   One and a half times natural size.

The navicular is comparatively short and wide, conformant with the characters of the astragalar head, on which, however, it covers only the navicular facet, leaving the supposed tibiale facet clear and distinct.   The ectocuneiform facet is smaller than the mesocuneiform facet.

The metapodials are of nearly the same length as in *Potos* but notably heavier, especially in the shafts.   The external process on the head of Mt. V is somewhat less developed proportionately, but the head of Mt. I is more massive.   The distal facets are somewhat less convex dorsad, as in the metacarpals. A proximal phalanx referred to the pes is decidedly longer and wider than in *Potos*, the distal end notably broad, suggesting that the hind claws were less compressed than in that genus.

From the above data it is evident that *Loxolophus* had the general skeleton proportions of the arboreal Viverridae or Procyonidae, but with even longer phalanges and less evidence of re-adaptation from a more terrestrial type.   It agrees well with other primitive creodonts in numerous details of construction.   The elongate ischium, third trochanter on femur, cnemial crest and flat oblique astragalar facet of tibia, heavy shaft of fibula, the wedging up of the head of the astragalus between tibia and fibula, the fibulo-calcanear facet, the broad, short, shallow astragalus with nearly flat trochlea, very oblique fibular facet, large astragalar foramen, short neck and wide, rather flat head with a supposed tibiale indicated; the calcaneum overlapping a little dorsad on the cuboid; the short, wide, shallow navicular, unreduced pollex and hallux and long phalanges may all be cited as in this category.   Nearly all of them are shared by *Claenodon*, most of them by the more primitive Miacidae, and the earliest Oxyaenidae and Hyaenodontidae are but little different in any of these particulars. It will be seen that the most primitive of the Mesonychidae, Condylarthra and Taligrada are not far removed from this type, and although diverse in adaptation appear to be fundamentally more nearly allied than are such primitive insectivoran types as *Prodiacodon* and *Onychodectes*.

It appears probable that the skeleton construction and most of the characters of skull and teeth of *Loxolophus* present the closest known actual approach to the common primitive type from which the placental mammals, or most of them, are derived.

Loxolophus priscus (Cope)

Chriacus priscus Cope, 1888, Trans. Amer. Phil. Soc., XVI, N.S., p. 337, fig. 6; (*Protochriacus*) Scott, 1892, Proc. Acad. Nat. Sci. Phila., XLIV, p. 296; Osborn and Earle, 1895, Bull. Amer. Mus. Nat. Hist., VII, p. 22; Matthew, 1897, *ibid.*, IX, p. 269.

Type: A. M. Cope Coll. No. 3108, skull crushed and incomplete, with p³–m³ left, damaged, m²⁻³ right, muzzle broken off in advance of teeth, part of lower jaw with right m₁₋₂ associated, also an upper(?) canine.

Horizon and Locality: Puerco formation, San Juan basin.   Collected by David Baldwin in 1887.

Author's Description: The superior molars are of the same size as those of the *C. pelvidens*, but are very different in form.   The absence of the strong internal angles of the crown at the inner extremities of the anterior and posterior cingula, is one character.   The straight outline with an open margination of the external side of the crown in *C. pelvidens* is in strong contrast with the two convexities, each following an external cusp which form the outline in the *C. priscus*.   The first premolar [p⁴] has a small internal cusp in the *C. priscus;* a large one in the *C. pelvidens*.   [The second premolar (i.e. p³) has a distinct internal cusp in *C. pelvidens*.²⁷]   It is represented by an angular cingulum in the *C. priscus*.   In this species the cingulum does not extend round the inner base of the crown, except weakly in the last molar.

The inferior molars have a relatively greater transverse diameter than those of any other species of the genus.   They also differ from those of the *C. pelvidens* and *C. stenops*, which they resemble in size, in the nearly equal elevation of the anterior and posterior cusps, and in the absence of an external basal cingulum. The fifth cusp is a small cone, and is not spread away from the fourth and connected with it by a crest, as in

²⁷ Some words in Cope's text are evidently omitted and are conjecturally supplied as above.—w.D.M.

MEASUREMENTS (IN MM.) OF *Lozolophus priscus*

Measurement	Type 3108	16356	3109	786a	3113	16358	16359	16361	3111	802	811
Diameters, p^2, ap.×tr.	6×4.9		4.9×4.2								
" p^3, "	5.5×5.7	5.2×5.9									
" p^4, "											
Upper molars, m^{1-3}	19.0	{16.6 / 17.8}	17.8		18.8						
Diameters, m^1, ap.×tr.	6.6×6.9	6.1×7.0	6.2×6.9		7.0×7.4						
" m^2, "	7.2×8.8	6.9×8.3	6.3×8.1		7.1×8.8						
" m^3, "	5.0×7.7	5.1×7.5	5.3×7.2		5.2×7.3						
Skull length, m^3 to condyle	42.7										
" width at postorbital constriction	21.4										
" across condyles	(e)* 20										
Basicranium across auditory prominences	24.0										
Palate, width across m^2	34.4										
Lower molar-premolar series		44.8					(e) 43.7	(e) 44.0			(e) 45.5
Diameters, p_1 (root)		4.0×1.8									4.0×2.2
" p_2, ap.×tr.		6.0×2.3		6.1×3.0							5.1
" p_3, "		6.0×3.3		5.8×3.4						5.8×3.7	5.2
" p_4, "		6.0×3.6		5.5×4.0			6.0×3.1	5.8×3.9	6.2×4.0	6.0×3.9	5.5×3.9
Lower molars, m_{1-3}		20.0	19.5	22.0		22.0			20.0	23.2	
Diameters, m_1, ap.×tr.	6.5×5.8	6.3×5.1	6.5×5.2	7.0×5.8		7.3×5.6	6.9×4.3	6.5×5.3	6.1×5.0	7.1×5.8	6.1×4.8
" m_2, "	7.1×6.3	6.3×5.7	6.3×5.7	7.5×6.7		7.0×6.5	7.1×5.2	6.7×5.9	7.0×6.5	8.2×5.5	7.3×4.8
" m_3, "		7.7×5.0	7.0×4.8	8.0×5.9		7.1×5.0			6.9×5.1		
Lower canine, length of crown	7.5										
Diameters, c_1, ap.×tr.	4.9×3.3										
Lower jaw, depth beneath m_2	12.1	12.3		13.1			13.0	13.8	12.2		14.0
" " " p_2							10.0	10.5			12.0

* (e) = estimated, in all such tables of measurements.

C. stenops. In specimens with the last inferior molar preserved, that tooth is seen to be of average proportions.

The brain-case is long, narrow and rather low, and the sagittal crest is low and thin. It does not in the least resemble that of Adapis, but is rather that of a Creodont. Postorbital region lost.

Revised diagnosis: One fifth larger than *L. hyattianus,* cusps of teeth lower and more massive; upper molars more quadrate, lower molars broader, premolars more robust, $m_{\overline{3}}^{3}$ less reduced. Skull behind orbits shorter, wider, occipital crest more prominent. Canine short, stout, little recurved, somewhat compressed and crested on anterior and posterior margins. Lower jaw scarcely deeper than in *hyattianus,* but much heavier. Cusps of upper and lower teeth distinctly striated, with vertical anastomosing ridges.

In addition to the type, a series of upper and lower jaws or parts of jaws is referred to this species, as follows:

No. 16356, upper and lower jaws, p^4–m^3 r. and l., p_3–m_1 and m_3 l.

No. 3109, upper and lower jaws, p^3 and m^{1-3} l., m^3 r., m_{1-3} l., p_4–m_3 r., and separate canine.

No. 16361, lower jaws and distal end of humerus, m_{1-2} r., p_4–m_1 l.

No. 786a, lower jaw, p_2–m_3 l.

Nos. 16358, 16359, 16360, parts of lower jaws; 3111, 3110, small variants.

No. 3113, upper jaw, m^{1-3} r.

No. 802, lower jaw, p_3–m_3 r.

Nos. 811, 813, 818, parts of lower jaws.

All the above are from the Puerco formation. The specimens collected in 1913, and probably these of the earlier collections, are from the upper or *Taeniolabis* fossiliferous level.

?Loxolophus interruptus (Cope)

Deltatherium interruptum Cope, 1882, Paleont. Bull. No. 35, Nov. 11, and Proc. Amer. Phil. Soc., XX, p. 463.

Type: a lower jaw with part of canine and p_4–m_1. Type not certainly identifiable in the Cope Collection. The species is indeterminate.

Horizon and Locality: Paleocene (?Puerco), San Juan basin, New Mexico.

Author's Description: The smallest species of *Deltatherium* is . . . represented by the anterior part of a right mandibular ramus, which supports the last premolar and the first true molar.

Protogonodon Scott, 1892

Scott, W. B., 1892, Proc. Acad. Nat. Sci. Phila., XLIV, p. 322; Earle, Charles, 1893, Amer. Nat., XXVII, p. 377; Osborn, H. F., and Earle, Charles, 1895, Bull. Amer. Mus. Nat. Hist., VII, p. 67; Matthew, W. D., 1897, *ibid.,* IX, p. 302

Type: Mioclænus pentacus Cope, from the Puerco formation, New Mexico.

Author's Diagnosis: To this genus I refer as a type species the *M. pentacus* which Cope provisionally incorporated with *Mioclænus,* though directing attention to the resemblance of its inferior premolars (superior unknown) to those of the phenacodont *Protogonia,* from which it differs in the simplicity of $p_{\overline{4}}$, which has no deuteroconid. Certain specimens, however, show rudimentary indications of it. I think there can be no doubt that this genus is referable to the *Phenacodontidæ.* A second species, *P. (Mioclænus) lydekkerianus,* the structure of $p_{\overline{4}}$ in which is not known, probably belongs to the same genus.

Osborn and Earle in 1895 provided a more adequate diagnosis:

Superior true molars tritubercular without a hypocone.   Both intermediate tubercles distinct, but tending to coalesce with protocone, forming an internal crescent.   Last inferior premolar simple in structure, and showing only in some specimens an indication of a deuteroconid.   Inferior true molars with trigonid not raised above talon, and with paraconid well marked.   Lower jaw long and slender.

Although retained in the Condylarthra, it was regarded as probably of artiodactyl affinities, as Earle has advocated in an earlier note.   Matthew in 1897 returned to Scott's view, although less positively expressed.

The resemblance to certain species of Oxyclaenidae, in particular to *Loxolophus priscus*, was by no means overlooked by any of the authors cited.   But this resemblance was weakened not only by the imperfection of the material available, so that it was by no means so apparent as now, but also by the uncertainty as to the true systematic relations either of the Oxyclaenidae in general or of *Loxolophus* in particular.   In the light of such comparisons as can now be made it seems reasonably certain that *Protogonodon* is quite closely related to *Loxolophus* and the skull and skeleton parts of the latter described in this memoir show it to be a primitive creodont.   Whatever the phyletic relations of *Protogonodon*, it is structurally an oxyclaenid creodont, not a condylarth.   Whether it is in any direct or approximate sense ancestral to *Tetraclaenodon* and *Phenacodus* is doubtful.   The resemblances, such as they are, are of a rather slight and superficial kind.   Nevertheless it would appear probable that *Tetraclaenodon* is derived from an ancestor not unlike *Protogonodon* in structure of teeth and like the Oxyclaenidae in skeleton characters, and that its relationship thereto was not in fact very remote, although probably hardly as near as the Puerco to the Torrejon.

The argument for artiodactyl affinities is quite unconvincing.   The lower jaw is not any more slender anteriorly than in most Oxyclaenidae or in the great majority of Creodonta and Carnivora.   The simplicity of the fourth premolar is also a common creodont character and affords no evidence for artiodactyl affinities of this more than many other creodont genera.   So, too, with the simple tritubercular molars—they are merely generalized and primitive characters of the Creodonta.   On the other hand, there are certain special characters that distinguish the early tritubercular Artiodactyla (*Diacodexis*, etc.) from other early primitive types: the peculiar elongation of the *third* premolar, the internal position of the paraconid and its twinning with the metaconid, the large m_3 with its peculiar heel, and other details of construction that enable us to distinguish jaw fragments of *Diacodexis* and its allies from other genera with superficially similar dentition.   Of these there is no foreshadowing in *Protogonodon*.   Nor is there any astragalus known from the Paleocene that shows any distinctive approach to the artiodactyl type, which in the Lower Eocene appears fully developed and fixed in type, much as in small modern artiodactyls.   Matthew in 1897 distinguished under the name of *Protogonodon stenognathus* certain specimens (upper and lower jaws) of smaller size with less robust and massive teeth.   In 1888 Cope described as *Chriacus rütimeyeranus* a lower jaw with m_{1-3} which Scott subsequently made the type of his genus *Paradoxodon*.   This jaw, we believe, may be a crushed specimen of *Protogonodon*.   The construction of the molars, if normal, would be wholly peculiar, but it is probable, from the thinness of the jaw as preserved and from the indications of crushing on the second molar, that these teeth have been deformed by the peculiar type of crushing noted on a preceding page of this memoir, although the enamel surface on m_1 and m_3 does not show any marks of breaking and crushing but has been narrowed uniformly throughout.

That this may occur is shown by another specimen in which m^{1-3} are preserved; m^{2-3} have retained their proportions but the cusps have partly collapsed and lost their fullness; m^1 has been reduced in transverse diameter and all the cusps narrowed transversely, throwing the tooth quite out of proportion with m^{2-3}; yet the surfaces of the enamel in all three teeth appear practically unbroken.   In *Loxolophus hyattianus* type the three upper molars have been crushed transversely in a corresponding manner.   If this interpretation be accepted, *Paradoxodon rütimeyeranus* is probably, although not certainly, the same as *Protogonodon stenognathus*.

Aside from any crushing of the teeth there is a rather wide range of variation in the width of outer and inner cingula of the upper molars, the basal cuspules at the outer angles of the upper premolars, size and massiveness of both upper and lower teeth, development of deuteroconid on p^4 (from wholly absent in the type to fairly strong in No. 16398), and other details.

The skull and skeleton are unknown; the reference of the genus to Oxyclaenidae rests upon the near resemblance to *Loxolophus* in all details of construction of the teeth.

Protogonodon pentacus (Cope)

Mioclaenus pentacus Cope, 1888, Trans. Amer. Phil. Soc., XVI, N.S., p. 325; (*Protogonodon*) Scott, 1892, Proc. Acad. Nat. Sci., Phila., XLIV, p. 322; Earle, 1893, Amer. Nat., XXVII, p. 377; Osborn and Earle, 1895, Bull. Amer. Mus. Nat. Hist., VII, p. 67; Matthew, 1897, *ibid.*, IX, p. 302.

Type (lectotype): A. M. Cope Coll. No. 3192, part of lower jaw.

Cotypes: Nos. 3193–7, 3948, part of lower jaws.

Horizon and Locality: Puerco formation, upper level as far as recorded.

Author's Diagnosis: The inferior molars are robust and Phenacodus-like, but the fifth cusp is well developed in all of them, forming an anterior triangle.   The three cusps of this triangle are only connected at their bases.   The heel is wide and supports a large external marginal cusp and a curved raised margin behind and within, which is notched at two points so as to produce two narrow areas on wear.   On the third true molar these are represented by distinct but rather small tubercles.   The second true molar is distinctly larger than the first or the third.   The external tubercles of all of the true molars stand within the base of the crown, so that their external faces are unusually oblique.   They have a crescentic section, and the posterior sends its anterior ridge obliquely to the base of the posterior of the anterior internal cusps, enclosing a basin as in Chriacus.   There is a delicate cingulum round the base of the crown on all except the internal sides.   The first premolar is a robust tooth, with a wide narrow heel, and a small anterior basal tubercle, besides the principal subconic cusp.   A delicate external basal cingulum.   The third premolar has a minute heel, and no external or anterior cingulum.   Fourth premolar one-rooted.   A short space between third and fourth premolars.   Enamel minutely rugose, but polished.

The ramus is not deep, but is robust.   Its inferior outline rises posteriorly from below the anterior part of the third true molar.   The anterior masseteric ridge is prominent.   The coronoid process rises gradually from the third true molar, without carrying that tooth with it.   The symphysis extends posteriorly to the line of the posterior border of the third premolar.   Mental foramina below the second and fourth premolars.

A number of lower jaws and an upper jaw (Nos. 949–956) were obtained in 1892 from the Puerco beds, most or all of them from the *Taeniolabis* zone.   In 1913 a series of upper and lower jaws, Nos. 16387–16395, were found in the upper level of the Puerco, all agreeing fairly well with *P. pentacus*, the species evidently being characteristic of this level and not found in the lower level or *Ectoconus* zone.   Nothing of the skull or skeleton is recognized.

Protogonodon stenognathus Matthew

Protogonodon stenognathus Matthew, 1897, Bull. Amer. Mus. Nat. Hist., IX, p. 302.

Type: A. M. Cope Coll. No. 3198, parts of the lower jaw with the three true molars.
Paratype: No. 761, upper jaw with p^4–m^3 and root of p^3.
Horizon and Locality: Puerco formation, San Juan basin, New Mexico.
Diagnosis: The teeth in this species are somewhat narrower and less massive than in
P. pentacus, but it appears to be closely related and of somewhat doubtful validity.   No
specimens from the 1913 collections are referable to it.

?Protogonodon kimbetovius, new species

Type: No. 16397, lower jaws with dp$_4$–m$_2$ of both sides and the point of a partly emerged
p$_3$.

Horizon and Locality: Puerco formation, New Mexico.   The type was found three
miles above Kimbetoh, in the San Juan basin.
Diagnosis: This is a slightly larger species than *Protogonodon pentacus,* with somewhat
different proportions between trigonid and talonid.   The trigonid on m$_1$ is of the same
width and on m$_2$ is wider than the talonid, whereas in *pentacus* the trigonid is narrower
than talonid on m$_1$ and of equal width on m$_2$.   The inference would be that m$_3$ is considerably smaller than in *P. pentacus.*   The dp$_4$ is unfortunately unknown in the type
species, so that further comparisons cannot be made.

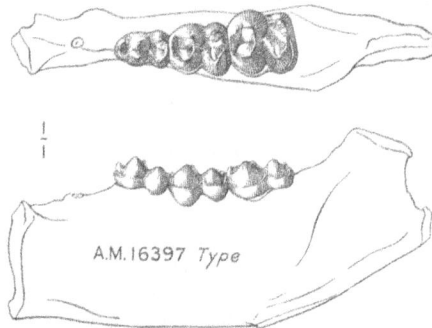

Fig. 9.   *Protogonodon kimbetovius,* new species, type mandible, A. M. No. 16397, crown view above, side view below.
Natural size.

?Protogonodon protogonioides (Cope)

Mioclaenus protogonioides (in part) Cope, 1882, Amer. Nat., XVI, Sept. 28, p. 833; 1884,
     Tertiary Vertebrata, p. 340, pl. xxv*f*, fig. 17; 1888, Trans. Amer. Phil. Soc., XVI,
     N.S., p. 329; (*Claenodon*) Scott, 1892, Proc. Acad. Nat. Sci. Phila., XLIV, p. 299;
     (*?Claenodon*) Matthew, 1897, Bull. Amer. Mus. Nat. Hist., IX, p. 291.

Type: A. M. Cope Coll. No. 3253, upper jaw fragments with m$^{2-3}$, right and left.
Horizon and Locality: Lower Paleocene, Puerco formation, New Mexico.

Author's Description: The largest species of the genus,[28] represented by the superior true molars. It is an exaggerated form of the *M. subtrigonus* [*Tricentes subtrigonus*]. The internal angle of the V, as well as the intermediate tubercles at the ends of its limbs, are distinct. Cingula extending entirely round the crown, the posterior with a small tubercle on the M.II, as in *A. subtrigonus;* none on M.III, which is .75 the area of the M.II. [Measurements follow.]

The type of this species is from the Puerco, and consists of a fragment of the upper jaw with m^{2-3}. Upper and lower jaw fragments from the Torrejon were referred to the species by Cope and Scott. Matthew in 1897 doubted the correctness of this reference. Specimens obtained in 1913 include associated upper and lower jaws which prove that the Torrejon specimens are quite distinct and agree with *Claenodon* in the vestigial character of the paraconids and some other characters, while the Puerco species is nearer to *Protogonodon*.

No. 16396, an upper jaw with p^3-m^3, agrees fairly well with the type of *protogonioides*. The hypocones of the molars are much more rudimentary than in *Claenodon*, and the premolars have the form of *Protogonodon* with the cusps at the external angles very little developed, inner cusp on p^3 distinct, and in various details it is nearer to *P. pentacus*.

No. 16398, a lower jaw with p_4-m_3, is provisionally referred to *P. protogonioides*. It agrees in size with No. 16396, and comes from the same level and locality, the lower level of the Puerco beds. The molars are like those of *P. pentacus* and *stenognathus* except for smaller size; the premolar is similar except for a fairly well-developed deuteroconid, stronger than in any specimen of *pentacus*.

The Torrejon specimens which have been referred to this species represent a species of *Neoclaenodon* described under that genus.

Carcinodon Scott, 1892

Scott, W. B., 1892, Proc. Acad. Nat. Sci. Phila., XLIV, p. 323

Type: Mioclaenus filholianus Cope, from the Puerco formation of New Mexico.

Diagnosis: Lower molars increasing in size from first to third, trigons relatively small, height intermediate between *Oxyclaenus* and *Loxolophus*. Premolars like *Loxolophus*, but less robust.

This genus is of doubtful validity. It is partly intermediate between *Loxolophus* and *Oxyclaenus*. No upper teeth are certainly referable to it.

Carcinodon filholianus (Cope)

Mioclaenus filholianus Cope, 1888, Trans. Amer. Phil. Soc., XVI, pp. 303, 329; (*Carcinodon*) Scott, 1892, Proc. Acad. Nat. Sci. Phila., XLIV, p. 323.

Type: A. M. Cope Coll. No. 3205, lower jaws with p_2-m_3 left, p_4 and m_{1-3} right.

Horizon and Locality: Lower Paleocene, Puerco formation, San Juan basin.

Diagnosis: The species is considerably smaller than *Oxyclaenus simplex*, from which it is readily distinguished by the long narrow molars and large size of m_3. The trigonid cusps have the usual height of *Oxyclaenus*, but the paraconid is more median, as in *Loxolophus;* external cingula are distinct, as in *Oxyclaenus*. The premolars are very like those of *Loxolophus attenuatus*, but the heel of p_4 is not broadened and m_1 and m_2 are much smaller, narrower and higher cusped.

---

[28] *M. ferox* and *corrugatus* were not described until later.—w.d.m.

The species is represented by the type and No. 3206, a lower jaw with m_{2-3} l. No. 16415, an upper jaw with part of the skull poorly preserved, is possibly referable to this species, but in absence of associated upper and lower teeth this is very doubtful. The upper molars provisionally referred here (No. 16415) are of appropriate size, tritubercular, with a rudimentary hypocone on m^1 and m^2, m^3 trigonal, equal to m^2 in size, the anterior border transverse but metacone unreduced.

FIG. 10. *Carcinodon filholianus*, type left $p_3 - m_3$, A. M. No. 3205, crown view. Twice natural size.

MEASUREMENTS (IN MM.)

	No. 3205 *Carcinodon filholianus* type	No. 16346 *Oxyclaenus cuspidatus*	No. 16352 *Oxyclaenus cuspidatus*	No. 3252 *Oxyclaenus cuspidatus*
Length, m_{1-3}	16.8	17.5		
Diameters, m_1 ap. × tr......	4.8×3.1	5.3×4.0		
" m_2 "	5.0×3.9	6.0×4.7	6.8×5.0	
" m_3 "	6.5×3.5	6.0×3.8	6.2×4.7	
" p_3 "	4.7×2.9	6.9×2.2		
" m^2 "			6.6×8.9	6.7×8.8
" m^3 "			4.2×7.3	4.2×6.3

Tricentes Cope, 1884

Cope, E. D., 1884, Paleont. Bull. No. 37, Jan. 2, and Proc. Amer. Phil. Soc., XXI, p. 315

Type: Tricentes crassicollidens Cope, from the Torrejon formation of New Mexico.

Author's Description: Char. gen. This genus is *Chriacus* with only three premolars in the superior, and probably inferior series. The canines are well developed, and lateral in position, leaving space for small incisors, thus differing from the genera of the Mixodectidæ,[29] *Mixodectes, Microsyops,* and *Cynodontomys,* on the one hand, and from *Necrolemur* on the other. It has, so far as known, the dental formula of several genera of typical Lemuridæ, but differs from these in the following points. The orbit is open posteriorly; the inferior molars have the anterior triangle of three cusps; and the fourth inferior premolar has an interior cusp. I have demonstrated the last mentioned characters on the type, *T. crassicollidens* only, but suspect its presence on some or all of the other species. In their details the superior true molars are like those of *Mioclænus,* as distinguished from those of *Pelycodus.*

To this genus belongs the *Mioclænus subtrigonus,* and probably, from the small size of its fourth premolar, the *M. bucculentus.* I add to these three a fourth, *T. inæquidens,*[30] and remark that it is yet uncertain how many premolars are present in the *Chriacus simplex.* Should the latter possess three only, it will be properly referred to *Tricentes.*[31]

[29] Cope at this time believed the enlarged front teeth of these genera to be canines. They have since been determined to be incisors in *Mixodectes.*—W.D.M.

[30] Type of *Ellipsodon* Scott, 1892.—W.D.M.

[31] Referred now to *Oxyclænus.*—W.D.M.

Scott in 1892 [32] refers the genus to the Oxyclænidæ, with the following description:

This genus is very closely allied to *Protochriacus*, but differs in the absence of p^1. The premolars are compressed, acute, very high and simple, except p^4 which has a small deuterocone. $M^{1 \text{ and } 2}$ have a nearly quadrate shape, produced by the well developed hypocone and are surrounded by a stout cingulum. M^3 is the smallest of the series, the hypocone is absent and the metacone reduced. The canine is large and separated from p^2 by a considerable diastema. Inferior dentition unknown. The face is very short and the anterior edge of the orbit is over the space between p^4 and m^1. The forehead is flat, the supraciliary ridges [postorbital crests] short and converging rapidly to form the sagittal crest.

Two species may certainly be referred to this genus: *T.* (*Mioclænus*) *bucculentus* Cope and *T. crassicollidens* Cope, Puerco. A third species is doubtful, viz., ?*T.* (*Mioclænus*) *subtrigonus* Cope, in which the number of upper premolars is not known, but the tooth structure agrees closely with *Tricentes*. Puerco.

Osborn and Earle in 1895 [33] define the genus as follows, placing it *incertae sedis*, but following *Protochriacus:*

Dentition: I?, $C_{\bar{1}}^1$, $P_{\bar{3}}^3$, $M_{\bar{3}}^3$. Premolars spaced and conic. Molars with rounded tubercles, hypocone well developed. Molars irregular in size; third molar reduced. Trigonid slightly elevated. Paraconid reduced.

Matthew in 1897 [34] places it under the Oxyclænidae, following *Protochriacus* [= *Loxolophus*], with the diagnosis here cited:

Dentition: $I_{\bar{7}}^?$, $C_{\bar{1}}^1$, $P_{\bar{3}}^3$, $M_{\bar{3}}^3$. Hypocone moderate, no protostyle [on molars], very rudimental deuterocone, or none, on p^3 and p_4. Cusps conical and blunted. Canines well developed in both jaws, incisors small or reduced. A considerable diastema behind the upper and lower canines, which are short and directed nearly vertically.

Tricentes differs from *Protochriacus* in the more conical form of the cusps, the loss of the first premolar and close setting of the remaining ones, and the reduction of the paraconid. From *Chriacus* it differs in the simpler premolars, less development of hypocone, absence of $pm_{\bar{1}}^1$, canines vertical instead of projecting forward, and in the much less trenchant and more rounded molar cusps.

A somewhat detailed description of *T. subtrigonus* follows, with figures of upper and lower teeth. *T. bucculentus* is made a synonym of this species, which is regarded as rather doubtfully congeneric with *T. crassicollidens*, the type of the genus.

The very imperfect figures and vague descriptions of the Cernaysian genus *Creoadapis* Lemoine [35] have recently been supplemented by the excellent photographs and figures and critical revision of Teilhard de Chardin. Père Teilhard refers the genus to the Oxyclaenidae + Arctocyonidae + Mioclaenidae, which he rightly regards as a group of nearly related genera. It appears to be nearly related to *Tricentes* and *Neoclaenodon* of the Torrejon and Fort Union and to *Thryptacodon* of the Tiffany (uppermost Paleocene) and Lower Wasatch (lowest Eocene). From *Tricentes subtrigonus* it may be distinguished by the different proportions of the lower premolars, p_{3-4}, p_3 being relatively smaller and the principal cusp not so high as that of p_4; the accessory cusps are better developed and the teeth more compressed. It agrees with *Tricentes* and differs from *Neoclaenodon* in the relative development and position of paraconid on m_1 and m_2 and in the size and proportions of these molars.

The genotype of *Tricentes*, *T. crassicollidens*, is a species known only from the type

[32] Scott, W. B., 1892, Proc. Acad. Nat. Sci. Phila., XLIV, p. 297.

[33] Osborn, H. F., and Earle, Charles, 1895, Bull. Amer. Mus. Nat. Hist., VII, p. 23.

[34] Matthew, W. D., 1897, Bull. Amer. Mus. Nat. Hist., IX, p. 270.

[35] Lemoine, V., 1893, Bull. Soc. Géol. Fr. (3), XXI, p. 361, pl. ix, fig. 1. Type *C. douvillei* from the Thanetian of France. The genus was named *Adapicreodon* by Lemoine a year earlier, *ibid.*, XX, p. cxlii, but with insufficient description to validate the name. Teilhard (Ann. Paléont. XI, p. 31) shows that both names are synonyms of *Arctocyonides* of 1891.—W.D.M.

specimen, a poorly preserved "skull" in which only the upper teeth are visible, and these have been so much damaged by attempts to clear them from the flinty matrix that their true characters are mostly uncertain.   It is of the same size as *Chriacus truncatus*, and is distinguishable from *Chriacus* by the supposed absence of p¹ (not very certainly demonstrated) and by different form of m³ clearly different from the trigonal shape in *C. truncatus*, but not very close to the equally characteristic form of that tooth in *T. subtrigonus*.   The characteristically thick rugose enamel of *T. subtrigonus* is likewise not clearly shown on the genotype.   There is, however, a sufficient general likeness in the teeth, as far as they are still preserved in the type, to warrant leaving it provisionally in the same genus with the much better-known and recognizable *T. subtrigonus*, which serves as a neotype for *Tricentes*, otherwise a practically indeterminate genus of oxyclaenid creodonts.

The generic characters as thus based are: Dentition $\frac{2 \cdot 1 \cdot 3 \cdot 3}{7 \cdot 1 \cdot 3 \cdot 3}$.   Canines of moderate size, round-oval, rather short and stout, a short but well-marked diastema behind them. Premolars rather short and high-pointed, not compressed, the basal cusps comparatively small; the molars low, rounded, quadrate with low subcircular cusps and more or less encircling cingula; m_3^3 unreduced.   Enamel of all teeth markedly rugose with short vertical wrinkles dominant.   P$\frac{2}{2-3}$ simple and high, p³ trigonal, p⁴ with large inner cusp and rudimentary basal cusps at external angles, p₄ with rudimentary inner cusp and small heel. Upper molars, m¹⁻², rounded quadrate, the hypocone and conules distinct but much smaller than the three main cusps; stylar cusps absent or very rudimentary.   M³ rounded trigonal, no hypocone or metaconule, the metacone reduced, sub-median.   M² considerably larger than m¹, m³ somewhat smaller.   Lower molars with paraconid well developed on m₁, smaller on m₂, minute on m₃, but sub-median on all and not merged with metaconid.   Entoconids and hypoconulids distinct and rounded on all three molars, hypoconids larger, also rounded.   M₂ a little larger than m₁, heel of m₃ comparatively short, the three molars of subequal length, but m₂ a little wider than the others.   Talonids moderately basined, larger than trigonids but notably wider on m₁ only.

Tricentes crassicollidens Cope

Cope, E. D., 1884, Paleont. Bull. No. 37, Jan. 2, and Proc. Amer.
Phil. Soc., XXI, p. 316

Type: A. M. Cope Coll. No. 3090, skull with broken teeth.

Horizon and Locality: Middle Paleocene, Torrejon formation, Gallegos Canyon, San Juan basin, New Mexico.

Author's Description: The *Tricentes crassicollidens* is about the size of the *Chriacus truncatus* and resembles it a good deal.   The latter has, however, a more transverse form of true molars, as compared with the present species, where the form is subquadrate.   In the present animal the premolars are smaller, and if the third (second present) has an internal cusp, it is much more insignificant than in the *C. truncatus*. These two species and the *Mioclænus opisthacus* resemble each other in the similar size, and in the true molars having the posterior inner cusp more distinct than in other species.   They differ in the dimensions of their premolars, those of the *M. opisthacus* being the largest, and those of *C. truncatus* being intermediate in size.   In the *T. crassicollidens* the anterior cingulum is also distinct.   The external cusps are conic, and are well separated, and the internal V is distinct.   The internal cusp of the fourth premolar is small and compressed, so as to be transverse.   The base of the third premolar is triangular and much longer than wide. All the superior molars, except the first premolar, are furnished with an internal cingulum, which rises into a more or less distinct apex at its anterior and posterior angles.   The first premolar is a simple cone.   The alveolus of the canine tooth is of large size.   The last true molar is not much reduced, and the first is as large as the second. This is not the case with the *T. bucculentus*, where the first is considerably smaller than the second. [Measurements follow.]

The original reference also describes a lower jaw found at the same time and at or near the same place, and "probably belonging to the same species if not to the same animal." It appears to have been referred to some other species, as it is not subsequently mentioned by Cope or identified in his collections as *T. crassicollidens*.

Tricentes subtrigonus (Cope)

Mioclaenus subtrigonus Cope, 1881, Paleont. Bull. No. 33, Sept. 30, and Proc. Amer. Phil. Soc., XIX, p. 491; 1883, Proc. Amer. Phil. Soc., XX, March 16, p. 555, and Paleont. Bull. No. 36; 1884, Amer. Nat., XVIII, p. 349, fig. 17.

Phenacodus zuniensis Cope, 1881, Paleont. Bull. No. 33, Sept. 30, and Proc. Amer. Phil. Soc., XIX, p. 492.

Mioclaenus bucculentus Cope, 1883, Proc. Amer. Phil. Soc., XX, p. 555.

Type: A. M. Cope Coll. No. 3227, part of skull with most of the palate and cheek teeth.

Horizon and Locality: Middle Paleocene, Torrejon formation, San Juan basin, New Mexico.

Author's Description: Fourth superior premolar with flattened external and conic internal cusp; . . . equilateral [in outline]; all cusps acute; true molars .0165. . . . The species differs from *M. turgidus* in the greater acuteness of all its cusps, and in the equilateral form of the fourth premolar. . . .

The inner borders of the molar teeth are shorter than the outer, especially in the last two molars. The last true molar is smaller than either of the others. The cusps are all subconical, but the internal is connected with the intermediate by ridges, which give it a triangular section. The latter form a V which is homologous with that in *Anisonchus*, but not so distinct, and the intermediate tubercles are not lost in its branches as in that genus. The posterior inner lobe of that and other genera, is represented [in "*M.*" *subtrigonus*] by a thickening of the cingulum. This cingulum extends entirely round the P-m.IV and M.I and M.II; the M.III is injured. The sides of the base of the P-m.IV are slightly concave. The enamel of all the molars is wrinkled.

Type Description of *Phenacodus zuniensis:* The least species of the genus, represented by the mandibles of two individuals. The first and second true molars are narrowed in front, and there is no distinct anterior ledge, only a minute anterior inner tubercle. The external cingulum is more distinct and the enamel is wrinkled. The fourth premolar has a short base and the inner cusp is much smaller than the principal one; it has a wide heel and an anterior basal tubercle.

Type Description of *Mioclaenus bucculentus:* Characterized by the remarkably small size of the fourth premolar, and large size of the second true molar. The first true molar is intermediate.

The fourth premolar consists of an external cone and a much smaller internal one. There is a weak posterior basal cingulum. The reduced size of the internal cone suggests the probability that the third premolar has no internal cusp, and that there may be but three premolars. In either case the species must be distinguished from *Mioclænus*.

The first and second true molars have conic, well separated external cusps, and a single pyramidal internal cusp. The intermediate tubercles are distinct. There is a posterior cingulum which terminates interiorly in a flat prominence. There is an anterior cingulum and a strong external one, which form a prominence at the anterior external angle of the crown. Enamel wrinkled. [Measurements follow.]

In 1883 Cope described the lower teeth of *Mioclaenus subtrigonus* and added a few particulars as to the upper teeth:

The second true [upper] molar is not so much longer than the first as in *M. bucculentus*, although the difference in size is very evident. The third is smaller than the first, and ovoid in outline, while the first and second are subquadrate. The external cusps are conic and widely separated and the intermediate areas are distinct. There is a cingulum all round the crown of the last two, and round that of the first except at the inner side, and at the anteroëxternal angle.

The last three inferior premolars are higher than long at the base, and are compressed and the apex acute. The posterior edge of the third and fourth is truncate, and simple. Each has a posterior cingulum which forms a narrow heel on the fourth. No other cingula. Of the true molars only the second is wanting. The form of these is like those of the *M. ferox*, with the cusps more prominent. The first only has trace of the anterior V; in the others, the two anterior tubercles are opposite and connected by a short anterior ledge. The heel of the first consists of a basin bounded by these tubercles, of which the external is pyramidal and largest. The median posterior is small. The heel of the third is narrow and prominent, and the internal lateral tubercle is represented by a short raised edge. The enamel of all the molars is wrinkled, and the inner side of the premolars is grooved with the height of the crown. A weak external cingulum on M.iii.

Matthew in 1897 (p. 270) gave the following diagnosis of the species, to which there is little to add:

Upper canines straight, rather short, directed downwards, strongly striate longitudinally, somewhat ridged posteriorly, with a considerable diastema behind. Second premolar two-rooted, high and trenchant, without cingulum. Third and fourth three-rooted, with strong cingulum all around the base, the fourth with large deuterocone. Molars with strong enveloping cingulum, except sometimes on inner side of protocone. M^1 and m^2 subquadrate, with small hypocone, m^3 triangular. Intermediates not large, but distinct.

Lower canines with sharply curved root of round-oval section, the cusp directed upward, not at all forward, and having a considerable diastema behind. Second, third and fourth premolars of nearly equal size, but successively wider and less trenchant, the third and fourth with cingulum around the base, developed into a small heel on the fourth. Lower molars with much reduced paraconid and an external cingulum more or less obsolete.

The second upper and lower molars are considerably larger than the first; the third is somewhat variable in size and shape, the upper one considerably smaller than the second, the lower never larger.

The premolar and molar cusps are marked by a characteristic interrupted striation or wrinkling longitudinal to the cusp; this is partly worn off on old individuals, but very distinct on unworn teeth.

Measurements [in meters], $m^{1-3} = .017$, No. 2399; $c-m_3 = .0455$, No. 4001. Lower diastema, .008.

A large number of upper and lower jaws and fragments of jaws may be referred to this species, but they represent a rather wide variation in size and not inconsiderable variation in relative proportions of the teeth, distinctness of stylar cusps on upper molars, amount of reduction of m_3^3, size and compression of premolars and other features. They are found in both the upper and lower levels of the Torrejon, but more abundantly in the upper beds. No constant distinctions are observed between the specimens from the upper and those from the lower level. The genus has not been found in the Tiffany or Clark Fork.

Two or three of the jaws have fragments of skull and parts of limb bones associated, from which nothing can be determined as to the affinities of the genus. The postorbital crests of the skull diverge sharply from the anterior end of the sagittal crest, with broad somewhat concave frontal surface between them, beneath which are the narrow parallel olfactory sinuses much as in *Deltatherium*. The glenoid fossa appears to be little excavated, and the postglenoid crest quite small, the postglenoid foramen of moderate size behind it. The distal end of the humerus shows a prominent entepicondyle and rather small foramen, prominent supinator crest and no signs of supratrochlear foramen.

Mixoclaenus Matthew and Granger, 1921

Matthew, W. D., and Granger, Walter, 1921, Amer. Mus. Novitates No. 13, p. 7

Type: *Mixoclaenus encinensis*, from the Middle Paleocene (Torrejon) of New Mexico.

Diagnosis: Upper molars resembling those of *Chriacus*, but wider transversely, and more triangular, external angles more prominent, hypocone less so, m^3 much reduced, transverse.   Premolars with blunt-pointed, somewhat inflated crowns, more as in Mioclaenidae and in the more primitive Anisonchinae.   Lower canine small, partly premolariform.   Jaw elongate, shallow, condyle not transverse.

This genus is provisionally placed with the Oxyclaenidae, as the characters of the molars are nearest to those of *Chriacus*, but the premolars lack the usual sharp cusps of the Oxyclaenidae and are suggestive of a very primitive anisonchine.   The rounded condyle and other characters of the jaw are likewise exceptional for an oxyclaenid.   The small premolariform canine is also difficult to reconcile with Oxyclaenidae.   But as none of these aberrant features shows any near approach to any known types of condylarth or insectivore it appears best to regard the genus as an aberrant oxyclaenid.

The molars are not unlike those of *Didelphodus* in many respects, but there is no resemblance in the premolars, which suggest Mioclaenidae in their simple construction and tendency to inflation, but in pattern are rather more like *Conacodon cophater*, *Anisonchus gillianus* and *Oxyacodon*.   The anterior basal cusp is usually absent in Mioclaenidae and there is a tendency to progressive molarization of the premolars from first to fourth.   Some resemblance to *Palaeosinopa* is to be noted both in molars and premolars.   The characters of the back of the jaw are not known in this genus; in the allied *Pantolestes* it is quite creodont-like, and unlike *Mixoclaenus*.

Mixoclaenus encinensis Matthew and Granger

Mixoclaenus encinensis Matthew, W. D., and Granger, W., 1921, Amer. Mus. Novitates No. 13, p. 7.

Type: No. 16601, upper and lower jaws of one individual, p^3–m^3 l., p_2–m_1 and roots of m_{2-3} r.

Paratype: No. 17074, lower jaw nearly complete with canine and p_2–m_3.   Torrejon formation, Bohannan Canyon.

Horizon and Locality: Torrejon formation, lower level, east fork of Torrejon Arroyo.

Author's Diagnosis: "C–m_3 = 26 mm.; m_{1-3} = 10 mm."

Description of Specimens: The type specimen shows the teeth moderately even and fairly well preserved.   The premolars are rather heavily worn, their principal cusps being worn down at the tips so that some of the detail is obliterated.   The paratype is somewhat less worn and very perfectly preserved, but the tips of the (lower) premolars are heavily worn in this specimen also.   Two other specimens, parts of lower jaws, show the same character of wear on the lower premolars.

All the specimens are from the lower level of the Torrejon formation.

The upper premolars are of moderate size, a little smaller than the first two molars; p^3 and p_4 are subequal in size and both are three-rooted, with rounded trihedral bases, the anteroposterior diameter being greater in the third and the transverse diameter in the fourth. Each tooth has an encircling basal cingulum, obsolete externally in p^3, rising internally to a good-sized inner cusp in p^4, and apparently broadened out into a large, low postero-internal heel in p^3, with the inner cusp rudimentary and postero-internal in position.

The first and second molars are like those of *Chriacus* in the prominent external angles and small but well-separated hypocone.   Their transverse diameters are about equal, but the anteroposterior diameter is considerably greater in m^1.   They have encircling cingula except upon the inner face of the protocone of m^1.   The paracone is somewhat larger than the metacone, both are round-conic, set somewhat in from the outer cingulum, the protocone has the usual crescentic form, and the conules are small and angular.

The third molar is much reduced, narrowed anteroposteriorly and transversely extended; the postero-external angle has disappeared, the metacone is vestigial and hypocone absent.

The lower premolars are of nearly equal size, of a moderately compressed oval in cross section, with minute anterior basal cusps, small, slightly basined heels, and the principal cusp oval, slightly crested in front and behind, moderately inflated, and always heavily worn on the tips.   The first premolar is small and one-rooted, but its crown is not preserved.

The lower molars decrease in size from first to third; their trigonids are somewhat larger and considerably higher than the talonids.   The protoconids decrease in relative size from m_1 to m_3, on the first it is much larger than the metaconid, on the third subequal.   The paraconids are distinct on all, but much lower than the other trigonid cusps.   The talonids are basined, the hypoconids being connected by a crest to about the middle of the posterior face of the trigonid; the hypoconulid is distinct only on m_3.

The lower canine is a little larger than the premolars and appears to have been partly premolariform, or at all events not laniary in type.   Its root is compressed oval, and the crown, as far as can be judged from its worn condition, was short, compressed, recurved, with anterior and probably posterior crest.   No diastemata.

The lower jaw is rather shallow and thick, elongate posteriorly, nearly half its length being behind m_3; the condyle is on a level with the crowns of the cheek teeth, rounded and not transversely extended, the angular process broad, prominent, flattened, somewhat incurved along its lower border; the coronoid process is only partly preserved, but appears to have been broad rather than high.

CHRIACINAE, new subfamily

(= Chriacidae) Osborn, H. F., and Earle, Charles, 1895, Bull. Amer.
Mus. Nat. Hist., VII, p. 20

Chriacus Cope, 1883

Cope, E. D., 1883, Proc. Acad. Nat. Sci. Phila., XXXV, p. 80, footnote

Type: Pelycodus pelvidens Cope, 1881,[36] of Upper Paleocene of New Mexico.

Author's Diagnosis: Two species of *Pelycodus* must be removed from this genus and family [Adapidæ] and be placed in the Creodonta with *Mioclænus.* They are the *P. pelvidens* and *P. angulatus,* which have the posterior inner tubercle of the superior molars, a mere projection of the cingulum. I place them in a new genus which differs from *Mioclænus* in the possession of an internal cusp of the fourth inferior premolar, under the name of *Chriacus;* type *C. pelvidens.*

In later publications Cope revised and redefined the genus, removing from it *Chriacus angulatus* (now *Cynodontomys*) and adding various other species, mostly new.   In 1888 [37] he distinguished it from *Mioclænus* by "the presence of an internal cusp of the first inferior premolar [i.e. p_4], which is however but little developed in some of the species.   The molars of the inferior series may be usually easily distinguished from those of Mioclaenus in the well-developed fifth cusp, and in the trihedral cusp of the external side of the heel,

[36] *Lipodectes pelvidens* Cope, 1881.   See p. 65.—THE EDITORS.
[37] Cope, E. D., 1888, Trans. Amer. Phil. Soc., XVI, N.S., p. 336.

which with the raised internal border gives the heel the basin-like character which is seen in Pelycodus.   The character of the inferior molars graduates into that of Mioclaenus through such species as *M. subtrigonus* [*Tricentes*] and *C. schlosserianus.*" Of the ten species referred at this time to the genus, *C. priscus*, *C. hyattianus* and *C. simplex* are now placed in *Loxolophus*, *C. inversus* in *Pentacodon* and *C. rütimeyeranus* provisionally in *Protogonodon*.

Scott in 1892 limited the genus to *C. pelvidens* and *C. truncatus*, provisionally placing with it *C. baldwini*, *C. stenops* and *C. angulatus*.   He defined it at some length, comparing it to the primitive lemuroids (i.e. *Pelycodus*).

Osborn and Earle in 1895 referred *Chriacus* to the Lemuroidea, making it the type of the new family Chriacidae.   In 1897 Matthew returned it to the creodont Oxyclaenidae with the following definition [38]: "Dentition: I$\frac{2}{7}$, C$\frac{1}{1}$, P$\frac{4}{4}$, M$\frac{3}{3}$.   First and second upper molars with strong spur-like hypocone projecting inwards and backwards, and smaller protostyle on m², giving them a quadrate outline with concave sides.   Intermediates small.   Cusps higher than in the two preceding genera [*Protochriacus* and *Tricentes*], trigonid more raised. Paraconid somewhat reduced, p³ and p₄ with well-developed deuterocone in the larger species." *Epichriacus* Scott, based upon *Chriacus schlosserianus* Cope, was considered a synonym of *Chriacus*, and four species were recognized, *C. pelvidens*, *C. baldwini*, *C. truncatus* and *C. schlosserianus*, the last doubtfully separable from *truncatus*.

Considerable additional material of *Chriacus* was secured by the expeditions of 1912 and 1913, but nothing to add materially to our knowledge of the genus.   The very evident relation of the larger species, *C. pelvidens* and *C. baldwini*, to *Deltatherium* leaves no doubt that the two genera should be nearly associated.   The smaller species, *C. truncatus* and *C. schlosserianus*, are similar in dentition to *Tricentes* and *Thryptacodon*.   Some parts of the skeleton of a species of *Chriacus* from the Wasatch of New Mexico were described by me in 1915 and confirm its primitive creodont affinities.

The more acute and angulate cusps of premolars and molars, raised trigonids and better-developed paraconids on the lower molars, more quadrangular, less rounded form of first and second upper molars and the sharply triangular m³ with more or less extension of parastyle, sufficiently distinguish the teeth from those of *Tricentes* and *Thryptacodon*.   From *Loxolophus* and *Protogonodon* the genus is distinguished by well-developed hypocone and somewhat reduced paraconids as well as by the higher trigonid and more acute cusps. From *Deltatherium* it is distinguished by the well-developed hypocones on m¹⁻² giving a quadrangular instead of triangular form to the teeth, by more simple p$\frac{4}{4}$, and less height of trigonids in lower molars, smaller canines and retention of p₁.

Chriacus pelvidens (Cope)

Lipodectes pelvidens Cope, 1881, Amer. Nat., XV, Dec. 3, p. 1019; *Pelycodus* Cope, Tertiary Vertebrata, p. 225 (part); *Chriacus pelvidens* Cope, 1883–4, Proc. Amer. Phil. Soc., p. 314 (part); *Chriacus pelvidens*, Matthew, 1897, Bull. Amer. Mus. Nat. Hist., IX, p. 273; *Chriacus stenops* Cope, 1888, Trans. Amer. Phil. Soc., XVI, N.S., p. 341; *Chriacus baldwini*, Osborn and Earle, 1895, Bull. Amer. Mus. Nat. Hist., VII, p. 21.

Type: A. M. Cope Coll. No. 3097, part of lower jaw, with left p₄–m₃.

[38] Matthew, W. D., 1897, Bull. Amer. Mus. Nat. Hist., IX, p. 272.

Horizon and Locality: Paleocene of San Juan basin, New Mexico, undoubtedly from the Torrejon formation, Middle Paleocene.

Author's Description: This species is about the size of *L. penetrans* and differs from it in the less carnassial character of the inferior molars. The anterior cusps are relatively smaller in every way, and are more distinctly separated by deeper emarginations. The heel is wider and has a less elongated external marginal cusp. The inner margin of the heel is elevated, enclosing a basin-like fossa and rises into a flat cusp posteriorly. There is a small median posterior marginal tubercle, which runs into a posterior cingulum, and is wanting from the *L. penetrans*. The tubercular has the three anterior cusps distinct as in *Didymictis* sp., while the heel is longer than in the known species of that genus. Its external border rises into a prominent cusp with triangular base. The fourth premolar has a small heel on the inner posterior side, and an acute anterior basal cusp. The principal cusp is robust and the basal portion is widely grooved posteriorly (apex lost). True molars with an external cingulum. Enamel obsoletely wrinkled. [Measurements follow.]

Revised Diagnosis, Matthew, 1897, p. 273: Third upper molar not reduced, being as wide transversely as the second. Lower jaw long and slender; canines strong, rather long, oval in section, projecting forward, striate longitudinally. First lower premolar one-rooted, second two-rooted, both spaced in front and behind. Third and fourth premolars considerably larger than the second, fourth with well developed deuteroconid.

The above diagnosis is based upon the types of *C. pelvidens* and *C. stenops* and a number of referred upper and lower jaws obtained by J. L. Wortman for the American Museum. Additional specimens of upper and lower jaws were secured by the expeditions of 1912 and 1913 from the upper and lower levels of the Torrejon formation.

Chriacus baldwini (Cope)

Deltatherium baldwini Cope, 1882, Paleont. Bull. No. 35, Nov. 11, and Proc. Amer. Phil. Soc., XX, p. 463; *Chriacus baldwini* Cope, 1888, Trans. Amer. Phil. Soc., XVI, N.S., p. 340 (part); Matthew, 1897, Bull. Amer. Mus. Nat. Hist., IX, p. 274; *Pelycodus pelvidens* Cope, 1885, Tertiary Vertebrata, p. 225 (all specimens except type).

Type: A. M. Cope Coll. No. 3114, part of right lower jaw with supposed p_3–m_2 (actually dp_{2-4}).

Horizon and Locality: Torrejon formation, New Mexico.

Author's Description: Differs from the *D. fundaminis* in its materially smaller size, and in the forms of the teeth. The first true molar is a more robust tooth, and the basis of the posterior or heel crest is more rounded, and less angulate. The anterior inner cusp projects less anteriorly. The fourth premolar has a distinct anterior basal lobe which is wanting in the *D. fundaminis*. Its heel is short and wide, and the posterior face of the principal cusp is flat, and there is a rudiment of an internal tubercle on its side. The second premolar is elevated and acute, has no anterior basal lobe, and has a very short wide heel, enamel slightly roughened. The animal was rather aged.

Matthew in 1897 revised this species, re-identifying the type specimen as milk dentition, removing from it a specimen of *Tricentes subtrigonus* which had served Cope as paratype, and including in it all the specimens except the type, which Cope had placed under *C. pelvidens*. As thus revised the species is close to *pelvidens* and is rather doubtfully distinguishable as follows:

C. baldwini is somewhat smaller than the preceding species [*C. pelvidens*], the last molar is reduced, and the second, third and fourth premolars are close together, of nearly uniform height, increasing successively in width. The deuteroconid on p_4 is small. The fourth milk molar in the type is smaller than the first true molar, but composed of the same elements, the paraconid more reduced, the other cusps too much worn for exact comparison.[39]

A few additional jaw fragments were obtained by the expeditions of 1912 and 1913.

Chriacus truncatus Cope

Chriacus truncatus Cope, 1884, Paleont. Bull. No. 37, Jan. 2, and Proc. Amer. Phil. Soc., XXI, p. 313; Cope, 1888, Trans. Amer. Phil. Soc., XVI, N.S., p. 336; Matthew, 1897, Bull. Amer. Mus. Nat. Hist., IX, p. 275, fig. 7 (type refigured); *Chriacus schlosserianus* Cope, 1888, Trans. Amer. Phil. Soc., XVI, N.S., p. 338; Matthew, 1897, Bull. Amer. Mus. Nat. Hist., IX, p. 275; *Epichriacus schlosserianus* Scott, 1892, Proc. Acad. Nat. Sci. Phila., XLIV, p. 296.

Type, A. M. Cope Coll. No. 3101, upper jaw with p^4-m^3 r., and roots of front teeth.

Horizon and Locality: Paleocene of San Juan basin, undoubtedly Torrejon formation, Middle Paleocene.

Author's Description: Posterior cingulum of superior molars with large tubercle; small species. In the *C. truncatus* the posterior singular (inner) tubercle reaches the largest development, but is not present on the cingulum of the last superior molar. The anterior cingulum is weak on that tooth and on the first true molar, but on the second it is thickened into a small anterior or inner tubercle. This with the posterior inner gives the crown a truncate internal outline, as is also the case in the *C. pelvidens*. The intermediate tubercles are distinct and the external cusps are separate at the base. An external cingulum. The fourth premolar has a triangular base; a single compressed external cusp, and a small acutely conical internal one. The internal tubercle is small and acute on the third premolar. The second premolar is small and probably one-rooted, and it is possible that there is no first premolar. The canine is directed vertically downwards, and the base of the crown is oval.

Besides the considerably smaller size, the posterior internal cusps are relatively larger than in *C. pelvidens*.

Matthew in 1897 refigured the type of *C. truncatus*, and questioned the distinctness of *C. schlosserianus*. Additional specimens fail to confirm even the specific validity of this type, which is here regarded as a synonym of *truncatus*. Besides a number of upper and lower jaws referable to Cope's species there are two or three specimens of more doubtful reference. Among these is a lower jaw, humerus and some fragments which seem to indicate a species of *Chriacus* of smaller size and more slender proportions than *truncatus* (or *schlosserianus*) but of interest chiefly as showing the characters of the humerus rather well (see pl. VIII, fig. 3).

Deltatherium Cope, 1881

Cope, E. D., 1881, Amer. Nat., XV, p. 337

Type: Deltatherium fundaminis, 1885, Tertiary Vertebrata, p. 277, from the Torrejon formation, New Mexico = *Lipodectes* Cope, 1881, Amer. Nat., XV, p. 1019; type: *L. penetrans*.

Author's Diagnosis (in type description of *D. fundaminis*): Family Leptictidæ, agreeing with *Ictops* and *Mesodectes* in possessing an internal tubercle of the third superior premolar, but differing from both in having but one external cusp of the fourth superior premolar.

The lower dentition of *Deltatherium* was not at first recognized as pertaining to it and was described as another genus of creodonts (*Lipodectes*) as follows:

The number of possible combinations of tubercular and tubercular-sectorial molar teeth is considerable, and many of them are represented in the genera of the Creodonta. A new one must now be added,

[39] Matthew, W. D., 1897, Bull. Amer. Mus. Nat. Hist., IX, p. 274.

in a genus which has, in the lower jaw, two tubercular sectorials and one tubercular posterior to them. The genus thus stands between *Stypolophus* and *Didymictis*, but is nearer the former than the latter, since it has three true molars. It differs further from both in having but three premolars and a wide diastema. The canine is well developed.

Cope based the genus *Lipodectes* upon two species, *L. penetrans*, shortly afterwards recognized as being the lower dentition of *Deltatherium fundaminis*, and *L. pelvidens*, later referred to *Pelycodus*, and subsequently made the type of the genus *Chriacus*. If the "method of elimination" be adopted, it would appear that *Lipodectes pelvidens* became the type of the genus after elimination of *L. penetrans*, antedating *Chriacus*. It is evident, however, from the diagnosis that *L. penetrans* was essentially the type of the genus, and *L. pelvidens* merely referred to it provisionally. Cope's subsequent procedure accords with this view, and Hay in 1902 [40] definitely specified *L. penetrans* as the type of the genus. It appears therefore in every way undesirable to revive the name *Lipodectes* to replace *Chriacus*.

Revised Description, Cope, Tertiary Vertebrata, p. 277: Dental formula: $I\frac{3}{7}$; $C\frac{1}{1}$; $Pm\frac{3}{3}$; $M\frac{3}{3}$. Superior premolars, the first and second with simple crowns, the third with one large external cusp and an internal small one. The fourth premolar with a large, simple external cusp and a prominent internal one. The first and second true molars with triangular bases, supporting two external compressed conic cusps and a subtriangular internal one. Last molar similar in its internal portions, the external part narrow. A wide diastema in the lower jaw. Inferior premolar simple, two rooted. True molars with anterior inner cusp well developed, forming with the anterior external a sectional edge, as in *Stypolophus*. Heels well developed, much produced, and supporting a special tubercle in the last molar.

The superior molars of this genus may be distinguished from those of *Pelycodus* by the absence of the intermediate tubercle and of the posterior internal tubercle. They differ from those of *Esthonyx* in the absence or weakness of the posterior inner tubercle, and in the absence of the ear-like expansion of the external angles.

The number of possible combinations of tubercular and tubercular-sectorial molar teeth is considerable and many of them are represented in the genera of the Creodonta. *Deltatherium* is a genus which has, in the lower jaw, two tubercular-sectorials, and a third with a long heel posterior to them. The genus thus stands between *Stypolophus* and *Didymictis*, but is nearer the former than the latter, since it has three true molars. It differs further from both in having but three premolars and a wide diastema. The canine is well developed. Although there is a tubercular tooth, the cutting apparatus is well developed, and indicates more than usually rapacious habits. There is as yet but one species known. It is allied to *Leptictis*, and agrees with *Ictops* and *Mesodectes* in possessing an internal tubercle of the third superior premolar, but differs from both in having but one external cusp of the fourth superior premolar, resembling in this respect the more typical *Oxyænidæ*.

Cope at this time referred the genus to the Leptictidae. Schlosser in 1887 rearranged the creodont genera, placing under the "Proviverridae" the division of Leptictidae to which *Deltatherium* belongs. He remarks [41]: Die zweite Gruppe bilden jene Creodonten, deren Höcker kantig erscheinen und deren untere *M* mit einem grubigen Talon versehen sind. Es schliessen sich diese Formen enger an *Didelphys* an als an die Raubbeutler.

Die primitivste Form ist *Deltatherium*, doch hat bereits die Zahl der *Pr* abgenommen, und trägt der obere *Pr₂* ebenfalls bereits einen wohlentwickelten Innenhöcker. Die oberen *M* sind noch echt tritubercular, ohne irgendwelche Modification; ein Fortschritt besteht jedoch insoferne, als der obere *M₃* dem *M₂* gleich geworden ist in Folge der Vergrösserung des unteren *M₂*. Auch hat die Grösse des Talons der unteren *M* ganz bedeutend zugenommen.

[40] Hay, O. P., 1902, Bibliography and Catalogue of the Fossil Vertebrata of North America, U. S. G. S. Bulletin No. 179, p. 751.

[41] Schlosser, Max, 1887, Die Affen, Lemuren, . . . p. 171.

Cope in 1888 [42] accepted Schlosser's classification in most particulars, and included it along with "*Mioclaenus* and probably *Triisodon* and *Onychodectes*" (presumably *Chriacus* as well) in the Proviverridae. His view of the relationships of the genera is thus expressed:

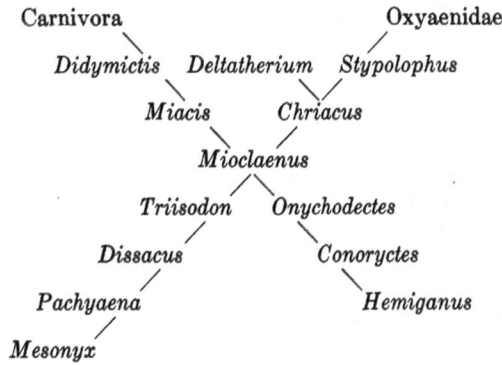

Carnivora   Oxyaenidae

Didymictis   *Deltatherium*   *Stypolophus*

Miacis   *Chriacus*

Mioclaenus

Triisodon   *Onychodectes*

Dissacus   *Conoryctes*

Pachyaena   *Hemiganus*

Mesonyx

Scott in 1892 [43] arranges *Deltatherium* under the Proviverridae, but places *Chriacus* and other related genera under the Oxyclaenidae. His description follows [44]:

    The Creodonta form an extremely heterogeneous group, very difficult to define and still more difficult to classify and subdivide. This difficulty arises partly from the imperfection of the available material, but more especially from the lack of diagnostic characters which are common to all the members of the order and from the minute steps of gradation by which they shade into other groups of allied unguiculates and even ungulates. Creodonta were among the earliest fossil mammals which were accurately studied and they were then referred to the carnivores. Laurillard, Pomel and others, however, regarded them as marsupials, and Aymard and Gaudry following this example, have called them Sous-didelphes. In 1875 Cope proposed the name of Creodonta for the group which he regarded as a suborder of the Insectivora, but in 1877 he named this comprehensive order the Bunotheria, referring to it as suborders, the Creodonta, Mesodonta, Insectivora, Tillodonta and Tæniodonta. The creodont division has not found universal acceptance, Filhol regarding them as Carnivora, Wortman as Insectivora and Lydekker as a suborder of the Carnivora. Nevertheless, they cannot be included among either the insectivores or the carnivores without uniting these groups, and it is therefore most convenient to regard them as an order.

Osborn and Earle in 1895 [45] adopt Scott's location of the genus and figure the upper cheek teeth and lower jaw of *D. fundaminis*.

Matthew in 1897 [46] retained the genus under "?Proviverridae," but in 1899 [47] transferred it to the Oxyclaenidae, on account of its near relationships to *Chriacus*, the best-known genus of that group. The affinities as then conceived were expressed in a diagram published in 1901.[48] There has been, however, no serious study of the genus since Cope (1885) and Scott (1892).

[42] Cope, E. D., 1888, Trans. Amer. Phil. Soc., XVI, N.S., p. 309; separata published Aug. 1.
[43] Scott, W. B., 1892, Proc. Acad. Nat. Sci. Phila., XLIV, p. 308.
[44] *op. cit.*, p. 291.
[45] Osborn and Earle, 1895, Bull. Amer. Mus. Nat. Hist., VII, pp. 39, 40, figs. 10, 11.
[46] Matthew, 1897, Bull. Amer. Mus. Nat. Hist., IX, p. 263.
[47] Matthew, 1899, Bull. Amer. Mus. Nat. Hist., XII, p. 29.
[48] Matthew, 1901, Bull. Amer. Mus. Nat. Hist., XIV, p. 21.

Deltatherium fundaminis Cope

Deltatherium fundaminis Cope, 1881, Amer. Nat., XV, p. 337; 1884, *ibid.*, XVIII, p. 352, fig. 20.

Type: A. M. Cope Coll. No. 3315, part of skull, with palate and cheek teeth well preserved.

Horizon and Locality: Middle Paleocene, Torrejon formation, San Juan basin, New Mexico.

Author's Description (p. 338): The second premolar is convex on the inner face. The base of the third is a nearly equilateral triangle. The bases of the true molars are triangles, with the bases external. The internal angle supports an acute cusp, and has a posterior basal cingulum, which is very strong in the last three molars. The two external cusps of the first and second molars are situated well within the base, which is folded into a strong cingulum. This cingulum develops strong anterior and posterior angles. This is the largest species of the family yet discovered. Extent of series of last six molars, m. .045; of true molars, .026; diameters of fourth premolar, anteroposterior, .0074; transverse, .0076; do. of second true molar, anteroposterior, .0087; transverse, .0100. This species was a fourth larger than the common opossum, and very much resembles it in dental characters.

The principal specimens now available for study are as follows:

No. 16610, skull in fine preservation, with most of the lower jaw and humerus. Expedition of 1913.

No. 16611, lower jaws and various parts of skeleton all buried in very difficult matrix. Expedition of 1913.

No. 3338, Cope Coll., skull and jaws, crushed and mostly buried in very hard matrix.

No. 3315, Cope Coll., type specimen, palate with upper teeth.

No. 783, Cope Coll., palate, jaws and limb-bones.

No. 17037, lower jaws. Animas valley, Colorado.

There are in addition various parts of upper and lower jaws which serve chiefly to prove the limits of individual variation and the characters that may be regarded as distinctive of the species.

The dentition was well described by Cope and Scott, except for the remarkable proportions of the canines. Of the skull only the general proportions were discernible in No. 3338, and nothing of the skeleton was known.

The teeth of *Deltatherium* have much the same construction as in *Chriacus*, but differ in their more triangular form, higher, sharper cusps, greater transverse and less anteroposterior extension of the upper molars, the sub-molariform p_4^4, in the loss of p_1^1, enlargement of upper and decrease of lower canines, the markedly concave postcanine diastema and slight flanging of the lower jaw. This associated group of characters constitutes a distinct approach to the machaerodont type of dentition, but neither in this instance nor in *Anacodon*, *Machaeroides*, *Phlaodectes*, etc., does this signify relationship. Possibly it is to some extent an inherited primitive character, but for the most part it must be regarded as a specialization, the upper canines being used more and more for stabbing and tearing instead of biting and holding.

The small brain-case, high sagittal and occipital crests, broad and strong zygomatic arches, backwardly-placed orbits without distinct postorbital processes, long nasals wide posteriorly, etc., are primitive creodont characters.

The upper incisors are quite small, pointed, not spatulate.

The upper canine is a large, powerful tooth, somewhat compressed-oval in cross section, with crested anterior and posterior margins, little curved, tapering regularly toward the point and with a moderate taper in the root.

A.M. 16610

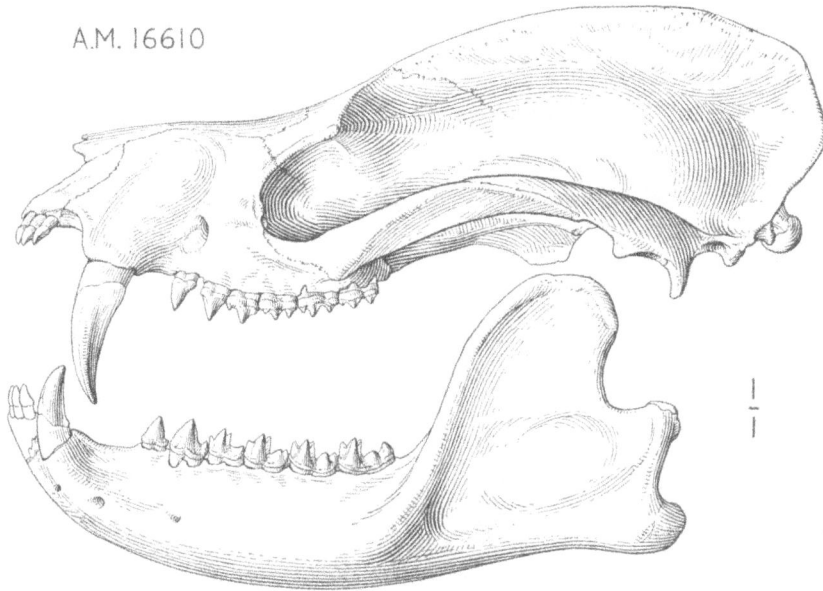

Fig. 11.   *Deltatherium fundaminis*, skull and lower jaw, A. M. No. 16610, side view, some crushing corrected.   Natural size.

The first premolar is absent. P^2 is two-rooted, high-pointed and compressed, but narrow anteriorly and somewhat broader posteriorly, with a diastema before and behind it. P^3 is triangular in form, with large internal root but no inner cusp. P^4 is widely trihedral, with large internal cusp and basal cingulum, and a prominent hooked parastyle. M^1 is widely triangular, with large, crescentic protocone, smaller paracone and metacone equal in size to each other, anteroposteriorly oval, with sharp external cingulum and a considerable external shelf lying within it, a distinct parastyle, no mesostyle, metastyle not separate, a rudimentary cingular hypocone.

M^2 is similar to m^1, but of slightly larger size, with rudimentary cingulum, hypocone and protostyle. M^3 has the construction of m^{1-2}, except that its postero-external portion is reduced so that the angle becomes somewhat obtuse instead of sharply acute, the metacone is reduced, there is a distinct metaconule, and the hypocone and protostyle, although rudimentary, are somewhat larger and farther apart than on m^2.

The palate is very broad and moderately concave, bordered posteriorly by a thickened margin.

The arches are wide, moderately heavy; the jugal extends back in the usual way, falling some distance short of reaching the glenoid fossa.   The glenoid fossa is little excavated, with moderate postglenoid and no preglenoid processes.   The postglenoid foramen is large.   The basioccipital is typically carnivore in form, rather long, broad, rectangular in outline.   There are weak postorbital processes on the frontals, and rather prominent postorbital crests extending inwards and only a little backwards till they come together at the median line in a sagittal crest.   The parietals overlap the frontals at the anterior portion of the sagittal crest, extending almost to the point where it forks into postorbital crests; the frontal bone is quite massive, especially heavy beneath the anterior end of the sagittal crest, where it overlies the anterior (olfactory) lobes of the brain.   The otic region is very well preserved.   The tympanic ring, if present, has been lost.   The auditory prominence of the petrosal is large, prominent, oval and rather uniformly convex, the *fenestra rotunda* situated at the top near the posterior end, the arterial tracks not discernible on the surface of the prominence.   Externally lies the deep, long mesotympanic excavation, posteriorly the rather small

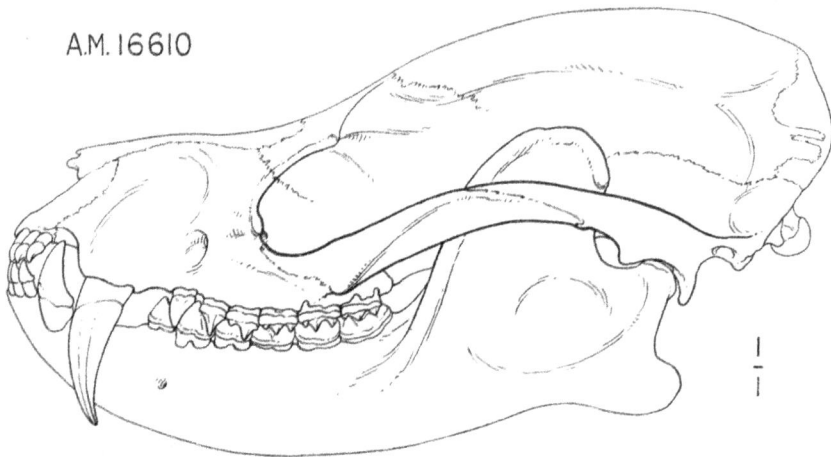

A.M. 16610

Fig. 12.   *Deltatherium fundaminis*, skull and lower jaw in articulation, A. M. No. 16610.   Natural size.

and almost slit-like *foramen lacerum posterius*, and postero-externally the rounded, prominent stylomastoid foramen.   The condylar foramen is small, a little behind the *f. l. p.* and some distance from the condyles.   The postglenoid foramen is large, situated behind the external end of the glenoid.   The alisphenoid canal is distinct, of moderate length.

The muzzle is short and wide.   The nasals are remarkably broad for a creodont, expanded a little at their posterior ends and slightly so at the anterior end, with very slight anterior notch, and apparently without contact with the premaxillae.

The parietals are long, narrow, remarkable for the great height of the sagittal crest.

The occipital bones are expanded into a high semicircular occipital crest.

The lower jaw is deep but not thick, slightly angulate at the symphysis, but not flanged, the angulation being much like that of *Nimravus* and *Archaelurus*, less than in *Dinictis*. The lower incisors are unknown except to Professor Scott, who has described them as three in number, the description being based upon some specimen in the Cope Collection not now discoverable.   The lower canine is much smaller than the upper and quite short, especially as to the crown.   A deep antero-internal flange curves backward and ends abruptly on the inner side of the crown about one-third from the base of the enamel.   There is a faint posterior flange extending to the base of the crown; otherwise the crown is round-oval in cross section, with a rather strong backward curve.

P_1 is absent. P_2 is large, two-rooted, trenchant, high, compressed, with only traces of posterior cingulum or heel. P_3 is similar, somewhat longer and wider, with more distinct heel and cingulum distinct except at the middle part of the inner face. P_4 has a larger heel and strong paraconid and metaconid, thus assuming a decided molariform type.

The molars are of very uniform construction.   All have a moderately high trigonid composed of three strong and widely separated cusps, of which the protoconid is the largest and the paraconid the smallest.   The heel is as large as the trigonid, composed of a large, high hypoconid, a small, low entoconid, and a median hypoconulid on the posterior margin.

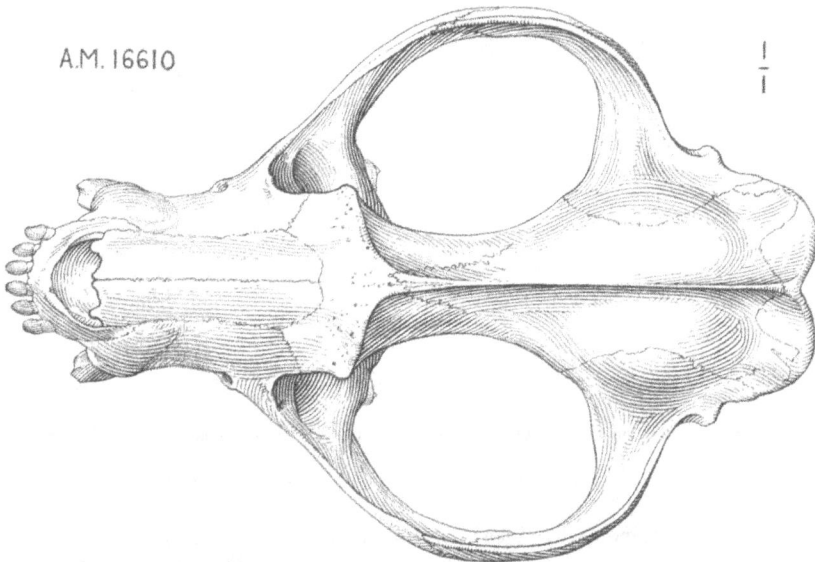

FIG. 13.   *Deltatherium fundaminis*, skull, A. M. No. 16610, top view, crushing corrected.   Natural size.

On m_{1-2} the hypoconulid is not larger than the entoconid, although considerably higher; on m_3^3 it equals the hypoconid in size and surpasses it in height, projecting backward to form a small third lobe to the tooth. M_2 and m_3 are of equal size except for the hypoconulid; m_1 is slightly smaller in the talonid and decidedly smaller in the trigonid.   The principal mental foramen is beneath p_3; a minor one is beneath p_2.   The condyle is nearly at the level of the alveolar border; the dental foramen is considerably below this line and about half-way between condyle and posterior border of m_3.   The coronoid process is wide, directed rather strongly backward; the angle is moderately deep, thick and heavy, but apparently is not prolonged into much of a spine.

The humerus is of about the same length as the lower jaw.   *Deltatherium*, therefore, is a moderately macrocephalic creodont, about midway between the extremes represented by such types as *Didymictis* on the one hand and *Claenodon* on the other.   The direct relation between macrocephaly in the skull and predaceous specialization in the teeth I pointed out some years ago in discussing the Bridger Creodonta.[49]   The greater trochanter is of moderate size, slightly overtopping the head; the lesser trochanter is quite small; the deltoid crest is high and ends abruptly a little below the middle of the bone.   The supinator crest is wide, thin, and curiously flat, the distal roll-facet is comparatively shallow and wide.

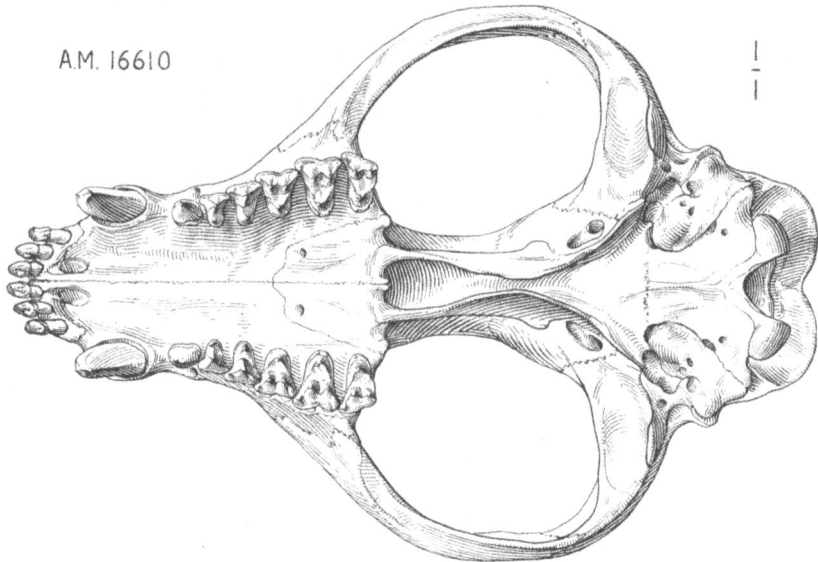

A.M. 16610

Fig. 14. *Deltatherium fundaminis*, skull, A. M. No. 16610, palatal view, crushing corrected.   Natural size.

The femur is of the usual creodont type, with high greater trochanter, lesser trochanter internal, small third trochanter.

The tibia has a somewhat reduced cnemial crest, obscure save near the middle of the shaft, where it stands out in a more prominent process.   The distal end is strongly oblique, with trochlear groove very slight.

TRIISODONTINAE Trouessart, 1904

Trouessart, E. L., 1904, Catalogus Mammalium, Quinquennale Supplementum, p. 161
    (= Triisodontidae) Scott, W. B., 1892, Proc. Acad. Nat. Sci. Phila., XLIV, p. 294

Author's Diagnosis: Superior molars tritubercular with low massive cusps, sometimes having a well developed hypocone on m^2; trigonid of lower molars much higher than talon, but not forming a shearing blade, paraconid reduced; premolars high and acute.

Scott included in the family *Triisodon, Goniacodon, Microclaenodon* and *Sarcothraustes*. This arrangement was left unaltered by Osborn and Earle in 1895 and by Matthew in 1897. The affinities to the Mesonychidae have been recognized in the teeth, which, however,

[49] Matthew, W. D., 1909, Mem. Amer. Mus. Nat. Hist., IX, pt. vi.

especially in the Puerco *Eoconodon*, retain the primitive type of tritubercular molar, peculiarly modified in the Mesonychidae.   In *Eoconodon* (*Triisodon* p. p.) the teeth are hardly distinguishable from those of certain Oxyclaenidae; in *Triisodon* (syn. *Sarcothraustes*) and *Microclaenodon* of the Torrejon they are partly intermediate between the primitive type and that of *Dissacus*, but with important differences that indicate a somewhat diverse line of specialization, not leading into the Mesonychidae.

The type species of *Triisodon* appears to be the same animal later called *Sarcothraustes* by Cope, and generically distinct from the several smaller forms later referred to it.   These species, all from the Puerco, are placed in the new genus *Eoconodon;* the horizon of *T. quivirensis*, the type of *Triisodon*, is unrecorded but reasons will be shown for the conclusion that it came from the Torrejon, like all the referred material.

Revised Subfamily Diagnosis: Molars above tritubercular, below tuberculo-sectorial, with basined heels, cusps massive, blunt-pointed, progressively conical.   Heel of p₄ bicuspid. Premolars backwardly pitched.   Skeleton much as in other members of the Arctocyonidae as far as known.   *Eoconodon:* Trigonid low, metaconids and protoconids slightly connate, subequal, paraconid strong, heel large, basined.   *Goniacodon:* Trigonid moderately high, metaconids and protoconids somewhat connate, paraconids much reduced, heel large, basined; m$\frac{3}{3}$ much reduced.   *Triisodon:* Trigonid moderate, protoconid overtopping metaconid, moderately connate, paraconid much reduced, heel bicuspid, the outer cusp much larger, partly crested.   Cusps massive, rounded.

The triisodonts have been regarded as a distinct family and also as a subfamily of Mesonychidae.   Their special relationship to this family has been recognized from the teeth but is not confirmed by the characters of the skeleton.

The relationship of the triisodonts to the mesonychid series corresponds, however, to the relationship of the Oxyclaeninae to the four remaining groups of creodonts—arctocyonines, miacids, hyaenodonts, oxyaenids.   In each case they represent a primary stock from which one or several specialized stocks are derived.   The known genera in each case are not the direct ancestors of the specialized phyla, but are related thereto in varying degrees of approximation.   They are all little side branches or twigs of the phylogenetic tree that have not survived.   In the triisodont group, the approximations are partly toward the mesonychid series, partly peculiar.

Triisodon Cope, 1881

Cope, E. D., 1881, Amer. Nat., XV, p. 667

Type: Triisodon quivirensis from Torrejon of San Juan basin.

Synonym: Sarcothraustes Cope, 1882; type: *S. antiquus* from Torrejon of San Juan basin.

Author's Description: Char. gen.: Derived from the lower jaw. Probably only three premolars. True molars alike, consisting of three anterior cusps and a heel.   The cusps are relatively small and the heel large.   Of the former the internal is much smaller than the external, and the anterior is rudimental, being merely a projection of the cingulum.   The cutting edges of the large external cusp are obtuse.   The heel is basin-shaped, and its posterior border is divided into tubercles, of which the external is a large cusp.   The fourth premolar has no anterior inner tubercle, so that the anterior part of the crown consists of a compressed cutting cusp.   The heel has two well-developed posterior cusps.   The third premolar has a similar principal trenchant cusp, but a smaller heel.   Canines large.

Type Description of Sarcothraustes: Char. gen.: We have in evidence of the characters of this genus, the last two superior molars, the last one lacking the crown; and parts of both mandibular rami which exhibit teeth as far posteriorly as the first true molar inclusive; all belonging to one individual.   A part of a skeleton of a second individual, which includes a fragment of lower jaw, belongs probably to this species.

Sarcothraustes resembles both *Amblyctonus* and *Mesonyx*, but it is probably to the latter genus that it is allied.   The last superior molar is transverse, much as in *Oxyaena*.   The crown of the penultimate is subtriangular and transverse.   It has two external subconic cusps and a single internal lobe, whose section on wearing is a V, each branch of the face extending to the base of the corresponding external tubercle.   There are three small inferior incisors, and a large canine.   There are probably only three inferior premolars, the first one-rooted.   The crown of the second has no heel.   The crown of the third has a short wide heel.   The crown of the first true molar consists of an anterior-elevated cone and a posterior heel.   The latter is wide, having a posterior transverse, as well as a longitudinal median keel.   The fragments of the supposed second individual include two large glenoid cavities with strong preglenoid crests as in *Mesonyx*.

As compared with *Mesonyx*, this genus differs in the V-shaped crest of the penultimate superior molar; in *Mesonyx* it is represented by a simple cone.   The last superior molar of *Mesonyx* is triangular and not transverse, but the composition of the crown of that tooth in *Sarcothraustes* must be known before the value of this character can be ascertained.   If the view that *Sarcothraustes* has but three inferior premolars be correct, this character distinguishes it from *Mesonyx*, as do also the transversely expanded heels of the molars.[50]

In 1888, Cope made *Sarcothraustes* a subgenus of *Mioclaenus*, placing under it a number of species which agree in having the "inferior true molars with anterior triangle of three cusps," [51] but are now known to be of very diverse affinities and are distributed among six genera of Creodonta and Condylarthra.   He retained *Triisodon* as a distinct genus, with the type species and a referred new species, *T. biculminatus*, and indicated it as intermediate between "*Mioclænus*" and *Dissacus*.[52]

Scott in 1892 revived *Sarcothraustes* as a distinct genus, and re-defined both it and *Triisodon*, placing under *Sarcothraustes S. antiquus* (type), *S. coryphaeus*, *S. bathygnathus* and *S. crassicuspis*, and under *Triisodon T. quivirensis* (type), *T. biculminatus* and *T. heilprinianus*.   His generic diagnoses are:

Triisodon: . . . The canine is large and of oval section, without cutting edges. $P_{\bar{3}}$ is small and $p_{\bar{4}}$ very large, with very high, acute and trenchant protoconid and a talon of two trenchant cusps, of which the external is much the higher and more acute.   Seen from the outer side, this tooth closely resembles the corresponding one of *Dissacus*, differing only in the presence of the tetartoconid.   In the molars the trigonid rises considerably above the talon and is composed of a high, sharp and massive protoconid, of a small, low metaconid, and a still smaller and lower paraconid; the two latter cusps are on the same anteroposterior line and, in $m_{\bar{3}}$ at least, are not visible from the external side.   The talon consists of a high and sharp hypoconid with trenchant anterior edge and internally three very much smaller cusps, representing the hypoconulid, entoconid, and a tubercle in front of the latter to which no name has been given.   This crenulate inner border of the talon is highly characteristic of the genus.[53]

Sarcothraustes: . . . The superior molars, so far as they are known, and the lower premolars agree closely with those of *Goniacodon*, the only differences being their larger size, more massive cusps, thicker and more prominent cingulum, especially at the antero-external angle of the crown. $M^{\underline{2}}$ is oval in shape and reduced in size, having lost the metacone.   The anterior lower premolars are remarkable for their small size and simple construction, but $p_{\bar{4}}$ is very much larger and higher and has a large talon, divided into inner and outer cusps.   The lower molars differ from those of *Goniacodon* in the composition of the trigonid; the protoconid is much the largest element, the para- and metaconids are greatly reduced and placed on the same fore and aft line, as in *Triisodon*, but the talon is very different, consisting of hypo- and entoconids and small hypoconulid, which may or may not be much enlarged on $m_{\bar{3}}$, the size of which tooth is very variable in the different species.   The skull has a very small cranial cavity and a very high occipital crest, which is

---

[50] Cope, 1882, Proc. Amer. Phil. Soc., XX, pp. 193–4.
[51] Cope, 1888, Trans. Amer. Phil. Soc., XVI, N.S., p. 320.
[52] Cope, 1888, *op. cit.*, pp. 309, 342.
[53] Scott, W. B., 1892, Proc. Acad. Nat. Sci. Phila., XLIV, p. 300.

arched from side to side and continued forward into an extremely prominent sagittal crest. The zygomatic arches are heavy and project strongly from the skull; the glenoid cavity is deeply concave, with prominent pre- and post-glenoid crest. The mandible varies much in size and proportions among the different species, being in some long and slender and in others very massive, but in all the ascending ramus is of remarkable antero-posterior extent, the distance from the condyle to $m_{\bar{3}}$ exceeding the length of the molar premolar series. The condyle is placed low down and there is a short hooked angular process. The masseteric fossa is large but shallow, especially so in the larger species.[54]

Scott's generic definitions are based upon both the types and referred species now distributed otherwise as to genera. They are largely the characters of the family Triisodontidae as here defined, many of them shared by the Mesonychidae; hardly any are generic characters of his type species.

Osborn and Earle in 1895 define *Sarcothraustes* in the following terms, including in it, as also in *Triisodon*, the species assigned by Scott:

Dentition: $I_{\bar{7}}^{2}$, $C_{\bar{1}}^{1}$, $P_{\bar{4}}^{2}$, $M_{\bar{3}}^{2}$. Superior true molars with paracone and metacone conical and equal in size. Last superior premolar not molariform, and same tooth of the lower series with talonid consisting of two cusps. Inferior true molars with trigonid raised above the talonid, the former consisting of three cusps with the protoconid much larger than the para- or metaconid. Metaconid distinctly separated from the protoconid and on the same fore and aft line with the paraconid.[55]

This definition is drawn chiefly from the referred species *S. coryphaeus*, now transferred to *Eoconodon*.

Matthew in 1897 revised the reference of the various species referred to *Triisodon* and *Sarcothraustes*, placing in *Triisodon* the type species *T. quivirensis, Goniacodon gaudrianus,* and *T. heilprinianus,* of which *Goniacodon rusticus, Sarcothraustes crassicuspis, S. coryphaeus, S. bathygnathus* and *Triisodon biculminatus* were regarded as synonyms. This necessitated a redefinition, as follows:

Dentition: $I_{\bar{3}}^{2}$, $C_{\bar{1}}^{1}$, $P_{\bar{4}}^{4}$, $M_{\bar{3}}^{3}$. Second upper premolar two-rooted, third and fourth three-rooted with well defined deuterocone. Upper molars wide transversely, subquadrate, with weak hypocone on $m^{\underline{1-2}}$. Second and third lower premolars small with simple minute talonids, fourth large with strong bicuspid talonid, the outer cusp larger. Lower molars with moderately high trigon, the protoconid and metaconid of nearly equal size, paraconid lower and placed partly in front of metaconid.[56]

This definition is also based chiefly upon wrongly referred species and does not agree wholly with the characters of the type.

Revised diagnosis: Teeth tritubercular, with robust, massive, well-rounded cusps. Upper molars somewhat wider than long, rounded trigonal with conic blunt-pointed cusps.

Paracones and metacones somewhat connate, the metacones reduced on m^{2-3}. No hypocones. M^1 decidedly smaller than m^2, larger than m^3. Premolars very robust, canines large and massive. Lower molars with paraconids much reduced, metaconids connate with protoconids. Heels large, equalling or exceeding trigonids, with rather high subequal hypoconid and metaconid, hypoconulids vestigial or absent. Jaw very massive and deep.

The construction of the teeth differs from that of *Eoconodon* in the high rounded connate metacone and paracone of the upper molars, and correspondingly of the protoconid and paraconid of the lower molars, the disappearance of hypocone and reduction of paraconid, change of the heels of the lower molars to a bicuspid rather than basined character, reduction

---

[54] Scott, 1892, Proc. Acad. Nat. Sci. Phila., XLIV, pp. 302–3.

[55] Osborn and Earle, 1895, Bull. Amer. Mus. Nat. Hist., VII, p. 28.

[56] Matthew, 1897, Bull. Amer. Mus. Nat. Hist., IX, p. 279.

of metacones, and in the generally more massive character of teeth and jaws.   The typical species is of larger size, the smaller referred species being to some extent intermediate.

The limb bones are of relatively small size, exaggerating the tendency seen in *Eoconodon*.

The number of species represented is uncertain, as the material does not allow of very satisfactory comparisons.   The genotype, *T. quivirensis*, is based on the lower jaws of a young individual, in which dp$_4$–m$_2$ are preserved with the lower canine partly emerged and p$_4$ preformed in the jaw.   To this species may be provisionally referred part of a lower jaw with m$_3$ unworn.   A second species, larger and more robust, with m$_3$ more specialized in the almost complete disappearance of paraconid and metaconid, is represented by the upper and lower jaw on which *T. conidens* was based.   The type of *Sarcothraustes antiquus* is very probably co-specific with *T. conidens*, and a number of upper and lower jaws may be referred to this species.   A third smaller species, with m$_3$ more primitive in retention of paraconid and metaconid, is represented by the type of *"Conoryctes" crassicuspis*, and with this may be associated the jaw fragment on which *T. rusticus* was based.

Triisodon quivirensis Cope

Triisodon quivirensis Cope, 1881, Amer. Nat., XV, p. 667, July 27; 1884, *ibid.*, XVIII, p. 257, fig. 1.

Type: A. M. Cope Coll. No. 3352, lower jaws, young, with canines, dp$_4$–m$_2$ in place, and p$_{3-4}$ and m$_3$ preformed in the jaw.

Horizon and Locality: Paleocene of San Juan basin, New Mexico, probably all from the Torrejon formation.

Author's Description: Probably only three premolars.   True molars alike, consisting of three anterior cusps and a heel.   The cusps are relatively small and the heel large.   Of the former the internal is much smaller than the external, and the anterior is rudimental, being merely a projection of the cingulum. The cutting edges of the large external cusp are obtuse.   The heel is basin-shaped, and its posterior border is divided into tubercles, of which the external is a large cusp.   The fourth premolar has no anterior inner tubercle, so that the anterior part of the crown consists of a compressed cutting cusp.   The heel has two well-developed posterior cusps.   The third premolar has a similar principal trenchant cusp, but a smaller heel.   Canines large. . . .

Char. specif.—Size about that of the wolf.   Inferior canine directed upwards, its section nearly elliptic; a faint posterior, no anterior cutting edge.   Fourth premolar rather large, with an anterior basal cingulum which is angulate upwards, and is not continued on the inner side of the crown.   Cusps of the heel each sending a ridge forwards, the internal lower, obtuse and descending to base of inner side of large cusp; the external larger, with an acute anterior cutting edge continuous with the cutting edge of the large cusp.   True molars with an external, but no internal basal cingulum.   Border of heel with one large and three small tubercles, the former with, the latter without, anterior cutting edge.   Enamel of all the teeth nearly smooth. All the cusps are rather obtuse.

Triisodon quivirensis was based upon the lower jaws of a young individual retaining the posterior milk molars and with the canines and first two molars in place, the permanent premolars and m$_3$ preformed but not erupted.   The jaws and teeth are uncrushed and their surface very perfectly preserved.   The character of the preservation and of the concretionary scale adherent in places to the specimen are very like the appearance of a number of specimens of various species from the Torrejon, and more or less different from anything in the Puerco collections.   It is probable, therefore, that this specimen is from the Torrejon formation, and not from the Puerco to which it was provisionally accredited by Matthew in 1897.   The various specimens which can be referred to *T. quivirensis* are certainly from the Torrejon.   Moreover, in describing the type of *Ptilodus mediaevus,* Cope states [57] that it was found in removing the matrix from the type jaw of *Triisodon quivirensis.*   This species is also strictly limited to the Torrejon, and neither it nor the genus *Ptilodus* has been found in the Puerco formation.

The immaturity of the type makes comparison with other species difficult.   It appears to be of smaller size, with m_{1-2} less specialized than in *T. conidens,* and for these reasons we associate with it No. 16559, part of a lower jaw with m_3 unworn, found with various fragments of the skeleton.   This specimen is either *T. quivirensis* or an undescribed species; the m_3 is clearly distinguished from *T. conidens* by retention of small metaconid and paraconid on m_3, as well as by smaller size; and from *T. crassicuspis* by larger size and the greater reduction of these cusps.   With this specimen were found associated various fragments of the skeleton, including a humerus and parts of the hind foot.   These would be of great interest if it could be made clear that they belonged to *Triisodon,* but they agree so closely with the corresponding parts of *Periptychus* that the association with *Triisodon* jaw must be regarded as an accidental intermixture.

Triisodon antiquus (Cope)

Sarcothraustes antiquus Cope, 1882, Paleont. Bull. No. 34, Feb. 20, and Proc. Amer. Phil.
    Soc., XX, p. 193; Tertiary Vertebrata, p. 347, pl. xxiv *d,* figs. 19–22.
Triisodon conidens Cope, 1882, Proc. Acad. Nat. Sci. Phila., p. 297; Tertiary Vertebrata,
    p. 274, pl. xxiii *d,* figs. 9–10.

Type: A. M. Cope Coll. No. 3172, anterior half of lower jaws, fragment of upper jaw and numerous fragments of skull and skeleton, all considerably weathered.   Type of *T. conidens:* A. M. Cope Coll. No. 3174, upper and lower jaw.

Horizon and Locality: Both from the "Puerco" (probably Torrejon) of the San Juan basin of New Mexico, exact locality and horizon unknown.

Author's Description (p. 194): *Char. specif.*—The penultimate superior molar has a strong posterior cingulum which commences within the line of the internal bases of the external cusps, and rises into considerable importance behind the internal cusp.   There is also an anterior cingulum which does not rise internally, and which is continuous with a strong external basal cingulum.   The latter passes round the posterior base of the posterior cone, and runs into the posterior branch of the internal V.   The posterior cone is smaller than the anterior cone, and its apex is well separated from the latter.   The appearance of this tooth is something like that of a carnivorous marsupial.

The symphysis mandibuli slopes obliquely forwards, and is united by coarse suture.   The ramus is stout and deep, as compared with the size of the molar teeth.   The roots of the teeth are relatively large, especially those of the first two premolars.   The crown of the canine is lost.   The first premolar points forwards, nearly parallel with the canine, and divergent from the second premolar.   The crown of the second premolar is small and subconic, and has a rudimental heel, and no anterior basal tubercle.   The first true molar resembles considerably that of *Mesonyx.*   There is a small anterior basal tubercle on the inner side of the principal cusp.   The expansion of the heel is transverse only, there being no longitudinal lateral edges or tubercles.   The enamel is obsoletely, rather coarsely wrinkled.   There are two rather large mental foramina; the posterior below the anterior root of the first true molar, and the anterior below the posterior root of the second premolar.

The type description of *Triisodon conidens* is a detailed one and is reprinted in Tertiary Vertebrata, pp. 274–7.

[57] Cope, E. D., 1881, Amer. Nat., XV, p. 921.

Matthew in 1897 regarded *T. conidens* as a synonym of *S. antiquus,* with the following diagnosis: In this species the paracone is larger than the metacone, while the protocone is much enlarged at the expense of the accessory cusps around it.   The hypocone has disappeared except on m^1, where a small remnant still holds out.   On m^1 the paracone is scarcely larger than the metacone, on m^2 the difference is considerable, while on m^3 the metacone is very small.   The great transverse width of the molars is the most prominent distinction from the Mesonychidae.   P^3 and p^4 have deuterocones.   In the lower jaw corresponding differences from *Triisodon* [*i.e.* = *Eoconodon*] are seen.   The protocone is larger than the other cusps of the trigon, progressively larger on the succeeding molars till on m$_3$ the *pa^d* and *me^d* are vestigial. In the heel the entoconid is not lost, but two high rounded cusps arise, internal and external.[58]

Besides the two type specimens the following from the Torrejon are referred to this species:

No. 3175, upper jaws, p^4–m^3 l., p^4–m^2 r., unworn.   Crocodile Kitchen Midden.

No. 758, lower jaw, p$_4$–m$_2$ l.   Escavada Canyon.

No. 3176, upper jaw, m$^{2-3}$ r.   Locality not recorded.

No. 3173, palate with p^4 l.   Crocodile Kitchen Midden.

No. 16747, upper jaw fragments, c^1, p^3, m$^{2-3}$.   Head of Kimbetoh Wash.

No. 17079, lower teeth, c$_1$, p$_3$–m$_2$.   Coots Canyon.

The additional specimens found in 1913 and 1916 confirm the identity of the two species and show the characters of the teeth a little more clearly.

Although partly paralleling the Mesonychidae in the specialization of the teeth, the short deep massive jaw with reduced premolars, the large basined or bicuspid heels of the molars and their much more massive construction are rather in contrast with the mesonychid type as shown in the contemporary *Dissacus.*   There is no evidence that the limbs and feet of *Triisodon* were of mesonychid type, and the affinities through *Eoconodon* and *Goniacodon* to the normal arctocyonid skeleton characters suggest that this animal was a gigantic and peculiarly specialized member of the Arctocyonidae.

Triisodon crassicuspis (Cope)

Conoryctes crassicuspis Cope, 1882, Paleont. Bull. No. 35, Nov. 11, and Proc. Amer. Phil.
  Soc., XX, p. 468; 1885, Tertiary Vertebrata, p. 201, pl. xxiiie, fig. 6.
Triisodon rusticus Cope, 1884, Paleont. Bull. No. 37, Jan. 2, and Proc. Amer. Phil. Soc.,
  XXI, p. 310.

Type: A. M. Cope Coll. No. 3178, part of lower jaw with m$_3$ and heel of m$_2$, l.   Type of *Triisodon rusticus:* A. M. Cope Coll. No. 3225, lower jaw fragment with damaged m$_{1-2}$.

Horizon and Locality: Paleocene of San Juan basin.   Probably Torrejon formation, Middle Paleocene, so far as one may judge from the matrix and preservation of the specimen.

Author's Description: A second and larger species of the genus [*Conoryctes*].   The ramus is one-half deeper than that of the *C. comma,* and the second true molar is much larger than in that species.   The last true molar is much smaller than the penultimate, and consists of three anterior cusps and a longer heel. The former are obtuse, the external the longer, the internal equal, the anterior on the inner edge of the crown. The heel sustains a low conic tubercle.

Author's Description of T. rusticus: The species is of the type of *T. levisanus,* but is much larger. . . . Cusps of inferior molars not compressed.   Anterior cusp very low. . . . In dimensions the *T. rusticus* is about equal to the *T. quivirensis,* thus exceeding the other species excepting the *T. conidens.* The interior anterior cusp is nearly as elevated as the exterior, and is united with it nearly to the apex; the anterior cusp is a tubercle which projects forwards from its anterior base.   The heel of the tooth is wide, and

---

[58] Matthew, 1897, Bull. Amer. Mus. Nat. Hist., IX, p. 283.

is rounded posteriorly, and supports three tubercles, an external, a posterior and an internal, all in contact with each other. On the second true molar the internal anterior tubercle presents a slightly projecting edge anteriorly and posteriorly, which bounds a shallow vertical groove of the mass which represents their united bodies. This is not apparent in the first. The enamel is smooth, but the animal is rather old.

No additional specimens of importance are referable to this species. It is considerably smaller than *T. quivirensis*, despite Cope's statement (in type description of *T. rusticus*), and is distinguished from that species by relatively larger metaconid on m_{1-2}, but the protoconid is decidedly larger than metaconid although worn down nearly to its level in the type as described by Cope. If No. 16559 be correctly referred to *T. quivirensis*, *T. crassicuspis* is also distinguished from that species by the subequal size of metaconid and paraconid on m_3; in No. 16559 the paraconid is smaller and lower than the metaconid.

Eoconodon Matthew and Granger

Matthew, W. D., and Granger, Walter, 1921, Amer. Mus. Novitates No. 13, p. 6

Type: Sarcothraustes coryphaeus Cope, 1888 = *Triisodon heilprinianus* Cope, 1882.

Horizon and Locality: Puerco formation of the San Juan basin, New Mexico.

Diagnosis: Dentition unreduced, $\frac{3 \cdot 1 \cdot 4 \cdot 3}{3 \cdot 1 \cdot 4 \cdot 3}$. Molars typically tritubercular, robust and massive. Upper molars wider than long, subquadrate. Paracones and metacones round-conic, protocone subcrescentic, conules and rudimentary hypocones on m^{1-2}, heavy encircling cingula except on inner face of m^3. M^3 somewhat reduced, metacone smaller than paracone, metaconule vestigial or absent. Lower molars with trigonid and heel of subequal size, heel somewhat wider, trigonid a little higher, protoconid slightly higher than metaconid, paraconid well developed but lower, the talon broadly basined with hypoconid largest, entoconid and hypoconulid distinct, sometimes a rudimentary cusp in advance of entoconid. Premolars with high cusps, somewhat blunted but not compressed. P^{3-4} three-rooted, with internal cusps, rudimentary on p^3, large on p^4, and small cusps at external angles, in addition to the principal cusp. P^1 one-rooted, simple and small, p^2 and p_{2-4} two-rooted, p_4 much larger than the preceding teeth, with high protoconid, sharp basal anterior cusp and bicuspid heel. The posterior premolars, above and below, pitched backward, as in the Mesonychidae. Canines large, stout, not compressed, moderately recurved and subequal. Lateral incisors fairly large, medial and second pair small, crowded.

Jaws long, deep, not thickened; condyles transversely extended, low, coronoid process projecting backward and upward from its base, little recurved. Symphysis moderately loose and long. Skull massively proportioned with heavy and stout zygomatic arches, stout postglenoid processes, very high sagittal crest and prominent occipital crest, wide above and below expanded in equally prominent lambdoidal crests which extend to the superior border of the zygoma.

Limb bones stout and short, the humerus with long deltoid crest not so prominent or so abruptly ending as in arctocyonids, entepicondylar bridge well developed, supinator crest less prominent than in arctocyonids. Ulna with long straight olecranon laterally flattened, shaft broader than radius, head of radius flattened. Cuboid with distinct proximal astragalar facet; astragalus with trochlea short, little grooved, heavily pitched inwardly, fibular facet oblique, foramen well developed, head not squared.

This genus is principally represented by *Eoconodon* (*Triisodon*) *heilprinianus*, fairly common in the Puerco formation, represented in the collections by one complete and two incomplete skulls, a series of upper and lower jaws and various parts of the skeleton. *Sarcothraustes coryphaeus* and *S. bathygnathus*, *Triisodon biculminatus* and *T. rusticus* of Cope are probably synonyms, and *Mioclaenus gaudrianus*, also from the Puerco, is provisionally referred to the genus.

Eoconodon is nearly related, and appears to be directly ancestral, to *Triisodon* of the Torrejon.

Eoconodon heilprinianus (Cope)

Triisodon heilprinianus Cope, 1882, Paleont. Bull. No. 34, Feb. 20, and Proc. Amer. Phil. Soc., XX, p. 193; 1885, Tertiary Vertebrata, p. 273, pl. xxiiid, fig. 11; (*Mioclaenus*) Cope, 1888, Trans. Amer. Phil. Soc., XVI, N.S., p. 321; (*Triisodon*) Scott, 1892, Proc. Acad. Nat. Sci. Phila., XLIV, p. 301; Matthew, 1897, Bull. Amer. Mus. Nat. Hist., IX, p. 279; 1901, *ibid.*, XIV, p. 30, figs. 10–12.

Sarcothraustes coryphaeus Cope, 1885, Amer. Nat., XIX, p. 386; (*Mioclaenus*) Cope, 1888, Trans. Amer. Phil. Soc., XVI, N.S., p. 323; (*Sarcothraustes*) Scott, 1892, Proc. Acad. Nat. Sci. Phila., XLIV, p. 303; Osborn and Earle, 1895, Bull. Amer. Mus. Nat. Hist., VII, p. 29; (= *Triisodon heilprinianus*) Matthew, 1897, *ibid.*, IX, p. 279.

Mioclaenus bathygnathus Cope, 1888, Trans. Amer. Phil. Soc., XVI, N.S., p. 320; (*Sarcothraustes*) Scott, 1892, Proc. Acad. Nat. Sci. Phila., XLIV, p. 303; (= *Triisodon heilprinianus*) Matthew, 1897, Bull. Amer. Mus. Nat. Hist., IX, p. 279.

Triisodon biculminatus Cope, 1888, Trans. Amer. Phil. Soc., XVI, N.S., p. 343; Scott, 1892, Proc. Acad. Nat. Sci. Phila., XLIV, p. 301; Osborn and Earle, 1895, Bull. Amer. Mus. Nat. Hist., VII, p. 28, fig. 7; (= *Triisodon heilprinianus*) Matthew, 1897, *ibid.*, IX, p. 279.

Type: A. M. Cope Coll. No. 3224, lower jaw fragment with damaged molar tooth. Type of *Sarcothraustes coryphaeus:* A. M. Cope Coll. No. 3181, an incomplete skull. Type of *Mioclaenus bathygnathus:* A. M. Cope Coll. No. 3177, a lower jaw, p_3–m_3, l. Type of *Triisodon biculminatus:* A. M. Cope Coll. No. 3354, part of lower jaw with unworn teeth, m_{1-2}, l.

Horizon and Locality: All the above types from the Paleocene of the San Juan basin recorded as from the Puerco.

Author's Description of Triisodon heilprinianus: This species may be readily recognized as smaller than the *T. quivirensis*, and as having the anterior inner cusp of the inferior true molar of larger proportions than in the corresponding teeth of the latter species. . . . The anterior cusp is very low, and is nearer the inside than the middle of the anterior border. The principal anterior cusps are opposite, and the external is a little the larger. The heel is larger than the basis of the anterior cusps, and has convex borders. Its internal border supports three tubercles, and the external border rises into a cutting lobe with lenticular section. Enamel smooth. No cingula, but the external base is injured.

The principal specimens referred to this species are:

No. 16329, skull nearly complete, with incomplete lower jaws. Expedition of 1913.

No. 3181, Cope Coll. (type of *Sarcothraustes coryphaeus*), incomplete skull.

No. 764, incomplete skull, also left humerus. Figured 1897 as *Triisodon heilprinianus*. Expedition of 1892.

Nos. 3177, Cope Coll. (type of *M. bathygnathus*), and 3354, Cope Coll. (type of *T. biculminatus*), lower jaws.

No. 16331, upper teeth and lower jaw.

Nos. 16333–5, 3187, lower jaws.

Nos. 3280, 773, fragments of skeleton.

In addition to these there are ten or twelve specimens with parts of the upper or lower dentition, which served to indicate the individual variability. This is very considerable in the size of canine and last molar, depth of jaw, and relative size and robustness of skull and skeletal parts, but there appears to be a considerable amount at least of intergradation between the two extremes, the larger form represented by No. 16331 and the types of *S. coryphaeus* and *M. bathygnathus*, the smaller by Nos. 764 and 3187. The diversity is about as great as that between *Claenodon ferox* and *C. corrugatus*, and it may be that these also represent two species, but it is impossible to say to which of them the type specimens of *T. heilprinianus* and *T. biculminatus* belong, and it appears better to leave them as arranged by Matthew many years ago in a single varying species, the differences being regarded as probably sexual. If necessary to separate them, the smaller form could be called *E. biculminatus*, with Nos. 764 and 3187 as neotypes.

The skull as shown in Nos. 16329 and 3181 is large in proportion to the teeth, with moderately long, heavy muzzle, zygomatic arches heavy, the glenoid region massive, depressed considerably below the palatal level, the orbits set well back, the anterior border above m^3, postorbital processes not very prominent, the postorbital crests less prominent and less transverse than in *Deltatherium*, the sagittal crest high and long with nearly straight superior border, the occiput wide, high-crested and pitched strongly backward. The brain-case is very small, making scarcely any swelling on the sides of the cranium; a number of large nutrient foramina perforate the outer table in its posterior upper portion near the angle between the bases of sagittal and occipital crests. The mastoid exposure is large, facing posteriorly and bounded externally by the prominent lambdoid crest, making a robust but not prominent mastoid process, and continuous inferiorly with the auditory prominence which projects inward and forward into the bottom of the very wide and deep mesotympanic fossa. Its surface does not project downward but faces outward and backward; near the posterior end of the outer face is the fenestra rotunda; the opening of the fallopian aqueduct is behind it and postero-external to that the stylomastoid foramen. The posterior lacerate foramen is rather small and widely separated from the condyloid foramen which lies at the bottom of a round fossa just in front of the inner part of the occipital condyle. The anterior lacerate foramen and carotid canal form a broad notch at the antero-internal border of the large mesotympanic fossa. The foramen ovale has its usual position and is rather wide apart from the posterior opening of the alisphenoid canal. The paroccipital process behind the mesotympanic fossa is rather small, projects outward and backward but not much downward. The occipital condyles are of moderate size and the foramen rather small. The basicranial bones are rather long and narrow, the posterior nareal canal is narrow elongate, not roofed over and open to a short distance behind the last molars, the palate being considerably thickened at its posterior border. There is no indication of any alisphenoid bulla, the conditions in the basicranial region being most nearly approached among modern mammals in *Nandinia*, but differing in the much narrower basicranial bones, deep and large mesotympanic fossa with the auditory prominence much more deeply buried, and the large projecting glenoid region of the squamosal, small exoccipital region and large exposure of the mastoid, and various other characters mostly correlated with the much smaller brain-case.

This small brain-case, comparable with those of other Paleocene mammals, is not nearly approached by any modern Mammalia. Even in *Thylacinus* and other marsupials the brain-case is greatly enlarged and capacious in comparison with the Paleocene mammals. It is smaller in *Eoconodon* and *Ectoconus* of the Puerco than in *Deltatherium* or *Claenodon* of the Torrejon. *Periptychus* of the Torrejon appears to retain more of the very primitive character; *Pantolambda* and *Tetraclaenodon* are somewhat more progressive. But the small and primitive brain is an outstanding character of all the Paleocene mammals. It has not been adequately examined in any of them, owing to the rarity of well-preserved skulls, but it probably was of much the same type as that shown in *Phenacodus* of the Wasatch described by Cope many years ago.

The sutures of the top of the skull are not very clear, but the construction was apparently much as in *Deltatherium*, with long, wide nasals somewhat expanded posteriorly, the frontals very short, the squamosal widely expanded over the side of the cranium. The anterior limits of lachrymal and jugal are not determinable, but presumably the lachrymal had a moderate facial expansion as in other creodonts. Posteriorly the jugal bar is wide but rather short, extending to 10 mm. in front of the anterior border of the glenoid fossa. This fossa is moderately wide transversely, rather flat with no preglenoid crest and a stout short postglenoid process not wide transversely.

The infra-orbital foramen appears to be well forward above p^3 or p^4, and far in front of the orbit. The anterior nareal aperture is large, the notch on the anterior border of the nasal bones is deep and narrow, the superior branch of the premaxilla is wide, extending backward some 25 mm. between nasals and premaxillae.

The principal characters of the teeth and jaws have already been given under the genus.

The characters of the skeleton are imperfectly known from specimens of rather doubtful · association. The limb bones have much the same proportions as in *Claenodon* but are relatively smaller in comparison with the skull. The humerus is less massive, the deltoid crest extends farther down on the shaft and ends less abruptly, and there is no opposite crest on the postero-internal face of the shaft. The distal end is broad, shallow and expanded in the usual way, with very prominent entepicondylar process, the foramen smaller and the bridge not so heavy. The ulnar shaft is correspondingly wide and heavy, the head of the radius somewhat more round-oval, the olecranon appears to be shorter and more backwardly pitched. The relative length of ulna and radius, tibia and fibula appears to be somewhat less, but in none of these bones is the complete length preserved. The distal end of the fibula appears to face somewhat more internad than in *Claenodon*, but to have a similar calcanear facet. The tuber-calcis is more narrow and deep; the cuboid has a facet for the astragalus about a third as wide as that for the calcaneum, but facing rather more internad than proximad. Fragments of metapodials are preserved, indicating a smaller foot, perhaps more slender; nothing to show the characters of the phalanges.

The characters of the skull are those of a very primitive carnivore, nearly related to *Deltatherium* and *Loxolophus*, much larger and more robust, but with a more omnivorous adaptation in the teeth. From *Claenodon* and *Arctocyon* it differs in the more robust and powerful canines and correlated characters of the skull and jaws, a somewhat more predaceous adaptation. All of them appear to be rather nearly related. No suggestion of any especial relationship to marsupials appears in any of them; they are typically carnivore, but extremely primitive, with a far lower type of brain-case than in any living mammal, whether marsupial or placental, yet the nearest comparison, in spite of the great difference in development of the brain, is to be found in *Nandinia*, the most primitive of the Viverridae.

There is, as will be seen, a great deal in common with the skulls of the most primitive Taligrada and Condylarthra, indicating that these two groups are not far from a common ancestry with the Creodonta.   Relationship to the primitive Insectivora and cognate groups of placentals is likewise indicated, although not so close as to the taligrades and condylarths. The conclusion seems unavoidable that none of them is very far from a common pro-placental ancestor.

If the contemporary ancestors of the marsupials were adequately known, they might make also a near approach to the pro-placental type, but as yet the characters of the skull in early Tertiary or Mesozoic marsupials are too imperfectly known for any satisfactory comparisons to be made.

Eoconodon gaudrianus (Cope)

Mioclaenus gaudrianus Cope, 1888, Trans. Amer. Phil. Soc., XVI, N.S., p. 326; (*Goniacodon*) Scott, 1892, Proc. Acad. Nat. Sci. Phila., XLIV, p. 302; (*Triisodon*) Matthew, 1897, Bull. Amer. Mus. Nat. Hist., IX, p. 280.

Type: A. M. Cope Coll. No. 3200, parts of upper and lower jaws and calcaneum.   No. 4029, upper and lower jaws and a few skeleton fragments, was referred by Matthew to this species.

Horizon and Locality: Both specimens are from the Puerco formation, Cope Collection, exact locality and level unknown.

Diagnosis: The species is distinguished from *Eoconodon heilprinianus* chiefly by the smaller size, about equal to that of *Goniacodon levisanus* of the Torrejon.   From *Goniacodon* it is distinguished by the greater transverse diameters of the molars and unreduced m$\frac{3}{3}$.   No additional specimens have been found by subsequent collectors.

Goniacodon Cope, 1888

Cope, E. D., 1888, Trans. Amer. Phil. Soc., XVI, N.S., p. 320

Type: Triisodon levisanus Cope, 1883, from the Torrejon formation of New Mexico.

Author's Diagnosis (as a subgenus of *Mioclaenus*): The fifth cusp is quite distinct, but is median in position and near the base of the crown (except in *M. heilprinianus*), forming an anterior angle in the outline of the crown. These species [*M. heilprinianus, assurgens, levisanus, rusticus*] I have referred sometimes to Triisodon and sometimes to Diacodon, but I think I have now found their proper position.

Scott in 1892 removed *heilprinianus* and *assurgens* to the genera *Triisodon* and *Microclaenodon* respectively, added *Mioclaenus gaudrianus*, and re-defined the genus, raising it to full generic rank:

The species are of moderate size, smaller on the average than those of either of the allied genera. The anterior upper premolars are small and simple. P$\frac{2}{}$ is implanted by three fangs (at least in *G. levisanus*) but has no distinct deuterocone. P$\frac{4}{}$ has a very high, acute protocone and well developed deuterocone. The upper molars are triangular in shape with low, conical cusps and m^1 has a fairly well developed hypocone. P$_4$ has a small talon, divided into minute outer and inner cusps. The lower molars are the characteristic feature of the genus; the trigonid is moderately elevated above the talon; the proto- and metaconids are of nearly the same size and very closely approximated, forming a twin cusp which is cleft but a short distance below the apex; the paraconid is very small, depressed and submedian in position, i.e., standing in front of the space between the proto- and metaconids.   The talon is basin-shaped and consists of hypo- and entoconids which may be of nearly the same size (*G. rusticus*) or the former may be much the larger (*G. levisanus*). A minute hypoconulid is also present. $M_{\overline{3}}$ is much reduced.[59]

Scott's diagnosis is a composite drawn from three species which have little in common save a certain resemblance in m_{1-2}, but no other corresponding parts were then known.

Matthew in 1897, with more complete specimens of the dentition, limited the genus to the type species and re-defined it as follows:

. . . Upper molars . . . much less quadrate than in *Triisodon* [i.e. *Eoconodon*], m^3 reduced and m^1 as large or larger than m^2, instead of smaller.   The third upper premolar is three-rooted, without well-developed deuterocone.   The paraconid is perhaps somewhat smaller than in *Triisodon*, and the proto- and metaconid higher and more equal in size.   The position of the paraconid is not entirely constant, but it is usually submedian on m_1, internal on m_2, as in *Triisodon*.   The third lower molar is reduced.   A character observed in two specimens, and perhaps a constant one, is the position of the mental foramen underneath the *second* premolar instead of between the third and fourth.   This is associated with the short deep symphyseal part of the jaw.   The symphysis is ovate and widest behind, extending back to beneath the third premolar, while in *Triisodon* it is widest anteriorly and pointed behind.[60]

A specimen of *G. levisanus* in the 1913 collection has the upper and lower dentition, better preserved than in any previously found, and various fragments of the skull and skeleton, including some parts of the feet.   These make it clear that the genus is not related to the Mesonychidae but to the Arctocyonidae.

Goniacodon levisanus (Cope)

Triisodon levisanus Cope, 1883, Proc. Amer. Phil. Soc., XX, Feb. 14, and Paleont. Bull. No. 36, April 17, p. 546; 1885, Tertiary Vertebrata, p. 273, pl. xxiv *f*, fig. 3; (*Mioclaenus*, sub. gen. *Goniacodon*) Cope, 1888, Trans. Amer. Phil. Soc., XVI, N.S., p. 321; (*Goniacodon*) Scott, 1892, Proc. Acad. Nat. Sci. Phila., XLIV, p. 301; Matthew, 1897, Bull. Amer. Mus. Nat. Hist., IX, p. 282.

Type: A. M. Cope Coll. No. 3217, right ramus of lower jaw with broken p_4 and m_{1-2}.

Horizon and Locality: Torrejon formation of San Juan basin.

Author's Description: The ramus is deep, and probably belonged to an animal of about the size of the red fox.   The molars have the structure most like that of the *T. heilprinianus*, especially anteriorly. The principal anterior cusps are united together for most of their elevation, while the anterior inner is much smaller and lower, and is situated between the middle and inner side of the anterior cusp.   The heel is rather wide, and has a raised border.   The external part of it is angular, and is somewhat within the vertical line of the base of the crown.   The fourth premolar differs from that of the type [of] the genus, *T. quivirensis*, in having two acute longitudinal tubercles situated close together on the heel.

The anterior masseteric ridge is very prominent.   The masseteric fossa is strongly concave, but shallows gradually inferiorly.   Its inferior border presents a low thickened ridge, which is recurved in front.   This may be an individual character only.   The inferior outline of the ramus is generally convex, and does not rise much below the masseteric fossa.

Revised Diagnosis: Upper canines large, stout, round-oval, without ridges or crests, premolars short, stout, without diastemata, p^4 triangular.

[59] Scott, 1892, Proc. Acad. Nat. Sci. Phila., XLIV, p. 301.

[60] Matthew, 1897, Bull. Amer. Mus. Nat. Hist., IX, p. 282.   The comparisons with "*Triisodon*" are throughout with *Eoconodon.*—w.d.m.

Upper molars subtrigonal, considerably wider than long, cusps massive, rounded, small hypocones, conules quite small, encircling cingula. M^3 much reduced, metacone vestigial or absent, no hypocone or conules.   Lower canines large, massive, oval, no posterior crest, premolars short, massive, without diastemata, p_1 one-rooted, p_{3-4} with well-developed heels. Lower molars with moderately high trigonids and basined heels, paraconid submedian, low on m_1, smaller on m_2, vestigial on m_3; metaconids and protoconids rounded, rather high and somewhat connate. M_3 much reduced.   Jaw deep and heavy, of uniform depth beneath dentition, the symphysis very stout and massive as in Oxyaenidae.

The genus is most readily recognized by the reduction of m^2 and the short, heavy anterior portion of the jaw with very massive symphysis and stout uncompressed canines.

Of the upper teeth the incisors are not known.   The upper canine is large, with robust root and rather short, pointed crown of round-oval section, the major axis of the cross section of the crown twisted obliquely to the axis of the root.   No anterior or posterior crests.   Four upper premolars, rather small and crowded, not notably compressed, p^1 and p^2 two-rooted, the second with rudimentary anterior and posterior basal cusps, p^3 and p^4 three-rooted, with trihedral crowns, rudimentary cusps at the angles on p^3, and on p^4 a fairly well-developed inner cusp and small basal cusps at the external angles.   The molars are tri-

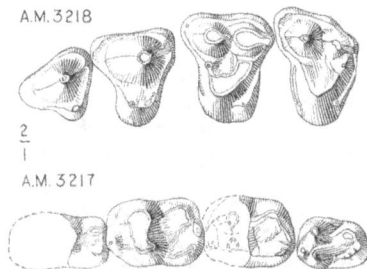

Fig. 15.   *Goniacodon levisanus*, upper and type lower teeth, A. M. Nos. 3218 and 3217, crown views.   Twice natural size.

tubercular, considerably wider (transversely) than long (anteroposteriorly), with rather massive cusps, paracone and metacone round-conical, protocone only a little crescentoid, protoconules absent but metaconules distinct, small rudimentary hypocones.   Cingula moderately strong except upon the inner face of the protocone. M^1 symmetrical, m^2 slightly smaller, and somewhat extended at the antero-external angle. M^3 much reduced and simplified, transversely suboval, about twice as wide as long, but the transverse diameter only about half that of m^1 and m^2, metacone vestigial, no metaconule or hypocone, cingula as on preceding molars.

Of the lower teeth the incisors are quite small and crowded; their alveoli are so indistinctly shown that it is uncertain whether they were reduced in number.   The lower canine is a large tooth with massive root and the crown round-oval in cross section, without crests in front or behind, longer and sharper than the upper canine, and lacking the outward twist. The lower premolars are crowded and rather small, the first one-rooted, the others two-rooted.   The crowns are not well preserved, except of p_4, which has the principal cusp (protoconid) rather high, acute, recurved, but not compressed, a small anterior basal cusp, and a fairly large bicuspid heel.   The tooth is throughout much like that of *Eoconodon heil-*

prinianus, except that the protoconid is rather recurved towards the tip than backwardly-pitched from the base.

The lower molars are constructed much as in *Oxyclaenus*, except that the protoconid and metaconid are more connate and much reduced, especially on m_3, and this tooth is much more reduced than in *Oxyclaenus cuspidatus*. The resemblance to *Eoconodon* is superficially greater than to *Oxyclaenus*, but not so close on a detailed comparison, as *Eoconodon* does not show the peculiar round-conic form of trigonid cusps of *G. levisanus* approached in *O. cuspidatus*, and the basin of the heel is more enclosed internally in *Goniacodon*. The last molar, somewhat reduced in *O. cuspidatus*, much reduced in *Goniacodon*, is wholly unreduced in *Eoconodon*. The reduction and low position of the paraconid is much nearer in *Goniacodon* to *Eoconodon* than to any of the Arctocyonidae.

The lower jaw is short, deep and heavy anteriorly, with a solid oxyaenoid symphysis; the mental foramina are remarkably far forward, situated under p_1 and p_2. The condyle was apparently set low, the coronoid process rises abruptly close behind the last molar, and the masseteric fossa is fairly well defined, its lower border much nearer to the inferior border of the ramus than in *Eoconodon*.

From the skull fragments the following data are determined: the proportions of the skull were somewhat as in *Loxolophus*, with a long but thin and low sagittal crest, the cranial bones not massively proportioned as in *Eoconodon*, but rather thin and delicate. The crest is by no means so high as in *Deltatherium* nor do the frontal and occipital regions appear to have been broadened out as in that genus, but the postorbital region was long and somewhat tubular, the brain-case small and low.

The jugal has less depth anteriorly than in *Deltatherium;* the superior branch is long and slender, the inferior branch being very small, while in *Deltatherium* it is somewhat expanded into a plate. It is typically carnivore in proportions and relations, resembling various modern Viverridae and with no approach to *Didelphis*. There is no postorbital process on the jugal. It is overlapped by the zygomatic process of the squamosal to a point about six millimeters behind the tip of the maxillary process on its inner face, a much closer approximation than in *Deltatherium;* and it does not approach the glenoid fossa, although its exact distance from it cannot be determined.

The glenoid region is much like that of *Deltatherium*, the transverse extension of the fossa being somewhat greater, and much greater than in *Loxolophus*. The postglenoid process is much wider transversely than in *Deltatherium* and somewhat more of a preglenoid crest is developed. The postglenoid foramen is of moderate size, somewhat smaller than in *Deltatherium*, much smaller than in *Loxolophus*. The foramen ovale is situated within the alisphenoid, but very close to its suture with the squamosal, an unusual position approximated in *Deltatherium* and *Loxolophus*, while in most modern Carnivora it has shifted to a varying degree towards the middle of the alisphenoid.

The occipital condyle is wider than in *Deltatherium*.

The humerus is of about the same length as in *Deltatherium*, but somewhat more slender, and the deltoid and supinator crests decidedly less prominent, the entepicondylar foramen and bridge apparently smaller.

The ulna and radius are of moderate length and of usual creodont proportions. The ulna has a rather long olecranon, as in *Vassacyon*, nearly in line with the shaft, pitched a little inward and ending in an obliquely truncated knob. The facet for the radius is concave and almost double, the posterior and internal faces almost separated by a deep notch and continuous only at the upper margin, which is well defined.   The shaft of the ulna is wide and flat, as in most primitive mammals, retaining most of its width to the distal end.   The radius has a flattened ovate head, bicipital tuberosity low and obscure, shaft irregularly rounded, by no means as wide as that of the ulna but thicker anteroposteriorly, the distal end trihedral, with the facet uniformly concave and no process on the inner side.

In *Miacis palustris*, a Bridger species of about the same size and proportions, the ulna has a much shorter olecranon, marked sigmoid curvature from end to end, the radial facet is flat, the head of the radius is rounded, and the ulna is less flattened in the shaft while the radius is considerably more so.

The tibia is of moderate length and proportions, and very like that of *Miacis palustris*, except that the astragalar facet is somewhat more excavated and the internal malleolus slightly heavier.   In *Vassacyon* the tibia is of relatively larger size, as it is in the majority of the early creodonts, but the structure does not differ to any notable degree.   The cnemial crest is more obscure than in *M. palustris*, in which it is somewhat definitely limited to the second fifth of the shaft.

The specimens referred to here are:

No. 16561, upper and lower jaws, fragments of the skull, and some parts of the skeleton, including humerus, radius, ulna, tibia, astragalus and calcaneum, parts of metapodials and phalanges.   Exped. 1913, from Alamito Wash.

No. 16562, upper jaw with p^4–m^2 and roots of anterior premolars.   Exped. 1913, West Fork, Torrejon Arroyo.

No. 3218, Cope Coll., parts of upper and lower jaws, p^4–m^2, m_{1-3}.   Locality not recorded.

No. 2396, lower jaw, m_{2-3} and alveoli or roots of all teeth in front of these.   Exped. 1896, Rio Torrejon.

No. 3217, Cope Coll., type, lower jaw, p_4–m_2 more or less broken.   San Juan basin.

No. 4018, Cope Coll., lower jaw, p_3–m_2 poorly preserved.   Locality not recorded.

Nos. 16728, 16729, lower jaws.   Exped. 1913, West Fork, Torrejon Arroyo.

The specimens obtained by the Expedition of 1913 were all from the upper level of the Torrejon.   The level of the others is not recorded, but presumably was the same.

MESONYCHIDAE Cope, 1875

Mesonychidae Cope, 1875, Paleont. Bull. No. 20, Dec. 22, p. 3; 1885, Tertiary Verte-
brata, p. 259; Scott, 1892, Proc. Acad. Nat. Sci. Phila., XLIV, pp. 294, 303; Wortman, 1901, Amer. Jour. Sci., XII, p. 285; Matthew, 1909, Mem. Amer. Mus. Nat. Hist., IX, pt. VI, p. 485; Matthew (and Granger), 1915, Bull. Amer. Mus. Nat. Hist., XXXIV, p. 84.

This family was originally distinguished by Cope by the trochlear ankle-joint, the non-shearing type of molar teeth with conic tubercles and heels, and the transversely extended glenoid articulation with preglenoid crest.   Subsequent studies have modified and extended these distinctions and shown the family to be a peculiar aberrant group of the Creodonta, apparently very early differentiated from the main stock.   Matthew defined the family in 1901 as follows:

No carnassials; teeth with high, round, blunted cusps; upper molars tritubercular; lower molars pre-molariform.   Claws blunt . . . hoof-like, resting on the ground.   No tendency to union of the carpals.[61]

The family is well distinguished by the peculiar characters of the teeth and feet.   The transfer of the triisodont group excepting *Microclaenodon* to the Arctocyonidae makes it practicable to define it more clearly as follows:

No carnassials, teeth with high round blunted cusps, upper molars tritubercular, without conules, paracone and metacone more or less connate, lower molars narrow and pre-molariform in type, the metaconid absent or closely twinned with protoconid, paraconid low, heel low, with strong median crest, no entoconid except in *Microclaenodon*.   Muzzle long, mesocranial region of skull and posterior portion of jaw elongate, basicranium shortened. Limbs and feet showing progressive cursorial specialization, the phalanges short and flattened, unguals hoof-like.

The Paleocene stages are, as might be expected, much more primitive than their later Eocene descendants.   *Dissacus* retains the complete dentition, metaconids on the molars, ungrooved astragalus and functionally five-toed feet, although the ungues are quite hoof-like and the tooth-pattern unmistakably mesonychid.   *Microclaenodon* is more primitive and may not be a member of the family, as the characters of limbs and feet are unknown and the upper dentition known only from one molar.

The later Mesonychidae are diversely specialized.   *Synoplotherium* (= *Dromocyon*) represents a cynoid specialization, as Wortman has shown; *Mesonyx* is similar but with loss of m³.   *Harpagolestes* is a gigantic short-faced type with short, stout limbs, *Andrewsarchus* a still more gigantic form, long-faced, and paralleling the later entelodonts in skull (presumably also in skeleton, although this is unknown).   *Hapalodectes* at the other extreme has a tiny jaw with much compressed teeth, perhaps a fish-eating or flesh-eating specialization.   *Dissacus* may be regarded as representing the common ancestry of these diverse specializations.   Its relations to the remaining Creodonta are not close.   In spite of the pentadactyl and more or less plantigrade feet, the phalanges are as much shortened and the ungues as hoof-like as in the Taligrada, and more so than in the contemporary Condylarthra, while the characteristic peculiarities of the teeth are quite as far from the primitive tritubercular pattern as are those of the Periptychidae or Pantolambdidae.   The skull characters are equally specialized.   *Microclaenodon* indeed appears to be intermediate, having the narrowed lower molars with high twinned protoconid and metaconid, low paraconid and heel, but retaining primitive basined heels.   But it is too imperfectly known to be sure of its exact relationship; it certainly is not ancestral.

The inference from these data would seem to be that the mesonychids are not more nearly related to the remaining Creodonta than are the taligrades, and not so near as are the condylarths.

The following table shows the geological distribution of the genera as now known:

[61] Matthew, 1901, Bull. Amer. Mus. Nat. Hist., XIV, p. 7.

	Paleocene					Eocene								
	Lower	Middle	Upper			Lower				Middle			Upper	
	Puerco	Torrejon	Fort Union	Cernaysian (France)	Tiffany	Lower Wasatch	Ageian (France)	Middle Wasatch / Lower Wind River	Upper Wind River / Lower Huerfano	Upper Huerfano / Lower Bridger	Upper Bridger / Lower Washakie	Irdin Manha (Mongolia)	Upper Washakie / Lower Uinta	Upper Uinta
Microclaenodon.......		×												
Dissacus...........		×	×	×	×	×	×							
Hapalodectes........						×		×				?		
Pachyaena..........						×	×	×						
Andrewsarchus.......												×		
Synoplotherium......											× ×	?		
Mesonyx...........										×				?
Harpagolestes.......										×			× ×	×

KEY TO GENERA

1. Heels of lower molars basined.
   Size small, metaconids distinct... *Microclaenodon*
2. Heels of lower molars crested.
   Metaconids distinct on m_{2-3}. M^3 reduced.
      Larger, teeth more robust. Tibio-tarsal joint oblique, without trochlea, pollex
        and hallux well developed.. *Dissacus*
      Smaller, lower teeth much compressed................................. *Hapalodectes*
   Metaconids vestigial or absent on all molars.
      M^3 unreduced.
        Size medium to large, astragalus with shallow trochlea, pollex and hallux
          vestigial... *Pachyaena*
        Size gigantic, muzzle elongate.................................... *Andrewsarchus*
      M^3 reduced.
        Size medium, deep astragalar trochlea, pollex and hallux absent. Cursorial. *Synoplotherium*
      M^3 absent.
        Size medium, astragalus with deep trochlea, pollex and hallux absent.
          Cursorial... *Mesonyx*
        Size large, muzzle shortened. Massive............................ *Harpagolestes*

Dissacus Cope, 1881

Cope, E. D., 1881, Amer. Nat., XV, Nov. 29, p. 1019

Type: Mesonyx navajovius, from the Middle Paleocene of New Mexico.

Author's Diagnosis (p. 1018): A fine series of specimens of *Mesonyx*[62] demonstrates the following points: (1) Pachyaena was founded on a superior molar of *Mesonyx*, and must be suppressed. (2) Apterodon

---

[62] These were specimens of *Pachyaena* from the Bighorn Wasatch, sent to Cope about this time by Wortman. Cope made *Pachyaena* a synonym of *Mesonyx* in 1882–5, but afterwards revived the genus as distinct.—W.D.M.

Fischer, is the same as *Mesonyx*. (3) *Mesonyx navajovius* Cope must be separated as a distinct genus, since the apices of the crowns of the last two molars have two cusps. This genus may be called *Dissacus*. (4) It results that there are four species of *Mesonyx: M. ossifragus* Cope, *M. lanius* Cope, *M. obtusidens* Cope and *M. gaudryi* Fisch.

Dissacus navajovius (Cope)

Mesonyx navajovius Cope, 1881, Paleont. Bull. No. 33, Sept. 30, and Proc. Amer. Phil. Soc., XIX, p. 484; (*Dissacus*) Cope, 1884, Amer. Nat., XVIII, p. 267, fig. 11.

Dissacus carnifex Cope, 1882, Amer. Nat., XVI, p. 834. (Not *D. carnifex* Osborn and Earle, 1895, Bull. Amer. Mus. Nat. Hist., VII, pp. 30–9, figs. 8, 9.)

Type: A. M. Cope Coll. No. 3356, jaw fragments, p_3–m_3, r; p_4–m_3, l. Type of *Dissacus carnifex:* A. M. Cope Coll. No. 3361, a lower jaw with p_2–m_3 l., recorded as from the "Upper Puerco" = Torrejon, San Juan basin, New Mexico.

Horizon and Locality: Paleocene of the San Juan basin, New Mexico. Both undoubtedly from the Torrejon formation, Middle Paleocene, but the type of *D. navajovius* is not specifically recorded as from that horizon.

Author's Description: Smaller than the two known species [of *Mesonyx*, i.e. *M. obtusidens* of the Bridger and *M. ossifraga* of the Wasatch], and with the crowns of the molars more compressed and the blades of the heels of the inferior series more acute. Molars seven, the first one-rooted. Last molar with a cutting heel like the others, and with the penultimate, with a rudimental anterior inner cusp [i.e. metaconid]. All the molars with an anterior basal tubercle except the first, second and third [premolars]. No basal cingula. Principal cusp elevated and compressed, as in the premolars of *Oxyaena*. Enamel minutely rugose. Mandibular rami and inferior canine teeth compressed, the angle of the latter not inflected. Length of inferior molar series M. .078; do. of premolar series .046; fourth premolar, length of base .010; elevation of cusp .008; second true molar, length .012, elevation .010; width of heel .005; depth of ramus at [?] .020; diameter of base of crown of canine, vertical .009.

Author's Description of Dissacus carnifex: This creodont differs from its only congener [*D. navajovius*] in its greater size, and in the presence of an anterior basal lobe on the third inferior premolar. . . . As compared with the latter the six inferior molars are as long as its seven, and the mandibular ramus is much deeper. Like it the P–m.IV and the true molars have an anterior basal tubercle; and the last two true molars have an internal supplementary cusp. After the *Sarcothraustes antiquus*, the largest flesh-eater of the Puerco.

The principal specimens referred are:

Nos. 3356 (type) and 3361 (type of *D. carnifex*), both lower jaws.

Nos. 3360 and 3362, palate and fragments of skull and skeleton (two specimens subsequently found to be the same individual).

No. 3357, upper molars.

No. 16557, upper and lower jaw.

No. 3359, upper and lower teeth and parts of skeleton. The last specimen is the one on which Cope based his description of the skeleton of *Dissacus* in 1888, but the specimen is now much more complete, the author of this memoir having spent considerable time in piecing together the fragments, having also found that Field Number 112 collected by Barnum Brown in 1896 is part of the same individual. It is of interest, however, to note how Cope built up from the fragments which he studied a correct concept of the characteristic features of the tarsals and certain limb bones.

Most of the lower cheek teeth and some of the upper are preserved in No. 3359, also the zygomatic arches complete, the greater part of the scapulae, humeri, radii and ulnae, parts of femora, tibiae, the tarsal bones except ectocuneiform of one or both sides, and parts of various metapodials including the heads of Mc. I, Mt. I–III, and parts of phalanges. These all accord with the better-preserved skeleton material of *D. saurognathus* except for much smaller size and much more slender proportions throughout.

The teeth have the characteristic mesonychid construction throughout, but the metaconid on m_{2-3} is distinct and well developed although closely connate with the protoconid. On m_1 it is much smaller, although distinct in this species (quite vestigial in *D. saurognathus*). P^3 and p^4 have internal cusps; the metacone is smaller than the paracone on m^1 and m^2, quite small or vestigial on m^3; the external angles of the molars and p^4 have low cinguloid parastyle cusps, and similar metastyles appear on m^{1-2}. The cingula are more or less obsolete

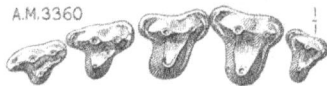

FIG. 16. *Dissacus navajovius*, upper teeth, A. M. No. 3360, crown view.   Natural size.

except at these external angles. The conules are absent and the wings of the protocones vestigial, the principal cusps having the characteristic rounded conic form of the family. The lower molars have the large crested square-based heels of the family fully developed, and these are progressively developed on p_{2-4}. The paraconids are of similar type but much smaller; a small anterior basal cusp is also present on p_4, rudimentary or absent on p_3. P_1 is one-rooted and simple-crowned. The principal cusp of the lower molars is the protoconid, high, rounded, blunt-tipped, twinned with a well-developed metaconid on m_{2-3}, but on m_1 the metaconid is vestigial. M_3 is about as large as m_1, m_2 a little larger than either. The canines are of moderately large size, long and rather slender, round-oval in cross-section, without posterior crests, upper and lower canines of equal size, both moderately recurved.

The lower jaw is rather deep but not thick, long, with loose symphysis, the anterior premolars more or less spaced, incisors apparently quite small. The posterior part of the jaw is unusually long, with wide, flat coronoid process directed backward from the base at an angle of forty-five degrees, the condyle prominent, strongly projecting, very convex and transversely expanded, the angle broad and flat.

The skull is very imperfectly known. The anterior border of the orbit appears to have been above or in advance of m^2; the infra-orbital foramen is set low down and above p^3. The zygomatic arches show the characteristic form seen in *Pachyaena*, the long and straight arches indicating an exceptionally long mesocranial region. The jugal, as in other Mesonychidae, has no antero-inferior branch, the superior branch being low, forming only the inferior border of the orbit, and heavily overlapped anteriorly and inferiorly by the maxilla; the lachrymal forms most of the anterior border of the orbit, extending considerably below the foramen which is just inside the orbital rim, and apparently had a considerable facial expansion. These conditions do not differ greatly from those found in other Creodonta except in the length, straightness, and backward extension of the zygoma. The glenoid articulation is probably set well back, and depressed, as in other Mesonychidae, and there is a high recurved postglenoid process, a well-developed preglenoid process, postglenoid foramen obsolete.

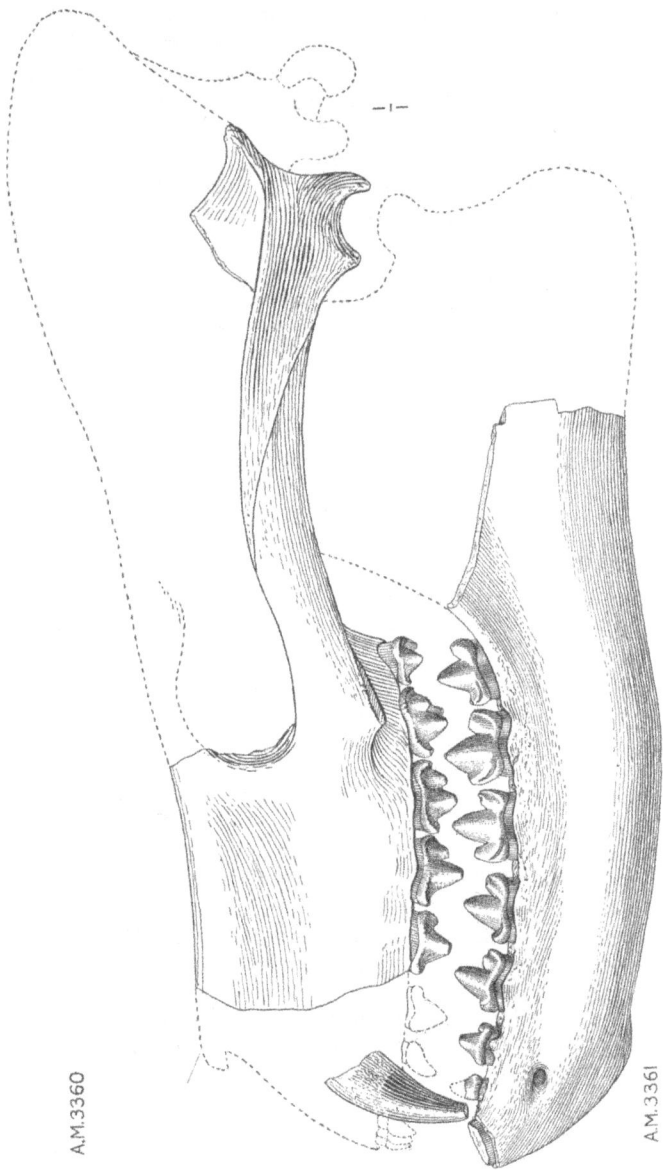

Fig. 17. *Dissacus navajovius*, skull and lower jaw, restored from A. M. Nos. 3360 and 3361, side view.   Natural size.

The vertebrae as observed by Cope show a long cylindro-conical odontoid process, and in other respects the atlas and axis appear to conform to the mesonychid type as described by Wortman in 1901, but they are too fragmentary for detailed comparison. The zygapophyses of the lumbars are closely interlocking, the posterior ones very convex but apparently not revolute.

A large part of the scapula is preserved, but the critical characters of coracoid and acromial processes are not determinable. The humerus is long and straight, not so much expanded distally as in Arctocyonidae, and the deltoid crest is not so high or abrupt. The distal facets are deeper antero-posteriorly and not so wide; the olecranar and anterior supratrochlear fossae are deeper and there may have been a supratrochlear foramen. In general the humerus shows a marked approach to that of *Mesonyx*, but the entepicondylar bridge is present. The ulna has a moderately long, straight olecranon, a wide and flattened shaft with a deep wide longitudinal groove on the anterior face, as described by Cope; the sigmoid cavity is rather shallow, the radial facet nearly flat. The bone is long and slender in comparison with *Claenodon*. The head of the radius is considerably expanded transversely with wide and only slightly convex ulnar facet and obscure bicipital process.

The manus is practically unknown in this species, save for the head of the first metacarpal and perhaps some other unrecognized fragments. The pollex is evidently well developed, although much smaller than the presumptive size of the other four digits.

The pelvis has the usual creodont proportions, with long ischia, the ischial spine far backward, the ilium rather short and broad. The femur is known only from a few fragments. The tibia has a heavy, wide internal malleolus, the astragalar trochlea is almost ungrooved and very oblique, and the fibula has a remarkably heavy shaft, the astragalar facet facing inwards and only a little downwards, the calcanear facet large and facing distally.

The pes as described by Cope is primitive mesonychid in character. The astragalus has an ungrooved trochlea, but the fibular facet is subvertical, and the whole tarsus is much narrowed and deepened as compared with Arctocyonidae. The astragalus has a very oblique and almost ungrooved trochlea and large astragalar foramen, so that the tibial facet is very short in comparison with its width. The cuboid facet on the head of the astragalus is distinct, forming a narrow uniform band facing more distally than externally; the navicular facet is deep and narrow; the cuneiform facets on the navicular are deep and narrow, that for the entocuneiform quite small but facing subdistally as far as the part of it preserved shows. The facet on the navicular for the head of the astragalus is strongly concave dorsoplantad, but is not transversely concave.

The metatarsals appear to have been rather long and compact, not so wide-spreading as in Arctocyonidae, and decidedly nearer in this respect to those of the later Mesonychidae; but the hallux was probably complete, both ends of the first metatarsal being preserved. They are considerably smaller than the other digits, and apparently the digit was more or less appressed against the inner side of the second metatarsal, although there is no facet for it on that bone.

The phalanges are short and flattened, although not so much so as in the later Mesonychidae, but are very different in appearance from those of Oxyclaenidae. The unguals are not known in this species.

Dissacus saurognathus Wortman

Dissacus saurognathus Wortman, in Matthew, 1897, Bull. Amer. Mus. Nat. Hist., IX, p. 285.
Dissacus carnifex Osborn and Earle, 1895, Bull. Amer. Mus. Nat. Hist., VII, p. 30.   (Not *D.*
   carnifex of Cope.)

Type: No. 2454, a complete lower jaw.

Horizon and Locality: Torrejon formation, New Mexico.

Diagnosis: It differs from *D. navajovius* in size, being about twice as large [lineally], in the presence of
a second internal cusp on p⁴ and a well defined anterior cusp on p₄.   The metaconid on m₁ is vestigial or
absent; on m₂ and m₃ it is well defined, as is the case in *D. navajovius.*[63]

To this species were referred the two fragmentary skeletons described by Osborn and
Earle in 1895 under the name of *D. carnifex*, also a humerus, ulna and radius, No. 2524.   No
important additional specimens were found in 1912 and 1913, but the two fragmentary
skeletons have been pieced together and restudied so that a revised and extended description
is advisable.

There is some uncertainty about the true succession of the teeth in No. 776, which is an
immature individual and may have retained some of the milk teeth; the tooth described as
p⁴ is possibly dp⁴, which would explain its having a second internal cusp; another tooth not
described or figured appears to be dp³.   Other specimens show p⁴ with the same construction
as in *D. navajovius.*   The disparity between the two species is not so great in size of teeth
as in the jaw and skeleton.

The skull is known only from a few fragments, which indicate a general resemblance
to the skull of *Pachyaena*, as described by Matthew in 1915.   This would mean a long
muzzle, an extremely long mesocranial region and a short basicranial region, the glenoid
articulations crowded back towards the condyles; a very small, narrow and low brain-case
with extremely high sagittal and occipital crests, deep heavy zygomatic arches of moderate
width and unusual straightness.   In *D. saurognathus* a few fragments of the top of the skull,
zygomata and other parts conform to this construction; in the smaller species, *D. navajovius*,
the arches are of the same type as in *Pachyaena*, allowing for its much smaller size and less
massive proportions.   The proportions of the jaw in *saurognathus* likewise indicate a skull
similar to that of *Pachyaena.*

The fore limb bones are preserved in No. 777 and in No. 2524.   The humerus is not
unlike that of *Claenodon*, the deltoid crest is less prominent but extends farther down the
shaft; the distal facets are somewhat differently set in relation to the shaft, the axis of the
shaft being about midway between the epicondyles instead of toward the ectepicondyle as in
Claenodon.   The entepicondylar foramen is large, the bridge rather heavy, the entepicondyle
less prominent than in *Claenodon*, the ectepicondyle weak as in that genus, supinator crest
prominent, internal crest as in *Claenodon*, anconeal fossa rather shallow, supratrochlear
foramen probably absent.

The ulna has a long straight olecranon, wide and flattened shaft, much as in *Claenodon*
but not so wide; it is deeply and broadly grooved on the anterior face, the groove extending
from a point external to the radial facet down to the lower end of the shaft; internal to it lies
the heavy elongate scar for the quadratus, as in *Claenodon*.   The radial facet is nearly flat.
The radius is moderately heavy and long, the head considerably expanded transversely,
more than in *Claenodon*, and its ulnar facet allowing of but little pronation.   The distal

---

[63] Cf. Matthew, 1897, Bull. Amer. Mus. Nat. Hist., IX, p. 285.

facet is concave anteroposteriorly but very little concave transversely, the scaphoid and lunar facets distinct, and the posterior border of the articulation deeply notched between them.

The pelvis has the usual elongate ischia of creodonts, with the ischial spine prominent and extending a considerable distance behind the acetabulum; the ilium is incomplete but appears to be short and broad.

The femur is comparatively long, straight and with broad shallow digital fossa, lesser trochanter internal, third trochanter prominent and situated well down on the shaft, as in Mesonychidae generally.   The patellar trochlea is rather short, wide and shallow.   The tibia is quite short in contrast to the femur, shorter than in *Claenodon*, although the femur is longer, and with high cnemial crest, prominent postero-internal crest, both fading out towards the middle of the shaft; the distal end is very oblique, with long, wide and heavy internal malleolus, the astragalar trochlear surface oblique and very slightly grooved.   The length of the tibia exceeds that of the radius only by the thickness of the proximal epiphysis.

In sum, the limb bones are those of a very primitive mesonychid, retaining many characters of Arctocyonidae.   The same relations are to be seen in the feet, which are described more in detail in comparison with those of *Claenodon* and of *Pachyaena*.

The fore foot of this species was described and figured by Osborn and Earle in 1895. Unfortunately, in their figure the metacarpals of the right and left side have been confused; the digits marked II, III and IV belong to the right manus, but those marked I and V to the left manus.   The specimen, No. 777, was more carefully pieced together and articulated by Matthew but no corrected description has been published.   The parts represented in No. 777 are the trapezoid, magnum and cuneiform and Mc. II, III, IV of the right side, the scaphoid, lunar, magnum, unciform, cuneiform and pisiform, metacarpals I and V complete, the proximal end of Mc. II and distal ends of Mc. II, III and IV, five proximal, three second and two distal phalanges.   Part of the hind foot is also present, consisting of the astragalus, calcaneum, cuboid, the three cuneiforms, and proximal ends of metatarsals I and II, together with several phalanges.

From the above material the feet can be reconstructed and compared with those of *Pachyaena* figured by Matthew in 1915, of *Synoplotherium* figured by Wortman in 1901, and of *Mesonyx* figured by Scott in 1892.

The metapodials are proportioned much as in the later Mesonychidae, but the carpus has less vertical height, the bones are broader and shorter and in all respects are much nearer in construction to those of Arctocyonidae.   The pollex is well developed, although not so large as the fifth metacarpal.   The distal ends of the metacarpals are "squared" as in the later Mesonychidae and the phalanges shortened and much flattened, although not so much so as in *Pachyaena*.   The unguals are broad and hoof-like but not so much so as in the later genera.

The scaphoid is wide and thin, with the radial facet strongly rolled over the top; it has quite a narrow contact with the lunar, which is as wide as high.   Distally it has facets for the centrale and trapezium; the centrale is not preserved but was evidently unreduced and almost excluded the trapezoid from contact with the scaphoid, quite as in *Claenodon* and unlike the conditions in the later Mesonychidae.   The trapezoid is small, low, with proximal facet for the centrale and a small contact with the scaphoid, an internal facet for trapezium and distal facet for Mt. II.   The trapezium is not preserved, but to judge from the form of the adjacent bones it had somewhat the same form as in *Claenodon*.   The magnum has a very small head, the usual proximal keel wedged between lunar and centrale; the unciform and cuneiform have less vertical height than in *Pachyaena*, but much the same relations as in this genus and in *Claenodon*.

The metacarpals conform to those of *Pachyaena*, except for the well-developed Mc. I, which has a wide head much as in *Claenodon* but less convex and lacking the peculiar notch that suggests a semi-opposable digit.   The second and third metacarpals have the same heavy overlap on the adjoining carpals as in *Claenodon* and *Pachyaena*, the fourth has no overlap but a peg-like joint externad for Mc. V, permitting some downward flexure of the fifth on the fourth digit.

The phalanges are intermediate in type between the normal carnivore form seen in *Claenodon* and the short flattened ungulate form seen in the later Mesonychidae, but considerably nearer to *Pachyaena* than to *Claenodon*.   The width of the ungues varies; the lateral ungues (figured by Osborn and Earle) are much narrower than the central ungues preserved in No. 776, which are nearly as wide and short as in *Pachyaena*.

The hind foot shows a correspondingly intermediate structure between the normal pes of primitive creodonts as shown in *Claenodon* and the specialized cursorial adaptations of the later Mesonychidae.   The astragalus is badly crushed, so that its true form is not well seen, but it evidently is as devoid of trochlea as is that of *Claenodon;* the astragalar foramen is large, and set well in from the border so that the tibial articulation is much limited posteriorly.   The fibular articulation is oblique but much flattened out by crushing; the neck is short and the head is wide, the cuboid facet obscurely separated from the navicular. The calcaneum has a wide fibular facet unlike later mesonychid conditions.   On the cuboid the navicular facet is sharply distinct but faces proximad, as in Mesonychidae, not so much internad as in other creodonts.   The cuboid is decidedly broader and shorter than in *Pachyaena*.   The navicular is not preserved, but must have been much shorter and wider than in *Pachyaena;* the ectocuneiform has a rather wide oblique facet at its proximal end for the cuboid; the entocuneiform is much wider and deeper distally than proximally, the reverse of its form in *Pachyaena*, and has a large concave facet for metatarsal I, which is about as large as metacarpal I.   The second metatarsal has a narrow head with much less overlap of trapezium and ectocuneiform on its inner and outer sides, respectively, than is seen in *Pachyaena*.

In all these characters of manus and pes *Dissacus* is evidently intermediate between the normal primitive creodont pes and the specialized conditions of the later Mesonychidae, in some particulars nearer to *Claenodon*, in others to *Pachyaena*.   The articulations with radius and tibia are evidently much more of the primitive type, indicating a marked angle between leg and foot and limitation of the straightening of this joint.   The pollex and hallux are well developed.   On the other hand, the character of the phalanges is evidently mesonychid.   The conditions show a good deal of parallelism to those of the Taligrada.

Microclaenodon Scott, 1892

Scott, W. B., 1892, Proc. Acad. Nat. Sci. Phila., XLIV, p. 302

Type: Triisodon assurgens Cope, 1884, from the Middle Paleocene of New Mexico.

This genus is very imperfectly known.   The type is part of a lower jaw with m_3 well preserved and the other molars mostly buried in flinty matrix; a second jaw fragment, No.

2473a, has m_1 well preserved; a third specimen, No. 15999, shows parts of the premolar and molar series mostly buried in matrix.   A fourth specimen, No. 3348, is a fragmentary skeleton in bad preservation.

Diagnosis: The lower molars are sharply distinguished from those of Arctocyonidae by their narrowness, agreeing in proportions with *Dissacus* but with narrow basined heels. The trigonids are high, with paired metaconid and protoconid, and the protoconid low but projecting forward.   The jaw is long and rather slender anteriorly, the premolars compressed and spaced, the symphysis loose, quite unlike the short deep jaw with massive symphysis of the triisodonts, but agreeing with *Dissacus*, to which it appears to be allied rather than to the Triisodontinae with which it has hitherto been associated.

The upper molars so far as known are trigonal with somewhat reduced inner half, approaching *Dissacus* in this respect, but an external cingulum and a small hypocone crest present, as in Arctocyonidae.   The skull is distinguished by a long slender muzzle, and the femur has a long narrow patellar trochlea; otherwise the characters of vertebrae and limb bones conform to those of Arctocyonidae of equal size.

Microclaenodon assurgens (Cope)

Triisodon assurgens Cope, 1884, Paleont. Bull. No. 37, Jan. 2, and Proc. Amer. Phil. Soc., XXI, p. 311; (*Microclaenodon*) Scott, 1892, Proc. Acad. Nat. Sci., Phila., XLIV, p. 302.

Type: A. M. Cope Coll. No. 3215, part of lower jaw.

Horizon and Locality: Paleocene of San Juan basin, undoubtedly Torrejon formation, Upper Paleocene.

Author's Description: This is the least species of the genus [*Triisodon*] and resembles in its inferior dentition the species of *Diacodon*. It is very much larger than the *D. alticuspis*, the larger species of that genus, which is found in the Wasatch formation.

The *T. assurgens* is known from a mandibular ramus which supports the last four molars, the last premolar having lost its principal cusp. The peculiarity of the true molars is seen in their generally more produced character; the anterior cusps are higher and the heels are longer. The anterior cusp is very small and basal; the principal anterior cusps are united to near their free summits. There are the usual low marginal tubercles on the heels. That of the fourth premolar is a short simple edge. [Measurements follow.]

Revised Description: Size about as in *Chriacus baldwini*, but distinguished readily by the narrow molars, high trigonids with connate round subequal pr^d and me^d, low small paraconid projecting forward at the base, heels long and narrowly basined, hy^d tending to become median and entoconid crest behind and within it; no distinct hypoconulid. M_1 and m_3 of subequal size, m_2 a little larger. P_4 long, compressed, with crested median heel,

A.M. 3215 $\frac{2}{1}$

Fig. 18.   *Microclaenodon assurgens*, type left m_3, A. M. No. 3215, crown view.   Twice natural size.

crowns of premolars otherwise not preserved, but p_3 appears to be similar, somewhat smaller, p_2 much smaller, two-rooted and spaced, p_1 small, one-rooted and spaced, canine of moderate size and somewhat compressed oval cross-section.   Symphysis long, loose, extending somewhat back of p_2.   Jaw elongate behind the tooth row much as in *Dissacus*.

The skull is long, quite slender in the premolar region, somewhat expanded at the canines.   Brain-case long, low, narrow.   Zygomatic arches appear to be moderately slender and glenoid region of squamosal is quite lightly constructed, postglenoid process prominent, glenoid fossa narrow and moderately extended transversely.   Postglenoid foramen quite small.

Cervicals moderately long.   Lumbars have the usual strongly convex peg-like posterior zygapophyses.   Caudals large and long.   Femur nearly equalling length of skull, rather long and slender, digital fossa wide and shallow, head projecting rather far internad, great trochanter probably nearly as high as head but surface broken, lesser trochanter internal, third trochanter rather slight and opposite lesser trochanter, the shaft rather widely flanged beneath it, the flange extending distally down nearly to the middle of the shaft.   Patellar trochlea very narrow and long, but distal end of femur of no great depth, the condyles rather small, facing posteriorly.   The tibia appears to be long and slender with very low cnemial crest.   Nothing is known of the feet.

MIACIDAE Cope, 1880

Miacidae Cope, 1880, Proc. Amer. Phil. Soc., XIX, p. 78; 1882, *ibid.*, XX, p. 157 (reprinted in Tertiary Vertebrata, 1885, p. 258); Schlosser, 1888, Die Affen, Lemuren, . . . p. 58 (282)—*Miacis* and *Uintacyon* included under Canidae; 1890, *ibid.*, pp. 60–2 (446–8)—*Didymictis* and *Miacis* as types of "Creodonta Adaptiva"; Scott, 1892, Proc. Acad. Nat. Sci. Phila., XLIV, p. 318; Matthew, 1909, Mem. Amer. Mus. Nat. Hist., IX, pt. vi; Matthew (and Granger), 1915, Bull. Amer. Mus. Nat. Hist., XXXIV, p. 16.   Canidae (in part) Wortman, 1901, Amer. Jour. Sci., XI, p. 338. Viverravidae Wortman, 1901, Amer. Jour. Sci., XII, p. 143.   Uintacyonidae Hay, 1902, U. S. Geol. Survey Bull. No. 179, p. 759.

Cope's definition of the Miacidae in 1882 is brief but sufficient, and is followed by a discussion which shows a very correct appreciation of their relationships to the post-Eocene Carnivora.

Schlosser in 1890 characterized the genera *Didymictis* and *Miacis* as "Creodonta Adaptiva," regarding them as approximately representing the group from which the Carnivora could be derived.   *Miacis*, he observed, is nearly related to *Cynodictis*, while *Didymictis* must be regarded as an extinct branch, unless possibly in the line of descent of the Subursidae (Procyonidae).   He considered the two genera as representing the group from which the true Carnivora arose, and derived in turn from very primitive creodonts, their relations being similar to that of *Paloplotherium* to the true equines on one hand and to *Phenacodus* on the other.   In the light of our present knowledge this comparison unduly stresses the affinities to the fissipede Carnivora, but at the time Schlosser wrote, *Phenacodus* was regarded as but little removed from the direct line of ancestry of the Equidae.   It is evident from his discussion that he regarded these adaptive creodonts as similarly near to the other creodont groups.

Wortman in 1901 referred *Miacis* (incorrectly synonymized with *Vulpavus*) to the Canidae, and *Didymictis* (wrongly regarded as a synonym of *Viverravus*) to the family Viverravidae, regarding both as "Carnassidentia" (= Fissipedia).   Matthew in 1909 revised the Eocene genera, reviving Cope's family name, and in 1915 added some further discussion of the affinities of the genera.

The Miacidae are a numerous and varied group in the Eocene, but in the Paleocene are represented by a single species of *Didymictis*, primitive in the genus but not so primitive as some of the Eocene miacid genera. It is quite sharply distinct from the rest of the Paleocene creodonta and presents an unfamiliar aspect.

Didymictis Cope, 1875

Cope, E. D., 1875, Report to Engineer Dept., U. S. Army, in charge of Lt. Wheeler, Apr. 17, pp. 5, 11

Type: Limnocyon protenus Cope, from the Lower Eocene Wasatch of New Mexico.

Diagnosis: Dentition $\frac{3\cdot1\cdot4\cdot2}{3\cdot1\cdot4\cdot2}$. Incisors small, canines of moderate size, slender, acute, premolars compressed, trenchant, with acute cusps and well-developed accessory cusps. P^4 carnassiform, metastyle blade much extended, protocone large, conical, antero-internal. M^{1-2} tubercular, trigonal, extended transversely, m^2 much smaller than m^1; parastyle much extended on m^1, less on m^2; protocones have anterior and posterior wings symmetrically developed; paracones larger than metacones. M_1 tuberculo-sectorial, with high trigonid and large basined heel; metaconid considerably overtopping both. M_2 tubercular with low obscure trigonid and elongate, shallow-basined heel.

Skull elongate, relatively large in comparison with skeleton. Brain-case low, small, occipital and sagittal crests very high, basicranial region long and narrow, glenoid articulations well forward. Limbs of moderate length, feet subdigitigrade, five-toed.

Didymictis is distinguished from most of the Miacidae by the sharp differentiation between the sectorial and tubercular dentition, the complete disappearance of the last molar and elongation of m_2, paralleling in this respect the bears, raccoons and tubercular Viverridae. From *Viverravus*, its nearest ally, it is distinguished by the larger size and more fully crushing character of the tubercular dentition, as expressed especially in the heel of m_1, large and basin-shaped instead of reduced and trenchant. This difference is more marked in the later species of the two phyla, *D. altidens* of the Wind River and Huerfano, and *V. sicarius* and *minutus* of the Bridger. In the Paleocene *Didymictis* the characters are to a great degree synthetic, so that it resembles *Viverravus* in details and in the "cut" of the teeth, although by definition a *Didymictis*. It is in truth rather a strain upon the reasonable limitations of genera to place the big, powerful, heavily-proportioned *D. altidens*, with its massive jaws and teeth, heavy molars of thoroughly crushing type, etc., in the same genus with the small, slender, sharp-toothed *D. haydenianus;* but it is not easy to specify any satisfactory definite specific differences worthy of ranking as generic distinctions. In the absence of these the Paleocene species may be retained for the present as a subgenus of *Didymictis* under the name of *Protictis*, as discussed below.

Cusps and crests robust, blunted. Accessory cusps well developed on p_{3-4}.
   M_2 completely tubercular . . . . *Didymictis* s.s.
Cusps and crests sharp, acute. Accessory cusps rudimentary on p_4, none
   on p_3. M_2 incompletely tubercular . . . . *Protictis*.

Didymictis (Protictis) haydenianus Cope

Didymictis haydenianus Cope, 1882, Paleont. Bull. No. 35, Nov. 11, and Proc. Amer. Phil.
  Soc., XX, p. 464; Amer. Nat., XVIII, p. 484, fig. 30 d, e, f; Matthew (and Granger),
  1915, Bull. Amer. Mus. Nat. Hist., XXXIV, p. 19, fig. 12.
Didymictis primus Cope, 1884, Paleont. Bull., No. 37, Jan. 2, and Proc. Amer. Phil. Soc.,
  XXI, p. 309.

Type: A. M. Cope Coll. No. 3368, upper and lower jaws. Type of *D. primus:* A. M.
Cope Coll. No. 3371, lower jaw.

Horizon and Locality: Paleocene of San Juan basin, undoubtedly from the Torrejon
formation, Middle Paleocene.

Author's Description: The arrangement of the superior molars is much as in *D. protenus*, the fourth
premolar being a true sectorial. The third premolar has no internal lobe, although the section of the base of the
crown is narrowly triangular. It has anterior and posterior basal lobes and a posterior lobe on the cutting
edge. In the sectorial the median lobe is a good deal more produced than the posterior, though the two form
together the usual blade. The anterior basal lobe is distinct; and the internal is larger and is conic. The
first true molar has the anterior external base of the crown produced. Its two external cusps are conic and
distinct. The internal part of the crown is rounded and supports a conic internal tubercle which is separated
from the external cones by two small concentric tubercles. The second true molar is considerably smaller,
and is transverse, its external border being very oblique. It has an acute internal lobe.

The character of the species is well-marked in the inferior true molars. The first has the form seen in
other species of *Didymictis*. The heel is large, and with a median basin between lateral cutting edges. The
two anterior inner cusps are of equal elevation and are near together; the external is much larger. The
last molar is elongate, but reduced in size. Its anterior three cusps, rudimental in other species, are here
elevated, forming the triangular mass seen in the first true molar. They are not so elevated, however, as in
that tooth, and thus not so much developed as in *Oxyaena, Stypolophus*, etc. The fourth premolar has a
median cutting edge on the short heel. . . . The peculiar characters of the last inferior molar distinguish this
species from its congeners. The last superior molar is relatively smaller than in the *D. protenus*. In size this
species is superior to the *D. dawkinsianus* [*Viverravus dawkinsianus*], and is smaller than the *D. leptomylus*.

Type Description of D. primus (p. 310): The inferior sectorial tooth is much like that of the
D. leptomylus, but the tubercular is only two-thirds as long, and is not only absolutely, but relatively narrower posteriorly. It has the usual three cusps in a reduced condition. In the first superior true molar the
external cusps are conical, and there is a small cusp between the anterior one and the produced anterior angle
of the crown. There is an anterior intermediate tubercle, but no posterior one. The cingulum does not
extend all round the inferior base of the crown, as it does in *D. protenus*. The sectorial has a distinct anterior
basal conic lobe. The internal lobe is in transverse line with the last named and is conical and not large. . . .

The fourth specimen is especially important as presenting almost the entire dentition including canines
and incisors, and the anterior part of the skull from the line of the coronoid process of the mandible. The
specimen shows that the species differs from the species of the Wasatch period with oval inferior tubercular,
in the absence of the posterior cutting lobe of the third, and probably fourth inferior premolar. The corresponding superior premolars are also simple. The first premolars in both jaws are one-rooted. The
canines are long and acute, and are directed vertically. Both have flat facets on their external (the only
visible) faces; on the superior canine I count four lateral, and one nearly anterior. On the inferior I see
three lateral and one nearly anterior. There are three small superior incisors, of which the first is the largest
and has a subconical crown. The infraorbital foramen is large and is above the anterior border of the
sectorial. . . .

In its simple premolars this species agrees with the *D. haydenianus*, and is more primitive than the
Wasatch species.

Revised Description: Lower dentition c–m_2 .045 as against .060 in *D. leptomylus*, .070
in *D. protenus*, and .090 in *D. altidens*. Jaw relatively shallow, teeth with more acute cusps
than in the nearest species, *D. leptomylus*. Trigonid of m_2 considerably higher than talonid;

paraconid distinct, unreduced, heel narrower than trigonid, with submedian hypoconid, entoconid-hypoconulid crest directed more anteroposteriorly than in other species. P_4 with two small cusps in the position of the posterior accessory cusp of the other species; of these the smaller anterior one appears to be its homologue, the posterior one corresponding to the normal heel cusp of p_4 in Creodonta. P_3 shows no trace of accessory cusp, and no distinct heel cusp; p_2 has not even the rudimentary heel.

2. ORDER TALIGRADA

Discussion

The Taligrada was defined by Cope as a suborder of Amblypoda distinguished by "astragalus with a head distinct from trochlea, with distal articular facets." [64] He included in it the single family Pantolambdidae.

In his revision of the Amblypoda[65] Osborn included the Periptychidae in the order, on the ground that the tarsus in *Periptychus* is non-serial and the molars "strictly trigonal" in symmetry. He further cited a long series of correspondences in structure between *Periptychus* and *Pantolambda* in support of this relationship. That such near relationship exists as between these two genera this structural similarity proves beyond question, but two facts should be considered in interpreting it:

1. *Periptychus* resembles *Pantolambda*, a primitive member of the amblypod phylum, or a marginal member of the order, but it does not resemble the typical Amblypoda as it does the typical Condylarthra. The points of resemblance to *Pantolambda* are mostly primitive characters, shared also by various Creodonta and Condylarthra.

2. The resemblances are not common to all Periptychidae, but only to *Periptychus*. *Ectoconus* shares many of them, the Anisonchinae some few. The trigonal molars cannot be called "persistently trigonal" in a fauna of which nine-tenths of the mammalian genera have trigonal molars. And *Anisonchus sectorius* does not have trigonal molars; they are fully tetragonal; nor can the molars of *Haploconus* be called "strictly trigonal" in symmetry. These facts suffice to show that the molars of Periptychidae are not persistently trigonal in the sense that they were unable to develop a tetragonal molar.

Osborn lists in the same paper an imposing series of characters of which he regards the "persistently primitive" characters as subordinal. Substantially this is equivalent to substituting the Amblypoda for the Condylarthra as the practical equivalent of the hypothetic Protoungulata. Cope was himself much inclined to take this view in his theoretical discussions of the origin of the ungulate orders. But in analyzing this list for the purpose of considering the relations of the Periptychidae one finds that (1) the seven characters marked persistently primitive are all shared by the Condylarthra; (2) of 37 characters marked primitive, 19 are shared by all Condylarthra, 9 by all Condylarthra except *Phenacodus*, 6 are found in taligrades and in *Periptychus* but not in all Periptychidae, while only three out of the thirty-seven are valid common characters of the Taligrada, including Periptychidae; (3) the single progressive character ascribed to the Taligrada is not found in any of the Periptychidae.

[64] Cope, E. D., 1883, Amer. Nat., XVII, Mar. 15, p. 406.
[65] Osborn, H. F., 1898, Bull. Amer. Mus. Nat. Hist., X, pp. 180–2.

Nearly all the truly primitive characters are equally to be found in the older and more primitive Creodonta, Insectivora, etc. They are not at all distinctive of Amblypoda or of Taligrada. They are the generalized characters of the primitive placentals. The characters common to Taligrada (i.e. *Pantolambda*) and *Periptychus* are not all primitive. Nevertheless the analysis brings out very impressively the primitive character of the *Pantolambda* skeleton; the single progressive feature listed is the crested teeth characteristic of the true Amblypoda.

That the true Amblypoda retain many of these primitive features is undoubtedly true; so do the Condylarthra, the creodonts, and all other Eocene groups. The only possible way to classify and distinguish them from other early Tertiary placental groups is by their special progressive features.

Gregory, in his essay on *The Orders of Mammals*, emphasizes the relations of the Taligrada to the Condylarthra by making them both suborders of the order Protoungulata. He gives a brief analysis of the characters of the families Phenacodontidae, Meniscotheriidae, Periptychidae, Pantolambdidae, Coryphodontidae and Uintatheriidae, and remarks: "The preceding brief analysis is sufficient to indicate that the Pantolambdidæ, Coryphodontidæ and Uintatheriidæ form an ascending series and that the Periptychidæ are connected on the one hand with this series at its base, and on the other hand with the condylarth families Phenacodontidæ and Meniscotheriidæ." [66]   He states the argument for the "phylogenetic" grouping favored by Osborn. This argument even as cautiously stated by Gregory seems open to grave criticism. For instance, his definition of the Condylarthra to include Phenacodontidae and Meniscotheriidae runs as follows: "Feet more slender, tridactyly progressive. Astragalus with slender neck; trochlea keeled. Lunar-unciform facet becoming reduced, carpals becoming serial. M^1, m^2 early becoming quadrate, with large hypocone. Premolars triangular with tritocone."

The above definition is almost wholly based only upon the phenacodonts. Very little of it applies to *Meniscotherium*, which is a contemporary of *Phenacodus* and at least equally specialized, although in a different manner; or to the hyopsodonts, which were the latest survivors of the Condylarthra, or to the older mioclaenids, from which the hyopsodonts were probably derived. In *Meniscotherium* the feet are no more slender than in periptychids of equal size; there is no trace of progressive tridactyly; the astragalar trochlea is no more keeled than in periptychids, nor is there any indication of reduction in the lunar-unciform facet or of progressive serialism elsewhere. The upper molars are quadrate indeed, but with a wholly peculiar pattern, as different from that of *Phenacodus* as are the equally quadrate molars of *Anisonchus*. Progressive molarization of the premolars is evident in both meniscotheres and phenacodonts—not in hyopsodonts. The characters cited are *not* the phylogenetic characters of the Condylarthra, but of the family Phenacodontidae. Similarly, the characters ascribed as progressive phylogenetic distinctions of the order Amblypoda apply only partially to the Periptychidae, although quite fully to the *Pantolambda-Coryphodon-Uintatherium* series. The really strong point of distinction between the groups as thus arranged lies in the serial or nearly serial tarsus and slender-necked, round-headed astragalus common to all phenacodonts, meniscotheres, mioclaenids, hyopsodonts, and not found in any periptychids, pantolambdids or Amblypoda. This, however,

[66] Gregory, W. K., 1910, Bull. Amer. Mus. Nat. Hist., XXVII, p. 358.

is not a progressive but a primitive character in my opinion. At all events it is shared by the primitive Creodonta and approached by the oldest and otherwise most primitive members of some other groups.

Gregory's own view at the time of writing this essay (1910) leaned perhaps a little too much to the other side, as did Matthew's in 1897. The placing of the Taligrada and Condylarthra together as suborders of "Protungulata" gives the much-needed weight to their close relationship and primitive character, but it tends to create a false concept. There is in fact no protoungulate type at all. The various ungulate groups are separately derived from various unguiculate groups. The relationship between Condylarthra and Creodonta in the Paleocene is a closer one than between Condylarthra and Taligrada. It is doubtful whether any wholly satisfactory classification can be devised for the relations as they stand. But it is at all events desirable to get away from clearly false and misleading concepts. It is not practicable perhaps to formulate wholly satisfactory diagnoses for Creodonta and Condylarthra which will be mutually exclusive. The distinction cannot be made upon the possession of hoofs and claws respectively, for one family of the creodonts, the Mesonychidae, did possess hoofs; and at least one family of the Condylarthra, the Hyopsodontidae, and probably the primitive members of all the others, possessed ungues that would be called claws rather than hoofs by any unprejudiced observer. At the same time the hoofed creodonts were not in other respects any nearer to the Condylarthra than their clawed fellows, nor did the clawed Condylarthra display any particular approach to the specialized characters of creodonts and Carnivora.

Distinctive Characters of the Amblypoda, Condylarthra and Taligrada

The progressive characters of the AMBLYPODA proper (Coryphodontidae and Uintatheriidae) consist in:

1. The very short five-toed feet, like those of Proboscidea in proportions, but with the carpus and tarsus alternating instead of serial. No centrale in carpus, no neck on astragalus.
2. Canines large and powerful, molar teeth with two curved or transverse crests and peculiarly patterned premolars.

In addition to these we have in the two known families of Amblypoda the massive skull, with tendency to develop horns or bony bosses, the short neck, straight limbs, broadly expanded ilium, etc., characteristic of all the subungulate mammals and of some other large mammals.

The Amblypoda retain a peculiarly small and primitive brain. They do not develop a proboscis.

The CONDYLARTHRA *as an order* display the following distinctive features:

1. Feet as in primitive Carnivora, of moderate proportions, digitigrade to subplantigrade, phalanges medium to short, ungues more or less flattened into hoofs or strictly claw-like. Carpus partly alternating with centrale, except in *Phenacodus*, tarsus serial, astragalus with distinct neck and convex head, tibial trochlea more or less grooved, obliquely pitched, the inner crest lower or rudimentary.
2. Teeth in continuous series without diastemata, incisors small, canines little enlarged, premolars comparatively simple, trenchant or inflated, molars bunodont,

except *Meniscotherium* and *Pleuraspidotherium*, which have peculiar lopho-selenodont patterns.

The Condylarthra are all small or of moderate size and have the proportions and characters associated with such size, with terrestrial habits and vegetarian diet. They have a small and primitive brain, no proboscis, no horns or other special means of defense.

The order TALIGRADA may be distinguished thus:

1. Feet moderately short, approaching the subungulate type, five-toed, with alternating carpus and tarsus, centrale present, astragalus with short neck, convex head.
2. Teeth in continuous series without diastemata, incisors small, canines large, premolars simple or progressively complex backward, molars tritubercular, bunodont or selenodont.

The best-known forms also show a more or less shortened neck and varying proportions of shortening in the limbs, but this apparently is not true of the less-known smaller forms included. They retain a very low and small brain, hornless skull and various other primitive features.

The relations of this group to the various specialized subungulate orders parallel the relations of the Condylarthra to the specialized ungulate orders, thus:

Paleocene	Eocene and later Tertiary
	Amblypoda
	Pyrotheria
Taligrada	Embrithopoda
	Proboscidea
	Sirenia
	Notoungulata
	Litopterna
Condylarthra	Hyracoidea
	Perissodactyla
	Artiodactyla

It is by no means meant to assert that the direct ancestry of all the later orders lies within the known families of Condylarthra and Taligrada. Save for the Litopterna. Notoungulata, Amblypoda and Pyrotheria, it probably or certainly does not. The other orders are all of Old World or of boreal origin. They are derived presumably from taligrade and condylarth families inhabiting those parts of the world in the later Cretaceous and early Tertiary. Perissodactyla and Artiodactyla had already evolved before the end of the Paleocene, and invaded the United States and western Europe at the opening of the Eocene. Proboscidea, Sirenia, Embrithopoda and Hyracoidea appear at the beginning of the Oligocene in Egypt, already specialized. With the relations of these groups we are not here concerned. The New World groups are more clearly connected with the Paleocene faunas. The Paleocene Condylarthra—*Tetraclaenodon* and the Mioclaenidae—are, as a group, at least as primitive as the Paleocene Taligrada—anisonchines, periptychines and *Pantolambda*. *Phenacodus* is a specialized Lower Eocene survivor, corresponding as to its degree of progressiveness with *Coryphodon*, but of course in a quite diverse sense. *Hyopsodus* is a peculiarly primitive survival in most respects, although lasting to the end of the Eocene. *Meniscotherium*, while specialized in teeth, is almost equally primitive in feet. The Eocene Condylarthra, therefore, while corresponding in relative geologic and phyletic position to the Eocene Amblypoda, do not differ so sharply from their Paleocene ancestral group and are not taxonomically separable as a distinct order.

The Notoungulata and Litopterna are clearly so separable, as at present known. It is quite probable that a knowledge of the skulls and skeletons of the numerous genera described by Ameghino and Roth from the *Notostylops* beds would make it difficult to draw any line between these and the Condylarthra. But at present these Eocene Patagonian genera are known chiefly from isolated teeth, few from jaws or parts of jaws, and still fewer from any part of the skeleton known to be associated with teeth or jaws. The better-known Patagonian ungulates of the Deseado, Santa Cruz and Pampean fall into readily definable specialized ordinal groups, some apparently derivations of the Condylarthra, others of the Taligrada.

PERIPTYCHIDAE Cope, 1882

Cope, E. D., 1882, Amer. Nat., XVI, p. 832

Author's Diagnosis: The brain is, as in *Phenacodus*, very small, with the olfactory lobes widely separated from the small hemispheres. The humerus has an epitrochlear foramen. The astragalus has no trochlear groove, and the neck is short. The head is convex, and presents a lateral face for contact with the side of the cuboid. Five digits on the posterior foot. The lateral ungues are rather narrow hoofs. Cervical vertebrae very short.

The absence of trochlea of the astragalus is a point of resemblance to *Meniscotherium*, and separates *Periptychus* [Footnote: *Catathlaeus* was established on the permanent dentition of *Periptychus*] from the *Phenacodontidae* as a family type, which I call the *Periptychidae*. With it must no doubt be associated *Anisonchus* Cope, *Haploconus*, and the following new genus [*Hemithlaeus*].

In 1884 and 1885 [67] Cope developed the above views somewhat more clearly in his discussion of the Condylarthra. His analysis of the genera at this time was (p. 794):

The characters of the genera are the following:
I. Three premolars.
 Fourth superior premolar like molars; inferior premolars without internal ledge........ *Hexodon*
II. Superior molars with intermediate tubercles, and tubercles anterior and posterior to the internal cusp; four premolars.
 Superior molars with an external cingular cusp; inferior premolars without internal ledge *Ectoconus*
 No supplementary external cusps, inferior premolars with internal ledges............. *Periptychus*
III. Intermediate tubercles wanting; four inferior premolars, without internal lobes.
 Superior molars with posterior internal cusp only, besides internal V; last two superior premolars with internal lobes....................................... *Anisonchus*
 Superior molars with internal V only, no other internal lobes; last two superior premolars with internal cusps.. *Hemithlaeus*
 Superior molars with posterior internal cusp only, besides apex of V; fourth superior premolar only with internal lobe............................... *Haploconus*
IV. Superior molars unknown; inferior premolar No. IV? with two opposed crescents and a heel.
 Inferior molars with one or two pairs of opposed crescents......................... *Zetodon*

[67] Amer. Nat., XVIII, pp. 790–805, and Tertiary Vertebrata, pp. 384–5.

Osborn and Earle in 1895 pointed out that the Periptychidae have a non-serial tarsus and in this respect do not conform to the (later) definition of the Condylarthra. They conclude that "*Periptychus* is quite as closely related in its pes to the Amblypoda as to the Condylarthra. *Periptychus* has the simple bunodont dentition of the Condylarthra, but it has the strictly trigonal molar of the Amblypoda." [68]

Cope in 1897, in a note in the American Naturalist (XXXI, p. 335), definitely transfers the family Periptychidae to the Amblypoda.

Matthew in 1897,[69] while recognizing the affinity through the Pantolambdidae to the typical Amblypoda, retains the concept indicated by Cope in 1888 that the Condylarthra include the ancestors of the Amblypoda, and re-defines the order so as to include the Periptychidae in it, pointing out that the serial carpus and tarsus of *Phenacodus* is a secondary character. The Condylarthra are regarded as practically equivalent to Protoungulata of various theoretical discussions on the origin of the ungulate mammals.

Osborn in 1898 [70] cites in detail the "primitive or protoungulate characters" of *Pantolambda*, and points out that "all these osteological characters are shared by *Periptychus*, so far as the skeleton of the latter genus is known." He includes the Periptychidae along with the Pantolambdidae in the suborder Taligrada of the order Amblypoda.

Gregory in 1910 discusses the relationship of the Periptychidae, which he places with the Pantolambdidae in the order Taligrada, distinct from the Condylarthra on one side and the specialized Amblypoda on the other. His view of their relationships is indicated by the following citation: [71]

The Basal Eocene families of the Condylarthra and "Taligrada" seem in fact to be not widely removed from each other, and there is evidence that the broad "horizontal" group of which they were doubtless a small part had about the same relation to certain of the more highly specialized ungulate orders that the most primitive Creodonta (including the Miacidae) had to the later Creodonta, Fissipedia and Pinnipedia. These "protoungulates" retained very many Creodont characters in the skull and skeleton and were separated from that group chiefly by the greater elaboration of the molar teeth.

The above citation very well expresses my present views as to the affinities of the Periptychidae and related groups. It is clear that the several genera of the family, while sharply distinguished in the dentition, and varying widely in size and proportions of skull and skeleton, are nevertheless rather nearly related to each other, and it is equally clear that they are related on one side to the Pantolambdidae and through that family to the typical Amblypoda, and on the other side to the Mioclaenidae and through that family to the typical Condylarthra. There is in fact no wide gap anywhere along the line from *Tetraclaenodon* to *Pantolambda*.

The periptychid genera all show certain characteristic peculiarities in the structure of the upper molar teeth. Underlying a great diversity in construction there is a certain fundamental resemblance, recognized by all who have had occasion to study them, but not very easily formulated. Osborn has stated it as consisting in a persistently tritubercular symmetry, but this does not express the peculiar type of specialization. The peculiarity seems rather to consist in this: in normal tritubercular teeth the protocone is a lingual cusp, internal in position, and becoming antero-internal when the hypocone develops in the

[68] Bull. Amer. Mus. Nat. Hist., VII, p. 47.
[69] Bull. Amer. Mus. Nat. Hist., IX, pp. 293–5, 321–3.
[70] Osborn, H. F., 1898, Bull. Amer. Mus. Nat. Hist., X, pp. 184, 186.
[71] Gregory, W. K., 1910, Bull. Amer. Mus. Nat. Hist., XXVII, p. 359.

quadritubercular tooth. In the Periptychidae the protocone tends to become central in position and either the hypocone alone or the hypocone and protostyle become the lingual cusps, either or both occupying a position more towards the lingual side of the tooth than does the protocone. In *Periptychus* and *Ectoconus* these two cusps occupy the normal posi-

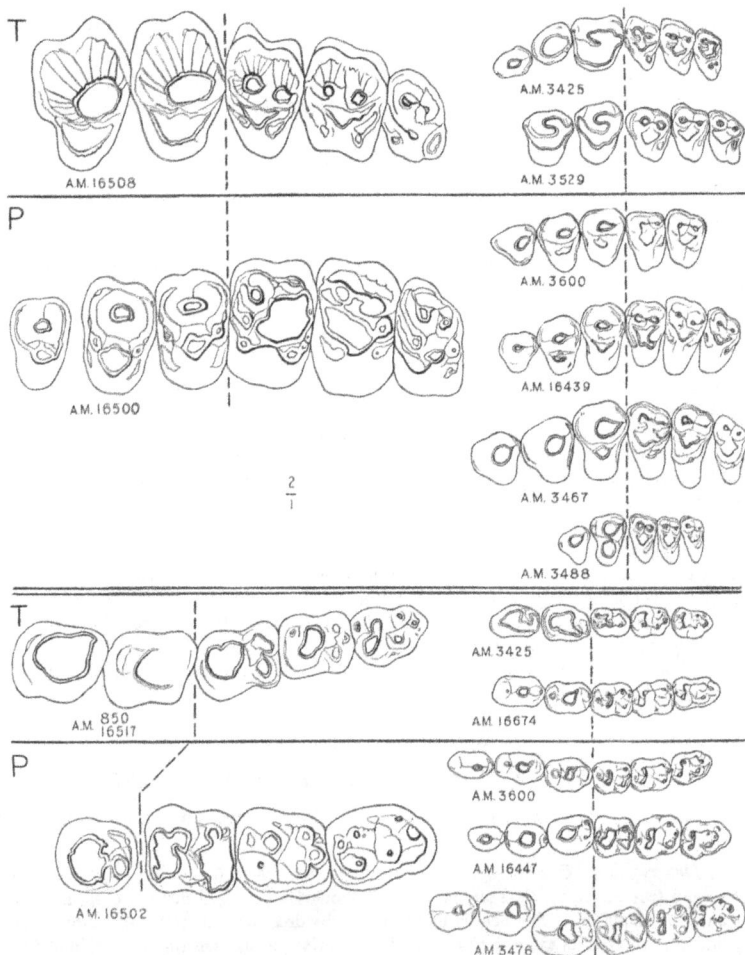

FIG. 19. Periptychidae, upper and lower teeth, crown views. All twice natural size.

Upper teeth: *Periptychus coarctatus*, A. M. No. 16508; *Ectoconus majusculus*, A. M. No. 16500; *Haploconus angustus*, A. M. No. 3425; *Anisonchus sectorius*, A. M. No. 3529; *Anisonchus gillianus*, A. M. No. 3600; *Hemithlaeus kowalevskianus*, A. M. No. 16439; *Conacodon entoconus*, A. M. No. 3467; *Conacodon cophater*, A. M. No. 3488.

Lower teeth: *Periptychus coarctatus*, A. M. Nos. 850 and 16517; *Ectoconus majusculus*, A. M. No. 16502; *Haploconus angustus*, A. M. No. 3425; *Anisonchus sectorius*, A. M. No. 16674; *Anisonchus gillianus*, A. M. No. 3600; *Hemithlaeus kowalevskianus*, A. M. No. 16447; *Conacodon entoconus*, A. M. No. 3476.

T = Torrejon. P = Puerco. Dentitions aligned on the anterior border of the first molar.

tion of the protocone and hypocone of the normal quadritubercular or sextubercular tooth, while the protocone has retreated towards the centre of the crown.   In *Haploconus* and *Anisonchus* the hypocone stands alone, postero-internal in position but more lingual than the protocone.   In *Conacodon* the hypocone is directly internal to the protocone.   In *Hemithlaeus*, which perhaps best represents the primitive condition, the protostyle and hypocone are small and flank the protocone in a more normal way, but their bases extended lingually, while the tip of the protocone is far in toward the centre of the tooth. *Anisonchus gillianus* and *Periptychus coarctatus* of the Puerco show less specialization than their successors in the Torrejon and serve to illustrate the progressive change in the teeth.

KEY TO THE PERIPTYCHIDAE

Family PERIPTYCHIDAE: Molars low-crowned, large-heeled, with tritubercular symmetry, usually with additional cusps near margin, premolars enlarged and inflated or else complicated, canines moderate-sized.   Limbs short or medium, feet short, astragalus with broad cuboid contact, short neck, trochlea flattened and oblique, fibular facet oblique, astragalar foramen well developed.

Subfamily PERIPTYCHINAE: Molars polybunous, with flattened crowns, very low cusps and prominent stylar cusps on the three sides of the trigon, conules distinct.   Feet very short and broad, pentadactyl, with lateral digits unreduced, astragalus with very short neck, flattened head, phalanges short, wide and flattened, unguals broadened into hoofs.

Premolars enlarged, inflated, not complicated....................................... *Periptychus*
Premolars smaller than molars, complicated but not inflated........................ *Ectoconus*

Subfamily ANISONCHINAE: Cusps relatively high and sharp, conules absent, stylar cusps variously developed, premolars enlarged and more or less inflated, simple or little complicated.

I. Teeth broad transversely, premolar cusps rounded.
  P³ with inner cusp.   Protostyle and hypocone small, subequal................ *Hemithlaeus*
  P³ without inner cusp.   No protostyle, hypocone large, internal to protocone..... *Conacodon*
II. Teeth narrower transversely, premolar cusps flattened.
  P³ with inner cusp.   Protostyle small to absent, hypocone large, postero-internal. Distinct paraconid on lower molars.................................... *Anisonchus*
  P³ without inner cusp.   No protostyle, hypocone more internal.   No paraconid on lower molars.. *Haploconus*

Periptychus Cope, 1881

Cope, E. D. 1881, Amer. Nat., XV, March 25, p. 337

Type: Periptychus carinidens from the Middle Paleocene, Torrejon formation, of New Mexico.

Author's Description: Gen. et sp. nov. Creodontium.   *Char. Gen.*   No distinct sectorial teeth, the first and second true inferior molars similar.   They support a principal median cusp, a broad heel and a prominent anterior cingulum.   The heel is more or less divided into tubercles; the anterior cingulum is on the inner side, and represents the anterior cusp of a sectorial tooth.   On the inner side of the principal cusp a cingulum rises, forming a flat internal tubercle.   Last molar not smaller than the others; premolars unknown.

This genus belongs to the *Amblyctonidae* with *Amblyctonus* and *Palaeonyctis*.   It differs from both in the rudimental character of the anterior cusp, and from the former, in the presence of the internal tubercle. In *Mesonyx* the heel has a median cutting edge.

In the above description, based upon an immature lower jaw with the last milk molar preserved and roots of the adjacent teeth, Cope entirely mistook the true affinities of the genus, also mistaking the milk molar for a true molar. He shortly afterwards described the permanent dentition of the same genus under another name, evidently without suspecting the identity of the two, as follows:

Catathlaeus rhabdodon gen. et sp. nov. *Char. gen.*—With this genus I commence descriptions of some genera with bunodont dentition, which has some resemblance to that of some of the hogs. The one above named, with *Mioclaenus*, remind one of *Tetraconodon* Falc. and Lydd. in the enlarged proportions of their premolar teeth. . . . Third and fourth superior premolars one or two-lobed externally, and with internal lobes . . . [and] one external cusp, enlarged; inferior fourth premolar with internal crest and cusp. . . . In the genus *Catathlaeus* the development of the premolars is remarkable, while the true molars are relatively small. The last three superior premolars have an elevated internal crescentic cingulum, homologous with the inner lobe of the fourth superior premolar of the ruminants. The general character of the true molars is that of *Phenacodus*.[72]

In 1882 Cope had secured additional material which enabled him to recognize the identity of *Catathlaeus* with *Periptychus* and to outline the family characters.[73] Early in 1883 he published a note on the brain-case of *Periptychus*, with a figure of the cast. The skull from which this cast was made is described in detail on page 113 of this memoir. In 1884 the characters of the genus are summarized in his article on the Condylarthra,[74] the dentition, parts of the skull and skeleton being described and figured.

Revised Diagnosis. Dentition $\frac{3 \cdot 1 \cdot 4 \cdot 3}{3 \cdot 1 \cdot 4 \cdot 3}$. Incisors small, pointed. Canines a little larger than incisors, and similar in type. Premolars enlarged, robust, larger than the molars, the protocones large and inflated with large crescentic inner cusps on p^{2-4}, the wings of the deuterocone set on each side with two or three small cusplets, small postero-internal and sometimes antero-internal basal cusps on lower premolars. Molars low-crowned, rounded, with tendency to polybuny; upper molars with protocone, paracone and metacone equal, rounded, conules small, hypocone and protostyle large, subequal, paired with respect to the protocone, and connected by an encircling cingulum around the posterior, external and anterior base of the molar. Lower molars with large, low, three-cusped trigonid and short, wide tricuspid talonid; the protoconid, paraconid and metaconid subequal, rounded, well separated; hypoconid, entoconid and hypoconulid also subequal, rounded, well separated, only slightly lower than trigonid cusps.

All cusps with strongly marked vertical ridges on the enamel, divaricating from tips of cusps and especially prominent on the premolars.

Upper molars decreasing progressively in size from m^1 to m^3; lower molars subequal. Premolars$\frac{2-4}{2-4}$ subequal, the $p\frac{2}{2}$ only slightly smaller; p_1^1 much smaller, one-rooted; the other upper premolars having three roots and the lower two roots.

Skull larger than that of *Ectoconus*, relatively to the size of the teeth, the sagittal and occipital crests high. Skeleton somewhat smaller and less massively proportioned than in *Ectoconus*, the construction corresponding in most details.

Periptychus is a genus very readily recognized by the peculiar inflated striated premolars, larger than the molars, and curiously suggestive of some of the bunodont artiodactyls (Leptochoeridae, *Tetraconodon*, etc.), as was noted by Cope. There is no question, however, of relationship to this group, nor to the Insectivora as advocated by Winge. The skeleton shows the genus to be related closely to *Ectoconus*, less closely to *Pantolambda*, and to fall with these genera into the Taligrada, of amblypod affinities.

[72] Cope, E. D., 1881, Amer. Nat., XV, p. 829, for Oct., published Sept. 22.
[73] Cope, E. D., 1882, Amer. Nat., XVI, p. 832.
[74] Cope, E. D., 1884, Amer. Nat., XVIII, pp. 790–805, figs. 2, 6–10.

The genus is represented by a large series of upper and lower jaws and incomplete skulls from both the Puerco and Torrejon and by incomplete skeletons from both horizons. In the Torrejon two closely allied species occur, *P. carinidens* (genotype) and the larger and more robust *P. rhabdodon*. In the Puerco a single very variable species occurs, *P. coarctatus*, differing so much from the Torrejon *Periptychus* that it might be considered sub-generically distinct. It is less specialized in many details of construction of molars and premolars, as also in the skull and skeleton. In many of these particulars it is nearer to *Ectoconus*.

KEY TO SPECIES OF Periptychus

1. Premolar cusps more conical, with more backward pitch; no antero-internal basal cusp (except sometimes on p₄), and postero-internal cusp less developed on lower series. Upper molars wider than long (tr. a.-p.), the hypocone and protostyle more internal than protocone. Lower molars with protoconid a little higher than metaconid and paraconid, talonid lower than trigonid..................................... *P. (Plagioptychus) coarctatus*
2. Premolar cusps more robust and inflated, with less backward pitch, antero-internal and postero-internal basal cusps well developed on p₂₋₄. Upper molars as wide as long; hypocone, protocone and protostyle forming a row antero-posteriorly. Cusps of lower molars of subequal height, and a small central seventh cusp usually present.
    Larger, premolars more robust, molars wider; milk molars with larger heels and only moderately compressed....................... *P. rhabdodon*
    Smaller, teeth less massive, molars narrower; milk molars much compressed, with smaller heels................................. *P. carinidens*
    Larger, premolars and milk molar more molariform, heel of m₃ more elongate.. *P. rhabdodon superstes*

Periptychus carinidens Cope

Periptychus carinidens Cope, 1881, Amer. Nat., XV, p. 337; 1881, Paleont. Bull. No. 33, Sept. 30, p. 484; 1885, Tertiary Vertebrata, p. 403, pl. xxiii *d*, figs. 14–15, pl. xxiv *g*, fig. 5, pl. xxv *a*, fig. 16; Matthew, 1897, Bull. Amer. Mus. Nat. Hist., IX, p. 297.
Periptychus brabensis Osborn and Earle, 1895, Bull. Amer. Mus. Nat. Hist., VII, p. 55. (*Not P. brabensis* Cope.)

Type: A. M. Cope Coll. No. 3620, jaw fragments, with right d₄ and left d₃.

Horizon and Locality: Paleocene, presumably Torrejon formation, San Juan basin, New Mexico.

Author's Description: See above (page 110), under the genus *Periptychus*.

This species is the type of the genus and was founded upon milk teeth not at first recognized as such. In 1885 Cope distinguished it from *P. rhabdodon* by smaller size and longer narrower m₃. This distinction was retained by Matthew in 1897, but the validity of the later species is much in doubt. The distinctions are much more apparent in the milk molars than in the permanent series.

Cope described and figured in Tertiary Vertebrata the jaws, incomplete skulls and various parts of the skeleton of *Periptychus carinidens* and *P. rhabdodon*. These fragmentary specimens have been more thoroughly pieced together by Matthew since they came into the American Museum collections and the acquisition of a number of additional specimens secured by the Museum parties in 1892, 1896, 1912, 1913 and 1916.

The principal specimens referred to these two species are:

No.		*P. carinidens*	*P. rhabdodon*
3620	Lower jaw fragments, dp₂₋₄	× Type	
15937	Lower jaw, milk dentition unworn	×	
3627	Upper and lower jaws. Permanent dentition		× Type
3636	Skull, limb bones, hind foot, etc. Permanent dentition.		× Fig'd
3637	Fragmentary skeleton, upper jaws with milk dentition..		× Fig'd
3720	Lower jaw, milk dentition unworn		×
3665	Skull and lower jaws		×
3669	Upper jaw, back of skull, etc.		×
2466	Skull, fragmentary but uncrushed	×	
837	Hind limb and part of tarsus		×
15936	Upper jaws, unworn dentition		×
16695	Lower jaw, teeth almost unworn	×	
16696	Upper jaws, teeth little worn		×
846	Upper jaws, teeth little worn (intermediate)	?	
3794	Palate, milk dentition		?
17075	Limb bones, manus and pes		× Fig'd

A considerable range of individual variation is noted in the series of dentitions referred to each species.   The metaconid on both the molars and premolars (= "tritoconid") may be simple or with a smaller or subequal cusp (metastylid) just behind it.   One or more cingular cusps may appear on the heel besides the usual three.   The width of the heel on m₃ varies considerably.   In no instance, however, is the heel elongate as in the *Periptychus* from the Tiffany beds.

The skull and skeleton material from the Torrejon beds vary within rather wide limits in size and robustness and in various minor details of construction, so that it has not been found practicable to distinguish the two species with certainty; they are therefore described below as one form.

The skull and jaws compare in size with those of the larger peccaries but have little resemblance otherwise.   The skull is massive, the muzzle moderately long, sagittal and occipital crests high, brain-case very small, zygomatic arches heavy and moderately spreading, canines small, no post-canine diastema, lower jaws deep and thick beneath premolars and molars, reduced anteriorly, condyles considerably above level of tooth-row, the angle widely expanded, flat, and rounded posteriorly, the coronoid process subvertical at base, large and strongly recurved.   In general the skull and jaws agree most nearly with those of *Ectoconus* and differ from *Claenodon* and *Arctocyon* chiefly in the narrower arches, the different form of muzzle, and other details of structure conditioned by the small canines and enlarged premolars.

The top of the muzzle is imperfectly preserved in all the specimens, but the fragments show that the nasals were long, reaching well backward and somewhat expanded posteriorly, as in all primitive mammals.   The frontals are expanded widely and reach backward in the median line to the anterior end of the sagittal crest, which is about opposite the middle of the zygomatic arch; forward and laterally they extend almost down to the end of the upper branch of the jugal, the lachrymal expansion on the face being small, but preventing the maxillary from reaching the margin of the orbit.   The orbit is well forward, its anterior border being above the front of m¹, and is of rather large size, facing upward more than laterally or anteriorly, without any well-marked postorbital process on either frontal or jugal. The jugal has no inferior branch, but the superior branch is long, outlining the lower half of the orbit with a heavy thickened border which forms a rather prominent shelf beneath the orbit. This shelf is supported beneath by the maxilla, which has a strongly concave surface above p^4–m^2. This infra-orbital shelf is not developed in *Ectoconus;* in *Pantolambda* a somewhat similar shelf is found, but it is more largely made up of the maxillary. Posteriorly the jugal extends as a long wide plate, thinning out to a trihedral tip just in front of the glenoid facet of the squamosal.

The parietal region is very thick and massive, with high sagittal crest, about as in *Ectoconus* or other primitive mammals, the brain-case small and indicating a very low type of brain. Anteriorly the parietals extend forward on each side of the frontal along the postorbital process in an overlapping triangle. Posteriorly they are plastered up on the anterior face of the high and broad occipital crest almost to its margin. Laterally the parietals are overlapped by the squamosals in the usual manner. The squamosals are large, the zygomatic portion massive and moderately wide, with nearly flat glenoid facet, small low postglenoid process, small postglenoid foramen (unlike the large foramen of *Ectoconus* or *Claenodon*), the post-tympanic border short and plastered against the face of the mastoid, widely separated from the postglenoid process, the superior border of the zygomatic crest fading out just before it reaches the lambdoid crest at the posterior border of the squamosal.

The mastoid exposure of the periotic bone is quite extensive, facing about as much laterally as posteriorly, but separated from the side of the face by a well-marked lambdoid crest along the suture with the squamosal. Inferiorly the mastoid is extended into a stout process flanked by the paroccipital and post-tympanic, which are plastered against its postero-internal and antero-external faces, but do not extend beyond it to form distinct processes. The mastoid foramen on the outer face of the mastoid bone is large and deep. The mastoid process itself projects laterally more than downward; on the under surface at the base is the stylomastoid foramen, and antero-internal to it the small prominent oval auditory prominence.

The condition of the basicranial region cannot be clearly seen in any of our specimens. Presumably it was not fundamentally different from that seen in *Pantolambda*.

The teeth in *Periptychus* have been briefly described in the diagnosis of the genus. In *P. carinidens* a series of well-preserved dentitions shows the characters in detail. This species is hardly separable, if at all, from *P. rhabdodon*, but is distinct throughout from the older and more primitive *P. coarctatus*. The incisors are small, round-pointed, rather characterless teeth. The canines are similar but of larger size, with rather reduced and blunted crown which shows the vertical grooving also seen in the cheek teeth. There is no post-canine diastema. The four premolars all have crowns of similar type, with a sub-circular base, a short blunt-conical principal cusp with a large crescentic inner cusp, its border raised into a series of six or eight rudimentary cuspules and the wings of the crescent extending around the anterior and posterior faces of the premolar (anterior wing rudimentary on p^1). The cusps, large and small, are all ornamented with the prominent vertical ridges characteristic of the genus (and more or less of the family). There are no basal cingula on the premolars. P^1 has a single root, the others are three-rooted. The principal cusps are pitched slightly backward and when unworn can be seen to have anterior and posterior crests at the tip and a small postero-internal crest, but these disappear with wear. The inner crescents are more nearly symmetrical, facing the midline of the skull; this gives a slightly skewed effect to the tooth that is more obvious in *P. coarctatus*.

The molars are smaller than the premolars, rounded and nearly quadrate in form, the cusps low, round, of subequal height, except that the conules are smaller and the hypocone and protostyle set a little lower.   These two inner cusps are almost in line with the protocone, being set only a little internal to it.   From the hypocone a basal cingulum extends around the posterior, external and anterior faces of the tooth to the protostyle, but it is sometimes interrupted or obsolete on the outer face of the paracone.   The molars decrease in size from m^1 to m^3, the third molar having the metacone considerably reduced in this species (less so in *P. rhabdodon*) with a consequent change in the form of this part of the tooth.   The cusps have the same vertical grooving as those of the premolars but less prominent on account of the smaller size of the cusps.

The lower premolars correspond to the upper in size and massiveness, but p_1 is relatively smaller and spaced apart from canine and p_2.   It is one-rooted, with short oval crown, a small posterior basal cusp.   The remaining premolars are two-rooted, robust, close-set, with massive, round-oval principal cusps and small subequal basal cusps antero-internal and postero-internal, and a semi-separate median internal cusp.   The main cusps when unworn are more or less prominently crested anteriorly and posteriorly.   They are very slightly pitched backward, much less than in *P. coarctatus*.

The molars have the six cusps of the genus all of subequal height, save that the protoconid is a little the largest.   A seventh cusp is sometimes seen at the junction of low ridges which connect protoconid and metaconid with hypoconid and hypoconulid, but it is rudimentary and not always present.   It is more generally present in *P. rhabdodon*, absent in *P. coarctatus*.   The lower molars are of nearly equal size, about three-fifths as wide as they are long, and the last molar in this species is slightly smaller and a little narrower than in *P. rhabdodon*.   The difference, however, is not so great as in the canines and premolars.

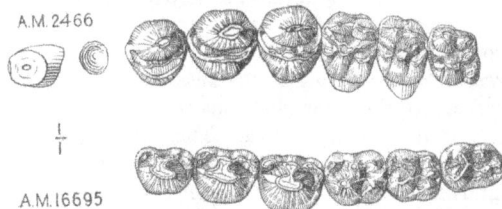

FIG. 20.   *Periptychus carinidens*, upper and lower teeth, A. M. Nos. 2466 and 16695, crown views.   Natural size.

The milk premolars of *Periptychus carinidens* have been described and figured by Cope in Tertiary Vertebrata, the species having been founded upon milk teeth.   They are retained unusually late, the permanent premolars replacing them after all the molars are fully emerged.   They are semi-molariform in type, the fourth fully molariform, the others progressively simpler and distinguished from the premolars by compressed form.

The type of *P. carinidens* consists of jaw fragments with dp_4 of the right side and dp_3 of the left.   In No. 3637, a fragmentary skeleton of *P. rhabdodon* partly figured by Cope, the upper and lower milk molars are preserved in place with the molars.   No. 15937 is a lower jaw fragment of *P. carinidens* with the milk molars and m_1 well preserved, m_2 partly preformed in the jaw.

The upper milk molars as shown in *P. rhabdodon* are distinguished by the small size and narrowness of the principal cusp. Dp^4 is completely like the molars in cusp construction, but the inner half of the tooth is reduced in all its dimensions. Dp^{2-3} are more like their permanent successors, but cusps smaller, protocone less robust and distinctly twinned at the tip; in front of dp^2 is a single large round alveolus, for dp^1 or for the permanent canine. Dp_1 is one-rooted and small; dp_{2-4} are like those of *carinidens* except for larger size and somewhat greater development of the heel-cusps.

In No. 15937 dp_{2-3} have large compressed principal cusps, twinned anteroposteriorly on dp_3 at the tip, and sharply crested in front and behind. A rudimentary inner cusp (deuteroconid) is formed by the rising of the inner cingulum toward the apex of the protoconid, forming on dp_2 a supplementary crest parallel to its posterior crest and rising at the apex to a closely twinned cusplet. In dp_3 this crest is more distinct, its apex between the two twinned apices of the protoconid, and the crest is continued forward to join the anterior basal cusp, which is distinct in dp_{3-4}, quite minute and rudimentary in dp_2. The posterior basal cusp is simple and of small size in dp_2, larger and bicuspid in dp_3, and in dp_4 has become quite large, tricuspid, and entirely like the talonid of the molars. The protoconid in dp_4 is much lower, smaller, less compressed, and has become almost like the protoconid of the molars, the deuteroconid has taken on largely the characters of the molar metaconid, while the small twinned cusp behind the protoconid has taken partly the form and position of the small seventh cusp already noted in the lower molars. The anterior basal cusp of the milk molars has approached to some extent the development and position of the molar paraconids, to which it obviously corresponds.

The progressive molarization of the milk molars in primitive mammals affords better evidence on the origin of the molar cusps than does the molarization of the premolars, since the milk molars belong to the same series as the true molars and their correspondence is closer. They lead to the conclusions that have already been drawn from the premolars as to the origin of the tritubercular molar and serve to strengthen these conclusions.

The skeleton of *Periptychus carinidens* is known from various fragmentary specimens which, so far as we are able to see, agree with the better-known and better-preserved specimens referred to *P. rhabdodon*, excepting in characters of smaller size and less robust proportions. The detailed description from skeletons referred to both species follows here.

The vertebrae are very imperfectly known. The cervical centra, described and figured by Cope, indicate a shorter neck than in *Ectoconus*, more as in *Pantolambda*. The construction of the atlas and of what is known of the axis was evidently similar, as also in *Pantolambda*. The resemblance to *Coryphodon* is much more striking than any proboscidean resemblances such as were suggested by Cope, nor is the resemblance at all close to *Uintatherium*.

Of the shoulder and pelvic girdles only an incomplete scapula and clavicle and indistinctive fragments of the pelvis are known. The head of the scapula, figured by Cope, agrees well enough with the scapula of *Ectoconus*, except for smaller size and less massive coracoid process. The full length of the scapula is preserved in No. 3636, but not the lateral borders. Its form was evidently much as in *Ectoconus*, but relatively more pointed at the superior border, with apparently a narrower blade and less prominent spine.

The humerus, best seen in Nos. 3636 and 17075, is smaller than in *Ectoconus* and somewhat less massive, especially in No. 3636; the supinator crest less prominent, the deltoid crest equally heavy and long, ending abruptly in the same manner, and the distal end resembles that of the Puerco genus on a somewhat smaller scale.

The ulna and radius are best preserved in No. 17075.   They are smaller and less robust than in *Ectoconus*, but with the same proportionate length.   The olecranon is much shorter (best seen in No. 3636); the ulnar shaft has a much shallower longitudinal groove on its

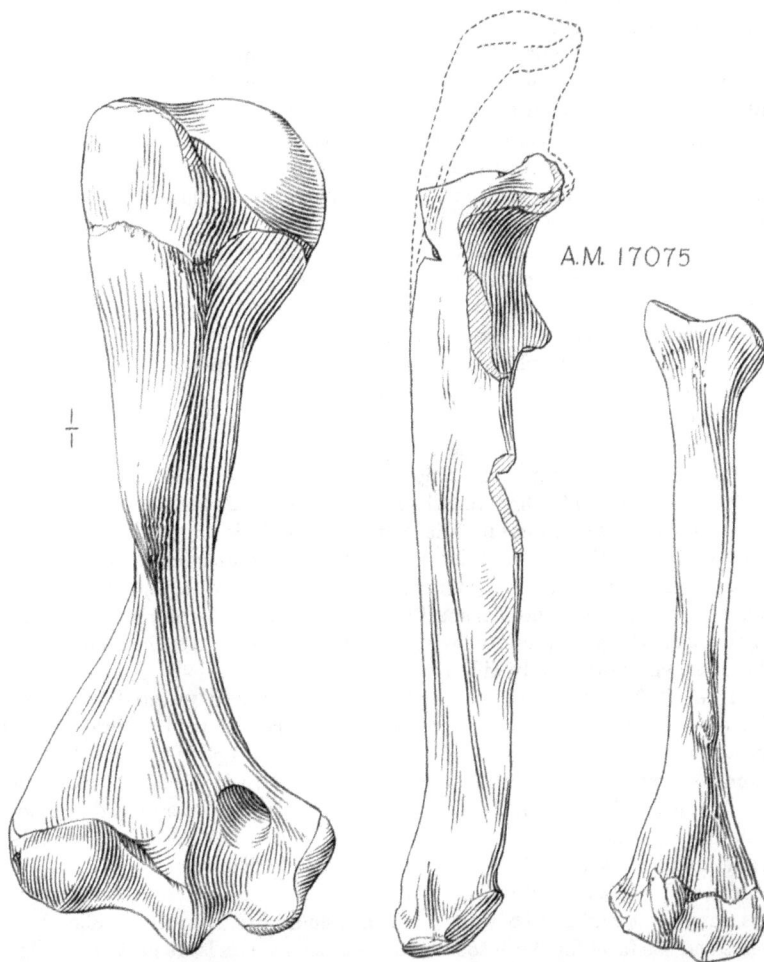

FIG. 21.   *Periptychus rhabdodon*, left humerus, anterior view; left ulna, internal lateral view; left radius, anterior view, A. M. No. 17075.   All natural size.

antero-external face, the head of the radius is somewhat more flattened, the bicipital tubercle similarly weak, narrow and elongate.   The distal part of the shaft has a rather prominent anterior low ridge with rugose surface, about in the position for the metacarpal extension of the first digit; this may indicate a concentration of the origin of this muscle, associated with the greater development of this digit and of its opposability in primitive placentals.

The distal end of the radius is broadly expanded and shows a single concave facet with obscure indications of the separation of lunar and unciform.   The distal end of the ulna is not expanded, carrying the width of the shaft, with an oblique distal-internal facet obscurely shown for the radius, and the usual convex distal facet for cuneiform and pisiform, but no styloid process on the external tip.

Of the manus the scaphoid, lunar, centrale, magnum, trapezium and pisiform are preserved in No. 17075, together with metacarpals II, IV and V and several phalanges.   These all indicate a construction of the manus very similar to that of *Ectoconus*, except as follows: the scaphoid is not so much reduced in width along the dorsal surface and has a smaller hook; the lunar has considerably less width, and its magnum facet is smaller, not reaching the dorsal surface; the magnum has a smaller dorsal surface and does not reach the lunar at the dorsal surface, the articulation of the keel on the lunar being also restricted by the shifting of the centrale; the centrale has a narrower scaphoid facet and wider lunar facet; the "peg" of the trapezium is more blunted, the scaphoid and trapezoid facets having a wider angle between them, and its internal border is shorter, shifting the pollex apparently to a somewhat more lateral position.   The metacarpals are very like those of *Ectoconus* except for somewhat smaller size; the phalanges show quite the same proportions, except the unguals, which are shorter and wider, quite distinctly more hoof-like and with the subungual process much reduced.

The hind limb and foot bones are preserved in a number of specimens, especially Nos. 3636, 17075, 837, 851, and have been figured by Cope and Osborn.   The femur is very like that of *Ectoconus* but only three-fourths as long, with shaft less flattened, third trochanter somewhat less prominent, digital fossa more restricted distad, patellar trochlea and distal end narrower.   The tibia is about five-sixths as long as in *Ectoconus*, with a more cylindrical shaft; the cnemial crest is less prominent and does not extend quite so far down.   The fibula has a much more cylindrical shaft than in *Ectoconus*, the prominent flanges characteristic of that genus being reduced to low ridges; the proximal end of the fibula is less expanded, but the distal end not notably different.

The pes of No. 3636 has been figured by Cope [75] and restored by Osborn,[76] and a number of other specimens supplement the characters.   It is in general very like that of *Ectoconus*, but smaller and a little less robust.   The astragalus has less depth in both the body and head, the facets of the head being a flattened oval; the cuboid facet is less distinct from the navicular facet, and the inner navicular facet is much reduced and not separable from the principal facet for the body of the naviculare.   The calcaneum differs chiefly in the reduction of the shelf on the external side which extends from the peroneal tubercle to the middle of the tuber calcis but is by no means so prominent as in *Ectoconus;* the fibular facet is somewhat wider relatively.   The navicular has a much smaller hook than in *Ectoconus* and the facet for the entocuneiform extends over a large part of it.   The cuboid is of less depth than in *Ectoconus*, the calcanear facet more convex dorso-plantad; the astragalar facet faces more nearly proximad and is somewhat more concave.

The three cuneiforms are preserved in No. 17075, with metatarsals II to V; in No. 3636 metatarsals II–IV, the proximal half of Mt. V and distal half of Mt. I are complete.   The cuneiforms are like those of *Ectoconus* except that the first is relatively somewhat reduced;

---

[75] Cope, E. D., 1885, Tertiary Vertebrata, pl. xxiii *g*.

[76] Osborn, H. F., 1890 [1889], Trans. Amer. Phil. Soc., XVI, N.S., diags. 6–7, p. 533.

the first metatarsal appears to be a little smaller in proportion than in *Ectoconus*, the others have about the same relative length but somewhat more slender in the shafts. The phalanges are indistinguishable from those of the fore foot.

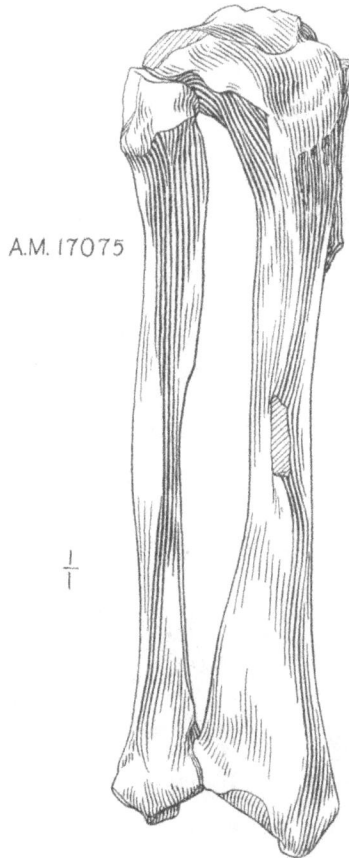

FIG. 22. *Periptychus rhabdodon*, right tibia and fibula, A. M. No. 17075, anterior view. Natural size.

The skeleton characters of *Periptychus* indicate an animal somewhat smaller than *Ectoconus*, although with a larger head. The neck is shorter, the limbs of about the same proportions but a little less massive throughout. The elbow joint appears to have been much flexed and considerably everted. The stout short limbs and short spreading plantigrade five-toed feet, the long heavy tail, are all much the same as in *Ectoconus* and *Pantolambda*. The toes are intermediate between the two in the broadening of the unguals and shortening of the first and second phalanges. The tarsal region shows a more flattened astragalus than either, approaching the form seen in the marsupials, and this approach is closer apparently in the older species *P. coarctatus* than in *P. carinidens*. This may be a

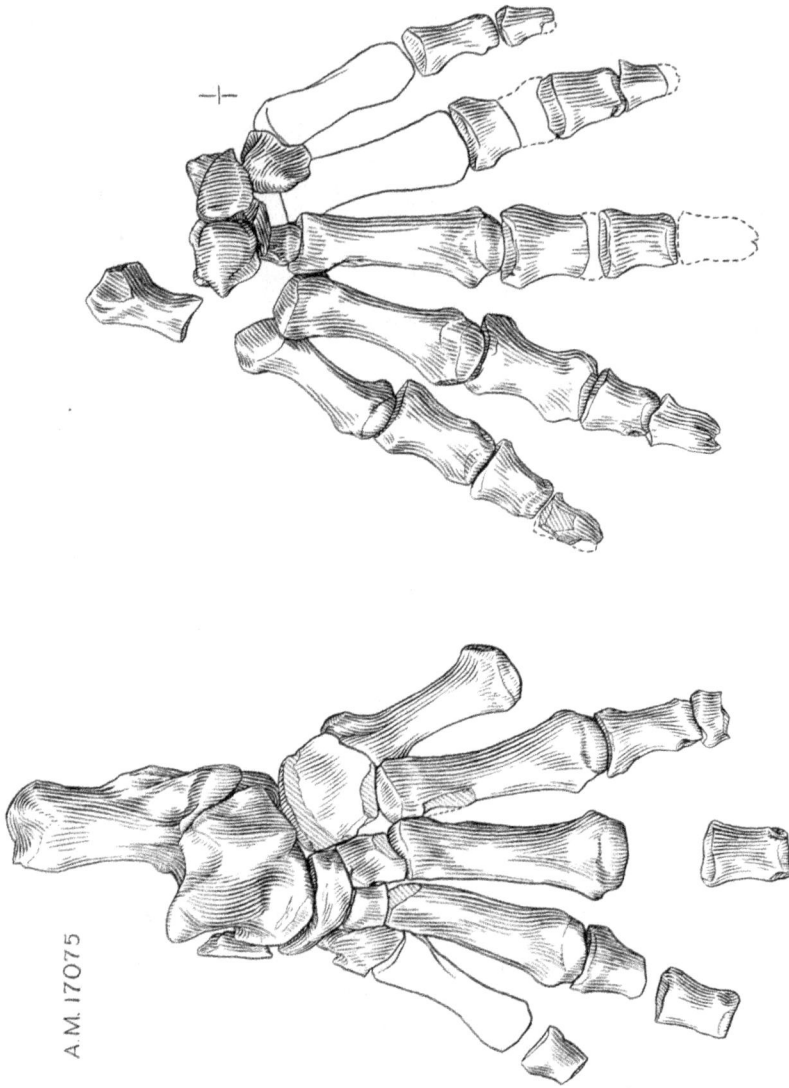

FIG. 23. *Periptychus rhabdodon*, right manus and left pes, A. M. No. 17075, anterior view.   Natural size.

primitive character. On the other hand the reduction of the hook of the navicular is an advance on *Pantolambda*, in which the hook is a separate bone; on *Ectoconus*, in which it is a large, semi-separate process, not yet invaded by the entocuneiform facet; and on *Claenodon*, in which it is reduced, but not so much as in *Periptychus*. There is no evidence in *Periptychus* of the "prehallux" seen in *Claenodon*, but its absence is not demonstrated. The reduction of the olecranon and shallowing of the ulnar groove are points of advance on *Ectoconus*.

No one modern animal has the proportions of the *Periptychus* skeleton. *Orycteropus* approaches it in proportions of neck, body and limbs, but has a much heavier tail and very different feet. The wombat has some resemblance in the proportions of the feet, aside from the inequality of its digits, but its lower limbs are much longer and straighter. The limbs and feet of *Sarcophilus* are much longer and more slender.

Periptychus rhabdodon (Cope)

Catathlaeus rhabdodon Cope, 1881, Amer. Nat., XV, Sept. 22, p. 829; (*Periptychus*) Cope, 1882, Amer. Nat., XVI, p. 832; Proc. Amer. Phil. Soc., XX, p. 564, pl. ii (description and figure of brain-case); 1884, Amer. Nat., XVIII, p. 792, figs. 1, 2, 6–9 (upper and lower teeth and skeletal bones).

Type: Amer. Mus. No. 3627, upper and lower jaws.

Horizon and Locality: Paleocene, Torrejon formation, San Juan basin, New Mexico.

Author's Description: Catathlaeus rhabdodon. Char. gen.—With this genus I commence descriptions of some genera with bunodont dentition, which has some resemblance to that of some of the hogs. The one above named, with *Mioclaenus*, remind one of *Tetraconodon* Falc. and Lydd., in the enlarged proportions of their premolar teeth. . . .

In the genus *Catathlaeus* the development of the premolars is remarkable, while the true molars are relatively small. The last three superior premolars have an elevated internal crescentic cingulum, homologous with the inner lobe of the fourth superior premolar of the ruminants. The general character of the true molars is that of *Phenacodus*. Parts of two or three individuals of this species have come into my possession, one of which includes nearly all of the molar dentition of both jaws. The external cusp of the superior premolars is compressed conic, and the internal cingulum extends to its *anterior* base in the second, third, and fourth. The crown of the last true molar is about as long as wide, while that of the first is wider than long. Each supports seven cusps; two subconic external; one large median internal, which is connected by ridges with a small anterior and posterior median. Then there are a small anterior and posterior internal, making three internal. The internal crest is distinct from the principal cusp in the inferior premolars III and IV, but unites with it in the II; it supports on the IV, an anterior, a median and a posterior cusp, the latter forming part of the rather narrow heel. The true molars I and II have seven tubercles, the four principal ones, and three smaller, one anterior, one posterior, and one median. On the third the posterior forms a large heel. All of the molars, but especially the premolars, have the enamel thrown into sharp parallel folds, in a manner I have not seen in any other mammal. Length of six superior molars, .067; length of three true molars, .029; length of base of third premolar, .012; width of do., .012; width of base of first true molar, .010; do. of third true molar, .009; length of do., .010. Length of base of fourth inferior premolar, .012; width do., .012; of third true molar, .0115; width of do., .009. The teeth indicate an animal of the size of the peccary.

Periptychus rhabdodon superstes, mut. nov.

Periptychus superstes Matthew, in Simpson, G. G., 1935, Amer. Mus. Novitates No. 817, pp. 25–7, fig. 14.

Type: A. M. No. 17181, lower jaws with p_4–m_3.

Paratypes: A. M. Nos. 17183, 17184, 17195, isolated teeth, p^1, m^2 and dp_4.

Horizon and Locality: Tiffany (basal Wasatch), Colorado.

A lower jaw and a few isolated teeth from the Tiffany horizon (basal Wasatch) represent a species of *Periptychus* close to *P. rhabdodon* but differing in a number of minor characters which lie outside the range of variation of the Torrejon specimens as far as known. In size it equals the larger individuals of *rhabdodon*, but the heel of m_3 is longer, the inner crescentic cusp of p^4 is larger and more widely separated, the inner cusps of the trigonid of dp_4 are higher than in Torrejon specimens, subequal in height to the protoconid and more widely separated from it.   These rather slight indications conform to expectation that the Tiffany *Periptychus* would be further advanced in specialization on the lines that are seen in comparing *P. rhabdodon* with *P. coarctatus*.   Better specimens might show it to be a distinct species; for the present it may be treated as a progressive mutation.

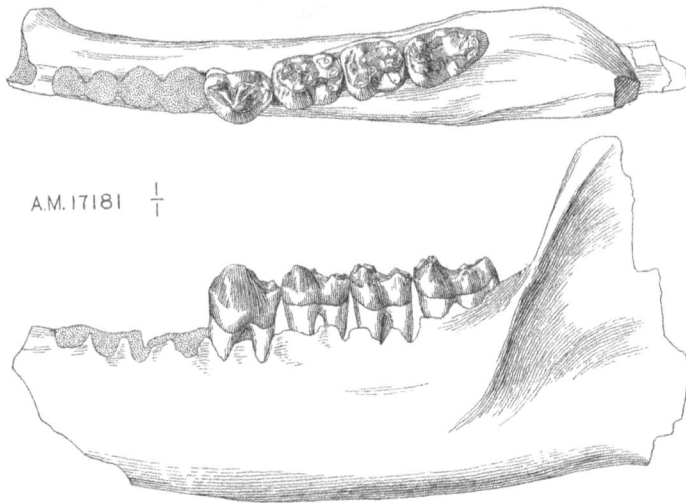

FIG. 24. *Periptychus rhabdodon superstes*, type mandible, A. M. No. 17181, crown view above, lateral view below. (From Simpson, 1935.)   Natural size.

Subgenus **Plagioptychus** Matthew

Matthew, W. D., in Simpson, G. G., 1936, Amer. Mus. Novitates No. 849, p. 9 [77]

Type: Periptychus coarctatus Cope.

Characters: Principal cusps of premolars pitched obliquely backward, the anterior basal cusps and deuteroconids of lower premolars lacking.   Upper molars wider than long, the basal portion of the crown extended inwardly, supplementary cusps not developed on molars.

Periptychus (**Plagioptychus**) coarctatus (Cope)

Periptychus coarctatus Cope, 1883, Proc. Acad. Nat. Sci. Phila., XXV, Sept. 18, p. 168; 1884, Amer. Nat., XVIII, p. 802, fig. 10; 1885, Tertiary Vertebrata, pl. XXIX d expl. and figs. 7–8; 1888, Trans. Amer. Phil. Soc., XVI, N.S., p. 354; Osborn and Earle, 1895, Bull. Amer. Mus. Nat. Hist., VII, p. 54; Matthew, 1897, *ibid.*, IX, p. 296.

[77] This name is given generic rank in Simpson's paper cited.   Its validity having been disproved, it is replaced by *Carsioptychus*. See p. 365.—THE EDITORS.

Periptychus brabensis Cope, 1888, Trans. Amer. Phil. Soc., XVI, N.S., p. 354; Matthew, 1897, Bull. Amer. Mus. Nat. Hist., IX, p. 296 (= *P. coarctatus*).   (Not *P. brabensis* of Osborn and Earle, 1895, which is *P. carinidens.*)

Type: A. M. Cope Coll. No. 3775, lower teeth.   Type of *P. brabensis:* A. M. Cope Coll. No. 3782, parts of upper and lower jaws, p⁴–m² right, p₄–m₂ right, etc.

Horizon and Locality: Both from the Lower Paleocene, Puerco formation, San Juan basin, New Mexico.

Author's Description: The characters of the species are well marked in the premolar and molar teeth.   The former lack the anterior and internal ledges of the *P. carinidens* and *P. rhabdodon*, having only a prominent ledge-shaped heel besides the principal conical cusp.   The true molars lack the small tubercle which is between the pair of threes which compose the crown.   The adjacent cusps of the threes are connected by low longitudinal ridges instead of oblique ones.   The cusps themselves are closer together than in the other species, especially those of the anterior three, which are closely approximated.   The anterior one is small and low.   The enamel is grooved as in the other species.   (Measurements follow.)

Matthew in 1897 (p. 296) distinguished *P. coarctatus* as follows: There is a wide difference between this species and its successors; the teeth are much less specialized.   The antero-internal cusp of the lower premolars is entirely wanting; it is always strongly developed in the Torrejon species.   The premolars, both upper and lower, are more pointed and less inflated.   The molar cusps are higher and rounder, and no subsidiary cusps appear besides the six normal ones.   Molars and premolars show more or less obsolete external cingula.

A series of upper and lower jaws of *P. coarctatus*, and most of the limbs and feet, are now available for comparison.   They show a considerable variation in size and proportions of most of the teeth, but all are distinguished by the characters cited above.   The premolars and molars are of more trihedral outline, their transverse diameters greater, the cusps higher and less inflated, the external cingula complete on upper molars and premolars, although partly obsolete on the premolars.   The protocones of the upper premolars are more backwardly pitched than in *P. carinidens* and *rhabdodon*, the peculiar "skewing" of these teeth more pronounced; their inner cusps are crescentic but less semicircular, the apices higher.   The upper molars are considerably wider transversely than long anteroposteriorly, the

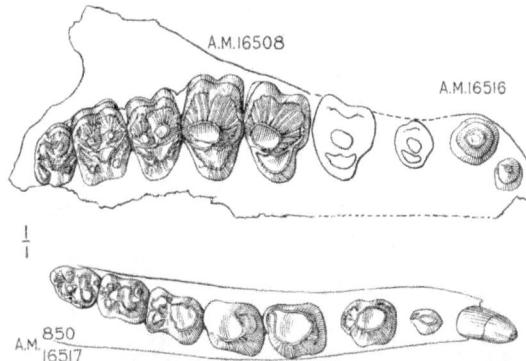

Fig. 25.   *Periptychus coarctatus*, upper and lower teeth, A. M. Nos. 16508, 16516, 850, 16517, crown views.   Natural size.

conules relatively small, paracone, metacone and protocone higher, hypocone and protostyle more internal in position.   The anterior teeth are in general form similar to those of *P. carinidens*, but the front of the muzzle is shorter, the teeth more crowded, the canines

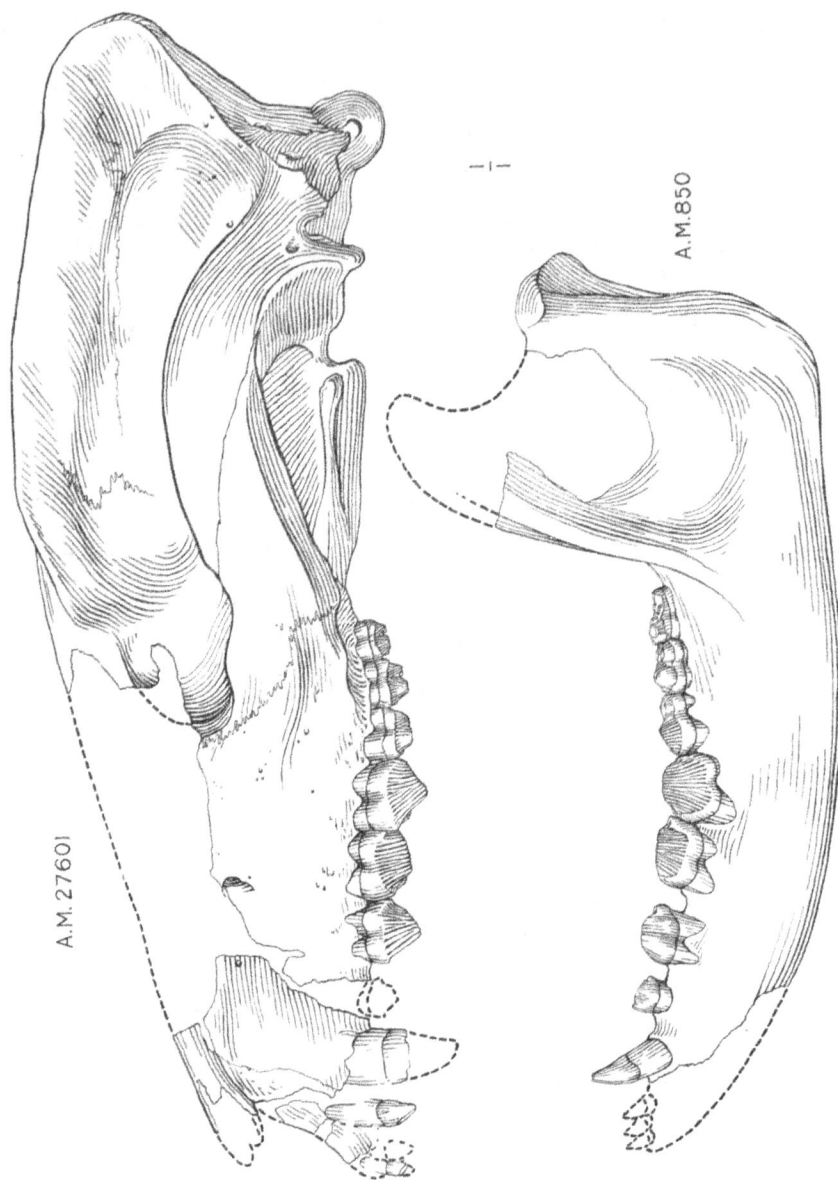

FIG. 26. *Periptychus coarctatus*, skull and jaw, A. M. Nos. 27601 and 850, lateral views. Natural size.

smaller and the lateral incisors relatively large, pointed, caniniform, about two-thirds the diameter and height of the canines, the first and second incisors being unknown but evidently small.

The lower teeth show corresponding differences from the Torrejon species. The lower canines are relatively small and slender, the front of the jaw reduced. The premolars have a decided backward pitch in the protoconids, which are higher and less inflated than in *P. carinidens* and *rhabdodon*, the anterior basal cusps are absent, and none of the specimens shows any trace of deuteroconid. The molars have a relatively high protoconid and metaconid and no trace of a seventh (central) cusp.

The skull is imperfectly known, parts of it being preserved in No. 16516 along with a large part of the skeleton. It appears to be relatively short, both in the facial and cranial portions. The length from canine to m^3 inclusive is only two-thirds that of *P. carinidens*, No. 2466. The anterior border of the orbit is in the same position, but the base of the zygomatic arch has a very different form, lacking the suborbital shelf and the excavated area beneath it. The postorbital crests come together far more abruptly than in *carinidens*, and the frontal between them is strongly concave. The postorbital constriction of the skull is very narrow, only two-thirds the width of that in *P. rhabdodon*, and the front part of the brain-case appears to be even narrower relatively than in that species.

The limb bones are of about five-sixths the length of those referred to *carinidens* and *rhabdodon*, the range in size being almost as wide as in specimens referred to the two Torrejon species. The coracoid process of the scapula is wider, though not so long as in *P. rhabdodon;* it approaches the stout and robust coracoid of *Ectoconus* to some extent. The humerus has a somewhat higher and longer supinator crest than in *P. rhabdodon*, much higher than in *carinidens*, and its deltoid crest is more compressed, probably equally high and long. The ulna has a somewhat longer olecranon, and the head of the radius is less flattened. The carpals appear to have the same form and proportions as in *P. carinidens*, but are of smaller size throughout. The scaphoid, lunar, cuneiform, part of the pisiform, the trapezoid, centrale and unciform are preserved in No. 16516, and an unidentifiable bone, possibly a "pre-pollex." The trapezoid, unknown in *P. carinidens*, is like that of *Ectoconus* except for smaller size and lesser depth proximo-distally. The undetermined bone is a round-oval nodule of about the same size as the trapezoid, with two facets on it meeting at right angles, the one shallow, concave, with a projecting dorsal lip, the other extending convexly backward from the line of meeting. This bone is also preserved in No. 16524. It cannot be any of the normal elements of manus or pes, all of which are well known and readily recognizable in *Periptychus* and its relatives; it is quite different in form from the compressed and flattened "pre-hallux" of *Periptychus* and *Ectoconus*, and it has not the character of a sesamoid. If not the "pre-pollex," it is an otherwise unknown element of manus or pes.[78] The second metacarpal, with the heads of the third, fourth and fifth, distal end of the third and the proximal phalanges of digits I–III are preserved partly articulated in No. 16516. The metacarpals are very much like those of the Torrejon species in their form and articula-

[78] It is perhaps advisable at this point to state that I do not regard this occurrence of the elements designated as "pre-pollex" and "pre-hallux" in these Paleocene mammals as evidence for a sixth digit in the Mesozoic mammalia. There is no trace of such additional digits in the pro-mammalian reptiles of the Permian, and it is likely that they are added elements of the nature of sesamoids; the same interpretation is generally, and, I believe, correctly, placed upon the more or less similar occurrences among recent mammals, to which the misleading names of pre-pollex and pre-hallux have been given.—W.D.M.

tions, except for somewhat greater slenderness; the phalanges are decidedly slenderer, their length twice their width and a little over half the length of the metacarpals, more relatively than in *Ectoconus* or *P. rhabdodon*, and approaching *Claenodon* and other creodonts to some degree.   The single ungual preserved has the narrow proportions of *Ectoconus* and is not flattened out as in *P. carinidens*.

The femur is about five-sixths as long as in *P. rhabdodon*, and of generally similar construction but slightly more slender.   The digital fossa is not reduced distally as in the Torrejon species, but agrees more nearly with that of *Ectoconus*.   The tibia has about the same relative size and proportions and does not differ in any significant characters from that of *P. rhabdodon*.   The fibula has much more prominent lateral crests on the shaft, again approaching that of *Ectoconus*.

The astragalus is shallower in the body and wider and flatter in the head than that of *P. carinidens* or *rhabdodon*.   The calcaneum has a relatively longer tuber and larger fibular facet; the crest along the external side of the tuber is longer and more prominent but not so much so as in *Ectoconus*.   The cuboid has the same form as in the Torrejon species, the astragalar facet less concave than in *Ectoconus*.   The navicular (No. 16524) has the reduced hook of the Torrejon species, and is otherwise similar, as are also the cuneiforms.

The metatarsals preserved in No. 16524 (II and IV) and No. 16516 (most of II, III and IV) correspond to the metacarpals in comparatively slender proportions, and the phalanges of the hind foot are so similar to those of the fore foot that no distinction can be made between them.

From the above data it appears that the Puerco species of *Periptychus* is decidedly and unmistakably more primitive in teeth than the Torrejon species.   In almost every particular it makes a nearer approach to the other periptychid genera, especially to the Puerco species of those genera.   In some particulars it approaches *Ectoconus*, in others it approaches *Conacodon, Hemithlaeus* and *Anisonchus*.   This is also the case in the characters of the skull and limb bones and in the foot structure.   The species is smaller and more slender in limbs and feet but with a shorter skull than the Torrejon form.

Ectoconus Cope, 1884

Cope, E. D., 1884, Amer. Nat., XVIII, p. 795

Type: Periptychus ditrigonus Cope, 1882, from the Puerco formation of New Mexico.

Author's Diagnosis: Superior molars with an external cingular cusp; inferior premolars without internal ledge . . . . *Ectoconus*

No supplementary external cusps, inferior premolars with internal ledges . . . . *Periptychus*

The affinity of this genus to the Periptychidae was recognized from the first by Cope, and was confirmed by his later studies and those of Osborn and Earle.   The correspondence in the arrangement of the molar cusps, a complex and very peculiar pattern, was regarded as decisive, and was confirmed, if further proof were necessary, by the characters of such skeleton parts as were known.   Winge [79] has referred *Ectoconus* to the Condylarthra and the remaining Periptychidae to the Insectivora, as nearly related to the Leptictidae, while *Pantolambda* is referred to the Amblypoda.   The only reason that he assigns for separating *Ectoconus* from the remaining periptychids is that its premolars are not enlarged and inflated.   There is a certain superficial resemblance between this genus and the larger

[79] Winge, H., 1923, Pattedyr-Slægter, I, pp. 196–7.

species of *Phenacodus* in that both have a tendency to polybunodont low-crowned teeth. The fundamental pattern of the teeth in *Ectoconus* is, however, the same as in *Periptychus;* in *Phenacodus* the pattern is that of *Tetraclaenodon.* The skull and skeleton characters of *Ectoconus* are very close to those of *Periptychus*, especially in the feet; while *Phenacodus*, as shown by Matthew in 1897, has a foot related, like the dentition, to the *Tetraclaenodon* type and widely different from any of the taligrade genera. The anisonchine genera are indeed indicated by the fragmentary evidence available as partly intermediate, but as they equally tend to link up with the oxyclaenid creodonts it is apparent that in this respect they are chiefly primitive survivals.

The jaws, skull and skeleton here figured make *Ectoconus* the most completely known of all the Puerco mammals, and from them its affinities can be definitely estimated.

The dentition was described by Osborn and Earle in 1895. They also described briefly a humerus in somewhat doubtful association.[80] Matthew in 1897 (p. 295) notes that "the foot was much like that of *Periptychus* but somewhat more primitive."

Cope in 1885 described and figured the cheek teeth.

In 1888 Cope added notes upon various skeleton fragments associated with the dentition, and in particular the tarsals, parts of scapula, centra of vertebrae, pelvis and femur.[81]

Although the description and figures of the dentition and fragmentary skeleton materials studied by Cope, Osborn, Earle, Gregory and Matthew have been very incomplete and inadequate, there has never been any question in the minds of those who have studied the specimens that *Ectoconus* was a member of the Periptychidae, and their authority has been accepted by all authors except Winge. Whether this family belonged in the Condylarthra or Amblypoda has been in question, but rather as a matter of definitions and the proper limitations of these orders than of any doubt as to the relative affinities. It was clearly recognized by Cope and subsequent writers that *Periptychus* was related to *Pantolambda*, in spite of obvious divergence in the tooth construction, and through *Pantolambda* to the more typical Amblypoda. On the other hand, all the Periptychidae showed unmistakable evidence of relationship to Mioclaenidae and Phenacodontidae in the dentition, evidence which no one who has studied the original materials could well ignore.

The arrangement adopted by Gregory in 1910, including the Periptychidae and Pantolambdidae in a distinct order Taligrada, appears to conform best with the evidence and is here adopted.

Winge, as pointed out above, referred the Periptychidae to the Insectivora, as near relatives of the Leptictidae, along with the Taeniodonta and Tillodontia. He excepts *Ectoconus*, however, which he refers to the Condylarthra. Following is a translation (from the Danish) of his statement:[82]

To the Periptychidae Cope referred also the Tertiary North American *Ectoconus* and herein he is followed by others. The genus is based upon a few fragments of jaws (see Cope, Tertiary Vertebrata, 1884, pl. xxix *d*, under name of *Conoryctes ditrigonus;* Amer. Nat., 1884, pp. 796 and 797, with figures; Trans. Amer. Phil. Soc., XVI, 1888, pp. 355–9; Osborn and Earle, Bull. Amer. Mus. Nat. Hist., VII, 1895, pp. 56–8). It shows nothing of the peculiarities in the anterior cheek teeth which are found in all the genera that in the present work are reckoned as periptychids. On the contrary, the two posterior premolars are much weaker than the true molars and have not the dagger form ("Dolk-Form"); the molars are of

---

[80] Amer. Mus. Nat. Hist., 1895, VII, pp. 56–8.

[81] 1888, Trans. Amer. Phil. Soc., XVI, N.S., pp. 355–9.

[82] Winge, H., 1923, Pattedyr-Slægter, I, pp. 196–7.

considerable size with quadrate outline, much suggesting *Phenacodus* yet with a relatively primitive character, a distinct remnant of the most posterior of the three outer cusps [*i.e.*, a metastyle]. *Ectoconus* belongs nearest to the Phenacodontidae, but it is too little known to decide rightly upon it.

Ectoconus ditrigonus (Cope)

Periptychus ditrigonus Cope, 1882, Proc. Amer. Phil. Soc., XX, p. 465; 1885, Tertiary
    Vertebrata, p. 404, pl. xxiii *g*, fig. 12; (*Conoryctes*) 1883, Amer. Nat., XVII, p. 968;
    1885, Tertiary Vertebrata, pl. xxix *d*; (*Ectoconus*) 1884, Amer. Nat., XVIII, p. 796,
    figs. 4–5; 1888, Trans. Amer. Phil. Soc., XVI, N.S., p. 355; Osborn and Earle, 1895,
    Bull. Amer. Nat. Hist., VII, p. 56.

Type: A. M. Cope Coll. No. 3798, part of right ramus of lower jaw with m_2 preserved.

Horizon and Locality: Puerco formation, San Juan basin, New Mexico.   Collected by
David Baldwin in 1882.

Author's Description: The second true molar presents very peculiar characters, and the mandibular ramus is shallower and thicker than in the two other species of *Periptychus* [*P. carinidens, P. rhabdodon*]. The former has a wide external cingulum which is not present in the other species, and there are only six cusps instead of seven.   These are peculiarly arranged.   The anterior three are much as in *P. rhabdodon*, the anterior being not quite so far internal as the posterior inner, close to it, and as large as the anterior external.   The posterior three, are a posterior inner and a posterior median as in *P. rhabdodon*, and a peculiarly placed posterior external.   This is not opposite the posterior inner, but is anterior to such a position and intermediate between the latter point, and the one occupied by the median tubercle in *P. rhabdodon*. It is as large as the anterior external tubercle.   All these tubercles are conical, and not connected by angles or ridges.   The posterior external cusp leaves the cingulum wide posteriorly, and its edge develops some small tubercles.   There are also some small tubercles at other points on the edge of the crown, but no other cingula.   The enamel is not regularly ridged as in *P. rhabdodon*, but has a rather coarse obsolete wrinkling.

This description is repeated in Tertiary Vertebrata, p. 404, but in the plate description of pl. xxix *d* of the same volume, the species is referred to *Conoryctes* in the following terms:

These specimens were received too late for insertion in the body of the work.   They indicate that the species must be referred to the genus *Conoryctes* Cope, and render it very probable that the genus belongs to the family of the Periptychidae.   The absence of ungual phalanges prevents absolute certainty.   The genus is near *Periptychus*, but differs in the one root and simple conic crown of the second true molar in both jaws, and the presence of cingular cusps of the superior molars, exterior to the external tubercles. *Conoryctes ditrigonus* has the molars of both jaws larger than those of the *C. comma*, and there is less difference in size between the posterior and anterior true molars than in that species.

The foregoing citation was copied almost verbatim from the American Naturalist for September, 1883 (published August 15).

In the American Naturalist for August, 1884, p. 796, Cope established the new genus *Ectoconus* for *P. ditrigonus* and gave an additional description and figures of the upper and lower jaws and parts of the skeleton:

Its upper lip, and probably muzzle, are prominent, since the premaxillary bone is produced, and the small conic incisor teeth are widely spaced. . . . The limbs were robust, and had the general character of those of *Periptychus*.   Thus the astragalus is flat, and the humeral condyles are wide and resemble those of a carnivorous animal.

In 1888 (p. 356) Cope gave a more detailed description of the skeleton parts, especially of the tarsal bones, concluding that "the foot was evidently entirely plantigrade and pentadactyle in this genus."   He recognized the distinct internal facet of the head of the astragalus as indicating "a large tibiale or 'internal navicular,' a bone well known in Rodentia, and in *Bathmodon* among the Coryphodontidæ."

In 1895 Osborn and Earle gave the following specific diagnosis: "Superior and inferior true molars with a strong external cingulum. Last superior molar nearly as large as first. Postero-external cingular cusp of superior true molars opposite metacone." They further described the upper and lower teeth in detail, comparing them throughout with *Periptychus*, as Cope had done.

Ectoconus majusculus, new species

Type: A. M. No. 16500, skull and nearly complete skeleton.

Horizon and Locality: Puerco formation, lower level, Kimbetoh Arroyo, San Juan basin, New Mexico. Found by Walter Granger, Amer. Mus. Exped., 1913.

Diagnosis: Somewhat larger than *E. ditrigonus*, teeth more robust, third molar relatively broad and massive.

This species is closely allied to *ditrigonus* but the distinctions run fairly constant in a large series of specimens. Of the specimens belonging to *Ectoconus*, the majority may be referred to *E. ditrigonus*; the inequality in numbers makes it unlikely that the difference is sexual, and as the two species occur together it is probably not mutational.

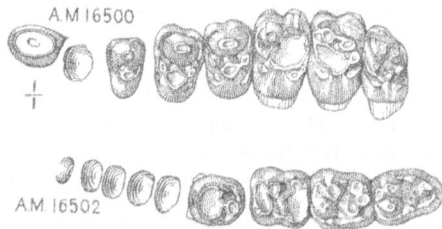

Fig. 27. *Ectoconus majusculus*, new species, type upper dentition and paratype lower dentition, A. M. Nos. 16500 and 16502, crown views. Natural size.

Description of *Ectoconus*

TEETH. The teeth of No. 16500 are considerably worn, so that the pattern of the molars is partly obscured; but all the upper teeth are preserved and most of them perfect.

The incisors are of moderate size, spaced, sharp-pointed and caniniform in type; they increase in size from the first to the third, which is but little smaller than the canine. The crowns of i^1 and i^2 are broken. I^3 is perfect and is a sharp-pointed laniary tooth, round in cross section and moderately recurved, the length of the crown nearly two and a half times the diameter of the base. The tooth is worn on the posterior face by attrition of the lower canine.

The canine is of the usual periptychid type, nearly straight, with round-oval root and crown projecting almost vertically downward.

The premolars are smaller than in any other of the Periptychidae. The combined length of p^{1-4}, less the spacing between them, equals the combined length of m^{1-3}; their width is considerably less. P^1 is a single-rooted tooth, of rather large size, with short robust crown, round at the base and hardly higher than the diameter of its base. The crown is slightly ridged anteriorly and posteriorly. The remaining premolars are of larger size, with two outer roots close together, stout, widely separated inner root and a large, low and rounded inner and outer cusp of nearly equal height. They increase a little in size and complexity from p^2 to p^4.

P² has no accessory cusps; the inner cusp (protocone) is somewhat smaller than the outer cusp (paracone) and from its apex obscure cingula extend downward and outward towards the external angles of the crown. In some other individuals the protocone is much smaller and the cingula are much stronger and extend around the entire outer base of the crown as well.

P³ is larger and of more quadrate outlines than p², the protocone fully equalling the paracone in height and bulk, and on its lateral wings are developed a strong pair of cuspules nearly in the position of conules, and a rudimentary pair external and lateral to them. The wings extend as cingula entirely around the external base of the crown, and short but well-marked basal cingula are also developed on the antero-internal and postero-internal flanks of the protocone.

P⁴ is almost exactly like p³ save for slightly larger size, more prominent cuspules and cingula and more fully quadrate outlines.

The molar pattern is peculiar, owing to the strong development of marginal cusps and conules. The protocone, paracone and metacone form the central triangle with the conules on the wings of the protocone prominent. The parastyle is also large and prominent and there are two small cusps in the position of mesostyle and metastyle. Between the two last is a large prominent cusp (ectostyle) directly external to the metacone, and there is also a pair of internal cusps, postero-internal and antero-internal to the protocone, which may be regarded as hypocone and protostyle, but are not in the normal position for these cusps.

M¹ and m² are of approximately equal size and very similar in pattern. M³ is decidedly smaller, lacks the ectostyle and has only a cingulum to represent the other postero-external stylar cusps.

All the upper cheek teeth from p² to m² are very nearly symmetrical. The premolars are not easy to distinguish as rights and lefts, owing to this symmetry.

Unworn premolars of *Ectoconus* show more or less distinctly the vertical ridges on the outer face which are so prominent a feature of *Periptychus* and more obscurely present in all Periptychidae. In the skull here described the ridges are obliterated by wear.

Variations in Upper Dentition. The range of variation is rather wide as to size, relative width of the teeth, development of the cingula and cingular cusps. In the skeleton specimen, No. 16500, the molars are unusually broad at the base, the premolars broader and more quadrate, the ectostyle and other external stylar cusps less prominent than usual.

In No. 16501, upper and lower jaw, p⁴–m³, p₂–m₂, the basal width of the teeth is less, the stylar cusps are more prominent throughout, so that the grinding surface of the crown is fully as wide as in No. 16500.

In No. 16504, p⁴–m², the transverse diameter of the base of p⁴ and m¹ is slightly less, of m² slightly greater, the inner cingula of p⁴ a little less prominent.

In No. 16505, upper and lower teeth, the upper molars are all from 10 to 15 per cent smaller.

In No. 16499, front of skull, the teeth are materially smaller, more spaced apart, so that the length of the upper cheek-teeth row is slightly greater; m³ is larger relatively, so that it equals m¹ and m² in size, p⁴ lacks the internal cingula and its conules are weaker so that it has a more trigonal outline, and p³ and p², although badly preserved, appear to have the protocones less developed.

In No. 16496, upper and lower jaws and many fragments of skeleton, the upper cheek teeth are somewhat more robust than in No. 16500, the premolars a little larger relatively. The transverse diameters of the teeth are nearly the same, but their anteroposterior diameters average 12 per cent greater.

All the above are from the lower level of the true Puerco.

No. 16490, maxillae, p^2–m^3 l., p^4–m^3 r., from the upper level of the true Puerco, has the teeth, especially the premolars, of less transverse diameter (p^3 is only 4/5 as wide) and the internal cingula on the premolars are lacking. In another specimen from the same level—a poorly preserved palate—they are apparently well-developed.

In No. 888, upper and lower jaws, etc., collected 1896, probably lower level, the upper cheek teeth are intermediate between No. 16496 and 16500.

In No. 882, upper jaws, collected 1896, also probably lower level, the upper molars are fully as robust as in No. 16496, but premolars are smaller, with the lingual cusps and cingula more reduced than in No. 16500; upper and lower jaws mostly buried in hard matrix, Cope Collection, exact level unknown; the teeth are nearly like those of No. 16499, except that the posterior cusps of m^3 are less reduced, and the ectostyle of m^2 much larger, the cingula on lingual side of p^4 less prominent but longer. The upper incisors are preserved on this specimen, similar in size and character to those of No. 16500.

Comparison of Upper Teeth with Other Periptychidae. *Periptychus* differs from *Ectoconus* in the larger size and different proportions of the premolars, the smaller molars, less extended transversely and wholly lacking the outer stylar cusps, and the prominence of the vertical grooving, especially upon the premolars.

In *Periptychus* the protocone and conules of the upper premolars are united into a wide crescentic crest which embraces the inner side of the paracone, and the paracone is compressed laterally and much higher, with somewhat of an inward and backward pitch. The backward pitch is greater in the Puerco species, but the transverse diameter of the tooth is greater and its inner cusps are less fully converted into a crescent. P^2 of *Periptychus* has the roots connate.

The molars of *Periptychus* have the same cusp composition as in *Ectoconus*, save for absence of external styles, but their transverse diameter is less, especially in the Torrejon species, in which the inner stylar cusps have moved outward so as to be nearly in line with the protocone.

The canines and incisors are much shorter and more blunt-pointed than in *Ectoconus*.

In *Hemithlaeus* the teeth are higher-cusped and less robust. The inner cusps of the premolars are less developed, the protocone of p^2 being practically absent and that of p^3 small. The conules are wholly absent on the premolars and relatively small on the molars. The external stylar cusps are very slightly developed, rudiments only of the parastyle and ectostyle being present. The molars are not so broad internally, their outlines less quadrate. The vertical grooving of the teeth is practically absent.

In *Conacodon* the premolars are still higher and their inner cusps (protocone) less developed. There is no inner cusp at all on p^{2-3} although p^3, like p^2 of *Hemithlaeus*, is triangular, with three roots. The inner cusp of p^4 is distinct, but smaller than the paracone. There are no traces of conules on the premolars or molars. The molars are triangular, very wide transversely, with a prominent crescentic internal stylar cusp nearly median in position. From its relations this is probably the same cusp as the "hypocone" of *Ectoconus*, but its position and character are very different. There are no external stylar cusps on the molars.

In *Anisonchus gillianus* (Puerco) the teeth are much as in *Hemithlaeus*, save that the "hypocone" is more prominent and the protostyle reduced to a cingulum.   The relations of this species to its successor in the Torrejon, *A. sectorius*, are almost exactly the same as of the Puerco and Torrejon species of *Periptychus;* in each case the Puerco species is nearer in every respect to *Ectoconus* than is its Torrejon successor.

In *Haploconus* p² and p³ have lost their internal cusps; the inner root as well is absent on p² and nearly absent on p³, while p⁴, like p³⁻⁴ of *Anisonchus*, has assumed much the same form as in *Periptychus*.   The upper molars are constructed as in *Anisonchus*, save that the position of the "hypocone" is more nearly internal (as in *Conacodon*) and the outline of the tooth is in consequence more triangular, less quadrate.   The vertical grooving of the cusps is more noticeable in the Torrejon *Anisonchus* and in *Haploconus* than in any of the other genera except *Periptychus*.

With all the clear-cut and marked generic distinctions between the different periptychid genera, the family likeness is apparent.   The tendency towards centralizing of the primary cusps and development of styloid cusps from the cingula; the correspondence of the lingual styloid and development of one or the other or both; the simplicity of the outer cusps and large size of the inner cusps of the premolars; the vertical grooving of both premolar and molar cusps more or less apparent in all the genera, but prominent only in *Periptychus*— these are the most obvious common features.

Comparison of Upper Teeth with Phenacodontidae.   The most obvious comparison is with *Phenacodus*, which has a marked superficial resemblance to *Ectoconus*.   In the upper premolars of *Ectoconus*, as in all other Periptychidae, there is no trace of metacone (trito-cone) or tendency to form one.   In *Phenacodus* the metacone is prominent on p⁴, rudimentary on p³, and a tendency to form it is observed even on p².   In the molars of *Phena-codus* there is a prominent true hypocone; the "hypocone" and protostyle of the Periptychidae and the cingula that represent them on the premolars are distinct cusps of different relations to the protocone, and no trace of them is seen in the Phenacodontidae.   Exter-nally the molars of *Phenacodus* have a normal mesostyle and a rudimentary parastyle, but no ectostyle or metastyle.   The conules are well-developed in both genera.   The surface of the teeth in the phenacodonts is rugulose, almost with a granular ornamentation, quite different from the vertical striation of the periptychids.

Tetraclaenodon is not any nearer to *Ectoconus*, save that the metacones of the premolars are less developed.   In other respects it shows the same differences as does *Phenacodus*, to which it is nearly related.

The larger species of *Phenacodus* tend to lower and more robust crowns and to the addition of supplementary cingular cusps to the wearing surface, but these cusps have no correspondence to the cingular cusps of *Ectoconus*, and the character is not a generic one and indicates only certain parallelism in adaptation.

SKULL.   The skull is of nearly the same size as in *Periptychus rhabdodon* and *Phena-codus primaevus*.   It is of moderate width, with heavy and rather short muzzle, deep zygomata, high sagittal and occipital crests and very small brain-case.   The animal was a mature individual, with well-worn teeth, and the cranial sutures are in consequence mostly indistinct.

The nasals are long, moderately wide, somewhat expanded anteriorly and considerably so posteriorly.   They extend backward a little beyond the line of the anterior margin of

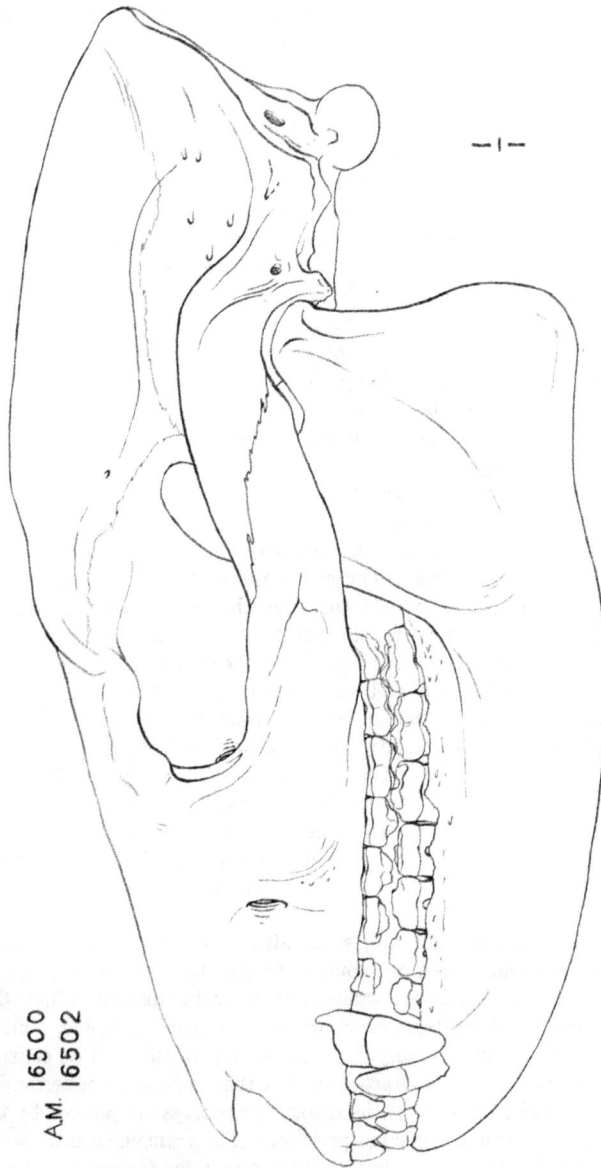

FIG. 28. *Ectoconus majusculus*, new species, type skull and paratype mandible, A. M. Nos. 16500 and 16502, side view, crushing corrected. Natural size.

the orbits and are separated from the frontals by a digitate transverse suture. Their lateral borders are overlapped by maxilla and premaxilla nearly to the anterior end.

The frontals are quite short, wide anteriorly between the orbits, narrowing posteriorly to a point at the postorbital constriction, where they do not quite reach back at the median line to the sagittal crest, and are overlapped laterally by the parietals. The postorbital process is represented by an obscure boss, from which the postorbitals sweep backward in the usual curve until they meet in the sagittal crest. The suture between parietal and frontal lies along the posterior portion of the postorbital crests and thence goes irregularly downward at about the narrowest portion of the postorbital constriction, but its lower course is obscure. Anteriorly the frontals articulate with nasals, maxillae and lachrymals upon the upper surface of the skull, but the inferior sutures cannot be clearly distinguished.

The parietals are long, rather narrow, the major part consisting of the high, thin sagittal crest. The parieto-frontal suture is obscure and recognizable only in its upper part; the parietals project forward in a broad tongue on each side of the postorbital crests, considerably in advance of the sagittal crest. The parieto-squamosal suture at the lower margin is marked by a somewhat obscure ridge, and above it the upper surface is notably concave over the brain-case, owing to the extremely small capacity of the latter.

The squamosals are exceptionally long and wide, extending forward far in advance of the zygomatic process, and even further backward, lapping up on the anterior flank of the occipital crests nearly to their summit.

Skeleton. The skeleton of *Ectoconus* compares in size and in most of its general proportions with that of *Orycteropus*. The tail is not so heavy and the feet are shorter, wider, less specialized and diversely adapted, but the proportions of neck, back, ribs, girdles and limbs are not very different. The skull is shorter and different in proportions, and the teeth are wholly diverse, but aside from the shape of the head the general appearance and proportions of the animal must have been more like those of *Orycteropus* than any other modern animal, only with a much smaller tail and small, flattened, cony-like hoofs. This general resemblance does not involve any near relationship, but indicates merely that the modern *Orycteropus* has retained with little alteration much of the proportions and structure that were common among primitive placentals of similar size. It is convenient, therefore, in describing the skeleton of *Ectoconus*, to make comparisons with that of *Orycteropus*.

Vertebrae. The vertebral series is preserved almost complete down to the proximal caudals, but considerably distorted by crushing. The ribs are also nearly complete but considerably crushed.

The cervical vertebrae are short and wide, although not so short as in *Periptychus*. The transverse processes and arches are comparatively short and weak, the centra large and neural canals spacious. A curious anomaly is noted in this individual: the vertebral artery pierces the roots of all the transverse processes on the right side, while on the left side the seventh cervical is imperforate. The significance of this will be considered later.

The atlas has the proportions of *Orycteropus* but the cotyli are smaller and not so wide apart. The neural arch and the base of the transverse process are pierced by the foramina for the vertebral artery in about the same position. The transverse process is somewhat longer but less massive; the facets for the axis are not quite so far apart and are continued inferiorly toward the median line so as to be rather pyriform than round.

The centra of the remaining cervical vertebrae are not so wide relatively as in *Oryc-*

teropus and are much more ridged inferiorly.   The axis is of nearly the same general proportions except for the high broad hatchet-shaped neural spine in contrast to the nearly round and backwardly projecting spine of *Orycteropus*.   The transverse processes have about the same size and shape, the centrum is not so wide, the atlanteal facets not so wide apart and nearly continuous with the more cylindrical inferior surface of the odontoid process.   The post-zygapophyses are smaller and not so wide apart.

The third, fourth and fifth cervicals are much alike, all with centra wider than long, ridged medially, and the facets between them very oblique, somewhat flattened, oval in outline, the transverse processes short and small, pierced at base for the vertebral artery, the superior lamella directed outward and backward, the inferior forward and somewhat downward, and neither lamella much expanded.

The sixth cervical differs on the two sides.   On the right side it is similar to the third, fourth and fifth in the character of the transverse processes, the inferior lamella is expanded into a broad plate extending backward into a spine, after the usual type of creodonts and other placentals.   The base of the transverse process is perforated on both sides by the vertebrarterial foramen.   The seventh cervical has on the right side a transverse process perforate at the base for the vertebral artery, and with superior and inferior lamellae well developed, the latter expanded into a broad plate which matches the plate of the sixth vertebra on the left side.   The left side of the seventh vertebra is imperforate and has no inferior lamella.   There is no rib-facet on either side of the posterior face of the centrum. Comparison with *Orycteropus* shows that the perforation of the transverse process of the seventh cervical by the vertebral artery in *Ectoconus* is not a normal character as it is in *Orycteropus*, many marsupials and some rodents, but is an abnormality associated with the shifting to the seventh vertebra on one side only of the peculiar transverse process characteristic of the sixth vertebra.   In *Orycteropus*, etc., the sixth vertebra has the normal type of inferior lamella, and the seventh is also normal in the absence of inferior lamella but is nevertheless perforate.   The failure of the head of the first rib to reach the centrum of the seventh cervical vertebra in *Ectoconus* is doubtless also associated with the abnormality of the skeleton here described; there is no reason to suppose it a normal character.

The neural spines of the cervicals are not perfectly preserved, but are evidently quite rudimentary, less developed than in *Orycteropus*.   The zygapophyses are similar to that genus, rather large, with nearly flat facets, the process not extending beyond the facet.

The first dorsal vertebra has a wide short centrum, nearly flat inferiorly, both anterior and posterior zygapophyses are wide apart, and the spine is quite high, thin, and rather wide anteroposteriorly; the transverse process is moderately stout and long, the facet for the tuberculum large, that for the capitulum small and situated on the side of the centrum, wholly on this vertebra.   The second dorsal is of similar type, but with the post-zygapophyseal facets nearer together; the spine a little higher and more trigonal, the centrum not so wide and more convex inferiorly.   The remaining anterior dorsals are of the usual type, with semi-cylindrical centra, moderately high trigonal spines directed backward, transverse processes of moderate weight, longer than in *Orycteropus* but not so stout, and projecting a little backward instead of directly outward.   The centra do not have the wide and flat form of *Orycteropus;* the spines are of about the same height and form, with somewhat less backward pitch.

On the ninth dorsal the spine has somewhat decreased in size but maintains its backward pitch; the tubercular facets on the transverse processes have moved inward and the process projects beyond them.   The tenth dorsal has a smaller and upright spine, thin and flat and no longer triangular in cross-section.   The arches of the eleventh and twelfth dorsal are lost; their centra are still small, semicircular, and medianly ridged beneath.

The series is broken behind this, and the number of dorsals is uncertain.   Two posterior dorsals are preserved, probably the last two.   They have larger and wider centra which, however, maintain a strong median cresting on the under surface.   The zygapophyses are quite of lumbar type, strongly convex behind and deeply concave in front, but not revolute. The anapophyses are large, stout, directed outward and backward; the neural spines are broken off, but appear to have been similar to those of *Orycteropus*, the most important differences being the greater convexity of the zygapophyses and the much less flattened centra.

Four lumbar vertebrae are preserved, which may have been in series with the two posterior dorsals; if so, four was the number of lumbar vertebrae.   The first has a well developed, the second a rudimentary, anapophysis; the transverse processes are broad and flat, directed outward and a little forward, increasing progressively in size towards the sacrum.   The anterior zygapophyses project but little beyond the facets, which are strongly convex and concave, much more than in *Orycteropus*, but not uniformly cylindrical, the lower part of the facet being flattened.   The spines are not preserved, but seem to have been stouter than in *Orycteropus;* the centra are progressively wider and flatter toward the sacrum.

The sacrum consists of four vertebrae well coössified, with the transverse processes completely grown together into a broad continuous plate on each side, with three small intervertebral foramina, the first and second vertebrae attached by their consolidated transverse processes to the ilium.   The centrum of the first is broad and flat, the length increasing to the third centrum, while the width decreases to the fourth.   The posterior end of the fourth is little more than half the width of the anterior end of the first centrum, quite unlike the sacrum of *Orycteropus*, which is almost as large behind as before, and consists of six vertebrae.   The neural spines are broken, but appear to be imperfectly united or separate, not consolidated into a single crest as in *Orycteropus*, and considerably more trace remains of the zygapophyses than in the modern genus.

Three proximal caudals are preserved, all of them with short centra (much shorter than that of the last sacral), wide transverse processes, short compressed neural spines, zygapophyses stout and curving forward with rather large and nearly flat facets comparatively wide apart; in general not unlike the caudals of *Orycteropus* save for smaller size, probably shorter transverse processes and the spines wider anteroposteriorly.

Ribs.   The ribs were not less than fourteen in number, but there may have been more. They are of moderate length, not strongly curved, the tubercle comparatively weak even upon the anterior ribs, but distinct except upon the last four in the series preserved.   In width and curvature and general form, they compare rather closely with those of *Orycteropus.*

Shoulder Girdle.   The shoulder girdle consists of scapula and clavicle, the coracoid reduced to a short process, as generally in eutherian mammals.   The scapula is of the size of that of *Orycteropus* but differs notably in form; the spine is continued distally to the border, which projects much in the same manner as in the Amblypoda and Proboscidea. The coracoid process is prominent as in *Orycteropus*, but more massive and more strongly recurved. The prespinous and postspinous portions are somewhat more widely expanded, especially towards the glenoid portion, so that there is even less of a "neck" than in *Orycteropus*. The acromion is broken off so that its character cannot be stated. The glenoid facet has the usual shape, somewhat wider than in the modern genus. The clavicle is a little shorter than in *Orycteropus*, equally heavy at the proximal end, but maintains its width and thickness throughout, and is even slightly expanded at the distal end. It agrees better with the clavicle of *Erinaceus*, but is stouter and more strongly curved. This probably represents well the primitive eutherian condition; the form of the scapula, on the other hand, is very suggestive of relationship to *Coryphodon* and the "subungulate" group.

Fore Limb Bones. The humerus is somewhat shorter than in *Orycteropus*, with broader head, more prominent deltoid and supinator crests. The entepicondyle projects further but is less massive; anteriorly the radial portion of the distal facet is less condylar, the ulnar portion more trochlear. It has much the same size and proportions as in *Claenodon*, but somewhat more robust, the supinator crest and entepicondyle a little heavier.

The ulna and radius show the same general proportions as in *Orycteropus*, but have straighter shafts and less expansion at the distal end, the olecranon is longer, straighter and more flattened, the head of the radius is considerably flattened, much as in *Claenodon*, with which these bones agree more nearly than with *Orycteropus*. The shaft of the ulna has a broad, shallow longitudinal groove, less than in *Dissacus*, more than in *Claenodon*. The broad, flat shaft of the ulna is characteristic of many primitive mammals; the considerable and slightly convex ulnar facet at the head of the radius is unlike the condition in *Orycteropus;* the bicipital tubercle is less prominent but extends further down on the shaft, the distal end is only moderately expanded, and the facets for lunar and scaphoid are quite shallow and indistinguishable one from the other. The distal facet for the ulna is gently convex anteroposteriorly, the pisiform portion not clearly distinct from the cuneiform portion.

The right olecranon of the *Ectoconus* skeleton has been very badly damaged during life, apparently as a result of accident and subsequent resorption of bone and false growth which have to a great extent destroyed the elbow joint.

Fore Foot. The fore foot has the general characters of the Taligrada. It is pentadactyl, with alternating carpus, the centrale well developed, five rather short metacarpals, short and peculiarly flattened phalanges, and small narrow, fissured unguals. It resembles throughout the fore foot of *Periptychus* and of *Pantolambda*. In most respects it agrees with the primitive creodont type and much more closely with the earliest Mesonychidae; resemblance to the condylarthran foot is much more distant.

The scaphoid is of considerable width but very small depth. Its upper border is thickest at the (inferior) process, thinning out to nothing before it reaches the external end of the bone. Proximally the surface shows a radial facet convex dorso-ventrally but not laterally, and below and internal to it, separated by a deep groove, is the thick round head of the (inferior) process. The convexity of the radial facet is strong inferiorly, but superiorly it fades out to a plane or even slightly concave surface. On the external side of the scaphoid is a narrow triangular facet, widest in the middle and thinning to nothing above and below, for the lunar bone. It faces somewhat distally as well as internally. On the distal face adjoining it is the large oval facet for centrale, concave dorso-ventrally, and internal to that a somewhat smaller oval facet for the trapezium, also slightly concave dorso-ventrally. The trapezoid just touches the scaphoid close to the dorsal surface between these two large facets.

The lunar also is wide and comparatively shallow, convex proximally and broad-wedge-shaped distally, resting about equally upon the unciform and magnum + centrale. The facet for the radius is of trapezoidal outline, broadest above, the superior and inferior borders convex, the lateral borders straight, strongly convex dorso-ventrally and slightly convex laterally. On the internal side is the narrow facet for the scaphoid facing internally and somewhat proximad; below that the much wider triangular-oval facet for the centrale, facing distal-internad, widest above and extending about three-fifths of the depth from the dorsum; it is moderately concave dorso-ventrally. Below this is the magnum facet, widest below, where it is concave both ways and faces distal-internad, and extending as a narrower strip with convex surface facing more distally at the dorsal surface. The facet for the unciform is broad and deep, widest at the dorsal surface, strongly concave dorso-ventrally and facing distal-internad. On the internal face of the lunar is a moderately large, round-oval facet for the cuneiform, slightly concave dorso-ventrally and limited to the dorsal half of the bone.

The cuneiform is about as large as the lunar, its height about the same, and its width about equalling that of the scaphoid. The ulnar and pisiform facets meet at a low angle, are of about equal area, but the ulnar deeper and the pisiform wider, both slightly concave. The lunar facet is round-oval, slightly convex, faces internally; the facet for the unciform is of larger size, oval, with horizontal diameter wider, slightly concave dorso-ventrally. The inferior process is a short, stout mass of bone, not developing into a hook.

The centrale has the usual diamond shape, with the transverse axis longer than the proximo-distal. Proximally it is about equally divided between the scaphoid and lunar, its facets for both being moderately convex dorso-ventrally. Distally the facet for the trapezoid is in a plane nearly parallel with that for the lunar, and is about twice as wide as that for the magnum which is parallel dorsally to the scaphoid facet, but under-runs the bone and pinches it out to a narrow edge so that its depth is less than that of the trapezoid and decidedly less than that of the magnum.

The trapezium has the usual primitive type found in the earliest creodonts, condylarths, taligrades, etc. It is a comparatively large robust bone, with a wide facet for metacarpal I, strongly concave both ways and facing distal-internad; proximally it presents a broad wedge inserted between the scaphoid and trapezoid; it is separated at the dorsal surface from the centrale by the point of the trapezoid which touches the scaphoid, but it just reaches the centrale towards the palmar surface. The scaphoid facet is oval, slightly convex and faces proximad and somewhat internad; the trapezoid facet is considerably wider than deep, strongly convex in a dorso-palmad direction, and faces more externad than proximad. Below the trapezoid facet, and facing externad and somewhat distad is a rather obscurely faceted, irregular surface for Mc. II.

The trapezoid is comparatively large, its dorsal surface nearly as extensive as that of the lunar, rudely trihedral in form, the distal facet for Mc. II, internal facet for trapezium and proximal-external facet for centrale constituting the three sides. Towards the palmar surface the bone decreases in size, pinched out between the three limiting facets. The trapezoid facet is strongly concave dorso-palmad, the metacarpal facet is nearly twice as wide and is slightly concave dorso-palmad, and the centrale facet is a little broader than the Mc. II facet and nearly flat. On the inner side of the trapezoid is a narrow, deep facet for the magnum, wider towards the palmar surface, while internally the bone just touches the scaphoid along the dorsal surface and has a slight facet for it.

The magnum has a very small dorsal surface, somewhat less than the centrale and much less than the trapezoid, but projects beneath in the usual heavy proximal keel, wedged in between lunar on the external side and centrale on the internal side. These two facets are strongly convex dorso-palmad, face obliquely, proximo-externad and proximo-internad and meet in a sharp, very concave crest. The sides of the magnum meet with vertical flat facets, much deeper than wide, against the trapezoid internally and the unciform externally; distally the facet for Mc. III is of moderate width, deep, quadrate, concave dorso-palmad, and slightly convex from side to side.

The unciform equals cuneiform or lunar in bulk, and exceeds either in the extent of its dorsal surface. It is comparatively broad and low, with wide saddle-shaped surface for the cuneiform facing proximo-externad, and a large facet for the lunar, strongly convex dorso-palmad, and facing proximo-internad. On the inner face of the bone are two flat facets united dorsally, separating palmad, but neither of them reaching the palmar surface. The proximal one for the magnum is narrowed dorsally, the distal one for Mc. III is widest dorsally. On the distal face of the unciform are two large facets, for Mc. IV and Mc. V. That for Mc. IV is trapezoidal, wider dorsad, concave dorso-palmad; the facet for Mc. V is concave both ways, not so wide dorsally as that for Mc. IV, and reduced in its width towards the palmar surface by the encroachment of the proximal facet for the cuneiform.

The pisiform is large, moderately long, with flat, transversely wide facet for the cuneiform facing distally and somewhat dorsad and a shorter oval, rather obscure ulnar facet, concave and facing dorsad.

There are five metacarpals, all well developed, stout, rather short and wide. The fifth is a little shorter and the first is shorter and more slender distally than the others. The distal ends tend to be wide and square-cut, conformad to the ungulate type of phalanges. The keels are not very prominent, even upon the inferior surface, obscure on the distal and do not extend at all upon the dorsal surfaces.

Metacarpal I has a wide convex head of no great depth, for articulation with the trapezium. It is somewhat divergent and does not touch Mc. II. The shaft is short, smaller than in its fellows and of little depth; the distal end, strongly convex from the dorsal to the palmar plane, has hardly any convexity from side to side, and no keel even upon the under side.

Metacarpal II is about one-half longer than Mc. I, and the distal part heavier in about the same proportion. The head presents a rectangular proximal facet, deeper than wide, moderately concave from side to side and convex dorso-palmar for the trapezoid; an irregularly plane internal facet for the trapezoid, a very narrow but deep facet internal and partly proximal, for the magnum, and on the internal face of the head, a larger dorsal and a small palmar facet for Mc. III.

The attachment for the extensor metacarpi (?digitorum) forms a prominent, obliquely transverse crest near the head of the bone, with a marked fossa proximal to it. The distal part of the metacarpal is like that of Mc. I, broad and rather flattened, the phalangeal facet wide and square-cut, with a shallow and rather obscure keel on the under surface. The keel runs somewhat obliquely.

Metacarpal III is the longest of the series, about one-eighth longer than Mc. II. It has a rectangular facet for the magnum, somewhat narrower and deeper than the proximal facet of Mc. II, more convex dorso-palmad and less concave transversely, and facing obliquely, proximo-internad. A narrower triangular facet for the unciform faces proximo-externad, is slightly convex either way, widest dorsally and narrows to a thin edge at the palmar surfaces. The internal facets for Mc. II correspond with their mates on that bone; on the external face distal to the unciform facet is a rather narrow, deep facet for Mc. IV, somewhat concave both ways and curving around at the palmar face and extending distally for a short distance. The attachments for the extensor tendons form two rather prominent rugosities near the proximal end on the dorsal surface. The distal part of the bone is very similar to that of the second metacarpal, but is nearly symmetrical.

Metacarpal IV equals Mc. II in length and has a somewhat narrower shaft but is deeper in the shaft and distal end. The facet for the unciform faces proximad and is convex dorso-palmad, faintly concave transversely toward the palmar surface, widest at the dorsal and narrowing towards the palmar border. The facet for Mc. III is internal and directed slightly proximad, a little convex both ways and conformant to its mate on that bone. The facet for Mc. V is much deeper than wide, moderately concave dorso-palmad, faces externally and somewhat palmad. The attachments for the extensor tendons are less prominent than on Mc. II and III, and the distal end of the bone is considerably less flattened. It shows a low keel and is more asymmetric than Mc. III but less than Mc. II.

Metacarpal V is intermediate in length and in size of distal end between Mc. I and Mc. II or IV. It is strongly asymmetric, corresponding to the first metacarpal. The unciform facet faces proximo-dorsad and is ovoid, narrower above, moderately convex both ways. The facet for Mc. IV faces inward and partly dorsad, is somewhat convex dorso-palmad and smaller than the unciform facet. The process for the ulnar flexor is short and stout, moderately prominent. The shaft is short and considerably flattened, the distal end wide with facet for the phalanx, more convex transversely than in the other metacarpals, and obscure keel underneath.

The phalanges are of thoroughly ungulate type, short, wide and flat. The proximal series are about one-half longer than their width, the proximal facets very moderately concave or almost flat transversely; the distal facets convex dorso-palmad, slightly concave transversely, facing distad and palmad but not at all reflected over the dorsal surface and about twice as wide as their depth. The distal ends are no wider than the shafts. The second phalanges are only one-third to one-fourth longer than wide, much flattened, the proximal facets moderately concave dorso-palmad, twice as wide as their depth, the ungual facets reflected over dorsal and palmar surfaces but with very little transverse concavity, the shafts wide and flat.

The ungual phalanges are comparatively small and narrow hoofs. The proximal portion is heavy, the distal portion rather short and weak, flat beneath, convex above, the lateral margins sharp, a little expanded, the tip somewhat fissured but not deeply. The proximal facet is moderately concave dorso-palmad, slightly convex laterally; the subungual processes are heavy but not prominent. The hoof is much like that of *Periptychus* but less flattened; it is more expanded than in *Euprotogonia* but much less than in *Phenacodus;* it is not nearly so short and wide as in *Pantolambda.*

Pelvis and Hind Limb Bones. The pelvis is mostly missing and its character only imperfectly seen in other specimens. It apparently had much the same characters as in *Periptychus* and *Pantolambda*. The femur is considerably longer and more slender than in *Orycteropus*, the head smaller, much more convex and set on a distinct neck, the great trochanter less massive, the notch separating it from the head much more distinct; the digital fossa has less depth but extends farther down on the shaft. The lesser trochanter in *Orycteropus* is separated into two processes quite wide apart, but in *Ectoconus* it is single, elongated, and intermediate in position on the postero-internal border of the shaft near the head. The third trochanter is almost as prominent as in *Orycteropus* but situated higher on the shaft. The distal condyles are not so broad as in *Orycteropus* but are more convex; the patellar trochlea is narrower, shallower and somewhat longer.

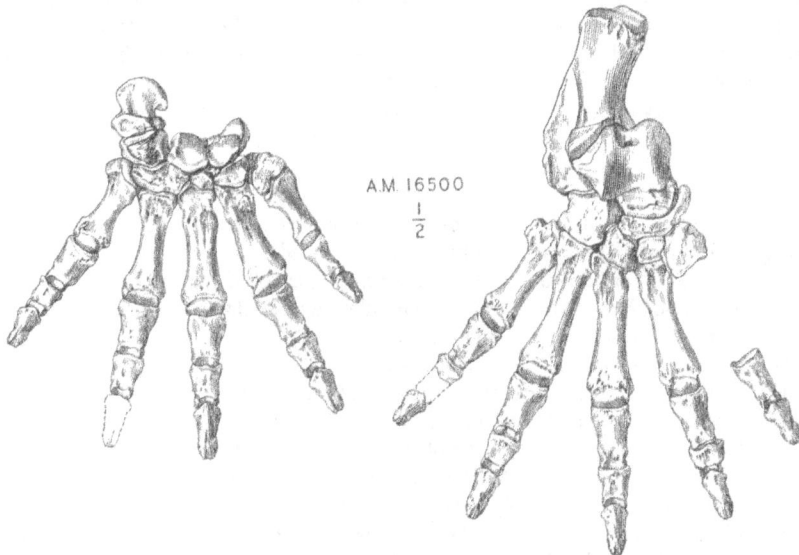

A.M. 16500
$\frac{1}{2}$

Fig. 29. *Ectoconus majusculus*, new species, right manus and pes of type, A. M. No. 16500, anterior views. One-half natural size.

The tibia and fibula are of quite primitive character, entirely separate, the fibula well developed, the tibia with simple oblique astragalar facet entirely lacking a trochlea, the astragalar head being wedged up between tibia and fibula in the same manner as in Arctocyonidae. The femoral facets of the tibia are both nearly flat, but slightly concave, and face proximad. The cnemial crest is very high and long, decidedly more so than in *Orycteropus*, extending down on the shaft two-thirds of its length. The postero-internal crest opposite it is very prominent, as in *Orycteropus*. The facet for proximal end of fibula lies directly under the outer femoral facet and faces principally distad, to a much less degree externad. The fibula has a straight shaft, considerably expanded at both ends, with a thin but wide antero-internal crest along its proximal half, and a prominent antero-external crest along the distal half of the shaft; distally it has an oblique facet for the astragalus, more than half as large as the astragalar facet of the tibia and almost as oblique, and a smaller nearly flat distally-facing calcanear facet. Apparently there is no distal facet between tibia and fibula. The flat oblique tibial and fibular facets for the astragalus, quite devoid of trochlea, are entirely unlike *Orycteropus*, but agree nearly with those of *Claenodon* and other primitive creodonts, as also with *Periptychus* and *Pantolambda*.

Hind Foot. The right hind foot in No. 16500 is preserved complete and parts of the left pes. Various other specimens include characteristic bones of the pes. The astragalus and calcaneum were described by Cope in 1888, but owing to the incompleteness of Cope's material he somewhat misunderstood the real construction and wrongly oriented the bones.

The astragalus has a heavy subquadrate body, short neck and wide, flattened head. The tibial facet faces obliquely, but more proximad than internad; the fibular facet is also oblique but more externad than proximad; the internal malleolar facet is nearly vertical. The tibial facet is wider than long, slightly concave transversely, moderately convex antero-posteriorly, limited behind by the astragalar foramen but extending a little beyond it on the inner side, and anteriorly extends down to and slightly over the neck. The fibular facet is large, slightly convex antero-posteriorly, widest anteriorly and ending abruptly on a line with the straight anterior margin of the tibial facet, curving backwards and ending at the back of the astragalo-calcanear facet. The angle between tibial and fibular facets is anteriorly somewhat less than ninety degrees; the crest between them is sharp at this end and gradually fades out towards the back. The neck is short but distinct, the head wider than deep, with three distinct facets, convex dorso-plantad, nearly flat in the other diameter. The external facet is for the cuboid, both the others for the navicular. The cuboid facet faces more externad than distad, and is triangular, widest dorsad. The distal navicular facet is as wide as the others put together, and is about as wide as deep. The internal navicular facet faces chiefly internad, separated by a distinct angle from the plantad portion of the distal facet. Cope supposed this facet to be for a separate "tibiale," but the navicular, which appears to be a compound bone, covers it. There is, however, upon the proximal margin of the internal navicular facet an obscurely faceted convex border, which may have carried a bone corresponding to the "pre-hallux" of *Claenodon*. The astragalo-calcanear facet on the astragalus is moderately concave, parallel to the tibial facet, and is about two-thirds as wide as long; the sustentacular facet is equally large, oval, most of it flat and horizontal (from side to side), but the external border bent upwards; distally the sustentacular is almost continuous with the cuboid facet, a little separated from the navicular facets. The groove for the interosseous ligament is deep, the two facets almost closing over it; it is continued backward to the astragalar foramen.

The calcaneum is massive, the tuber expanding towards its end so as to be as wide as deep, and supported on the external side by a wide shelf-like crest that extends along the outer border from the peroneal tubercle nearly to the end of the tuber. On the plantar surface of the calcaneum a rather prominent process projects plantad and somewhat proximad from the base of the shelf on the superior surface of which lies the sustentacular facet; another rather heavy process facing distad lies below the cuboid facet, separated from it by a deep groove. On the upper surface the astragalo-calcanear facet is set obliquely, facing as much distad and internad as upward, and is moderately convex from side to side; the sustentacular facet is rounded, facing partly distad, and with a narrow extension that reaches to the cuboid facet. The cuboid facet is wide from side to side, somewhat concave, overhanging dorsally so that it faces a little plantad and considerably entad.

The cuboid is comparatively wide and massive, with astragalar facet about two-fifths as wide as the calcanear facet, but facing more internad than proximad. A narrow navicular facet and a broader ectocuneiform facet lie distal to it, their form and relations comparing with those in *Claenodon* (and *Ursus, fide* Gidley) except for the partly proximad direction of the astragalar facet.

The navicular is broad and rather shallow, with a large and prominent hook extending proximad and internad. On the astragalar facet the portions for the hook and for the body of the navicular are separated by an obscure angulation and somewhat notched at the margins; distally the facets for the three cuneiforms are all on the body of the navicular, the hook having on its distal-internal surface what appears to be an obscure convex facet separated by a notch from the entocuneiform facet on the body. These conditions are of interest because in *Pantolambda* the navicular hook is a separate distinct bone. In *Ectoconus* the traces of its consolidation appear to be unmistakable. In *Claenodon* the hook is further consolidated and reduced, and in this genus we have preserved another inner tarsal bone which is indicated in *Ectoconus* by an obscure facet, but is not preserved (perhaps not ossified). The conditions in *Periptychus* appear to be much like those of *Ectoconus*, except that the hook of the navicular is somewhat more reduced and consolidated. It would appear therefore that two extra bones may be present on the inner side of the tarsus in these Paleocene mammals, more or less in the position of the so-called "pre-hallux" and "tibiale" found in various modern mammalia. One of them is the hook of the navicular; the other may be one of the "pre-hallux" bones, if indeed the bones so identified in various modern mammals are really homologous and not accidental analogies. So far as would appear from what is known of Permian theromorph reptiles, neither is a survival of primitive reptilian structures. In theromorph tarsi we find a bone that corresponds in form and in relations to the astragalus on one side and to the three cuneiforms on the other side, to the *navicular minus the hook* of Paleocene mammals—that is to say, to the navicular proper of *Pantolambda;* and there are no known tarsals internal to these. There is therefore nothing to support the idea of a sixth digit or pre-hallux. Whether the navicular of the theromorph reptiles is the tibiale, as believed by Broom, is a question to be decided on the evidence of primitive reptilia; it appears probable in any event that it is the navicular of ordinary mammals *minus* the hook, which may be regarded as a subsequent addition.

The three cuneiforms are, as already stated, articulated to the body of the navicular. The facets are of subequal size, the ectocuneiform widest, the mesocuneiform deepest, the entocuneiform moderately convex and facing almost as much entad as distad. The ectocuneiform is the largest of the three cuneiform bones, has a subquadrate body with a stout inferior process and the usual facets for adjacent bones, the second metatarsal having a considerable overlap along the inner border. The mesocuneiform is also quadrate, about two-thirds as large in either diameter, and has no inferior process. The entocuneiform is compressed and deep, longer proximo-distad than the ectocuneiform but rather less than the mesocuneiform in width, the facet for the mesocuneiform distinct but narrow, that for Mt. II twice as wide proximo-distad but rather obscure; the facet for Mt. I a narrow oval, two and a half times as deep as wide.

The metatarsals are complete except the first, which is missing from both feet. It was evidently considerably smaller than the other four, which are of subequal size, the fifth a little shorter than the others. The second has a considerable overlap on Mt. III, and on the internal side is heavily lapped by the entocuneiform but with a very obscurely marked facet.   The metatarsals all have somewhat flattened oval shafts and the ends broadened and "squared," being converted to a considerable degree into hinge-joints; the keels however are weak and limited to the plantar side.   The metatarsals are decidedly shorter and stouter than in *Claenodon*, the shafts more flattened, the ends more squared, the inferior keels much weaker; the first metatarsal, judging from the trapezium, digit II and size of its phalanges, was smaller and less divergent.   The metatarsals are somewhat stouter and relatively a little shorter than in *Periptychus*, and are rather longer than in *Pantolambda*, the fifth somewhat less reduced, the first apparently more so.

The phalanges are short, wide and flattened, with squared ends allowing but limited flexure, especially between the first and second row.   The unguals are flattened into narrow hoofs much as in *Periptychus* and *Tetraclaenodon*, not so wide as in *Dissacus*, not nearly so wide as in *Pantolambda*.   The phalanges compare on the whole rather nearly with those of *Periptychus*.   Compared with *Dissacus saurognathus* the distal ends of the metatarsals are somewhat more squared, the intermediate phalanges are much alike, the lateral unguals are broader in *Ectoconus* but the middle unguals not so wide.   *Claenodon* has very different phalanges, of the usual carnivore type, not shortened or flattened, and with large, high, compressed claws, quite unlike those of any taligrade (and equally unlike those of Mesonychidae).

The pes of *Orycteropus* differs widely from that of *Ectoconus*; the astragalus has a deep trochlea, moderately long neck and very convex head, the tarsus is serial, the metapodials are not flattened but strongly hinge-jointed, the keels extending over the dorsal surface; the phalanges are longer, not flattened, and the ungues conform only in general proportions.

SUMMARY.   The skeleton of *Ectoconus* represents altogether an early stage in the evolution of "subungulate" mammals, with especial relations, through *Pantolambda*, to the Amblypoda.   The structural agreement with the Arctocyonidae appears in innumerable details, in spite of a considerable adaptive divergence manifest in teeth and feet.   This points to derivation from a common source not very remote.   Most of these common characters are shared, as might be expected, by such primitive modern mammals as *Orycteropus*, again in spite of wide adaptive divergence in teeth and feet.

Ectoconus was of about the size and general bodily proportions of the aardvark, except for the smaller tail, but the shape of the head was quite different, not nearly comparable to any modern mammal, and the feet are equally different from any existing Mammalia.   The animal was probably sub-plantigrade, with short wide foot considerably padded beneath, wide spreading digits and small tapiroid hoofs.

The near relationship to *Periptychus* is obvious throughout the skeleton, although the premolars show a diverse line of specialization.   The skeleton is larger and more massively proportioned, although the skull is somewhat smaller.   The tarsals are heavier in *Ectoconus* and the astragalus deeper; the external calcanear ridge is not present in *Periptychus*.   The metapodials are somewhat shorter and heavier in *Ectoconus*.   The astragalo-cuboid facet is more lateral than in *Periptychus*.

The skeleton of *Periptychus coarctatus* so far as known agrees with that of *P. rhabdodon* except for somewhat smaller size and more slender proportions.   Hence the *Ectoconus* skeleton does not represent an older stage in the specialization of the Periptychidae than does *Periptychus*.   Probably the smaller genera of the family are more primitive structurally than either large genus, but they are very imperfectly known.

Conacodon Matthew, 1897

Matthew, W. D., 1897, Bull. Amer. Mus. Nat. Hist., IX, pp. 297–8;
Hay, O. P., 1902, U. S. Geol. Surv. Bull. No. 179, p. 695

Type: Haploconus entoconus Cope, as fixed by Hay, from the Puerco formation of New Mexico.

Author's Diagnosis: With round premolar cusps; third upper premolar simple.

The above diagnosis merely follows the lines of generic distinction indicated for the Anisonchinae by Osborn and Earle in 1895 and no further description of distinctions has been published, except for a note by Matthew on the page cited as to the distal ends of the tibia and fibula.   *Conacodon* is, however, a distinct and characteristic genus; the dentition of both species is admirably shown in a large series of upper and lower jaws, although very little is known of the skull and skeleton.

Revised Diagnosis: Molars trigonal, very wide transversely, with hypocone very large, high and sharp, directly internal to protocone, no protostyle.   Protocone central, acute, high, crescentic, paraconule absent, metaconule involved with wings of protocone to a variable extent but usually disappearing at an early stage of wear.   Paracone and metacone subcircular, equal on m^{1-2}, metacone reduced on m^3.   Encircling cingula on molars extending from apex of hypocone; anterior and posterior cingula on premolars.

Premolars rather larger than molars, especially in *C. entoconus*, with round, high conical cusps, backwardly pitched.   Internal cusp on p^4, none on the others.   Lower molars with distinct paraconid.   Cusps robust.   Lower premolars with small single-cusped heels on p_{2-4}, otherwise much like premolars of *Hemithlaeus*.   The lower molars also have much the proportions and cusp construction of those of *Hemithlaeus*, but protoconid somewhat higher than metaconid.

The genus is readily distinguished by the widely transverse trigonal molars, with large internal hypocone, directly median instead of in the usual postero-internal position.   The premolars are simple as in *Haploconus*, but robust as in *Hemithlaeus*, and the enlargement and backward pitch is very suggestive of *Periptychus coarctatus*.   The lower molars resemble those of *Anisonchus* and *Hemithlaeus* in retaining the paraconid.

Conacodon entoconus (Cope)

Haploconus entoconus Cope, 1882, Amer. Nat., XVI, July 28, p. 686; 1884, Amer. Nat., XVIII, p. 802, fig. 11 *a, b*; 1885, Tertiary Vertebrata, pp. 421–3, pl. xxv *f*, figs. 4, 5; 1888, Trans. Amer. Phil. Soc., XVI, N.S., p. 348; Osborn and Earle, 1895, Bull. Amer. Mus. Nat. Hist., VII, p. 63; (*Conacodon*) Matthew, 1897, *ibid.*, IX, p. 298.
Anisonchus coniferus Cope, 1882, Amer. Nat., XVI, Sept. 28, p. 832; 1884, Amer. Nat., XVIII, p. 803, fig. 12 *c*; 1885, Tertiary Vertebrata, p. 409, pl. xxiv *g*, fig. 6.

Type: A. M. Cope Coll. No. 3462, upper jaw, right, with p^3–m^3.   Type of *Anisonchus coniferus:* A. M. Cope Coll. No. 3551.

Horizon and Locality: Puerco formation, San Juan basin, New Mexico.   All the specimens obtained in 1913 were from the lower horizon of the Puerco, to which the species is probably limited.

Author's Description: The largest species of the genus. . . . The peculiarity which distinguishes it from the *H. lineatus* is the conical form of the internal lobe of the fourth premolar, which is in the *H. lineatus*, flat and concentric in section.   Further, the posterior inner or cingular cusp of the true molars is extended further inwards than in that species, giving the crowns a greater transverse extension. The posterior molar has the posterior external angle less developed than the other molars. The third premolar is a robust cone with subtriangular base. [Measurements follow.]

Type Description of Anisonchus coniferus: . . . of larger size than the *H. kowalevskianus.* This species differs materially from the last in the larger development of the cingular internal cusp of the superior true molars, so that the transverse diameter of the latter exceeds that of any of the species of this group. The apex of the median V is not very prominent. Third superior premolar with a rudiment of the anterior and posterior basal lobes; internal lobe not large, conic. Fourth not wider than first true molar, which equals the second and exceeds the third in size. An external cingulum on the true molars, none on the premolars. Probable inferior true molars with anterior and posterior median cusplets. [Measurements follow.]

There is a large number of specimens of *C. entoconus,* chiefly upper and lower jaws, sometimes in association, and a few fragments of the skeleton associated with some of them, but in no case is the association of the skeleton parts wholly certain. They represent about sixty-five individuals, of which thirty-seven are catalogued. The most important are:

No. 3462, Cope Coll., type specimen, upper jaw with p³–m³ r.

No. 3551, Cope Coll., type of *Anisonchus coniferus,* upper jaw, p³–m² l., part lower jaw, teeth broken off.

No. 3476, Cope Coll., lower jaw, part of upper jaw, p³, p₂–m₃ l., astragalus.

No. 3503, Cope Coll., upper teeth, parts of lower jaw, calcaneum and distal ends of tibia and fibula.

No. 3552, Cope Coll., parts of upper and lower jaws. Figured by Cope as *Anisonchus coniferus.*

No. 3467, Cope Coll., upper and lower jaws, p³–m² r., p²–m¹ l., p₂–₄ r.

No. 3473, Cope Coll., upper and lower jaws, p⁴–m³ l., p₃–m₃ r.

No. 16418, upper and lower jaws, p³–m³ l., p₂–m₃ l.

No. 16420, upper and lower jaws, c–m³ r., p³–m¹ l., p₃–₄ r.

No. 16430, upper and lower jaws, p³–m¹ r., m₂–₃ l.

No. 16423, upper and lower jaws, p³–m² r., p₃–₄ and m₂–₃ l.

No. 16433, lower jaw, p₃–₄ r., m₁–₂ l., p².

Nos. 16425, 16424, 16427, 16431, lower jaws.

Upper Teeth. The incisors are unknown but cannot have been large. The canine (No. 16420) is of moderate size, round-oval in cross section, nearly straight, conical, pointed. A faint internal cingulum, and obscurely indicated anterior and posterior ridges. The enamel is limited to the distal third of the tooth; its base does not extend as low as the wearing surfaces of the premolar crowns. P¹ is much smaller than the canine, round, conical, pointed and quite simple. It is slightly spaced from the canine, and a little further spaced from p².

The second, third and fourth premolars are large, robust teeth, principally composed each of a large, round-conic, massive cusp, pitched somewhat backward. There is a slight posterior heel on each, and on p⁴ is a large internal cusp with cingula extending from it to encircle the base of the main outer cusp, almost meeting on its outer side. P² and p³, although they have well-developed inner roots, have no trace of inner cusp, only a broadened base to the main cusp internally. The inner cusp of p⁴ does not share the backward pitch of the outer cusp, but projects upward and outward, giving the same curiously skewed effect to this tooth that one sees in the upper premolars of *Periptychus coarctatus.*

The upper molars are much less massive in appearance than the premolars.   They are chiefly distinguished by their sharply trigonal form, extreme transverse extension, and the singular position assumed by the hypocone.   This cusp is almost median or lingual in its relations to the protocone, equally large, and occupies practically the normal position of the protocone, which in turn has retreated to the centre of the crown and takes almost the position of a protoconule.   A small metaconule is present in the unworn teeth on the posterior wing of the protocone crescent, but soon becomes obliterated by wear.   The hypocone is likewise subcrescentic, its wings extending as cingular crests to the base of the protocone wings, thence as basal cingula encircling the entire outer half of the crown.   The outer cusps are subequal, slightly connate, the paracone round, the metacone slightly flattened externally.   The postero-external angle of m^2 is slightly that of m^3 much drawn in, and the metacone of m^3 is noticeably reduced.   No basal cusps at the external angles of either molars or premolars.   The enamel of the premolars and of the outer cusps of the molars bears, sometimes but not always, an obscure ventral striction suggestive of *Periptychus*, but by no means so distinct or regular.

Lower Teeth.   The lower incisors and canines are unknown. P_1 (No. 16424) is a small, simple tooth, very like the corresponding upper tooth. P_2, p_3 and p_4 are large, robust teeth, the principal cusp round-oval, conic or slightly inflated, moderately high, distinctly pitched backward.   A minute posterior heel on p_2, larger on p_3, distinctly larger, basin-shaped, with inner, median and outer cusps clearly foreshadowed, thus assuming the characters of the talonid of the true molars.   There are no anterior basal cusps on any of the premolars.   Faint antero-internal basal cingulum often present.

The molars are slightly smaller than the premolars, much less robust and more complex in structure.   They are subequal in size, m_3 being longer but narrower than the others.   The trigonid is of about the same width as the talonid, but it occupies the greater moiety of the tooth, at the base.   The principal trigonid cusps are the metacone and paracone, the metacone being more posterior in position and somewhat smaller.   The paraconid is much smaller and lower, submedian in position and connected with the protoconid by a crest.

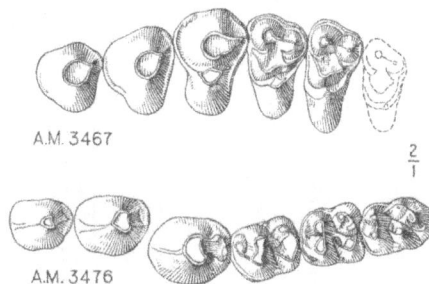

A.M. 3467

$\frac{2}{1}$

A.M. 3476

FIG. 30.   *Conacodon entoconus*, upper and lower teeth, A. M. Nos. 3467 and 3476, crown views.   Twice natural size.

A well-defined internal cingulum encircles the base of the metaconid, and a more obscure cingulum encircles the external and anterior base of the trigonid.   The three talonid cusps are subequal in size, surrounding a somewhat contracted basin.   They are fairly distinct, the hypoconulid somewhat smaller and considerably to the rear of the other two.   In m_3

MEASUREMENTS (IN MM.)

	Type 3462	3551	16420	3467	16418	16422	3473	3476	16433
Upper teeth, c–m³			45.3						
canine, ap. × tr.			4.4						
			4.0						
premolars, p¹⁻⁴			27.3						
" p²⁻⁴	18.9		21.3	18.6		19.3			
P¹, diameters, ap. × tr.			3.0						
			2.7						
P² " "			6.2	5.9		5.8			6.2
				5.5		5.3			6.9
P³ " "	6.7	7.0	7.0	6.4	7.0	7.3		7.0	
	7.2	7.4	8.2	7.1	7.6	7.3		7.3	
P⁴ " "	6.3	6.5	6.9	6.4	6.4	6.0	6.4		
	8.1	9.1	9.5	8.6	9.0	8.8	8.0		
molars, m¹⁻³	13.8		14.3		13.6	14.0	13.7		
M¹, diameters, ap. × tr.	5.2	5.1	5.3	5.2	5.2	5.4	5.4		
	8.5	9.3	9.4	8.8	9.0	8.7	8.7		
M² " "	5.0	4.8	5.0	5.0	5.0	4.7	4.8		
	8.0	10.0	10.7	9.7	10.2	10.0	10.0		
M³ " "	3.6		4.0		3.5	3.9	4.0		
	7.9		9.0		8.0	8.2	8.2		

	16425	16431	16420	3467	16418	16424	3473	3476	16433
Lower teeth, c–m₃	47.2							e 46.0	
premolars, p₁₋₄								23.1	
" p₂₋₄	19.1			19.5	19.3	19.0		19.0	
P₁, diameters						3.1		3.1	
						2.6		2.7	
P₂ "	5.8		6.3	6.0	5.9	5.9		5.5	
	4.0		4.5	4.4	4.8	4.5		4.0	
P₃ "	6.7	6.2	6.8	6.9	6.9	6.6	7.0	6.5	7.5
	4.7	5.0	5.5	4.7	5.4	5.0	5.0	4.8	5.9
P₄ "	6.4	6.4		6.6	6.9	6.7	6.5	6.5	7.0
	4.7	5.3		5.0	5.4	5.0	4.9	5.1	5.8
molars, m₁₋₃	16.7	16.1			17.1	16.5	15.5	17.7	
M₁, diameters	5.7	5.3			5.5	5.6		5.7	5.6
	4.4	4.5			5.0	4.4		4.4	4.8
M₂ "	5.2	4.7			5.3	5.5	4.9	5.2	5.4
	4.5	4.8			5.0	4.6	4.6	4.8	5.0
M₃ "	6.1	5.6			5.9	5.9	5.1	6.2	
	4.0	4.1			4.2	4.0	3.6	4.1	
Depth of jaw beneath p₂	12.2			11.6			11.9	11.8	
" " " " m₂	13.7	15.8					13.6	14.0	

the hypoconulid is set considerably farther to the rear and is larger and higher than the other heel cusps.

There is considerable variation in size and proportions of the teeth among the specimens compared but little variation in cusp structure. The relative position of the hypocone is not always exactly the same; sometimes it is a little farther back. Apparently there is some variation in the height and robustness of the premolars and in the transverse width of the molars, especially those of the upper jaw. The type of "*Anisonchus*" *coniferus* has larger and more robust teeth than the type of "*Haploconus*" *entoconus*, and No. 16433 has distinctly larger premolars than any other measured specimen, although not exceptional as to the molars. All these differences are apparently individual.

Conacodon cophater (Cope)

Anisonchus cophater Cope, 1884, Paleont. Bull. No. 37, Jan. 2, and Proc. Amer. Phil. Soc., XXI, p. 321.

Type: A. M. Cope Coll. No. 3486, lower jaw fragments, p_3 r., m_{2-3} r.

Horizon and Locality: Paleocene of San Juan basin, New Mexico, undoubtedly Puerco formation, Lower Paleocene.

Author's Description: Its proportions are the same as those of the *A. agapetillus* [*Oxyacodon agapetillus*], that is, much smaller than the *A. gillianus*, and the single premolar is much more like that of other species of the genus. The true molars differ from those of the *A. agapetillus* in two strong characters. First, the internal posterior cusp is inside the rim of the heel of the crown, that is, outside the bordering edge, and is therefore very distinct from the posterior median cusp. It is a sharp cone; secondly, there

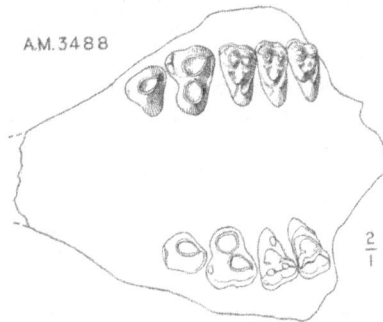

FIG. 31. *Conacodon cophater*, upper teeth, A. M. No. 3488, crown view with outlines of palate. Twice natural size.

is a cingulum extending from this cusp round the internal base of the internal anterior cusp. There is also one at the base of the external anterior cusp, which continues to the heel only on the last inferior molar. The posterior heel is relatively wider, and the anterior V relatively more contracted, than in the *A. agapetillus*. The anterior tubercle is moderately developed at the anterior base of the anterior V. The third or fourth premolar is equilateral, and larger than the true molars. It has a short apiculate heel, and a rudimental anterior basal tubercle. [Measurements follow.]

Revised Description: This species is of about two-thirds the lineal dimensions of *C. entoconus*, and the premolars are much less robust. There is a distinct antero-external cusp on p^4 and anterior basal cusps on p_{3-4}. The proportions and construction of the molars are the same throughout as in the larger species.

Hemithlaeus Cope, 1882

Cope, E. D. 1882, Amer. Nat., XVI, Sept. 28, p. 832

Type: Hemithlaeus kowalevskianus Cope, Lower Paleocene, Puerco formation, New Mexico.

Synonym: Zetodon Cope, 1883.

Author's Diagnosis: Char. gen.—Dentition of the type of *Anisonchus*, but there is but one internal tubercle of the superior true molars, which is the apex of the V, the posterior cusp being absent, no intermediate tubercles. Last and penultimate premolars with internal cusp. Last inferior true molar with heel.

Type Diagnosis of Zetodon: The true molars consist of narrow crescents in two pairs, which are both concave towards each other, embracing a fossa. The posterior crescents soon unite on attrition, closing the fossa, while the anterior are well separated, and only unite by their anterior apices. Each has a small, columnar heel. Fourth premolar with the posterior pair of crescents only, which soon unite. The anterior pair is represented by a part of the external one which forms a narrow lobe. The heel is larger than in the true molar.

The position of this genus it is impossible to determine from the specimens obtained. It may be Marsupial or Condylarthrous, and, if the latter, one of the Meniscotheriidae; but if not of these groups, its position is not likely to be in any known order of the Tertiary periods.[83]

Cope's original diagnosis of *Hemithlaeus* is in direct contradistinction to the characters of his "type" specimen, No. 3576, a palate with $p^2–m^3$, which shows quite unmistakably the hypocone and protostyle flanking the two sides of the worn protocone on all the upper molars. But this specimen is not the type, although it was so designated in Professor Cope's collection. The type, No. 3587, and paratype, No. 3581, specimens are figured in Tertiary Vertebrata, pl. xxv *f*, figs. 6 and 7. They are not so well preserved, so that the true character of the upper molars is less clearly indicated; they belong, however, unmistakably to *Hemithlaeus*. A great number of upper and lower jaws and jaw fragments of this genus are in the collections, and although there is much individual variation they may all be referred to a single species. Some parts of the skull are known but not enough for reconstruction. Except for the astragalus and fragments of the limb bones the skeleton is unknown.

Revised Diagnosis: Upper molars tritubercular, with vestigial conules and moderately large subequal hypocone and protostyle joined externally by encircling cingula, with rudimentary mesostyle. Base of tooth extended lingually, making the protocone subcentral in position. Cusps low, round-conical. P^3 and p^4 with well-developed internal cusps, the external cusps round-conic; p^2 trigonal, with inner root and rudimentary inner cusp, the principal cusp rounded trigonal; p^1 small, one-rooted, anterior teeth unknown. Lower premolars with short, round-conic simple cusps, small heels on $p_{3–4}$, molars wide and short, round-cusped, protoconid and metaconid subequal, elevated a little above the rest, paraconid low, distinct, talonid wide with three subequal cusps, m_3 with hypoconulid enlarged and set back forming a rudimentary third lobe. Enamel of the teeth finely rugose, without distinct vertical grooving.

The teeth are not widely different from those of *Periptychus coarctatus* and *Ectoconus* in cusp construction, but show no trace of the peculiar grooving of the enamel prominent in *Periptychus* and the tendency to polybuny and round-quadrate form of both the larger genera. The upper molars tend to assume the trigonal form more prominent in other small periptychids.

[83] Amer. Nat., XVII, p. 968. *Zetodon gracilis* is figured in the article on the Condylarthra in the American Naturalist for 1884, XVIII, p. 802, fig. 11 *d*.

Zetodon, as pointed out by Matthew in 1897, appears to have been based upon badly crushed teeth of *Hemithlaeus*.

Hemithlaeus kowalevskianus Cope

Hemithlaeus kowalevskianus Cope, 1882, Amer. Nat., XVI, Sept. 28, p. 832; 1884, *ibid.*, XVIII, p. 802, fig. 11 *c*; Osborn and Earle, 1895, Bull. Amer. Mus. Nat. Hist., VII, p. 60, fig. 18; Matthew, 1897, *ibid.*, IX, p. 297, fig. 11.

Type: A. M. Cope Coll. No. 3587, p²–m² r., p₁₋₂, m₁₋₂ l.

Paratype: A. M. Cope Coll. No. 3581, upper and lower jaw fragments associated.

Neotype: A. M. Cope Coll. No. 3576, skull, p³–m³ r., m¹⁻³ l.

Horizon and Locality: Lower Paleocene, Puerco formation, San Juan basin, New Mexico.

Author's Description: Char. specif. The internal lobes of both third and fourth premolars are conic. The true molars are distinguished from the species of *Anisonchus* and *Haploconus* in that the posterior cingulum does not develop an internal cusp. Instead of this, the apex of the median V forms the internal angle of the crown, and an anterior and a posterior cingula of equal size rise to meet it. The inferior molars have anterior and posterior median cusps, and the internal anterior cusp is not compressed. Length of P-m. IV with true molars, .0185; diameters P-m. IV, anteroposterior, .005; transverse, .007; do. of M. II, anteroposterior, .0046; transverse, .007. The last true molar is smaller than the first or second.

Revised Description: Cope's original description is incorrect in stating that the hypocone is absent. The species is of medium size among the Puerco Anisonchinae, smaller than *Conacodon entoconus*, larger than *C. cophater* or *A. gillianus*, distinguished from the two former by deuterocone on p³, form of molars, etc., and from the last by the more obese premolars and better development of protostyle. It is confined so far as known to the lower horizon of the Puerco.

The canines are of moderate size, pointed, simple, little recurved. The first upper premolar is quite small, one-rooted, with simple pointed-spatulate crown. P²⁻⁴ are larger, three-rooted, inner cusp rudimentary to absent on p², well developed on p³⁻⁴ and more or less rounded, the wings of the deuterocone being weakly developed, basal cingula on the

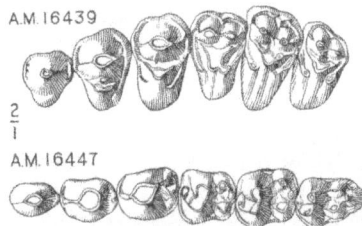

FIG. 32. *Hemithlaeus kowalevskianus*, upper and lower teeth, A. M. Nos. 16439 and 16447, crown views. Twice natural size.

anterior and posterior sides of p²⁻⁴, obsolete on their external faces. The molars are considerably widened by extension of the lingual side at the base of the crown, and between trigonal and oval in form, the outer half being wider, m² a little larger than m¹ or m³. The first lower premolar is quite like p¹, p₂₋₄ larger and two-rooted with round-conic short protoconids, small heels on p₃₋₄, the heel of p₄ having three small cusps. There are no traces of anterior basal cusps on any of the premolars, except sometimes on p_4. The first and second molars are a fifth longer than wide, the paraconid on a level with the talonid cusps, lower and smaller than metaconid and protoconid, the metaconid postero-internal to the protoconid and somewhat twinned with it. The entoconid and hypoconid are equal and opposite, the hypoconulid a little behind and somewhat smaller. On m_3 the hypoconulid is enlarged, set back and raised to a level with the protoconid, this tooth being narrower and longer than the others and wider in front than behind, the m_{1-2} having trigonid and talonid of subequal width.

A few fragments of skull and skeleton preserved with different specimens show that the skull is short and wide, the sagittal crest much less developed than in the larger periptychids, the orbit in about the same position with equally rudimentary post-orbital processes above and below, infra-orbital foramen above p^3, no trace of jugal shelf beneath the orbit. The limbs and feet are small compared with the skull; the astragalus and calcaneum, figured by Matthew in 1897, are not unlike a diminutive copy of *Periptychus coarctatus* except for somewhat longer neck, the size being about four-ninths that of the larger form, the teeth being a little more than half as large lineally.

A few specimens, especially No. 3611, show teeth somewhat more compressed throughout, with more distinct though quite rudimentary anterior basal cusps on the premolars, paraconids vestigial on m_{2-3}, and other characters that are regarded as varietal or at most subspecific; these specimens may be named var. *compressus* with No. 3611 as type, consisting of lower jaw with p_3-m_3 and upper jaw fragment with p^4-m^1.

Anisonchus Cope, 1881

Cope, E. D., 1881, Paleont. Bull. No. 33, Sept. 30, and Proc. Amer.
Phil. Soc., XIX, pp. 487, 488

Type: Mioclenus sectorius Cope, 1881, Amer. Nat., XV, Sept. 22, p. 831; of the Middle Paleocene of New Mexico.

Author's Diagnosis: Third and fourth superior premolars . . . with internal lobes. Superior premolars enlarged . . . with one external cusp. A posterior internal cusp of superior molars; . . . intermediate tubercles wanting, replaced by branches of an internal V; no cusp on inner side of last inferior premolar. . . . The inner posterior lobe [of the upper molars] is more prominent in this genus than in any of the others, and has a V-shaped apex. It projects further inwards than the anterior inner lobe. It is represented by a mere tubercle of the cingulum in *Mioclaenus.* In the lower jaw the last premolar is quite simple, consisting of a principal cusp, and a non-cutting heel. The second true molar has intermediate anterior and posterior cusps.

Revised Diagnosis: Molar construction as in *Hemithlaeus*, except that the protostyle is vestigial or absent. Premolars much less robust, more or less compressed, well-developed anterior cusps on p_{3-4}. Distinguished from *Conacodon* by less transverse width of upper molars and postero-internal position of hypocone, also by greater compression of lower premolars with well-developed anterior cusps. Distinguished from *Haploconus* by the more quadrate form of upper molars, distinct paraconids on lower molars, premolars ^{3} and ^{4} with well developed inner cusps, premolars $_3$ and $_4$ with distinct anterior cusps; premolars less enlarged relatively to molars.

The affinities of the genus are evidently between *Hemithlaeus* and *Conacodon*, but nearer to the former, as represented by the Puerco species of the three genera. The Torrejon species of *Anisonchus* is a more specialized type.

Anisonchus sectorius (Cope)

Mioclaenus sectorius Cope, 1881, Amer. Nat., XV, Sept. 22, p. 831; *Anisonchus* Cope, 1881, Paleont. Bull. No. 33, Sept. 30, and Proc. Amer. Phil. Soc., XIX, p. 488; 1884, Amer. Nat., XVIII, p. 803, fig. 12, *a, b.*

Mioclaenus mandibularis Cope, 1881, Amer. Nat., XV, Sept. 22, p. 831.

Type: A. M. Cope Coll. No. 3527, upper and lower jaws, containing right p_4^4–m_2^2.

Horizon and Locality: Middle Paleocene, Torrejon formation, San Juan basin, New Mexico.

Author's Description: Cusps of last premolars compressed in both jaws. . . . Last interior molar larger than penultimate; true molars, .014; p.m. III, .006.

The fourth superior premolar covers a larger base than either of the true molars. The external cusp has a base extended anteroposteriorly, but the apex is conical, and there are no basal tubercles. The inner cusp has a crescentic base as in *Catathlaeus* [= *Periptychus*], but the apex is narrowed and compressed conic. The external tubercles of the true molars are subconic, and do not develop any external ridges. They are connected by the crescentic, slightly angular crest, whose apex forms the inner anterior boundary of the crown. This crest is not divided into parts homologous with the intermediate tubercles. The crowns of the M. I and II are surrounded by a basal cingulum, which in the M. I develops a tubercle at the anterior external angle. No internal or external cingulum on P–m. IV. Enamel nearly smooth.

The ramus of the mandible is rather slender anteriorly. The P–m. IV is robust, and the cusp is behind the middle of the base of the crown. The heel is short and narrow, and has a raised border, connected with the base of the main cusp. The cusps of the second true molar are elevated and conic, the anterior external the highest, the others subequal. The base of the posterior pair is a little lower than that of the anterior pair. There is no central tubercle as in *Catathlaeus* [= *Periptychus*] *rhabdodon*, and no basal cingulum on either tooth.[84]

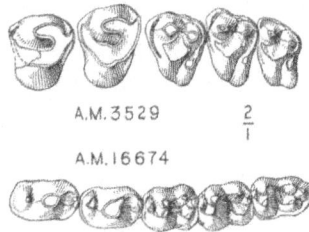

A.M. 3529 $\frac{2}{1}$

A.M. 16674

FIG. 33. *Anisonchus sectorius*, upper and lower teeth, A. M. Nos. 3529 and 16674, crown views. Twice natural size.

The original diagnosis was evidently based only upon the type lower jaw. The extended description, published a few days later, and distinguishing the genus *Anisonchus* as new, clearly resulted from identification of additional specimens as belonging to the species and showing the construction of the upper teeth, widely different from those of *Mioclaenus.*

Revised Diagnosis: Upper molars subquadrate, about one-fourth wider than long, the hypocone postero-internal, its base not projecting farther inward than the base of the much larger protocone, but its apex a little farther in. Conules absent, paracone and metacone subequal, not connate, metacone somewhat smaller on m^3. Premolars slightly larger than molars, round-trigonal, p^4 somewhat wider than long, p^3 about as wide as long, p^2 of less

[84] First paragraph is from Amer. Nat., *loc. cit.*; second and third paragraphs from Paleont. Bull. No. 33, *loc. cit.*

width than its (antero-posterior) length, and with no inner cusp. P^1 smaller, simple, moderately compressed, somewhat spaced, canine somewhat larger than p^1. Lower premolars compressed, with large anterior cusps on p_{3-4}, small on p_2, small inner cusp on p_4, and good-sized single-cusped heels. Size nearly a third larger than *A. gillianus.* Astragalus differs from that of *Hemithlaeus* and *Conacodon* in the longer neck and more vertical fibular facet, and greater extension backward of the tibial trochlea, encroaching upon the astragalar foramen to some degree. An incomplete femur is of rather slender proportions, much as in the smaller Artocyonidae, with long, narrow patellar trochlea, weak third trochanter, lesser trochanter internal and somewhat prominent, digital fossa unreduced.

The species is abundant in the Torrejon formation and characteristic of both the upper and lower levels. No constant distinctions can be made between specimens from the two levels.

Anisonchus gillianus Cope

Haploconus gillianus Cope, 1882, Amer. Nat., XVI, July 28, p. 686.
Anisonchus (Hemithlaeus) apiculatus Cope, 1885, Tertiary Vertebrata, pl. xxv *e*, plate description and fig. 7;[85] 1888, Trans. Amer. Phil. Soc., XVI, N.S., p. 352.

Type: A. M. Cope Coll. No. 3543, fragments of upper and lower jaws and a few skeleton fragments. Type of *A. apiculatus:* A. M. Cope Coll. No. 3600, parts of upper and lower jaws.

Horizon and Locality: Puerco formation, Lower Paleocene, San Juan basin, New Mexico.

Author's Description: Size of the *H. angustus,* but having the same peculiarity of the fourth superior premolar as the *H. entoconus, i.e.,* with the internal lobe conic. That tooth is also relatively smaller than

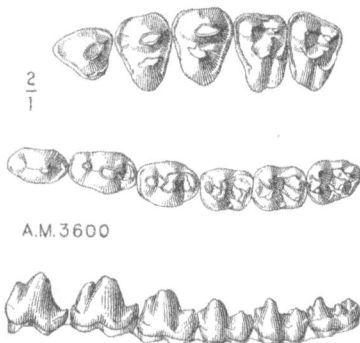

Fig. 34.   *Anisonchus gillianus,* type upper and lower teeth, A. M. No. 3600, crown view of upper teeth, crown and side views of lower teeth.   Twice natural size.

in the *H. entoconus,* and has an anterior basal external tubercle not found in that species. The inferior true molars have an anterior median tubercle which is not found in the *H. angustus* and *H. xiphodon,* and the external anterior cusp has not the compressed form general in the genus. [Measurements follow.]

Revised Diagnosis: Smaller than *A. sectorius,* averaging about three-fourths as large lineally. Upper molars much wider, about one-half wider transversely than long antero-

[85] This figure is captioned "*Anisonchus sectorius. . . .* Erroneously marked *A. apiculatus* on the plate in some of the editions."—THE EDITORS.

posteriorly, the hypocone more internal in position and vestigial protostyle usually present but reduced to a cingulum. Metaconule minute or vestigial. Upper premolars more triangular and less rounded than in *A. sectorius*, lower premolars somewhat less compressed, the anterior cusp set lower and less connate.

Anisonchus gillianus is common in the lower horizon of the Puerco, in occurrence with *Hemithlaeus* and *Conacodon*, but it is also found in the upper levels of the Puerco, the specimens showing no constant distinction from those of the lower levels. It is replaced in the Torrejon by the very distinct species *A. sectorius*, equally common, and constantly showing a very considerable progressive change, comparable to the amount of difference between the Periptychidae of the Puerco and of the Torrejon.

Haploconus Cope, 1882

Cope, E. D., 1882, Amer. Nat., XVI, April 24, p. 417

Type: Haploconus lineatus = Mioclaenus angustus Cope, 1881, from Upper Paleocene of New Mexico.

Author's Diagnosis: Char. gen. The same as *Anisonchus*, excepting that the crown of the third superior premolar is a simple cone, wanting the large crescentic crest of the inner side seen in that genus and *Catathlaeus* [= *Periptychus*]. It is more nearly allied to the two genera named than to *Phenacodus*.

Revised Diagnosis: Upper molars obliquely trigonal, the hypocone forming the posterointernal angle of the triangle, prominent, sometimes with a small hypostyle; protostyle small or vestigial, protocone central, conules small, variable. Premolars more enlarged than in any other Anisonchinae, large inner crescentic cusp on p^4, none on preceding premolars. Canines large and stout, much larger than in other anisonchines. Lower premolars relatively large, moderately compressed, with progressively large heels on p_{2-4} but no anterior cusp. Molars without paraconid, otherwise much as in the other genera.

Haploconus is found only in the Torrejon, where it is as abundant as *Anisonchus sectorius*. It is apparently related to *Conacodon*, and might be regarded as a specialized descendant of *C. cophater*, but the difference is considerably greater than the distinctions between the two species of *Anisonchus*, and the change in proportions of the molars and premolars, the shifting of the position of the hypocone, and other modifications, make it doubtful whether the known Puerco species can be directly ancestral. It is clearly not derivable from *Hemithlaeus* or *Anisonchus*.

The skull is somewhat better known in this genus, affording some points of comparison with the better-known larger periptychid genera. It shares many of their primitive characters, such as the wide posterior expansion of the nares, lack of postorbital processes, narrow postorbital constriction, and small and narrow brain-case (at least in the anterior part). In general, however, the skull is much less massive and agrees in many ways with *Loxolophus*. It appears to be considerably longer and narrower than that of *Hemithlaeus*, the maxilla expanded around the enlarged canine as in the creodonts and unlike *Hemithlaeus*, *Conacodon* or *Anisonchus*. The lack of lachrymal expansion on the face is probably characteristic of all the Periptychidae.

The few fragments of the skeleton known do not clearly differentiate *Haploconus* from other Anisonchinae, but indicate a rather slender-limbed animal as large as a raccoon, showing a distinct approach toward the primitive phenacodont type, apparently subcursorial in tendencies.

Haploconus angustus (Cope)

Mioclaenus angustus Cope, 1881, Amer. Nat., XV, Sept. 22, p. 831; 1881, Paleont. Bull. No. 33, Sept. 30, and Proc. Amer. Phil. Soc., XIX, p. 491.

Haploconus lineatus Cope, 1882, Amer. Nat., XVI, April 24, p. 417; 1884, *ibid.*, XVIII, p. 804, fig. 13 a-b.

Haploconus xiphodon Cope, 1882, Paleont. Bull. No. 35, Nov. 11, and Proc. Amer. Phil. Soc., XX, p. 466; 1884, Amer. Nat., XVIII, p. 804, fig. 13 c.

Type: A. M. Cope Coll. No. 3477, lower jaw, p_4–m_3 r. Collected by D. Baldwin Aug. 29, 1881, near Huerfano Peak. Type of *Haploconus lineatus:* A. M. Cope Coll. No. 3425, part of skull and jaws, p^3–m^3 r., p^4–m^3 l., p_3–m_3 r., m_{1-3} l; collected by D. Baldwin, March 22, 1882. Type of *H. xiphodon:* A. M. Cope Coll. No. 3481, lower jaw, young, dp_2–m_2 r.

Horizon and Locality: All the above are from the Paleocene of the San Juan basin, undoubtedly from the Torrejon formation, Middle Paleocene. *H. lineatus* positively so recorded.

Author's Diagnosis: Cusps of last premolars compressed in both jaws. . . . Second and third lower true molars subequal; cusps, especially the internal, elevated; anterior inner confluent into an edge; true molars, .013.

This brief diagnosis was supplemented in the Bulletin published a few days later by a more extended description, as follows:

The least species of the genus [*Mioclænus*], with the teeth about the size of *Hyopsodus paulus* Leidy, but with more robust jaw. The molar teeth diminish in size regularly posteriorly from the P–m. IV. They all have three subequal posterior cusps which are less elevated than the anterior ones. The median is enlarged into a heel on the last tooth. The anterior are opposite, and the external is larger than the internal. There is no anterior internal. The external wears into an anteroposterior narrow grinding surface, which looks like a combination with an anterior median. The latter is, however, not separate on the least worn molars. The anterior outer cusp increases in size anteriorly, and is the large cusp of the P–m. IV. It sends a branch backwards on the inner side of the crown which forms the edge of the narrow concave heel. There are no cingula except a short one on the anterior corners of the base of the crown of the P–m. IV. Enamel obscurely wrinkled.

Author's Description of *H. lineatus:* It is about the size of the *Anisonchus sectorius*, and differs from it in several details besides in the generic characters. In the *H. lineatus* the base of the posterior inner tubercle of the superior molars is more distinct, and projects further inwards. The fourth premolar is relatively larger, and the enamel is delicately plicate, remotely approaching the condition of the surface seen in *Catathlaeus rhabdodon*. In the inferior molars, the anterior marginal tubercle is wanting. The first premolar has but one root; the second and third have a posterior but no anterior basal lobe. The canines of both jaws are rather large, are acute, and flat on the inner side, and vertical in direction. [Measurements follow.] A second species of the genus is probably the *H. angustus* . . . which I described as a *Mioclaenus.*

Author's Description of *H. xiphodon:* Represented by a mandibular ramus, and perhaps by three rami. The one on which the series rests contains five molars, the middle one of the series broken, so that its form cannot be positively ascertained. It is probable that it is the first true molar, so that the animal exhibits the last true molar not entirely protruded, and is therefore nearly adult, but there are some reasons for suspecting it to be young. Thus the last inferior molar does not exhibit more of a heel than the second usually does, and the third supposed premolar is smaller than that tooth is in the other species, having nearly the proportions of the second premolar. The teeth present may then be supposed to be the molars from the second to the sixth inclusive. But opposed to this view is the fact that the supposed third premolar has more the structure of that tooth in details, than that of the second, and the specimens accompanying, which have the temporary dentition apparently of the same species, present premolar teeth of a very different character. In any case the present specimen represents a third species of the genus, and I describe it at present as an adult.

The third premolar has a simple compressed crown, about as high as the length of its base, and without anterior basal tubercle. It has a narrow triangular posterior face which is concave, and truncated by a cingulum below; no heel proper, nor lateral cingula. The fourth premolar is an elongate tooth consisting of a compressed principal median lobe, an anterior lobe connected with it, and a heel. The latter has elevated posterior and interior borders. A rudiment of an exterior border is seen in a narrow ridge on the external side of the posterior face of the principal lobe of the tooth.

The sides of the premolars present rather distinct ridges, as in *Periptychus carinidens*. The second true molar has two anterior and three posterior tubercles; the latter close together, pointed and of about equal size. Of the anterior tubercles, the external is much the larger and more elevated. It is compressed and has a curved subacute anterior edge, which extends much in front of the internal tubercle. There is no anterior inner tubercle, nor are there any cingula. The enamel of the sides of the crown presents a few vertical ridges. The last inferior molar only differs from the second, in the greater size of the median posterior lobe, which is nevertheless smaller than in the two other species of *Haploconus*.

There is a mental foramen below the posterior edge of the second inferior premolar. [Measurements follow.]

The two rami with the temporary premolars, exhibit the last true molar enclosed in the jaw. The third and fourth premolars are much like the fourth premolar of the specimen above described, but the fourth is a little more robust than that of the latter, which is very much like the third of the deciduous series. The space occupied by the supposed first premolar of the type specimen is too short for the fourth premolar of the deciduous series, otherwise it might be supposed to have occupied that position. The two true molars resemble those of the type, excepting that the last one does not extend so far into the base of the coronoid process, and is in accordance with the position as number two in the series.

In 1888 [86] Cope reconsidered the status of *H. xiphodon*, and concluded that it was, as originally supposed, the permanent dentition of a species distinct from *H. lineatus* (= *angustus*). Matthew, however, in 1897 [87] regarded it as based upon the milk dentition of that species. Re-examination of the types and other specimens confirms this conclusion. The type jaw of *H. xiphodon* is in the first place obviously an immature jaw, as indicated by the form of the coronoid process and character of the surface bone. The molars preserved are unworn and the second one does not have the characters of m_3 in the heel. They are probably therefore m_1 and m_2. The tooth in front is broken, so that its characters cannot be seen; it would be dp_4 according to Matthew, but m_1 according to Cope. The two teeth in front of it are narrow, much compressed, and well worn; they would be dp_{2-3} according to Matthew, p_{3-4} according to Cope. The amount of wear contrasted with the unworn character of the molars indicates that they are milk molars. A second specimen, No. 3483, confirms these characters. It shows dp_3–m_3, the last molar not yet emerged but preformed in the jaw. The milk molars are more worn than in the type of *xiphodon*, the dp_3 agrees in construction and form but appears a little less compressed, partly on account of greater wear, partly owing to a slight crushing of the type specimen. The molars are practically unworn, and dp_4 is clearly more fully emerged and of earlier eruption than m_1. Cope in 1888 interpreted this specimen as being the permanent dentition of *H. lineatus* (= *angustus*) but this is impossible, as shown by the above data and also by the much greater compression of dp_{3-4}, dp_3 being longer than dp_4, while p_3 and p_4 are of the same length in *H. angustus*. Associated with this specimen, No. 3483, is an upper milk molar, probably dp^3. It is to be noted that the specimen was a little more carefully cleaned by Matthew and the preformed m_3 exposed, in connection with his restudy of the matter. A third lower jaw, No. 3482, identified by Cope as *H. xiphodon*, shows dp_2–m_1 with the crowns well preserved and in about the same stage of wear as No. 3481.

[86] Cope, E. D., 1888, Trans. Amer. Phil. Soc., XVI, N.S., pp. 348–9.

[87] Matthew, W. D., 1897, Bull. Amer. Mus. Nat. Hist., IX, p. 298.

The most important specimens referred to *Haploconus angustus* are the front parts of six skulls, Nos. 3425 (type of *H. lineatus*), 3426, 3427, 16679, 16680, and 16689, in addition to numerous upper and lower jaws more or less complete. In No. 16680 the surface of the skull is best preserved, the others being more or less buried in refractory matrix so that the skull characters cannot be studied in detail. The teeth, however, are well shown in all of them.

The upper incisors are evidently quite small and two or three in number, to judge from what remains of the roots. The canine is separated from the incisors by a short diastema; it is large and creodont-like, but the crown is not preserved. The first premolar is small and simple, much as in *Anisonchus*, rather close to the canine, and with a short diastema behind it. The remaining cheek teeth are set close together. P^2 is two-rooted, p^{3-4} and the molars three-rooted, the internal root of p^4 enlarged. The crown of p^2 is simple, oval, pointed, without accessory cusps or crests save for a slight posterior basal cingulum. The crown of p^3 is a little larger, ovate-trigonal, the long axis oblique to that of the jaw, with a cingulum around most of the base but no other cusps than the protocone. P^4 is much larger,

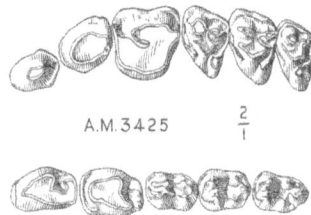

A.M. 3425 $\frac{2}{1}$

Fig. 35. *Haploconus angustus*, upper and lower teeth, A. M. No. 3425. Twice natural size.

with large outer and inner cusps (protocone and deuterocone), the outer compressed antero-posteriorly, especially toward the tip, and considerably pitched backward, the inner crescentic and symmetrical, the wings of the crescent sweeping around to the external angles at the base of the tooth. The molars are smaller than p^4, of subequal size and very similar construction. The outer cusps are slightly connate, the metacone considerably crested antero-posteriorly. The protocone is median, somewhat crescentic in its form, with obscure and variable remnants of the conules on its flanks. The hypocone is large and makes a prominent postero-internal bastion projecting from the otherwise normally trigonal molar. The protostyle is variable, sometimes a small but distinct cusp, more often an abrupt ending to the cingulum which sweeps around the front, outer and hinder faces of the base of the tooth to join the apex of the hypocone. On m^3 the metacone is variably reduced and the tooth a little smaller than the others; m^1 is to a variable extent smaller than m^2.

The lower canines are acute, recurved, slightly crested behind, quite like those of some of the small arctocyonids; the incisors are unknown. The first premolar is small, one-rooted, with a simple, rather short crown, and a short spacing before and behind. The other premolars are of much larger size, two-rooted, with large heels, progressively larger on p_{2-4}, no anterior cusps, the heels single-cusped. The protoconid of p_2 is simple, moderately compressed, with anterior and posterior crests somewhat internal making the inner face of the tooth nearly flat. On p_3 and p_4 the protoconid is moderately compressed and crested, with the posterior crest doubled, the inner crest extending down to the postero-internal corner of the heel, while the outer crest is separated from the heel by a distinct notch.   This structure is best seen in No. 16682; generally it is lost by wear of the teeth.   The lower molars are smaller than p_3 or p_4, of subequal size and very similar construction, narrower than in other Anisonchinae and with no paraconids.   The protoconid is larger and higher than the meta-conid, and more anterior in position; the heel cusps are a little lower than the trigonid; hypoconid and entoconid are of the same height and size and are opposite, the hypoconulid smaller and postero-median on m_{1-2}, but on m_3 considerably larger and set farther back so as to form a rudimentary third lobe.

The supposed dp^3 (No. 3483) has the construction of p^4, so far as the worn specimen shows, but is much narrower transversely and of more trigonal form, the proportions and relations of the inner and outer cusps being much as in p^4.   In the lower jaw dp_2 is a simple tooth very like its permanent successor but much more compressed and with minute heel. Dp_3 is longer than p_3, much more compressed, and has a well-developed anterior accessory cusp, nearly as high as the principal cusp, which is much lower than the protoconid of p_3, strongly compressed and crested with a double posterior ridge running from the apex of the protoconid to the posterior angles of the heel. Dp_4 is as compressed as dp_3 but not so long; it lacks the anterior cusp, the heel is larger and distinctly bicuspid, the protoconid also twinned with crests running from its doubled apex to the heel cusps.   This construction shows in various points an approach toward the molar construction, but it is much closer in every way to dp_4.   The resemblance to the premolars is much more remote in any of the milk series; they are much more suggestive of a progressive complication from the simple crested p_2 leading towards the true molar construction.

The skull differs considerably from that of the other Anisonchinae in its greater length and less width of muzzle, contrasting especially with *Hemithlaeus* in this respect.   From the larger genera it differs in the reduction of the region of the anterior premolars, and prob-ably in the much less prominent sagittal and occipital crests and lighter zygomatic arches.

The superior branch of the premaxilla is wide but short, ending quite abruptly (as in *Phalanger*) instead of in a long splint between maxilla and nasal.   It has an extensive nasal contact, chiefly because the nasal extends far forward, almost to the tip of the muzzle. Backwardly the nasals extend to a point opposite the postorbital process, and are broad-ened out in a diamond-shaped expansion much as in the opossum.   The fronto-maxillary suture extends from the widest part of the nasals postero-externad to the middle of the orbital rim half-way between the lachrymal tubercle and the postorbital process.   The lachrymal sutures are not very clearly preserved, but the bone had evidently little if any expansion upon the face; the lachrymal foramen lies within the orbital rim.   The superior limit of the jugal cannot be certainly traced, and the zygomatic arches are broken off on all our specimens.   The postorbital process on the frontal is very slightly indicated; the postorbital crests are of moderate prominence, with a marked concavity between them. The postorbital constriction appears to be narrower than in *Hemithlaeus*, much narrower than in *Ectoconus* or *Periptychus*, but it is difficult to say how much of this is due to crushing. The back of the skull is unknown.

The few fragments of the skeleton preserved in No. 891 were described by Osborn and Earle in 1895.[88]   Some additional piecing adds a little, but not much, of importance, to their notes.   The humerus and femur are completely known save for the distal ends, the

[88] Osborn, H. F., and Charles Earle, 1895, Bull. Amer. Mus. Nat. Hist., VII, pp. 59–60.

proximal half of the ulna, the astragalus and calcaneum badly broken, and several vertebrae.

The vertebrae indicate an animal of larger size than might be expected from the proportions of other Anisonchinae, about the size of a raccoon, with limb bones of somewhat more slender proportions, approaching *Cryptoprocta* or some of the Felidae. The humerus is of moderate length, with very weak deltoid crest, quite in contrast with the larger periptychids, distal end little expanded, and supinator crest slight, the greater tuberosity high and prominent, somewhat overtopping the head, the shaft round-oval, long and little curved, the entepicondylar bridge weak. The ulna has a rather long, straight, compressed olecranon, moderately wide shaft, much flattened and with shallow groove on

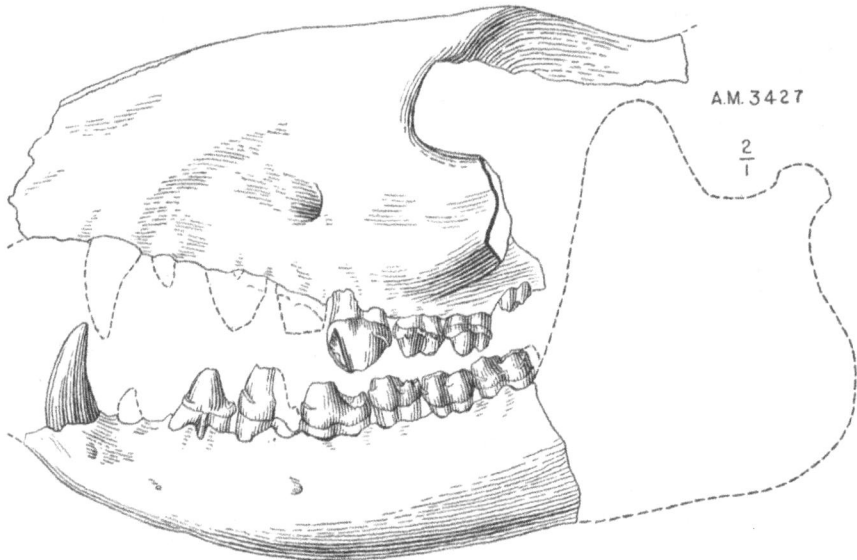

FIG. 36. *Haploconus angustus,* front of skull and jaw, A. M. No. 3427, side view. Twice natural size.

the front, a flat radial facet. The femur has a straight, long shaft of oval section without third trochanter, the second trochanter prominent, posterior rather than internal in position, the head rounded, facing sub-proximad, the digital fossa short, widely open and facing backward. The astragalus is like those of *Anisonchus* and *Hemithlaeus* except for transverse narrowing and grooving of the tibial facet and its extension downwards on the neck, somewhat deeper body and perhaps a longer neck. So far as can be judged, it makes a considerable approach to the astragalus of *Tetraclaenodon*. The calcaneum shows a rather limited facet for the fibula, but is not otherwise notable.

The above characters indicate rather an approach toward the semi-cursorial adaptation of *Tetraclaenodon* than the arboreal adaptation suggested by Osborn and Earle. Although compared by them to the lemur, the limb bones are very unlike those of modern lemurs and approach much more the characters of primitive terrestrial Carnivora. The skeleton throughout shows a marked approach to *Tetraclaenodon*, which is—rightly or wrongly—considered as a primitive "cursorial" adaptation, and is, at all events, a terrestrial, not an arboreal, form.   While rather suggestive than conclusive, the evidence points to a light-limbed terrestrial form, and it is probable that the other Anisonchinae were more or less similar.

Haploconus corniculatus Cope

Haploconus corniculatus Cope, 1888, Trans. Amer. Phil. Soc., XVI, N.S., p. 349.

Type: A. M. Cope Coll. No. 3489, front of skull, buried in refractory matrix.

Horizon and Locality: Torrejon formation, San Juan basin, New Mexico.

Author's Description: Five more or less complete crania and a set of jaws represent this species. Its characters are to be seen in its peculiar superior molar teeth, and in its superior size.

The species presents the general characters of the *H. lineatus*, especially as to the form of its first premolar, which has the internal cusp an elevated concentric cingulum, and its enamel vertically striated with shallow grooves.   The anterior cingulum of the true molars however terminates at its interior extremity in an acute erect cusp which is wanting in the *H. lineatus*, and the cusp of the posterior cingulum is isolated by a notch of the latter, which develops a second lower cusp immediately posterior to the first mentioned. This second posterior cusp is seen on the posterior molar of the *H. lineatus*, but not on any of the others. The dimensions of the *H. corniculatus* are constantly superior to those of the *H. lineatus*, as the following measurements will show.   The lengths on the superior molar series to the canine tooth, in three specimens of each species, are as follows:

H. corniculatus	.045	.043	.040
H. lineatus	.034	.033	.035

The skull is elongate and is narrow in the cerebral region.   The sagittal crest is low, as is also the inion. The orbit is small and lateral.   The canines are directed vertically downwards.

Measurements of Skull

	M.
Total length	.125
Length from occipital condyle to last molar	.058
Width between last molars, inclusive	.037
" " canines, inclusive	.026

The inferior dentition is similar to that of the *H. lineatus*, but is more robust.

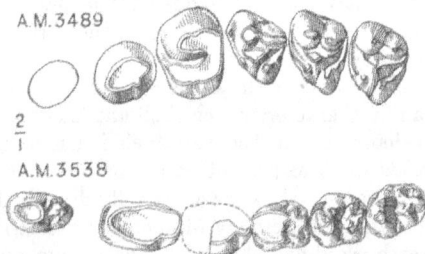

Fig. 37.   *Haploconus corniculatus*, type upper and referred lower teeth, A. M. Nos. 3489 and 3538.   Twice natural size.

Revised Description: The distinctive characters cited by Professor Cope are the superior size, development of protostyle into a distinct cusp, greater isolation of the hypocone and presence of a rudimentary hypostyle.   All these characters are more or less variable in

H. angustus, the size being the most constant distinction. This, however, is so considerable that the species may be held as valid. The "five more or less complete crania" are all so deeply buried in refractory matrix, and the substance of the bone so altered by concretionary action, that there is little left of them save indications of the general form and proportions of the skull; the crowns of the teeth, however, are well preserved in two or three of them and constitute all that is really known of the species.

Three additional specimens, No. 16692, palate, No. 16693, lower jaw, and No. 16672, two upper molars, may be referred to this species. They are from supposedly lower beds of the Torrejon formation, two miles above Chico Springs, at the head of the left branch of Arroyo Gallegos. The matrix is similar to that on Cope's specimens of *H. corniculatus*, and it is quite likely that they were secured by his collector at the same locality. Some are recorded as from "Gallego," i.e. one of the heads of Gallegos Arroyo, others have no especial record. The recorded specimens of *H. angustus* appear to be mostly or all from other localities in the Torrejon.

PANTOLAMBDIDAE Cope, 1883

Cope, E. D., 1883, Amer. Nat., XVII, March 15, p. 406

Author's Diagnosis: Superior and inferior molars with the cusps developed into Vs. Postglenoid process present; posttympanic and paroccipital not distinct. All the vertebrae with plane articulations. Humeral condyles without intertrochlear ridge. Femur with third trochanter. Digits of posterior foot probably five. Metapodial keels small and posterior.

Of this family *Pantolambda* is as yet the only known genus.

Osborn in 1898 defined the family as follows: "Molars selenodont, primitive triangle less compressed [than in Periptychidae]. Lower molars without hypoconulid. No secondary internal cusps." [89]

Revised Diagnosis: Molar and premolar cusps selenodont; upper molars simple, tritubercular, with protocone internal.

The essential distinction from the Periptychidae lies in the sharply crescentic cusps of premolars and molars as against the polybunous tendencies of the periptychids. To the well-known genus *Pantolambda* should be added *Titanoides* Gidley, of the Fort Union, known only from lower cheek teeth. [90] In *Pantolambda* the paraconids of the lower molars are reduced; in *Titanoides*, they are strong.

As the family is known chiefly from the single genus *Pantolambda*, it is not practicable to separate generic and family characters in the skull and skeleton. There is in fact more diversity in skull and skeleton among the periptychid genera than difference between *Periptychus* and *Pantolambda* so far as present data show.

Pantolambda is the most completely known and fully described of the Paleocene mammals, and its relationships to the later Tertiary Mammalia are clearly defined. Comparison of the characters of the teeth, skull and skeleton shows throughout more primitive stages

---

[89] Osborn, H. F., 1898, Bull. Amer. Mus. Nat. Hist., X, p. 181.

[90] As a result of an expedition conducted by the Field Museum in western Colorado during the summer of 1933, the genus *Titanoides* is now known from fairly complete skull and skeletal material. The reader is referred to p. 184 below for citations of recent papers by Bryan Patterson descriptive of this new material.—THE EDITORS.

of the various specializations seen in *Coryphodon*, but it seems equally clear that the molars of *Coryphodon* are not derived from those of *Pantolambda* in the same direct and immediate manner that the teeth of *Phenacodus* are derivable from those of *Tetraclaenodon* or *Dissacus* from *Pachyaena*. *Coryphodon* should be derived from some as yet undiscovered genus of Pantolambdidae in which the asymmetry of the outer cusps of the upper molars was more advanced than in *Pantolambda*.

Pantolambda Cope, 1882

Cope, E. D., 1882, Amer. Nat., XVI, April 24, p. 418

Type: Pantolambda bathmodon of the Middle Paleocene of New Mexico.

· *Author's Diagnosis:* The characters of these teeth remarkably resemble those of *Coryphodon*. *Char. gen.* Crowns of molars supporting two Vs, of which the posterior wears lower than the anterior. Premolars III and IV, crowns consisting of one V and a short median longitudinal crest, as in *Coryphodon;* II and I, unknown. The character which indicates that the genus is distinct from *Coryphodon* is the elevation of the anterior branch of the anterior V of the true molar, which is much more elevated than the posterior branch. In *Coryphodon* it is much less elevated. The type species is smaller than any known Coryphodontid.

A little less than a year later Cope revised this diagnosis on the basis of more complete specimens, including considerable parts of the skeleton. He erected the family Pantolambdidae and the suborder Taligrada to contain this genus (see above). His revised diagnosis of the genus was as follows:

Canine teeth large;[91] dental series continuous. Superior molars all triangular, that is with a single internal cusp. External cusps of premolars unknown; of molars two. Internal cusp V-shaped, sending its horns externally as cingula to the anterior and posterior bases of the external side of the crown, without intermediate tubercles. Inferior true molars with a crown of two Vs, the anterior the more elevated. Premolars consisting of one open V, with a short crest on a short heel, as in *Coryphodon*. Dental formula I$\frac{73}{3}$, C$\frac{1}{1}$, P-m$\frac{74}{4}$, M$\frac{3}{3}$; the last inferior with a heel. A strong sagittal crest. Auricular meatus widely open below. Large postparietal, postsquamosal and mastoid foramina.

The brain case indicates small and nearly smooth hemispheres, extending with little contraction into a rather large cerebellum. The olfactory lobes are produced anteriorly at the extremity of a rather long isthmus.[92]

In "First Addition to the Puerco Fauna" Cope adds[93] to the foregoing a diagnosis of the skeletal characters of the genus:

Cervical vertebrae rather short; other vertebrae moderate, the lumbars not elongate. A large tail. Humerus with large internal epicondyle. Femur with all the trochanters large. Ilium with the anterior inferior spine well developed. Metacarpals short, plantigrade. Phalanges of second series flat, and of subquadrate outline. The astragalus has a wide head, but no neck, as it is not separated from the trochlear portion by a constriction. It is as wide as the trochlear portion, but about one-third of its length extends within the line of the malleolar face of the trochlear portion. The navicular face is flat, that of the cuboid bone is convex vertically, and one-half as long horizontally as the navicular, and only half as deep. These two facets are continuous with the sustentacular below. Interior to all of these, on the internal tuberosity of the head is a sub-round facet looking inwards, like that characteristic of the genus *Bathmodon*, but relatively larger. A continuous facet is seen on the adjacent edge of the navicular. The use of these facets is unknown.

[91] In Paleont. Bull. No. 36, p. 558, this reads, "Canine teeth distinct."—w.d.m.

[92] Cope, E. D., 1883, Amer. Nat., XVII, March 15, p. 407.

[93] Cope, E. D., 1883, Proc. Amer. Phil. Soc., XX, March 16, p. 558, and Paleont. Bull. No. 36, April 17.

In 1898 Osborn defined the genus as follows:[94] "Dentition typical. First upper premolar one-rooted. Second, third and fourth three-rooted, with internal cones. Canines rounded." He gives also a long list of characters ascribed to Taligrada in comparison with those of Pantodonta and Dinocerata, which are based upon his study of the *Pantolambda* skeleton. His very admirable summary of the "persistent primitive or Creodont characters of *Pantolambda*" is very largely confirmed by our present knowledge of the Paleocene mammals. Some of the characters cited as primitive now appear to be taligrade or graviportal specializations; some are rather generally present in mammals whether primitive or not, but the greater part are primitive characters common to most of the Paleocene Mammalia, as will appear by comparing the following résumé of the generic characters of *Pantolambda* with those of *Periptychus*, *Ectoconus*, *Claenodon*, and other Creodonta so far as they are known.

Generic Characters of *Pantolambda* based chiefly upon *P. bathmodon*

TEETH. Dentition unreduced, $\frac{3.1.4.3}{3.1.4.3}$. Incisors moderately large, pointed conical, not expanded at the tip; canines enlarged, subvertical, and nearly circular in cross-section, oval only toward the tip. First premolar single-rooted with oval pointed crown, not crescentic, remaining upper premolars three-rooted with single outer crescent and progressively larger inner crescent. Molars with two outer crescents and larger inner crescent, conules vestigial, protostyle and hypocone lacking except as basal cingula on the two flanks of the protocone. Paracone crescent somewhat extended outwardly; metacone crescent set inward and reduced on m^3. Lower premolars with single inwardly-facing unsymmetric crescent and progressively larger transversely crested heel. Lower molars of two unsymmetric crescents, the paraconid forming the reduced anterior wing, the protoconid and metaconid the higher posterior wing of the first crescent. Similarly the hypoconid and hypoconulid form the posterior wing of the second crescent, with a strong crest falling antero-internad from the apex of the hypoconid. Entoconids wholly absent. Posterior wing of second crescent of m_3 more oblique.

SKULL. Short and wide, with heavy muzzle and deep, wide, massive symphysis; postorbital constriction narrow and long, brain-case small, sagittal and occipital crests high, occiput wide, zygomatic arches wide and moderately heavy. Nares terminal. Nasals long, wider anteriorly than in known Periptychidae, moderately expanded posteriorly. Jugals less extended upon the face than in Periptychidae but reaching farther backward. Lachrymal with very small facial expansion, lachrymal foramen within orbital rim. Occiput semicircular, squamosals more widely expanded than in *Periptychus*, mastoid exposure and otic region generally similar in construction.

SKELETON. Cervicals short as in *Periptychus* and *Ectoconus;* lumbars not less than five, sacrals three (four in mature specimens?), caudals large and numerous. Characters of vertebrae much as in *Ectoconus*.

Scapula with wide anterior blade, much narrower posterior blade, spine very high and prominent almost to the apex, long slender acromion, long stout recurved coracoid process. Clavicle large, long, not strongly curved. Ribs short, with little curvature, round-oval in section, with very little flattening. Sternal bones rather narrow anteriorly, flat and broad posteriorly; pre-sternum remarkably long, xiphisternal thin and flat.

[94] Osborn, H. F., 1898, Bull. Amer. Mus. Nat. Hist., X, p. 183.

Pelvis much as in *Phenacodus*, with long, strongly decurved ilium, flattened and expanded along the superior border into a thin plate, widest toward the anterior end; ischium long, expanded, and the lower border twisted inwardly toward the posterior end, ischial spine not prominent.

Limb bones rather short, heavy humerus with very prominent deltoid and supinator crests, entepicondyle wide, abruptly expanded, foramen large.   Ulna with wide flat shaft, long straight olecranon, no distinct styloid process.   Radius with flattened ovoid head, prominent anterior crest on distal part of shaft, distal end moderately expanded with a single concave facet for lunar and scaphoid.   Femur with prominent third trochanter at middle of shaft, lesser trochanter prominent, internal, set rather low, great trochanter high and widely separate from head, digital fossa wide and open, facing nearly backward.   Shaft curved, its line being concave anteriorly and interiorly, patellar trochlea rather wide and short, subdistal; condyles not especially deep or wide, facing mostly backward.   Tibia short with very prominent long cnemial crest, shaft curved backward, distal end with very oblique flat astragalar trochlea and wide short internal malleolus.   Fibula with straight and comparatively stout shaft, quadrate or round, expanded at upper and lower ends, the proximal facet oblique, the distal facet facing more inward than in *Ectoconus* or *Periptychus*.

Fore and hind feet five-toed, with short, flattened, spreading phalanges, the unguals small, round, flat hoofs.   Manus and pes of subequal size and proportions.   Carpus wide, short, "alternating," with wide lunar-unciform facet, large rectangular centrale set obliquely between lunar and trapezoid, magnum with very small dorsal surface, trapezoid small, shallow, trapezium somewhat reduced and flattened in comparison with creodonts; five well-developed, short, and somewhat flattened metacarpals, the third slightly larger than II and IV, the fifth somewhat and the first considerably smaller.   Phalanges somewhat shorter than in *Periptychus*, considerably shorter than in *Ectoconus* or *Dissacus*, unguals flattened but twice as long as their width, as long as, or longer than, the proximal phalanges.

The tarsus is also "alternating," the astragalus having a considerable facet on the cuboid, as in Periptychidae and Mesonychidae.   Astragalus with almost no neck, head wide, moderately deep, with three distal facets on it for cuboid, navicular, and navicular hook.   Astragalar trochlea flat, short and limited backward by the foramen, as in Periptychidae; a deep pocket in front of the trochlea, and the more internal position of the cuboid facet, facing more distally and widening the neck to a marked degree, are the most important differences from that family.   The character of the internal and external malleolar facets is much as in *Periptychus*.   The calcaneum has a fibular facet, a crest extending from the peroneal tubercle along the external side of the tuber, and other characters as in *Periptychus*.   The cuboid has a much broader astragalar facet than in Periptychidae.   The navicular hook is a completely separate bone articulating with the body of the navicular by a well-marked facet.   The three cuneiforms articulate to the body of the navicular; the entocuneiform is reduced and flattened as compared with Periptychidae. The five digits of the pes are similar in proportions to those of the manus, the first considerably, the fifth comewhat, reduced in comparison with the others, much as in Peritychidae, but to a slightly greater degree.   The ungual phalanges are somewhat shorter than those of the manus, the others quite similar.

The above characters appear to be common to the two species of *Pantolambda*, the chief difference of *P. cavirictus* being a greater massiveness throughout, shorter ungues, heavier entocuneiform and first digit, and in the skull a greater massiveness of the facial region and somewhat deeper jaw.

Comparison with the Periptychidae shows a near agreement throughout, especially with *Periptychus* and *Ectoconus*, in teeth, skull and skeleton. *Pantolambda* is diversely specialized in the selenodont cheek teeth as against the polybunodont tendencies of *Ectoconus* and *Periptychus* and the enlarged premolars of *Periptychus*. It carries further the tendency in the larger periptychids to enlargement of canine and incisors, deepening and broadening of symphysis and front of muzzle, early stages in coryphodont specializations. The broadening of the nasals and the backward shifting of the jugals are presumably associated with this advance. It retains the short, wide upper branch of the premaxilla much as in Periptychidae, and the primitive placental characters of the skull seen in the rudimentary postorbital processes, narrow long postorbital construction, long nasals expanded behind, small brain-case, the high sagittal and occipital crests, exposed mastoid, open auditory meatus, undeveloped tympanic region.

The skeleton throughout is closely and evidently related to the larger periptychids. The short, stout limbs and short, wide, amblypod type of feet are a little more specialized in the direction of *Coryphodon*, but retain much of the primitive characters of the Periptychidae. It has gone a little further in displacement of carpals and tarsals, loss of the primitive semi-opposability of pollex and hallux, shortening and flattening of the phalanges, but retains the large separate centrale and various other details of primitive composition; the form of the astragalus is intermediate between periptychids and *Coryphodon*.

In one feature, the separate navicular hook, the tarsus is more primitive than the known periptychids or creodonts. This bone was *inferred* by Cope and Osborn from the apparently separate facet on the inner side of the astragalar head in *Bathmodon* (= *Coryphodon*), *Ectoconus*, *Periptychus* and *Pantolambda*. It was identified as the tibiale of reptiles, and has figured rather extensively in theoretical literature concerning the origin of the tarsus. Its existence has been somewhat uncritically accepted or rejected and variously interpreted by anatomists quite innocent of any practical acquaintance with the evidence and unable to judge of its adequacy. Matthew in 1897 expressed some scepticism as to the presence of a "tibiale" in any of the Paleocene mammals:[95] "The presence of a tibiale has not been shown, although an apparent internal facet is more or less distinguishable on the head of the astragalus in several species; but that this really indicates a bone I am by no means certain." The reason for this scepticism was that the facet is not uncommonly distinguishable in some modern Carnivora, which certainly do not have any ossified "tibiale."

The material now available shows that the bone is really present as a separate ossified element in *Pantolambda*, that it is united to the navicular in *Ectoconus*, united and reduced in *Periptychus* and in *Claenodon*, although its facet is still distinct from the navicular facet in *Ectoconus* and *Periptychus*, obscurely separable in *Claenodon*. In these and later mammals it becomes the navicular hook, and the facet for the entocuneiform, confined to the true navicular in *Pantolambda* and still retaining this position in *Ectoconus*, is shifted over to the distal surface of the navicular hook in *Periptychus* and *Claenodon*. These various stages in its union with the navicular and further reduction may readily be paralleled among modern rodents, marsupials, and in *Hyrax* among the ungulates. The conditions in the

[95] Matthew, W. D., 1897, Bull. Amer. Mus. Nat. Hist., IX, p. 321.

Paleocene mammals show that the "tibiale" really is an element of the tarsus among the oldest Tertiary mammals.   Whether it is an original element of the tarsus in Mesozoic mammals and the pro-mammalian reptiles is another question.

In *Claenodon*, and less certainly in *Periptychus*, another small bony element appears in somewhat the same position but a little more proximad.   It quite evidently is *not* the navicular hook but a distinct ossification perhaps of a sesamoid nature, which may or may not have any correspondence with the "pre-hallux" or "tibiale" of modern rodents and marsupials.

If the navicular be the mammalian tibiale, as Broom believes, then all of these ossifications are best interpreted as sesamoids subsequently incorporated with the original tarsal elements.

Pantolambda bathmodon Cope

Pantolambda bathmodon Cope, 1882, Amer. Nat., XVI, Apr. 24, p. 418; 1883, *ibid.*, XVII, March 15, p. 406; 1884, *ibid.*, XVIII, pp. 1112–13, figs. 3–5 (teeth, skull, jaws and parts of skeleton).

Type: A. M. Cope Coll. No. 3956, part of the lower jaw with p_3–m_1 preserved.

Horizon and Locality: Paleocene of the San Juan basin, undoubtedly from the Torrejon formation, Middle Paleocene.

Author's Description: Char. specif.   The bases of the P–m. III and IV are subquadrate, the inner side rounded, that of the IV relatively the wider.   On the III the median keel constitutes the heel; on the IV, the keel is in the center of a wide heel.   No cingula.   The first true molar has an anterior cingulum, but no external one.   The enamel is wrinkled where not worn.   [Measurements follow.]

The principal specimens of *Pantolambda bathmodon* available for study are:
No. 3956, type specimen, lower jaw, p_3–m_1.
No. 3957, incomplete skull with parts of skeleton (described and figured by Cope).
No. 3958, incomplete skull with parts of skeleton.   Cope Coll. (but not put together or studied until recently).
No. 2546, fragmentary skeleton, weathered.   Exped. 1896.
No. 2547, fragmentary skeleton, weathered.   Exped. 1896.
No. 16663, skeleton nearly complete, articulated.   Exped. 1913.   No. 16665, a skull.
No. 16664, incomplete skull and jaws and fragmentary skeleton.   Exped. 1913.
No. 16666, lower jaws and part of skull, with part of skeleton.   Exped. 1913.

Besides the above there are various parts of upper and lower jaws which serve to show the range of variation in the teeth.

Cope's descriptions of *Pantolambda bathmodon* were based upon Nos. 3956 and 3957. Osborn's revised and extended description in 1898 was based chiefly upon Nos. 2549–51, 2545–47 and 3957.   The skeleton, No. 16663, discovered by W. J. Sinclair in 1913, confirms the characters of this composite in most particulars but is deserving of a more detailed description than the condensed summary of the characters of the genus and species given by Osborn.

Description of No. 16663 (*P. bathmodon*)

Upper Teeth.   The incisors are of moderate size, pointed, subconical, not at all spatulate, and considerably spaced apart.   All of them are round convex on the anterior face, flattened posteriorly, with postero-lateral marginal crests not very prominent.   The first and second are subequal, the third a little larger.   This type of incisor, resembling a canine in form, is remarkably similar to the teeth of thecodont reptiles, and in particular suggests some of the Permian cynodonts and related types from the Karroo formations of South Africa.   It is also shown in *Periptychus* and *Ectoconus*, and may be characteristic of all Taligrada so far as the evidence goes.

The canines are almost round in cross section, stout with conical crowns, the anterior face worn to a smooth, flat, vertical surface by the lower canine, the posterior face worn down to a certain point, partly by the lower canine, partly by the first lower premolar. This gives the half-worn canine a transverse trenchant edge in some cases, while in others the wear results in shaving away the posterior surface, affecting the anterior surface very little.   While much individual variation is thus evident, it is equally clear that the wear of the canines always involves a marked vertical shearing action between the upper and lower canines.   This shearing movement evidently controls to a great extent the shearing interaction of the premolar and molar teeth.   When the lower canine bites in front of the upper canine, the anterior shearing edges of the lower teeth, as the jaws close, are held pressed against the posterior shearing edges of the upper teeth and the slight curving obliquity of the molar and premolar crests tends to bring about a transverse movement of the jaw, the shearing crests moving transversely past one another as the shear closes vertically.   Similar relations on the opposite crests or sides of the cheek teeth result when the lower canine bites behind the upper one, as it apparently does almost as often as in front. The type of canine in *Coryphodon* is much the same, probably serving the same purpose.

The premolars are rather crowded behind the canine, the first premolar being close to it, but somewhat more internal than posterior in position.   It is a small tooth, single-rooted and simple-crowned, considerably compressed, with a moderately high apex and sharp anterior and posterior crests.   The second, third and fourth premolars are three-rooted teeth, with triangular base, increasing progressively in transverse width, each with an inner and an outer crescent, the inner crescent being quite small on p^2, hardly more than a cingulum, decidedly larger on p^3 and much larger on p^4.   The outer base of the tooth is deeply notched between the anterior and posterior roots, and the wings of the outer crescent extend into the prominent angles of the crown directly over the two outer roots.

These premolars are very like the corresponding teeth in *Coryphodon*.   They differ in the more compressed and trigonal form of the crown, and less development of the lingual crescent, especially upon the anterior premolars.   In *Coryphodon* it is distinct, in some species at least, even upon p^1, and quite large upon p^2, and its lateral wings extend outwards and downwards upon the anterior and posterior sides of the premolar much broader and more prominent than in *P. bathmodon*.   *P. cavirictus*, as will be seen, is intermediate.

The premolars of the uintatheres differ much more from those of *Pantolambda*, consisting in *Uintatherium* of two crests joined internally to form a single very deep V open externally.   This might be regarded as representing the outer crescent only of *Coryphodon* and *Pantolambda*.   Examination of the more primitive genera *Bathyopsis* of the Lost Cabin (Lower Eocene) and *Probathyopsis* of the Clark Fork (Uppermost Paleocene) indicates that this is not the case, but that the premolar V represents three cusps of a more or less molariform premolar, the apex of the V being therefore the protocone and not the para-metacone as it is in *Pantolambda*.   The uintatheres apparently did not pass through any *Pantolambda*-like stage in the premolars.

The fourth premolar of various Artiodactyla shows much the same construction as the premolars of *Pantolambda* and *Coryphodon*.   The resemblance is perhaps closest among the oreodont and anthracothere genera.

The upper molars of *P. bathmodon* are comparatively wide transversely and have the three principal cusps strongly crescentic.   The metacone and paracone are of equal size on m^{1-2}, with the anterior wing of the paracone and posterior wing of the metacone crescent extended into prominent external angles; the adjacent wings of the two outer crescents are shorter and come together with a very rudimentary mesostyle pillar at their junction. The protocone crescent is much larger than either of the outer pair, its wings sweeping down into basal cingula which rise up again at the external angles to join the wings of the outer crescents.   The paraconule, situated upon the anterior wing of the protocone, is distinct, though small; the metaconule is obscurely indicated on the posterior wing of the protocone of m^1, absent on m^2 and m^3.   A strong basal cingulum on the posterior side of the protocone represents the hypocone and a weaker one on the anterior side the protostyle.   There is no cingulum on the inner face of the protocone.   A narrow basal cingulum along the outer border of the molar, curving up at the external angles to join the tips of the wings from paracone and metacone.

The third upper molar differs from the others in the reduced metacone, with posterior wing smaller than the anterior, instead of being much larger as it is in m^1 and m^2.   The anterior wing of the paracone is slightly more extended than in m^1 and m^2, and the external end of the wing-crest is turned more backward, curving around more broadly to join the external basal cingulum, and more clearly distinct from the external end of the anterior cingulum.   The hypocone cingulum is somewhat less developed than in m^{1-2}, being equal to the protostyle cingulum, and the protocone crescent is distinctly narrower, its apex facing a little backward instead of directly inward (linguad) as it does in m^{1-2}.

The above described molar construction is very different from that of *Coryphodon*, and much more primitive.   In the coryphodonts, as Osborn has shown,[96] the paracone has become vestigial or wholly disappeared, the posterior wing of the protocone and the posterior wing of the metacone are degenerate in varying degree in different species and in different teeth of the upper molar series.   The tooth is essentially reduced to a large antero-internal crest and a smaller postero-external crest.   It is quite obvious that the *Coryphodon* molars may be derived from the pantolambdid type, but by no means from the genus *Pantolambda* and still less from *Titanoides*.   There is no tendency in *Pantolambda* to reduction of the paracone or of the anterior wing of the protocone, while on its third molar there is a very marked reduction of the metacone and of the posterior wing of the protocone, which would have to be completely reversed in order to progress towards *Coryphodon*.

The upper molars of the uintatheres are so entirely different in construction that no direct relations can be seen between them and those of *Pantolambda*.   Their structure can perhaps be best interpreted as a specialized derivative of the *Coryphodon* molar, but this would need to be verified by a careful and critical study of the upper teeth of all the Uinta-theriidae (Eocasileidae), which has not yet been made.

On the other hand, the relations of the upper molars of *Pantolambda* to the primitive tritubercular pattern are very clear and very close.   The chief change from the type presented in the Oxyclaenidae consists in the crescentic specialization of paracone and metacone

[96] Osborn, H. F., 1898, Bull. Amer. Mus. Nat. Hist., X, p. 173.

and the more marked emphasis on the crescentic character of the protocone, already distinct in the primitive trituberculates.

The teeth of *Pantolambda* are in every respect nearer to those of Oxyclaenidae than to any of the Periptychidae, which have departed from the primitive tritubercular pattern in a diverse line of specialization.

LOWER TEETH.   The lower ·incisors correspond in size to those of the upper jaw; the first is the smallest, the second slightly larger, and the third is larger than the second. Their crowns are decidedly shorter and more crested than the upper series, especially i_3, which has a crown almost as wide as high, the crest rising medially to a pointed apex and slightly expanded at the base of the crown; it approaches p_1 in construction and proportions and is but slightly smaller.   The lower incisors of *Coryphodon* have a much shorter crown, flat posteriorly, with compressed, crested margins, expanded at the base and rising to a central apex.   They are relatively larger and stouter teeth.   The third incisor of *Pantolambda* shows some approach to the coryphodont type.

The lower canine is a large, stout tooth of round cross-section and convex-conical form, equal in size to the upper canine, and very similar in shape, with weak anterior crest.   The principal wear of the tooth is a flat vertical shear on its posterior or postero-internal face. The tooth is set almost vertically in the jaw, with a marked backward pitch, especially towards the tip, which is corrected by the vertical shearing off of the posterior face of the tooth.   The lower canine of *Coryphodon* is shorter-crowned and more of the pointed-spatulate type of the incisors and anterior premolar.   It is hardly as large relatively as in *Pantolambda*, and is set much more laterally and not in any degree backwardly; a very marked posterior basal heel is present, which is wholly wanting in *Pantolambda*.

The lower incisors and canine of *Uintatherium* are of very peculiar type.   Each has a posterior heel which has grown up into a pointed cusp almost equalling the primary cusp in size, both cusps being pointed, convex-conical and moderately high.   Marsh's restorations of the lower incisor and canines of various uintatheres in the plates of his monograph upon the "Dinocerata" are far from being correct.   They were apparently based upon two or three scattered teeth figured in the text of the memoir.[97]   There was little excuse for this, as Osborn's accurate drawings of the lower incisor-canine series in "*Loxolophodon*" had been published at least three years previously.[98]

The lower premolars of *Pantolambda bathmodon* follow directly behind the canine, without diastema, but p_1 is partly internal, its alveolus lying directly behind the inner margin of the canine alveolus. P_1 and p_2 are one-rooted, although in some specimens p_2 has the root incompletely divided.   On p_3 and p_4 there are two distinct roots. P_1 has a nearly simple crown, compressed-conical, with pointed tip, strongly anterior crest, slight and

[97] Marsh, O. C., 1884, "The Dinocerata," U. S. G. S. Monograph No. 10, p. 42, text fig. 43.   This volume bears date 1884 on the title page, but apparently the author's separates were not published until January or February, 1885.   Among the reasons for this conclusion is that on p. 237 of the author's separates is a citation giving page and figure numbers of the abstract of the monograph published in the U. S. Geol. Surv. Annual Report, which did not come out until 1885, and could not well have been in page proof until the end of 1884.   The monograph is reviewed in 1885, in the Amer. Nat., Amer. Jour. Sci. (abstract before publication), Nature, Pop. Sci. Monthly, Geol. Magazine, Neues Jahrb. Mineral., and Ann. Sci. Géol. In the latter four publications there are abstracts.—W.D.M.

[98] Osborn, H. F., 1881, A Memoir upon *Loxolophodon* and *Uintatherium*.   Contr. E. M. Mus. of Geol. and Arch., Coll. of N. J., I, No. 1, p. 30, pl. iii, figs. 1, 3, 4.

sub-parallel posterior and postero-internal crests. P_2 is about twice as large, robust, nearly as wide as long. The three crests are much more prominent, the posterior crest shifted over to a postero-internal position and the postero-internal to an internal position. The internal crest is separated from the sharp anterior crest by a broad, uniformly concave valley; a more contracted valley separates the internal from the posterior crest; but these two crests unite about half-way up to the tip of the crown. P_3 is very like its predecessor, except for larger size and more prominent crests. The anterior crest here sweeps from the antero-internal corner of the crown outward and backward to the tip of the protoconid, thence curving around to a transverse crest, which, when it reaches the inner border of the tooth a little behind the middle, is deflected backward and drops rapidly to the inner border of the base of the crown. The posterior crest, springing from the flank of the transverse crest just mentioned, curves around the heel margin until it meets the deflected crest at the inner anterior corner of the heel and near the base of the crown. P_4 is somewhat larger than p_3, the anterior or paraconid crest is somewhat more transverse, the heel is larger and is buttressed from outside by a rather massive crest sweeping up obliquely from the postero-external base of the trigonid; from this inner border of the transverse crest (the metaconid) a crest extends down and back, partly closing in the heel basin from within.

The lower premolars of *Coryphodon* differ chiefly in that the metaconid and especially the paraconid ends of the principal crest are low, the crest falling away from the high protoconid apex nearly to the base of the crown within. The heel-cusps of p_3 and p_4 differ in a corresponding sense, the posterior crest falling away rapidly at the posterior border of the heel to the base of the crown internally, and buttressed externally on both p_3 and p_4 by a crest like that of p_4 in *Pantolambda*.

The effect of these changes is to reduce the premolar heels in *Coryphodon* to simple normal crested heels. In *Pantolambda* they soon assume this character on wear, but the pattern of the unworn tooth appears to be as described above.

The molars in *P. bathmodon* are chiefly made up of two sharp V's opening inward. The anterior V, or trigonid, is a little higher but slightly smaller than the posterior V, or talonid; when unworn these V's are seen to be composed of semi-distinct angulate cusps connected by sharp, high crests. Of the trigonid cusps the external or protoconid and internal or metaconid are of equal height, the paraconid (antero-internal) a little lower. The heel has a high external cusp (hypoconid), a smaller postero-median (hypoconulid) imperfectly distinct, but no postero-internal (entoconid) cusp. The two external cusps of each tooth (protoconid, hypoconid) have the outer faces compressed-convex and nearly vertical, the inner faces flat, falling rapidly to a contracted valley opening inwardly. The metaconids have, in addition to the crest connecting them with the protoconids, a strong postero-internal crest falling rapidly along the inner border of the heel to the bottom of the talonid basin. Thence the inner marginal crest of the heel rises to the hypoconulid, which is connected in turn with the hypoconid by a broadly notched crest. Sharp and moderately strong anterior and posterior basal cingula, none on inner or outer face of molar.

On m_3 the hypoconulid is much stronger, equal in size to the hypoconid and slightly higher. It has a flat anterior face and a strongly convex posterior face, recurved forward to a marked degree. As in Amblypoda generally, there is no distinct third lobe on the third molar.

The above description is based upon unworn teeth. The characters are best seen in Nos. 2551 and 2552. With wear many of the details are soon obliterated.

The molars of *Coryphodon* differ in the great reduction of the paraconid crest and of the anterior wing of the hypoconid V; also that on m_1 and m_2 the inner posterior cusp is in the normal position of an entoconid, instead of being submedian as it is in *Pantolambda*. On m_3 the position of the cusp is the same, but its inner crest, low in *Pantolambda*, is higher than the broadly-notched crest which connects it with the hypoconid. This is the condition in the more primitive species of *Coryphodon*, but in others the inner crest of m_3 may be constricted off to form a true entoconid, and in still other species there appears to be no hypoconulid, but only an internal and external pillar connected by a high crest, as in m_1 and m_2.

In *Uintatherium* the molar pattern is further altered, derived apparently from *Coryphodon* and with very little traceable to the primitive pattern of *Pantolambda*. The posterior crest has shifted to an oblique position, the valley between it and the anterior crest opening outwards instead of inwards as is usually the case, a protostylid develops in place of the metastylid so commonly seen, and in other respects it agrees with the upper teeth in reversing the usual construction as between lingual and buccal sides of the tooth. The posterior cingulum of *Coryphodon* has become much larger and more prominent.

The *Pantolambda* molars, on the other hand, show no very wide departure from the oxyclaenid pattern. The specialization consists chiefly in the sharp cresting of the teeth. The molars of *Deltatherium* have much the same pattern, but the cresting is less and is less transverse, the entoconid is distinct, the hypoconulid more prominent.

Summary of Teeth Characters. As a whole the teeth of *P. bathmodon* have a very primitive aspect. They are much nearer to the primitive placental pattern as seen in Oxyclaenidae than to the specialized amblypod pattern as seen even in *Coryphodon;* to the uintatheres the resemblance is still more remote. At the same time they do unquestionably show a marked advance towards the amblypoda—but so far as the molars are concerned, a partly divergent specialization, as noted above. It is in the premolars and canines that the nearest approach to *Coryphodon* is seen. To the uintathere teeth there is hardly any recognizable approach in constructive detail, and it must be remembered that this pattern is not a late Eocene development but is found already quite fully specialized in the Clark Fork beds, only a little later than the horizon of *Pantolambda*. The teeth indicate therefore that *Pantolambda* is in a broad way a collateral ancestor of the Amblypoda; it certainly cannot be anything like a direct ancestor of *Coryphodon* and can have but distant relations to the uintathere phylum. It is a very primitive type, and its amblypod affinities have been unduly stressed.

SKULL. The skull of *P. bathmodon* is rather short and wide as compared with the majority of Paleocene mammals. It has the usual primitive features of small brain-case, high continuous sagittal and occipital crests, absence of postorbital processes, long nasals expanded posteriorly, inter-orbital constriction narrow, elongate, giving this portion of the skull an almost tubular appearance. In the best skull, as shown in the photographs, the natural shortness of the muzzle has been accentuated by buckling due to pressure which has pushed the nasals, frontals and entire back portion of the skull forward about half an inch over the premaxillary-maxillary-jugal group. The overlap can be easily seen on the zygomatic arches and the premaxillo-nasal suture. The skull figured by Cope is now considerably more nearly complete than when he figured it, Matthew in 1897 having fitted many additional fragments to the broken edges, but it is still very incomplete and distorted. Two other skulls, less distorted but also incomplete, were secured in 1896, and two partial skulls, besides the fine one here figured, were obtained in 1913. In addition to these there is a fragmentary skull and jaws in the Cope Collection.

The muzzle is comparatively short and wide, the nares nearly terminal. The premaxillae are rather large, stout bones with wide and short ascending process, not extended far backward, but cut off abruptly. The two premaxillae meet in the middle line, although there is a considerable gap between the median incisors. The maxillo-premaxillary suture is almost vertical, but with a strong sigmoid curve, bending backward in the lower half until it is somewhat behind the vertical above the anterior border of the canine, then forward a little, and then backward to the nasal suture.

The maxillary is short, wide and deep. The suborbital crest is very prominent and massive, the anterior border of the orbit far forward, being situated above p⁴. The infraorbital foramen is above p³, and is rather large. It is larger than in *Periptychus* but its position is the same; the orbit is somewhat more posterior, the suborbital rim thicker, projecting more laterally and less vertically. The jugals are not clearly distinguished from

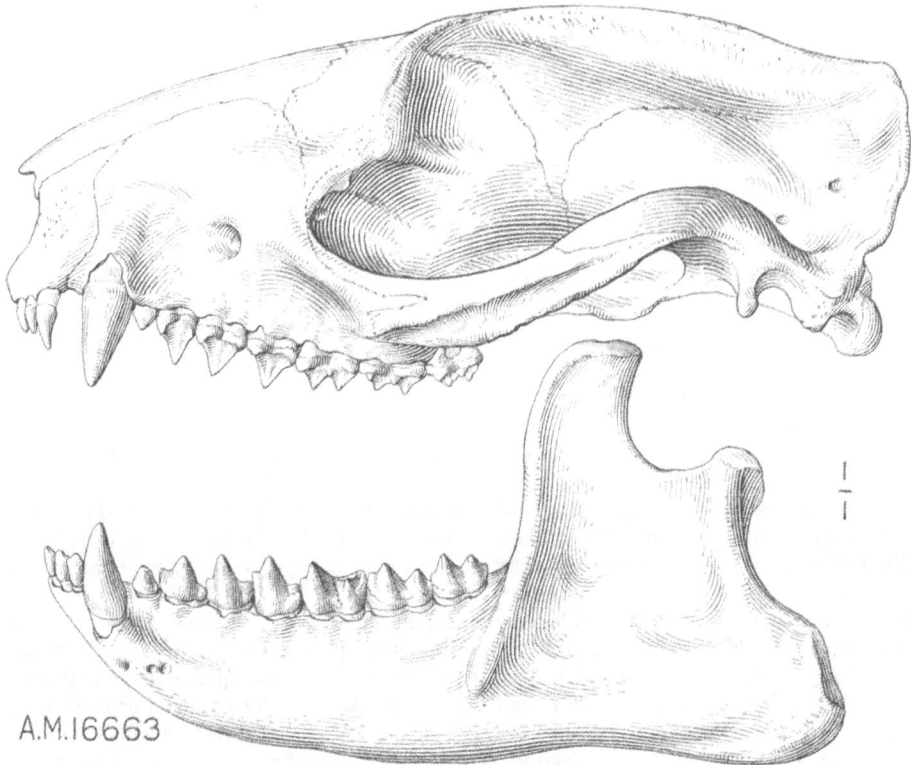

FIG. 38. *Pantolambda bathmodon*, skull and jaws, A. M. No. 16663, side view. Natural size.

the maxillae in any of the specimens. They are long, moderately stout, with an inferior crest and an obscure lateral ridge that runs parallel to it; the latter becomes more distinct and passes obliquely downward above the molars. In *Periptychus* this crest is less distinct, but there is a deep, round pocket in the maxilla just under the anterior end of the jugal that is wanting in *Pantolambda*. The nasal bones are rather wide and long, extending from a point above i³ to a point considerably behind the anterior rim of the orbit. They are expanded slightly at the anterior end and more considerably at the posterior end, but not so much relatively as in marsupials, the proportion being much as in *Deltatherium*, but the relative length of the nasals is considerably greater. The two nasals meet at a slight angle so that their conjoined surface is "obtusely roof-shaped." The frontal bones are very

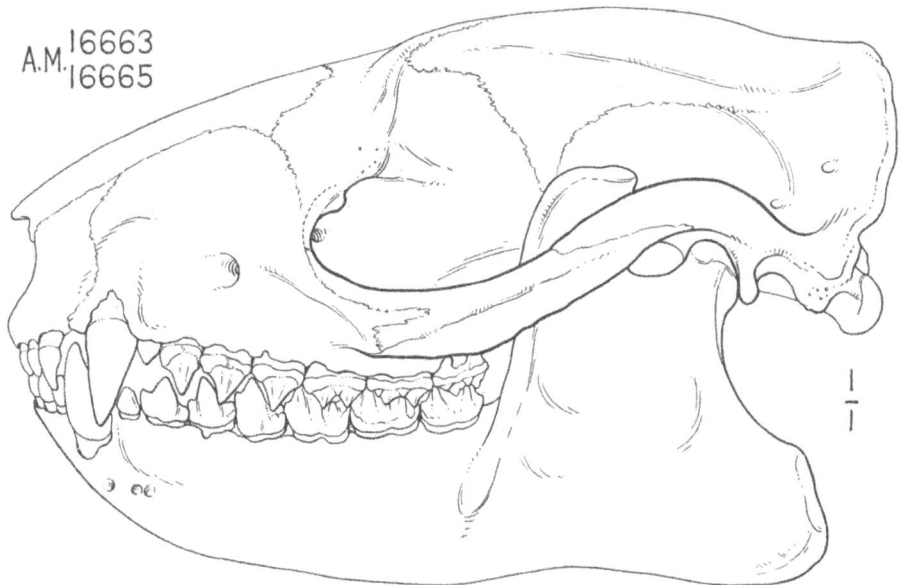

FIG. 39. *Pantolambda bathmodon*, skull and jaws in articulation, A. M. Nos. 16663 and 16665. Natural size.

short in the median line but extend forward laterally above and in front of the orbits, embracing the posterior ends of the nasals. Their suture with the parietals is transverse, situated at the point where the orbital crests join in the sagittal crest. The orbital crests are quite low, forming an obtuse angle on the surface of the frontals. In three of our skulls there is no excavation or concavity between them, in contrast to most creodonts; in a fourth (No. 16665) there is a shallow concavity. The parietals are long, narrow, forming the upper part of the tubular postorbital region, overlapped posteriorly by the squamosals so that they are not broadened posteriorly until close to the occipital crest. The sagittal crest is high, extending along the whole median line of the parietals; the occipital crest is equally prominent, high and wide. The squamosal is large, its cranial plate extensive, with a very considerable lateral expansion of the cranium which brings the bases of the zygomatic arches rather far apart. The zygomatic processes are moderately heavy, their superior crest continuous, but not so prominent as in *Periptychus*.

The occiput has no great amount of overhang, as it has so commonly with creodonts, but tends rather to be of insectivore type, broad and semicircular in its superior outline. The occipital crest is continuous down to the post-tympanic angle of the squamosal, which is hardly prominent enough to be called a process.   The mastoid has a rather small lateral exposure, but a more considerable inferior one, the mastoid process large and robust. The mastoid exposure is by no means as large as in *Periptychus*, nor is the process so promi-

A.M. 16663

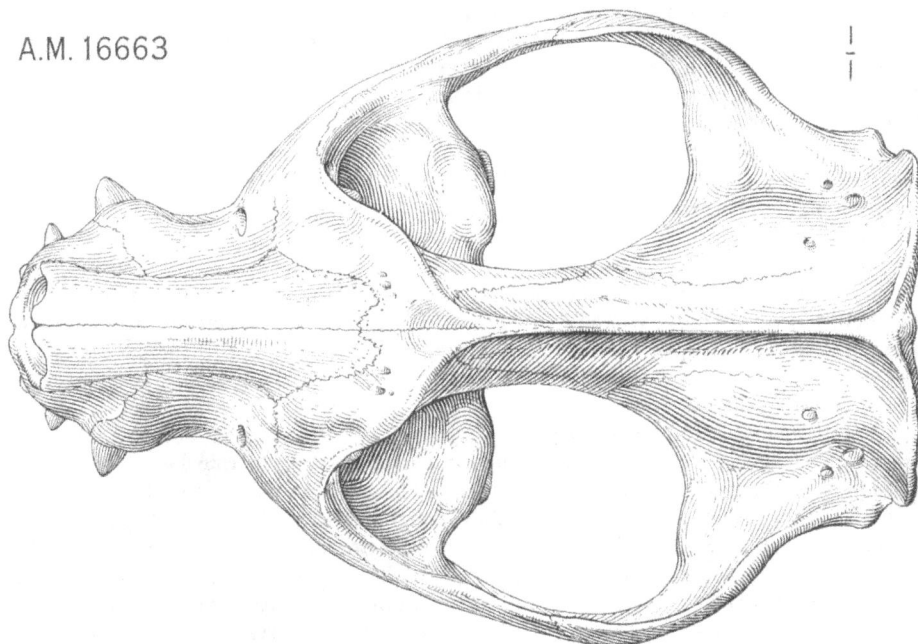

Fig. 40. *Pantolambda bathmodon*, skull, A. M. No. 16663, top view. Natural size.

nent. The condyles are comparatively large, moderately wide apart, not in any way remarkable.

The lower jaw is of rather characteristic shape, comparatively short, deep and massive, especially in front.   The chin is almost as deep as any other part of the jaw, and very thick, the symphysis solidly coössified in adult animals, sutural only in immature individuals. The inferior surface slopes upward at about 45°, making a broad, flattened plane on both sides of the median line beneath the incisors.   Two small mental foramina are beneath the median incisors, a much larger pair below the canines and another beneath the anterior part of p_2. The last is sometimes doubled, or is replaced by two foramina, beneath p_1 and p_2 respectively; the subcanine foramen is sometimes absent.   None of our specimens shows any mental foramina farther back than p_2.   This forward migration of the mental foramina, as compared with their primitive relations in the Creodonta and Condylarthra, seems to be also characteristic of *Periptychus*, as it is of *Coryphodon*, but in neither genus so marked as in *Pantolambda*. The coronoid process rises abruptly, almost at right angles to the alveolar line, is short and thick, not greatly recurved. The masseteric fossa is shallow, with the anterior border defined by a heavy crest, but it fades out inferiorly. The condyle is at a considerably higher level than the tooth-line; the angle is wide, flat, prominent backwardly.

In *Periptychus* the symphysis of the jaw is much less deep or abrupt, though equally massive, the depth of the jaw less uniform behind it; the jaw is thicker, the coronoid process projects more backwardly, is more recurved and slender toward the tip, the angle usually more prominent inferiorly, less so posteriorly.

Skeleton. The cervical vertebrae are preserved complete and articulated but mostly buried in very hard matrix. They are not so short in the centra as those of *Periptychus carinidens*, but distinctly shorter than in *Ectoconus*, and the construction appears to be much the same as in these two genera. The atlas and axis present in all details a close resemblance to those of the somewhat larger *Ectoconus* skeleton and are equally close to *Periptychus*, with which they agree better in size. The remaining cervicals, so far as they can be seen, present no peculiarities worthy of note. The sixth has the usual wide dependent inferior lamella of the transverse process, and the transverse process of the seventh appears to be imperforate. The dorsal vertebrae are so buried that, if all preserved, they are not visible. The three last dorsals are much as in *Ectoconus*, except that the centra are rounded beneath without median ridge and the anterior zygapophyseal process is less prominent and is compressed into an oblique plate instead of its rounded and more massive form in *Ectoconus*, and the facets are somewhat less convex on the posterior or concave on the anterior zygapophyses.

Five lumbar vertebrae are preserved, very similar to the posterior dorsals except for progressively larger size of the centra toward the sacrum and for the wide flat lumbar transverse processes, apparently increasing in width and probably in length from the first to the seventh. (The true sequence of the vertebrae is not certain in their present state of preservation and preparation.) The transverse processes are directed a little forward, chiefly outward, about as in *Ectoconus*. The sacrum consists of three coössified vertebrae instead of four as in *Ectoconus*, the sacral vertebrae being somewhat shorter relatively as well as smaller, and their transverse processes less fully united into a lateral plate. The transverse process of the first sacral forms most of the connection with the pelvis, but the anterior part of the second transverse process also takes part in it; behind this suture is a free border with no evidence of any connection of the transverse plate with the pelvis behind the middle part of the second sacral. The neural canal decreases rapidly in size from the first to the third sacral, the sciatic foramina, especially the first, being large. The first two caudals have transverse processes much like those of the last sacral but wholly free and somewhat shorter; the transverse processes decrease in width in the three following caudals but without much decrease in the size of the centra. The remaining caudals are imperfectly known, but the tail is long and heavy, as shown by various other specimens of *Pantolambda*.

The ribs are short and weak, not much curved, and even the anterior ones show a relatively small amount of flattening.

Shoulder Girdle. The scapula is remarkable for its very high spine, long acromion and well-developed metacromion, long stout recurved coracoid process, and the wide anterior blade and posterior blade narrow at the base.   This is the only Paleocene mammal in which the scapula is perfectly preserved, and its remarkably primitive characters are more clearly evident than in any of the other genera, in which incomplete scapulae have been preserved.   So far as the evidence goes, the other genera may be as primitive in the shoulder girdle as is *Pantolambda*.   The spine is certainly very high and the acromion long in *Ectoconus*, but this genus has the posterior blade wide at base as well as toward the apex. The clavicle is a long and heavy bone, moderately curved, of round-oval cross-section somewhat widened at the sternal end, widening into an oblique-ending expansion at the acromial end.   The bone is quite as heavy as in the beaver or in man, and has none of the sigmoid curvature of the primate scapula.   It does not differ materially from the clavicle of *Periptychus* or *Ectoconus*.

Pelvis.   The pelvis is articulated with the sacrum, and complete except for the pubis, most of which is missing.   In form and proportions it agrees rather nearly with that of the phenacodonts.   The ilium is a long, trihedral rod, decurved toward the anterior end, with a flange on the superior border.   The flange is widest towards the anterior end, decreasing gradually toward the neck of the ilium; in the phenacodonts it carries the width back to the neck and ends abruptly.   The decurvature of the rod and width of flange is less than in *Phenacodus primaevus*, rather more than in the small *P. astutus* or in *Tetraclaenodon*. The ischium is long, narrow, flattened, expanded toward the posterior end and progressively twisted into an oblique plane.   The pubis is a rather stout rod, apparently directed partly backward instead of wholly downward and inward as in *Phenacodus*, but this may be a result of crushing.   The pelvis of *Periptychus* and *Ectoconus* is of this same general type so far as fragmentary material shows.   The pelvis of *Meniscotherium* appears to be much like that of *Tetraclaenodon*.   The pelvis of creodonts has somewhat the same general proportions as in these taligrade and condylarth genera, but, so far as known, none of them shows the decurved tip of the ilium.   Unfortunately no Paleocene creodont has the tip of the ilium complete.   The ilium of *Coryphodon* and *Uintatherium* is wholly different, expanded into a broad, flat plate as in the Proboscidea.   To a large extent probably this is correlated with size, but it forms a striking difference between taligrades and amblypods as at present known, and emphasizes the relationship between taligrades and condylarths.

Limb Bones.   The limb bones of *Pantolambda bathmodon* have been excellently figured and described, partly by Cope in 1885-8, more fully by Osborn in 1898.   Some comparisons and minor emendations may now be added.   The humerus of *P. bathmodon* is like that of *Ectoconus* except for a higher and more compressed deltoid crest.   The supinator crest is equally prominent, much more so than in *Periptychus*, the distal end is very similar in its sudden broadening on the entepicondylar side, the large entepicondylar foramen and the character of the articulations for ulna and radius.   It is not so close to *Periptychus*, and differs from the creodonts much more in its short, robust proportions and greater development of deltoid and supinator crests.   In *Coryphodon* the greater tuberosity is longer and more prominent, the deltoid more massive but equally long and prominent, the supinator crest somewhat reduced and the entepicondyle greatly reduced, and the distal end of the bone much deeper as well as less expanded laterally.   The ulna has the same broad, flat shaft, long olecranon and absence of styloid process as in periptychids and primitive creodonts, and is intermediate in proportions between *Ectoconus* and *Periptychus*, but with the anterior groove of the shaft more vestigial than in either.   The radius is also much as in these two genera; the head is more flattened than in *Ectoconus*, but ovate instead of oval as in *Periptychus*, and the anterior crest of the shaft is prominent as in *Periptychus*. In *Coryphodon* the radius and ulna are more massive, straight and rod-like but of much the same proportions otherwise. The femur of *P. bathmodon* has the lesser and third trochan-

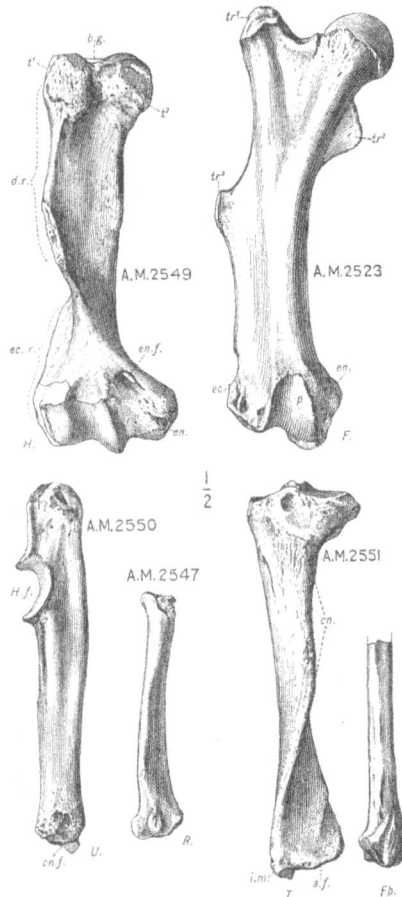

Fig. 41. *Pantolambda bathmodon*, right humerus, A. M. No. 2549, anterior view; ulna, A. M. No. 2550, lateral view; radius, A. M. No. 2547, anterior view; right femur, A. M. No. 2523, anterior view; right tibia and fibula, A. M. No. 2551, anterior views. All one-half natural size.

ters more prominent and more distally placed than in *Ectoconus* or *Periptychus* and the upper part of the shaft is bent inward and forward instead of its axis being a straight line as in these genera or in creodonts and most other primitive mammals. In these particulars it resembles the *Coryphodon* femur; otherwise it is much as in the Periptychidae. The tibia is somewhat shorter relatively than in *Ectoconus* and *Periptychus*, the cnemial crest longer and more prominent than in either genus; the astragalar facet faces less outward and more distally than in *Ectoconus* and somewhat more distally than in *Periptychus*. The shaft of the tibia is noticeably concave behind, not so straight as in *Ectoconus* or *Periptychus*. The shaft of the fibula lacks the prominent antero-internal and antero-external crests of *Ectoconus*, and is much more like that of *Periptychus carinidens* and *rhabdodon*. The tibia and fibula of *Coryphodon* have short, straight, more or less cylindrical shafts, the cnemial crest limited to the upper end of the tibia, not extending far down upon the shaft as it does in the taligrades.

The fore and hind feet of *Pantolambda* have very much the same construction and proportions throughout as those of *Periptychus* and *Ectoconus*. In some significant particulars they make a slightly nearer approach to the feet of *Coryphodon*. The carpals are very like those of *Periptychus*, with shallow scaphoid, large, rectangular, obliquely set centrale, small oval trapezoid, dorsal surface of magnum quite small, lunar with a broad unciform facet and its magnum facet scarcely if at all reaching the dorsal surface, trapezium with its scaphoid and trapezoid facets meeting at a broad angle as in the periptychids (not so much of a "peg" as in *Claenodon*). The metacarpals are similar to those of *Ectoconus* and *Periptychus*, the phalanges somewhat shorter and more flattened than in *Ectoconus*. The manus of *Coryphodon* is very greatly modified from the primitive status shown in *Pantolambda*. The carpals as a whole are broad, short and nodular; the centrale has disappeared and the proximo-distal height of the scaphoid, trapezoid and dorsal portion of the magnum is much increased. The unciform has shifted over on the proximal end of the third metacarpal, covering nearly half of it; the magnum has acquired a footing on the second metacarpal, the trapezium has become more quadrate but still overlaps the inner side of the second metacarpal. The metacarpals are all much shortened, not more than half as long in proportion to their width, while the phalanges are even more shortened up, the unguals being twice as wide as long, instead of half as wide as long, as they are in the taligrade genera.

The pes of *Pantolambda bathmodon* is a little more compact and rounded, with shorter phalanges, than in either *Ectoconus* or *Periptychus*, and of somewhat smaller size than in *Ectoconus*. The astragalus is most like that of *Periptychus*, but differs from all periptychid, creodont or condylarth genera in the shifting of the cuboid facet to face more distally, broadening out at the same time to a width on the cuboid about equal to that of the calcanear facet, and filling the sharp angulation that defines the neck and head from the body of the astragalus. This obviously is a step toward the complete loss of the distinct head of the astragalus that characterizes the Amblypoda. In *Pantolambda* a deep pocket in front of the tibial facet has expanded inferiorly on the external side so as to become continuous with the sustentacular facet; the astragalo-calcanear facet is slightly narrower than in *Periptychus*, decidedly narrower than in *Ectoconus*. The separate "tibiale" or hook of the navicular is a condition more primitive than in *Ectoconus*, in which it is united, and much more than in *Periptychus*, in which it is united and reduced. This bone has not been found as a separate element in Coryphodontidae, but the reduced size of the navicular in some members of that family, especially in *C. singularis* (see Osborn, 1898, p. 214), and its shifting over so as to allow the ectocuneiform to reach the astragalus suggest that the "tibiale" was, and remained, separate in these animals. The condition in *Pantolambda cavirictus* appears to have been partly intermediate (see under that species). The entocuneiform in *P. bathmodon* is considerably reduced and flattened as compared with that of *Ectoconus* or *Periptychus*. The metatarsals have somewhat straighter and more uniformly oval shafts, the lateral digits a little smaller relatively to the three central metatarsals. The proximal and medial phalanges are only two-thirds as long as in *Ectoconus* or *Periptychus* although of nearly the same width; the unguals are much wider and flatter than in *Ectoconus* but differ little from those of *Periptychus*.

The hind foot bones of *Coryphodon* are much more specialized, correspondingly to those of the fore foot. The astragalus has no neck or head, the navicular is much reduced and the distal tarsals are displaced on the metatarsals. The metatarsals are only from half to a third as long relatively, the phalanges are shortened and broadened as much as in the

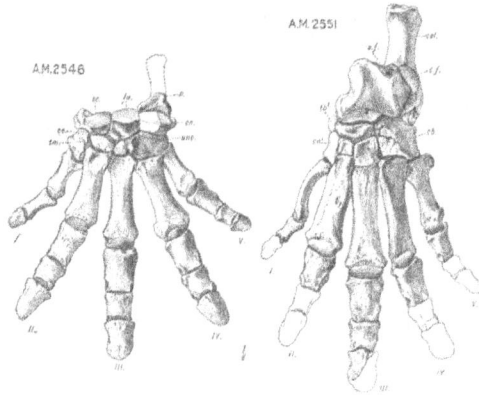

G. 42. *Pantolambda bathmodon*, left manus, A. M. No. 2546, left pes, A. M. No. 2551, anterior views. One-half natural size.

fore foot. Nevertheless the foot of *Pantolambda* shows in an incipient degree many of the characteristics of the amblypod foot and is structurally an excellent ancestral type, at least for *Coryphodon*. In most respects the still more specialized feet of the Dinocerata may also be derived from *Pantolambda*. *Pantolambda* cannot be regarded as a genetic ancestor, even of *Coryphodon*, and still less of *Uintatherium*, as fully specialized *Coryphodon* is found in the Clark Fork and Tiffany Upper Paleocene, and characteristic Dinocerata (*Prodinoceras*) in the Clark Fork and in the Gashato Paleocene of Mongolia. The structure of limbs and feet of these early coryphodonts and Dinocerata is not yet known.

Pantolambda cavirictus Cope

Pantolambda cavirictus Cope, 1883, Proc. Acad. Nat. Sci. Phila., XXXV, p. 169 (read July 3rd); 1884, Amer. Nat., XVIII, p. 1114, fig. 6 (lower jaw).

Pantolambda bathmodon Osborn and Earle, 1895, Bull. Amer. Mus. Nat. Hist., VII, pp. 43–6, figs. 14–16. (Not *P. bathmodon* Cope.)

Type: A. M. Cope Coll. No. 3961, a nearly complete lower jaw.

Horizon and Locality: Middle Paleocene, Torrejon formation, San Juan basin, New Mexico.

Author's Description: The characters are seen, first in the large size, the teeth having twice the linear dimensions of those of the *P. bathmodon;* and second, in the lateral prominence of the inferior edge of the ramus, which produces a concavity of the side of the jaw posterior to the canine teeth.   It is the largest mammal known from the Puerco formation.

The inferior canines are strongly curved, so that the crown is directed upwards and a little backwards. Both root and crown have a round section, but the apex of the crown cannot be described, as it is greatly worn by use in the specimen.   The incisive border is regularly convex, and the three incisors are not of large size, the first being least, and the third largest.   The premolars and molars have the form of those of the *P. bathmodon.*   The latter present two V's, the anterior narrower and more elevated.   In the former the posterior V is represented by a short crest.   The last molar is produced into a heel, which supports the posterior branch of the posterior V, and no cusp.   The first premolar is one-rooted, and is separated from the second premolar by a moderate diastema.   The symphysis is not long, is regularly curved upwards and has a flat inferoanterior face.   The canine alveoli create a marked prominence on each side.   [Measurements follow.]

The jaw of this species is about the length of that of a large tapir, but is deeper and more robust. The flare of the inferior edge in front is suggestive of the structure seen in the *Dinocerata,* and of the probability that the Taligrada (to which *Pantolambda* belongs) are the ancestors of that suborder as well as of the Pantodonta.   The flare is related to the flange of *Uintatherium,* exactly as the similar ridge in *Nimravus* is to the flange in *Machaerodus.*

In 1888 [99] Cope described the hind foot of *P. cavirictus* from a referred specimen (A. M. Cope Coll. No. 3963) as follows:

In the posterior foot the cuboid of the *P. cavirictus* differs from that of any of the Coryphodontidae which I have seen in the greater mutual obliquity of the two proximal facets.   That for the astragalus is a wide concave fossa; that for the calcaneum is a hook-shaped band, the convexity proximad, and the longer arm, or stem, of the hook anterior, and the shorter posterior to a ligamentous fossa.   The anterior band-like facet turns transversely distally.   The position of the cuboid is oblique in the foot, giving the digits which arise from it a divergent direction externally   The astragalus of this species closely resembles that of *P. bathmodon.*   The ectocuneïform is much like that of *Coryphodon,* but is not so depressed, the anterior faces being square.   The mesocuneïform has only two-thirds the longitudinal depth in front.   The entocuneïform is narrower transversely than in *Coryphodon,* and approaches the form of some of the unguiculates.   It indicates a smaller internal digit than in *Coryphodon.*   The above-described bones all belong to one individual.

Osborn and Earle in 1895 described and figured the skull and humerus of *P. cavirictus* under the name of *Pantolambda bathmodon.*   In 1898 Osborn corrected this reference and figured the hind foot described as above by Cope.

The principal specimens at present available for study are:

No. 3961, Cope Coll., type, lower jaw.

Nos. 963 and 964, skull, lower premolar, ulna, part of scapula, right humerus.

No. 2555, Cope Coll., upper teeth and fragments of skull.

No. 16723, upper teeth and premaxilla.

No. 3963, Cope Coll., right pes, part of left pes and a few fragments of fore foot, also one upper molar tooth.

Nos. 3964, 3967, 3986, Cope Coll., small lots of foot bones.

No. 965, parts of tibia, and patella.

No. 962, lower jaw, young, with milk premolars poorly preserved.

No. 2556, Cope Coll., atlas, axis and third cervical, parts of other vertebrae and of fore limb.

[99] Cope, E. D., 1888, Trans. Amer. Phil. Soc., XVI, N.S., p. 360.

The teeth and skull are about two-fifths larger lineally than those of *P. bathmodon*. The teeth agree in construction quite closely with that species, except that the incisors are relatively more robust, as shown by their roots in the type jaw and alveoli of the upper series in No. 16723. The first upper premolar is relatively smaller than in *P. bathmodon*, the second larger and more like its successors (see No. 2555). The canines are more crested posteriorly, and are set farther apart from each other in both upper and lower jaws, leaving more space for the incisor series. The molars and premolars apparently agree closely in

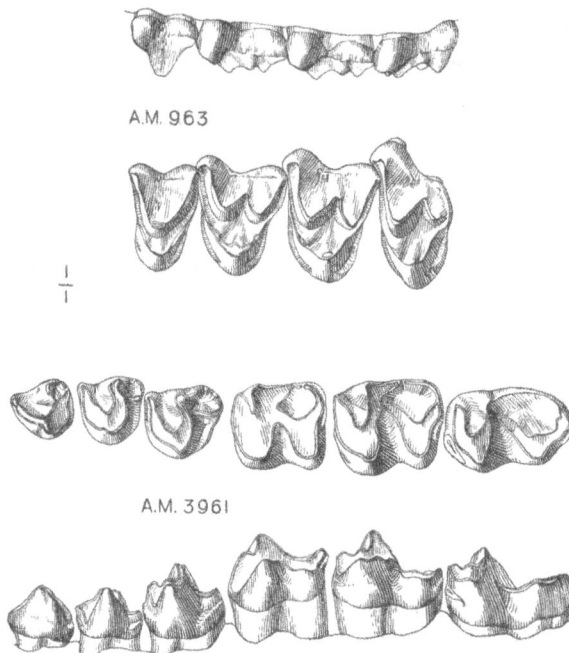

Fig. 43. *Pantolambda cavirictis,* upper teeth, A. M. No. 963, side and crown views; lower teeth, A. M. No. 3961, crown and side views. Natural size.

size and construction with the smaller species, but show some individual variation in size and in development of cingula.

The skull, as figured by Osborn, has been restored from several separate pieces and is too narrow in the palatal region and muzzle. The actual width, judging from the width of the symphysis in the type lower jaw, should give almost twice the width between the canines shown in the skull as figured,[100] the muzzle in *P. cavirictus* having somewhat of the hippopotamoid flare that is so characteristic of *Coryphodon*. The premaxilla is heavy and wide with evidently a short and wide superior branch, as in both *Pantolambda bathmodon*

[100] Approximately 100 mm. This is based on the proportionate width between upper and lower canines at the base of the crown, which is 44 mm. and 33 mm. in Nos. 16663 and 16664. In No. 3961, type of *cavirictus*, the width between lower canines is 75 mm., the width between upper canines should be 100 mm. In the *cavirictus* skull as at present restored it is only 54 mm.—w.d.m.

and *Coryphodon*. The nasals are not preserved. The zygomata are comparatively slender, relatively somewhat more slender than in *P. bathmodon;* the suborbital shelf is more prominent anteriorly; the jugal does not reach so far back. The cranial portion of the skull appears to be very similar, as far as preserved; the sagittal crest begins opposite the middle of the zygoma, but its height is not known; the occiput appears to have the same broad, semicircular form, but the height of the crest is not known. The entire basicranial construction appears to be much the same as in *P. bathmodon*.

The anterior cervical vertebrae are like those of *P. bathmodon* except for larger size, more prominent transverse processes and heavier, longer spine on the axis.

The scapula, as far as known, and the humerus have the same proportions and construction as in *P. bathmodon*, but are about one-half larger in lineal dimensions. The ulna is more than a half larger and is nearly twice as wide in the shaft. The proximal and distal ends of the tibia show much the same construction as *P. bathmodon* but three-fifths larger size.

The hind foot has been described and figured by Cope and Osborn. It is over a half larger than the smaller species and more robust throughout; the lateral digits are not reduced, being as large as the central digits although not so long, and the first metapodial only half as long as the second. The trapezium is correspondingly larger and less compressed than in *P. bathmodon;* the metapodials are one-half longer but their width is three-fifths greater. The astragalus has nearly twice the dimensions of that of *P. bathmodon* and the angle between the tibial and fibular facets is broader, but this may be due to crushing. The construction of the bone appears to be otherwise much the same. The navicular is not preserved, but the articulation of the adjacent bones indicates apparently that it was reduced on the external border so as to lose its cuboid facet and allow the ectocuneiform nearly but not quite to reach the astragalus. The mesocuneiform is increased in depth (proximo-distad), especially at the dorsal surface. Both features, and also the larger size of the entocuneiform, constitute an approach towards *Coryphodon*. The first row of phalanges is nearly as wide as long, the second row a little wider than long; the single ungual preserved is two-thirds as wide as long.

As a whole this species shows considerable approach to *Coryphodon* in proportions of the limbs and feet, the enlarged incisors and broader, more flaring muzzle. However, the cheek teeth do not show any approach, but are perfectly typical of *Pantolambda*, and it is probable that the approach seen in proportions of limbs and feet is merely associated with the larger size of the species and that it is not an intermediate stage connecting *Pantolambda* with *Coryphodon* but simply a giant species of *Pantolambda*. Nevertheless, it may serve to show that a great part of the difference between the two genera is merely due to the difference in size and does not indicate any wide diversity of origin.

Affinities of *Titanoides* Gidley, 1917

This genus described by Gidley [101] from the Fort Union beds is based upon four lower teeth, p_4–m_3 of an animal intermediate between *Pantolambda* and *Coryphodon* in size, and in my opinion closely related to *Pantolambda*. The teeth differ from those of *P. bathmodon* in two principal points: (1) the paraconid is more prominent and is a partly separated cusp, instead of being united to the protoconid by a continuous crest; (2) the entoconid of m_3

---

[101] Gidley, J. W., 1917, Proc. U. S. Nat. Mus., 52, pp. 431–5, pl. xxxvi, 1 text fig.

is a semi-separate and prominent cusp, instead of being merged in the inner crest of the hypoconulid as it is in *Pantolambda*.

Aside from these two characters the teeth agree with those of *Pantolambda*, and also resemble, though less closely, the teeth of *Coryphodon*. The fourth premolar is of the very characteristic amblypod type, identical with that of *Pantolambda*, and the heel of the last molar is equally characteristic, identical in pattern with certain species of *Coryphodon* and differing from *Pantolambda* only in the character cited. Both these teeth differ widely from any perissodactyl.

Gidley was disposed to see in the prominent paraconids of the molars, and in the bicrescentic structure thereby conditioned, an indication of affinity to the titanotheres. Undoubtedly the later and more specialized titanotheres have the trigonids of similar type; the older and more primitive titanotheres of the Middle and Lower Eocene have small, low paraconids and in every way are more different from *Titanoides*, while approaching in every particular the *Eohippus* or *Systemodon* type of dentition. The other resemblances to the titanotheres cited by Gidley—the relative proportions of the molars to each other in the series, the similarity in development of cingula, heavy, massive jaw, broad symphysis and position of the canine and submental foramina—are not points of resemblance to titanotheres any more than to Amblypoda and numerous other primitive Herbivora. The supposed shallowness of the symphysis is comparable to that of *Coryphodon*, although not to that of *Pantolambda;* the lack of coössification of the suture is perhaps an age character; at all events it is no more characteristic of titanotheres than it is of Amblypoda; immature individuals, especially in certain species in both groups, have the symphysis sutural.

It is perhaps unnecessary to point out that there is no warrant for the hypothetical upper teeth of titanothere pattern that Gidley ascribes to the genus. The lower teeth differ from those of *Pantolambda* in two points only: the more prominent and distinct paraconid should be correlated, as it is in other primitive dentitions, with a corresponding reduction of the hypocone element of the upper molars, i.e. a sharply trigonal protocone, without postero-internal cingular crest or cusp; the separate entoconid of the heel of m_3 makes this tooth like that of *Coryphodon* and the last upper molar should be modified from that of *Pantolambda* so as to approach the *Coryphodon* type, as the protocones would be modified in an opposite sense. Save for these points the close correspondence in the pattern of the lower teeth involves an equally close correspondence in the pattern of the upper teeth to those of *Pantolambda*. It is quite possible that the upper teeth of *Titanoides* differed from the type thus indicated, but that they were in the least like those modeled by Gidley appears wholly incredible.

It is unfortunate that Gidley seems to have been but little acquainted with the structure of the teeth in *Pantolambda*, perhaps from not having seen them in an unworn condition. His statement that the paraconid is small and low in *Pantolambda* is inaccurate as to *P. bathmodon* and appears to be equally or more inaccurate as to *P. cavirictus*, although no unworn lower molars of this species are known.

Titanoides is therefore referred to the Pantolambdidae.[102]

[102] The skull and skeletal material of *Titanoides*, recently discovered in western Colorado and now in the Field Museum, have been described in the following papers:

Patterson, Bryan, 1934, "A Contribution to the Osteology of *Titanoides* and the Relationships of the Amblypoda," Proc. Amer. Phil. Soc., LXXIII, pp. 71–101, pls. I, II.

Patterson, Bryan, 1935, "Second Contribution to the Osteology and Affinities of the Paleocene Amblypod Titanoides," Proc. Amer. Phil. Soc., LXXV, pp. 143–62.—THE EDITORS.

3. ORDER CONDYLARTHRA

GENERAL DISCUSSION

The Paleocene genera from the San Juan basin referred to this order are *Tetraclaenodon* (= *Protogonia* or *Euprotogonia*), *Mioclaenus*, and a number of less familiar generic groups. The Periptychidae were referred here by Cope and by Matthew in 1897, but in 1898 Cope transferred this family to the Amblypoda (Taligrada) on the ground of the non-serial tarsus, and this arrangement was followed by Osborn in 1898 and subsequent publications. The various Paleocene genera are, however, so closely allied on the one hand to the oxyclaenid creodonts and on the other hand through the smaller periptychids to the Taligrada that it is very difficult to decide upon their real affinities, especially where the skeleton characters are unknown or imperfectly known.

The order as originally defined by Cope comprised ungulates with distinct neck and condyloid head to the astragalus, serial tarsus and carpus and five-toed feet. The type of the order was *Phenacodus*. Osborn revising the order in 1898 distinguished it as cursorial in adaptation, as opposed to the graviportal Amblypoda. Cope's definition, however, applies in its entirety only to the typical genus, Osborn's only to the phenacodont family and these are cursorial only to a very limited extent.[103] *Tetraclaenodon* is only semi-ungulate and its carpus is non-serial. *Meniscotherium* is in no degree cursorial and has a non-serial carpus. The little that is known of carpus and tarsus in the smaller periptychids and the mioclaenids does not differentiate them from each other or from the oxyclaenids, *Meniscotherium* or *Tetraclaenodon*, by any clean-cut distinctions. The only real distinctions that can be drawn are of a negative character. They do not display the tendency towards a predaceous adaptation of the teeth characteristic of the Creodonta, nor the tendency towards an amblypod shortening of the foot characteristic of the Taligrada. None of them shows any approach, as far as known, toward the special construction of the feet characteristic of the primates nor to the shortening of the jaw, reduction of incisors, or any other of the specialized characters of this order. Such primate resemblances as any of them display consist merely in the extreme and primitive simplicity of molar construction often seen in that order. None of them shows the characteristic peculiarities of tarsus seen in all insectivores, including the contemporary Leptictidae. The teeth have a certain general resemblance to Insectivora but lack any special resemblance to any of the insectivore groups, nor are there any resemblances to Menotyphla sufficient to suggest near relationship.

The Paleocene genera grouped under Condylarthra appear to be a series of very primitive forms intermediate between the creodonts on one hand and the taligrades on the other, forming with these a group somewhat more distinct from the primitive insectivore and primitive primate stocks, but not far removed from the common ancestral stock of these and other placental orders. They left two surviving branches in the Eocene, *Phenacodus* descended from *Tetraclaenodon* and *Hyopsodus* more doubtfully descended from one of the mioclaenid genera. The first is somewhat specialized in the feet, rather primitive in the teeth, the second a little specialized in teeth but wholly primitive in feet. The entire group of Creodonta, Condylarthra and Taligrada retains what must be regarded as the primitive construction of the ankle, the astragalus wedged between tibia and fibula, a

---

[103] The feet of *Phenacodus* are less specialized in a 'cursorial' direction than are those of the tapir, yet the tapir is not regarded as a cursorial specialization!—W.D.M.

construction progressively lost in the later Carnivora and to some extent in *Phenacodus*, but lost at an earlier date in the insectivore group, as will be more fully discussed in a later chapter.

The validity of the order Condylarthra might well be called into question if these are the relationships of its members. It is hardly to be regarded as a specialized group comparable with most other orders of mammals. They might be united as Cope proposed in an order Bunotheria, including the primitive stocks of the Carnivora, the Insectivora and

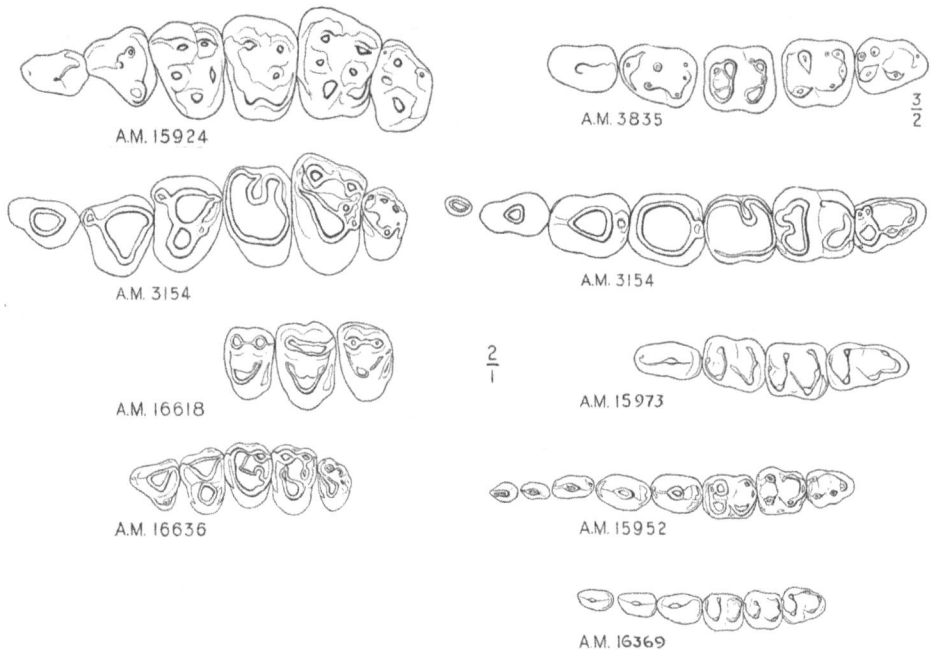

Fig. 44.   Upper and lower teeth of Condylarthra, crown views.   All twice natural size.
   Upper teeth (left): *Tetraclaenodon puercensis*, A. M. No. 15924; *Mioclaenus turgidus*, A. M. No. 3154; *Protoselene opisthacus*, A. M. No. 16618; *Ellipsodon lemuroides*, A. M. No. 16636.
   Lower teeth (right): *Tetraclaenodon puercensis*, A. M. No. 3835; *Mioclaenus turgidus*, A. M. No. 3154; *Protoselene opisthacus*, A. M. No. 15973; *Ellipsodon lemuroides*, A. M. No. 15952; *Oxyacodon apiculatus*, A. M. No. 16369.
   Dentitions aligned on the anterior border of the first molar.

other orders of which the specialized members are well distinguished, but such an arrangement would involve equal difficulties and inconsistencies in associating the various groups and in drawing the line between those worthy of separate ordinal distinction and the numerous early branches of the phylogenetic tree. It seems better on the whole to accept the current taxonomy and adapt the known data as well as we can to its outlines.

The Condylarthra from this viewpoint comprise primitive placentals of omnivorous adaptation lacking the predaceous adaptations of the creodonts and the graviportal adaptations of the Taligrada.

PHENACODONTIDAE Cope, 1881

Cope, E. D., 1881, Amer. Nat., XV, p. 1018

Author's Diagnosis: Molar teeth tubercular; the premolar teeth different from the molars; five digits on all the feet.

Revised Diagnosis: Dentition unreduced, bunodont, canines progressively enlarged. Last premolar partly molariform. Molars sexicuspid above, quadricuspid below. Skull long, with narrow occiput. Feet pentadactyl, with progressive tendency to tridactylism, becoming compact, narrow and digitigrade. Astragalus developing a rounded head, narrow and deep trochlea, with sharp keels and losing the astragalar foramen. Phalanges short, flattened, hinge-jointed, ungues progressively broadened into small flat hoofs.

The distinctive characters of the family are less fully developed in the Torrejon *Tetraclaenodon* than in the Wasatch *Phenacodus* and *Ectocion*. It shows, however, a marked approach toward the phenacodont type in all respects, in comparison with the mioclaenids, sufficient to conform with the above diagnosis. In the Meniscotheriidae the feet retain a very primitive character, while the teeth and skull are specialized in a diverse direction.

Tetraclaenodon Scott, 1892

Scott, W. B., 1892, Proc. Acad. Nat. Sci. Phila., XLIV, p. 299

Type: Mioclaenus floverianus Cope, 1888 = *Phenacodus* (*Protogonia, Euprotogonia*) *puercensis* Cope, 1881, Middle Paleocene of New Mexico.

Synonym: Protogonia Cope, 1881; type: *P. subquadrata.*

Author's Diagnosis: Superior dentition unknown; the inferior molars are like those of *Claenodon*, but the premolars are very different. The anterior ones are relatively larger and more massive; $p_{\overline{3}}$ is a stout, compressed cone and has a minute anterior basal cusp (paraconid) and a small heel, which forms two basin-like depressions, divided by a median ridge. $P_{\overline{4}}$ has all the elements of a molar, with a massive protoconid, and small para- and deuteroconids; the heel is low and composed of two cusps (meta- and tetartoconids). The molars are constructed as in *Claenodon*, but are less rugose than in either of the undoubted species of that genus. The humerus has a broad and flattened head, small tuberosities and wide, shallow bicipital groove. The deltoid ridge is very prominent and runs far down the shaft, which is stouter than in *Arctocyon*. The trochlea is higher, thicker and narrower than in that genus, and the supinator ridge less prominent; the entepicondyle is very large and is perforated. The distal end of the radius is narrow and flattened, and the facets for the scaphoid and lunar are separately marked, a very unusual feature among creodonts. The ilium is strongly trihedral and very little expanded; the inferior surface is broad and the spine prominent.

Professor Scott did not recognize the true affinities of the species on which this genus is founded and placed it in the Arctocyonidae, although he corrected the initial error in the description of *Mioclaenus floverianus*, which had led Cope astray as to its relationships, namely, the identification of p_4 as m_1. This, however, is not surprising, as the very creodont-like character of the limb bones in the type would not have suggested comparison with *Phenacodus*, and the skeleton of "*Protogonia*" was not then known.

The first phenacodont from the Paleocene was described by Cope under the name of *Phenacodus puercensis*. He had only the molar teeth of this type, but shortly afterwards recognized a somewhat smaller species in which the fourth upper and lower premolar were preserved with the molar, and based the new genus *Protogonia* upon the clearly more primitive characters of these teeth. He subsequently referred to it *P. subquadrata* and *calceolata* and the previously described species *P. puercensis* and *zuniensis*. The generic name unfortunately was preoccupied, and Cope in 1893 [104] substituted *Euprotogonia*. Scott in the meantime had based the new genus *Tetraclaenodon* upon *Mioclaenus floverianus*, although without recognizing its affinities. Matthew in 1897 placed this species as a synonym of *P. puercensis*, but retained the name *Euprotogonia* on the ground that Scott's genus was based on error, an untenable procedure by the rules of nomenclature, whatever its merits as a matter of justice. Matthew concluded that all these species except *zuniensis* were variants of a single very variable species for which the earliest name, *puercensis*, was used; the type of *P. zuniensis* he placed in the synonymy of *Tricentes subtrigonus* and certain specimens referred by Cope to *zuniensis* were renamed *Euprotogonia minor*.

The collections made in 1912 and 1913 include a large series of jaws of this genus, nearly all from the upper level of the Torrejon. A restudy of Cope's types in connection with this referred material, new and old, confirms the conclusions reached in 1897, except that *Protogonia plicifera* appears to be a distinct smaller and more primitive species, agreeing with the later-described *Euprotogonia minor*. It is a comparatively rare form, limited to the lower horizon of the Torrejon, where it occurs in company with the larger *puercensis*. This very abundant species shows a considerable latitude in size and a great deal of individual variation in proportions and cusp-construction of the teeth, but no varietal or specific differences can be found to support the validity of the species *subquadratus*, *calceolatus* or *floverianus* as distinct from *P. puercensis*.

Type diagnosis of Protogonia: Fourth superior premolar with one external and one internal lobe. True molars with two external, two internal, and two intermediate lobes, both the latter connected with the anterior internal by a ridge. Supposed inferior true molars with two Vs with weak anterior branches; last true molar with heel.[105]

Revised Diagnosis of Tetraclaenodon: Dentition $\frac{3 \cdot 1 \cdot 4 \cdot 3}{3 \cdot 1 \cdot 4 \cdot 3}$. Incisors small, canines large, laniary; anterior premolars simple, trenchant; protocone small on p^3, large on p^4; metacone usually present on p^4, but never so large as paracone, sometimes present on p^3; p$_4$ with well-separated metaconid, rudimentary paraconid, and large heel imperfectly bicuspid. Upper molars sexitubercular, hypocone and intermediates large and well separated, but not equalling the trigonid cusps in height; no mesostyle; m^3 without hypocone. Lower molars four-cusped, usually with vestigial paraconid on anterior flank of metaconid, always with an anterior crest curving around from protoconid or metaconid; talonid imperfectly basined, a little lower than trigonid; hypoconulid distinct on m$_{1-2}$, large on m$_3$.

Skull narrow, elongate, with high sagittal crest, high, narrow occiput, zygomatic arches of moderate depth but little thickness, glenoid fossa shallow, flattened, postglenoid process much reduced.

Limbs and feet slender, creodont-like, humerus with entepicondylar foramen, radius with distinct scaphoid and lunar facets, carpus alternating; the small, keeled magnum supporting scaphoid (scapho-centrale) and lunar, the lunar resting about equally upon magnum and unciform; tarsus serial; astragalus with shallow trochlea, imperfect inner crest, distinct astragalar foramen, head moderately flattened. Metapodials slender, the lateral pair moderately reduced, distal ends very slightly square (i.e. hinge-jointed).

---

[104] In letter to Charles Earle, Amer. Nat., XXVII, April 5, p. 378, footnote. For *Protogonius* Hübner see Palmer, T.S., 1904, Index Generum Mammalium, p. 582.

[105] Proc. Amer. Phil. Soc., XIX, p. 492; type: *P. subquadrata*. Name preoccupied by *Protogonius* Hübner, 1816; replaced by *Euprotogonia* Cope, 1893.

Phalanges narrow, long and creodont-like in comparison with *Phenacodus*, but somewhat shortened. Unguals narrow, almost like claws rather than hoofs.

The skeleton characters of *Tetraclaenodon* were described by Matthew in 1897 and some further particulars were added by Osborn in 1898. A few additional specimens supplement the characters of the skull, fore limbs and feet, but they are mainly based upon No. 2468, found by Walter Granger of the Museum expedition of 1896.

It differs more from *Phenacodus* in the skeleton than one would expect from the resemblance of the teeth, and the differences are all in the direction of approach to the Creodonta, especially the Arctocyonidae. In part, however, the lack of specialization is associated with the smaller size, and the smaller species of *Phenacodus* from the Wasatch are more like *Tetraclaenodon* than is the large *P. primaevus*.

The teeth are readily distinguished from any of the species of *Phenacodus* or *Ectocion* by the lack of mesostyle on the upper molars and by the small size or absence of metacone on p^4. The conules are relatively larger than in *Phenacodus*, much larger than in *Ectocion*.

The skull characters are drawn chiefly from No. 16653, which consists of parts of the skull, including the upper jaws, the occipital region with most of the brain-case and zygomatic arch, and are supplemented in some particulars by No. 15927, a fragmentary skeleton.

The general proportions are not widely different from those of the smaller species of *Phenacodus;* the differences from *P. primaevus* are chiefly due to the much larger size and robustness of the Wasatch type. The orbit, however, is considerably farther forward, its anterior rim being above the anterior end of m^1, while in *P. primaevus* it is above the posterior end of m^2. The posterior expansion of the nasals is by no means so great, nor are the frontals so reduced in antero-posterior length. The infra-orbital foramen is above p^3; in *Phenacodus* it is above p^4. The frontals are transversely convex above the orbits, as in *Phenacodus*, but they form a somewhat concave surface medially in front of the postorbital crests, while this surface is convex in *Phenacodus*. The postorbital processes are somewhat less prominent than in *Phenacodus;* behind them the crests are drawn together much more sharply to meet in a sharp, thin sagittal crest. The position of the fronto-parietal suture is the same as in *Phenacodus;* it commences medially a little in front of the anterior end of the sagittal crest, runs forward and outward just in front of the postorbital crest for a third of its length, then crosses the crest, turning backward and then downward on the side of the cranium, reaching the sphenoidal bones apparently at about the narrowest part of the postorbital constriction. The posterior and lateral sutures of the parietal are not clearly shown. The sagittal crest is narrow and sharp-edged, of moderate height, rising posteriorly to the occipital crest and spreading quite abruptly in front of the postorbital constriction into the postorbital crests.

The occipital crest is high, sharp and thin, projecting upward and backward and considerably overhanging the condyles. Laterally it passes into a thick, sharp and moderately prominent lambdoidal crest, forming the posterior border of the squamosal; the posterior branch of this crest behind the mastoid exposure formed by the lateral border of the occipital bones is little developed, and in consequence the exposure of the mastoid is more posterior than lateral.

The inferior exposure of the mastoid is small and the post-tympanic process of the squamosal is not prominent and is well separated from the postglenoid by a broad and deep tympanic notch. In *Phenacodus* the mastoid and post-tympanic processes are much

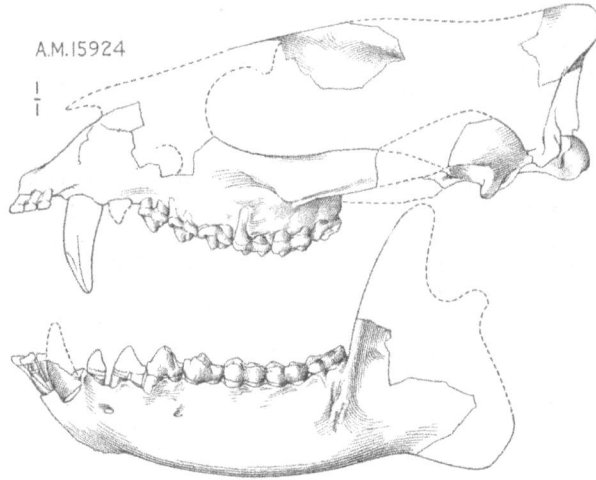

FIG. 45. *Tetraclaenodon puercensis*, skull and lower jaw, A. M. No. 15924, side view; dotted outlines from other specimens. Natural size.

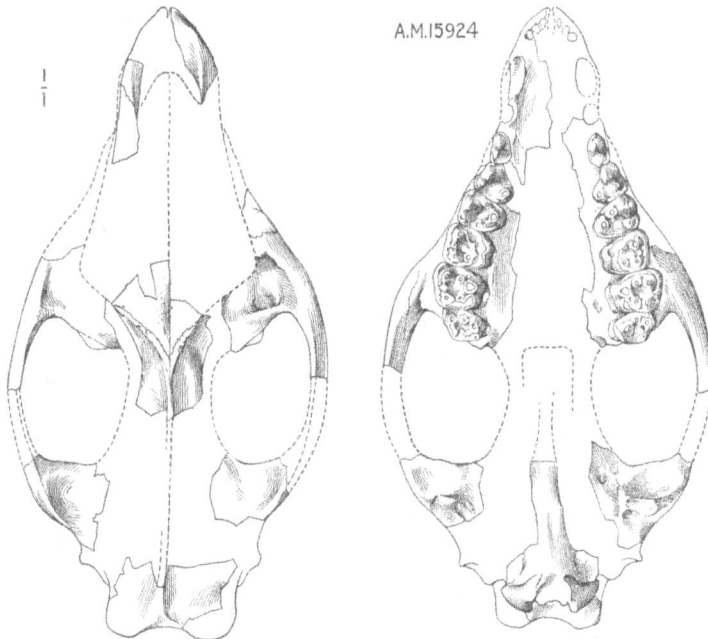

FIG. 46. *Tetraclaenodon puercensis*, skull, A. M. No. 15924, top and palatal views. Natural size.

more prominent, projecting strongly downward and curving forward at the tip towards the postglenoid, the occiput being much wider at this point. The postglenoid process is smaller, and the glenoid fossa less concave than in *Phencaodus*, the zygomatic process less massive. The jugal, however, has about the same proportionate length, although of less thickness. The postorbital process of the jugal, as in *Phenacodus*, is represented only by a slight rise and thickening of the upper border of the jugal; the antero-inferior branch is

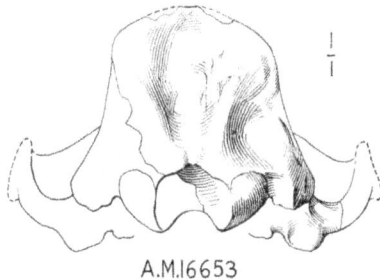

FIG. 47. *Tetraclaenodon puercensis*, back of skull, A. M. No. 16653. Natural size.

also absent; the superior branch is received into a deep slotted suture of the maxilla beneath the orbit. The relations of the lachrymal are not clearly shown in our specimens.

The maxillary is rather short, the premolar teeth following directly behind the canine without diastema; it is rather abruptly contracted in front of the infra-orbital foramen to a short and narrow muzzle; the suture for the premaxilla is nearly vertical. Parts of the premaxilla preserved in No. 15924 show that it is both short and small, the dentigerous branch curving sharply inward and bearing three small incisors, the superior branch ending in a short point slotted into the maxillary border.

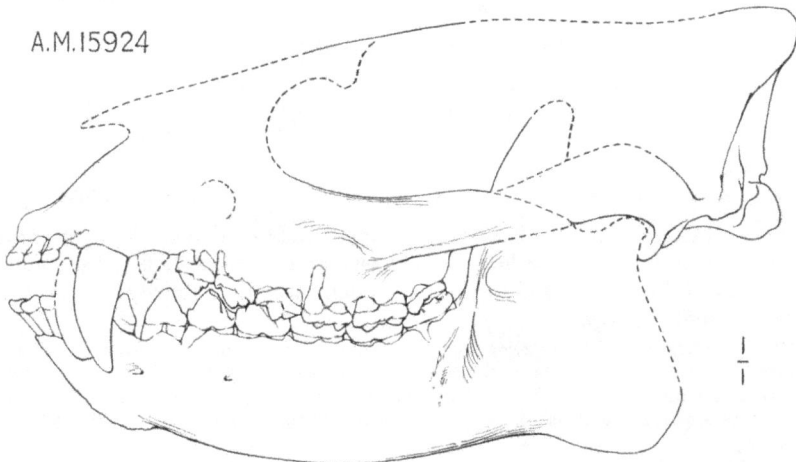

FIG. 48. *Tetraclaenodon puercensis*, skull and jaws in articulation, A. M. No. 15924, side view. Natural size.

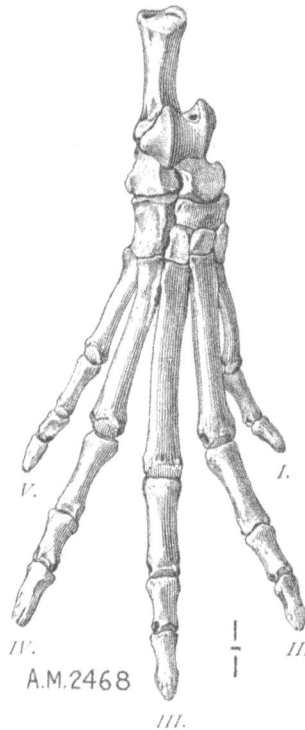

FIG. 49. *Tetraclaenodon puercensis*, hind foot, A. M. No. 2468, anterior view. Natural size.

Tetraclaenodon puercensis (Cope)

Phenacodus puercensis Cope, 1881, Paleont. Bull. No. 33, Sept. 30, and Proc. Amer. Phil. Soc., XIX, p. 492; 1884, Amer. Nat., XVIII, p. 900, fig. 22.

Phenacodus calceolatus Cope, 1883, Proc. Amer. Phil. Soc., XX, March 16, p. 561, and Paleont. Bull. No. 36, April 17.

Mioclaenus floverianus Cope, 1888, Trans. Amer. Phil. Soc., XVI, N.S., p. 330.

Type: A. M. Cope Coll. No. 3832, parts of upper and lower jaws. Type of *Phenacodus calceolatus:* No. 3947, upper and lower cheek teeth, fragments of skull and skeleton.

Horizon and Locality: Paleocene of the San Juan basin, undoubtedly from the Torrejon formation, Middle Paleocene.

Author's Diagnosis: Last superior molar smallest; first and second true molars with six tubercles, two external, two median and two internal. A strong basal cingulum except on inner side. Inferior true molars besides the usual five tubercles, furnished with an anterior ledge with a tubercle at its interior extremity. A weak external basal cingulum. A little larger than the *P.* [*Phenacodus*] *vortmani*. [Measurements follow.]

Type Description of Phenacodus calceolatus: The teeth are of the size of those of the *Phenacodus puercensis*, and like that species, there is no median external cingular cusp of the superior molars.

In these teeth the external basal cingulum is weak, but there is a strong anterior cingulum, distinct from any of the cusps. No internal cingulum. External cusps conical, well separated; intermediate cusps rather large; internal cusps rather large, close together, but deeply separated. The last superior molar is reduced in size. It has well developed anterior and posterior cingula, a weak external, and no internal cingula. The intermediate tubercles are rather large, and there is one large internal tubercle.

The heel of the last inferior molar is short, wide and rounded. The posterior tubercle is but little behind, opposite the posterior internal tubercle. The latter is separated from the anterior inner by a deep fissure, while the opposite side of the crown is occupied by a large median external cusp, which has a semi-circular section. The large anterior cusps are confluent on wearing. No anterior cingulum in the worn crown. The crowns of the first and second true molars of the specimen are rather worn. They show that the posterior median tubercle is very indistinct and probably absent. The bases of the smaller inner cusps are round, and on wearing unite with the larger external cusps. Of the latter the posterior is the larger. Anterior cingulum rudimental or wanting. No lateral or posterior cingula. The principal peculiarity of the lower dentition of this species and the one from which it is named, is the form of the third or fourth (probably third) premolars, both of which are preserved. They have a compressed apex, which descends steeply to the anterior base, without basal or lateral tubercle. The base of the crown spreads out laterally behind, and is broadly rounded at the posterior margin, so as to resemble the toe of a wide and moccassined foot. It is depressed, the surface rising to the apex from a flat base. [Measurements follow.]

Tetraclaenodon subquadratus (Cope)

Protogonia subquadrata Cope, 1881, Paleont. Bull. No. 33, Sept. 30, and Proc. Amer. Phil. Soc., XIX, p. 492.

Type: A. M. Cope Coll. No. 3876, right p^4–m^2.

Horizon and Locality: Paleocene of the San Juan basin, probably from the Torrejon formation, Middle Paleocene.

Author's Description (p. 493): *Char. specif.* . . . The animal was about the size of the red fox The external cusp of the fourth superior premolar is flattened externally, and has a small lobe on its posterior edge. The inner tubercle is conic and is separated by a tubercle from the anterior base of the external. True molars without external ridges. The external cusps of the true molars are lenticular in section. The posterior inner cusp is in nearly the same antero-posterior line with the anterior, its section about equalling that of the intermediate cusps. The first and second molars have an external, an anterior and posterior, but no internal, basal cingula. The enamel is somewhat wrinkled where not worn.

The heel of the last inferior true molar is elevated, and its worn surface forms the extended posterior branch of the posterior V. The posterior edge of the penultimate molar is elevated and curved forwards on the inner side of the crown. The anterior cusp forming the angle of the V of this tooth, is higher than the posterior angular cusp, but the anterior limb descends rapidly as in *Coryphodon*. A weak antero-external and postero-external cingula. Enamel wrinkled where not worn.

Tetraclaenodon pliciferus (Cope)

Protogonia plicifera Cope, 1882, Amer. Nat., XVI, Sept. 28, p. 833; 1884, *ibid.*, XVIII, p. 893, fig. 14.

Type: A. M. Cope Coll. No. 3900, upper and lower jaws, containing left p^4_4–m^2_2.

Horizon and Locality: Torrejon formation, San Juan River, New Mexico.

Author's Description: Differs from its congener, P. [*Protogonia*] *subquadrata* in that the internal cusp of the fourth superior premolar is connected with the anterior and posterior cingula by strong ridges, becoming thus the apex of a V. In the *P. subquadrata* it is a simple cone. Antero-external basal lobe distinct, intermediate lobe obsolete. The true molars are like those of the *P. subquadrata*, but all the molars are of smaller size. [Measurements follow.]

HYOPSODONTIDAE Nicholson and Lydekker, 1889

Hyopsodidae Schlosser, 1887. Hyopsodontidae Nicholson and Lydekker, 1889. Hyopsodontinae Trouessart, 1897. Including as a subfamily Mioclaenidae Osborn and Earle, 1895.

Dentition unreduced, bunodont; canines not differentiated, all anterior teeth uniformly set and regularly graded. Upper molars tritubercular (Mioclaeninae) or sexitubercular (Hyopsodontinae). Lower molars quadritubercular. Skull short, occiput wide.

Feet pentadactyl. Astragalus of very primitive type, the body wedged in between tibia and fibula, trochlea very oblique, flat, without inner crest, astragalar foramen distinct, neck distinct but short, head convex, flattened oval. Metapodials rather short, distal articulations somewhat hinge-type, phalanges short, unguals small but of claw rather than of hoof type.

The genera here referred to the Hyopsodontidae are *Mioclaenus* and its allies. *Hyopsodus*, the type of the family, is abundant in all the Eocene formations, but unknown in the Paleocene. *Haplomylus*, a somewhat doubtful relative, is common in the lower Wasatch, but is also unknown in the Paleocene.

Hyopsodus was originally described by Leidy as a " primitive suilline mammal," but a few years later Marsh referred it to the Primates, and this disposition was accepted by all writers until Wortman in 1903 transferred it to the Insectivora. Matthew in 1915 referred it to the Condylarthra.

Mioclaenus has been generally regarded as a condylarthran. Cope placed the genus among the Creodonta indeed, but this was evidently based upon the characters of the referred species, *M. ferox* and *corrugatus*, the only ones of which he had any characteristic parts of the skeleton. These species were separated under the name of *Claenodon* by Scott in 1892,[106] and with various other species which had been included under *Mioclaenus* by Cope were referred to the Creodonta, but Scott pointed out that *M. turgidus* was more probably condylarthran. In 1895 Osborn and Earle [107] placed it as a condylarthran in the family Mioclaenidae but suggested possible artiodactyl affinities—a suggestion prompted probably by the considerable resemblance of the lower jaw and teeth to those of "*Pantolestes*" *etsagicus*, of the Wasatch. Matthew [108] was at one time disposed to list the family among the Insectivora, on account of the evidences in the dentition of affinity to *Hyopsodus*, at that time regarded as insectivore. Winge in his recent revision of the Insectivora includes the genus as a member of the Periptychidae among the Insectivora.[109] " If the Periptychidae," he says, " despite all, are ungulates, they may be descended from one of the earliest ungulate types, specialized in opposition to all other ungulates, assuming the manner of life of insectivores and rodents. But, in fine, there has not been proven anything which would justify the belief that they are derived any more nearly from ungulates than directly from the Insectivora." The only apparent reason for this arrangement would seem to be the robustness of the premolars, for he compares it with *Hemithlaeus*, which has little else in common, and remarks that Matthew is " certainly wrong " in referring to

[106] Scott, W. B., 1892, Proc. Acad. Nat. Sci., Phila., XLIV, p. 299.

[107] Osborn, H. F., and Earle, Charles, 1895, Bull. Amer. Mus. Nat. Hist., VII, p. 48.

[108] Matthew, W. D., 1909, Mem. Amer. Mus. Nat. Hist., IX, p. 508.

[109] Winge, H., 1917, Udsigt over Insektædernes indbrydes Slægtskab. Vidensk. Medd. fra Dansk naturh. Foren, LXVIII, pp. 83–203.

the family "*Protoselene*" *opisthacus* and a number of smaller species (*M. lemuroides*, *acolytus* and *turgidunculus* presumably) which differ from *M. turgidus* only in the lesser degree of size and robustness of these teeth. Throughout his discussion, in fact, it is quite evident that this is Winge's real reason for referring the Periptychidae to the Insectivora instead of to the Condylarthra. He evidently thinks (quite erroneously) that enlargement of the premolars and reduction of the molars is something unknown among ungulates. The genera are as follows:

Hyopsodontinae:
    Hyopsodus . . . . .Wasatch, Bridger and Uinta

Mioclaeninae:
    Mioclaenus . . . . .Torrejon—2 species
    Ellipsodon . . . . .Puerco and Torrejon—5 species
    Protoselene . . . . .Torrejon—1 species
    Oxyacodon . . . . .Puerco—3 species
    ?*Haplomylus* . . . . .Lower Wasatch—1 species

MIOCLAENINAE Osborn and Earle, 1895

(= Mioclaenidae) Osborn, H. F., and Earle, Charles, 1895,
Bull. Amer. Mus. Nat. Hist., VII, p. 48

Mioclaenus Cope, 1881

Cope, E. D., 1881, Amer. Nat., XV, Sept. 22, p. 830

Type: Mioclaenus turgidus of the Upper Paleocene of New Mexico.

Author's Diagnosis: This genus differs from *Catathlaeus* [= *Periptychus*] in the greater simplicity of the structure of the inferior premolars, which are without internal crest or cusp. The inner lobe of the superior premolars is less developed than that genus. In the *M. turgidus* the characters of *Mioclaenus* are best seen in the subconical tubercles of the premolars, particularly that of the heel of the fourth inferior premolar. In the other three species this heel is more of a crest, and is connected with the principal cusp by a low ridge.

Revised Diagnosis: Dentition $\frac{1 \cdot 4 \cdot 3}{1 \cdot 4 \cdot 3}$; incisors unknown but quite small, canines small, premolars and molars very robust, low-crowned, with simple cusp construction, premolars large, much inflated, the cusps convexly conical, molars simple, with hypocones, paraconids, entoconids and basal cingula more or less obsolete or wholly absent; last molar considerably reduced.

The cusp construction in this genus is extremely simple. The low, robust crowns are devoid of ornamentation save for an obscure irregularly vertical wrinkling of the enamel surface. The premolars are large, with much inflated cusps and rudimentary heels. The posterior premolars do not show any tendency to become molariform, as in *Tetraclaenodon*, nor to the crescentic protocones of the periptychid genera. The molars contrast with those of the periptychids in their very low, massive cusps and simple construction, and differ from the phenacodont molars in lack of quadritubercular or sexitubercular symmetry.

The extreme robustness and simplicity of the teeth appear to be a specialization and distinguishes *Mioclaenus* from the smaller genera *Ellipsodon* and *Oxyacodon*.

Cope referred at one time or another more than thirty species to this genus, nearly all of them from the Puerco and Torrejon, and constituting about half the entire fauna. Some of these species he later removed to various distinct genera or subgenera; others were removed by Scott in 1892 and Matthew in 1897. In the present revision I retain in the genus only the type species, *Mioclaenus turgidus*, with the closely allied and doubtfully distinct *M. lydekkerianus.*

As thus limited, *Mioclaenus* is still very imperfectly known. Numerous specimens of *M. turgidus* are in the collections, but no considerable parts of the skeleton that can be positively identified by association with the teeth, and of the skull very little more is known than the palate and lower jaws. Various specimens show the dentition in all stages of wear.

Mioclaenus turgidus Cope

Mioclaenus turgidus Cope, 1881, Amer. Nat., XV, p. 830; 1885, Tertiary Vertebrata, p. 325, pl. xxv *e*, figs. 19–20, pl. LVII *f*, figs. 3, 4; Osborn and Earle, 1895, Bull. Amer. Nat. Hist., VII, p. 50, fig. 17; Matthew, 1897, *ibid.*, IX, p. 312.
Mioclaenus zittelianus Cope, 1888, Trans. Amer. Phil. Soc., XVI, N.S., p. 334; (= *M. turgidus*) Matthew, 1897, *loc. cit.*

Type: A. M. Cope Coll. No. 3135, parts of upper and lower jaws with the teeth unworn, p^4–m^2 l., p_4–m_3 l., p_4–m_1 r.

Horizon and Locality: Middle Paleocene, Torrejon formation, San Juan basin, New Mexico.

Author's Diagnosis (p. 831): Cusps of last premolars conical in both jaws. . . . Last lower molar disproportionately small; cusps low, two anterior inner distinct; true molars, .018.

Extended Description (Proc. Amer. Phil. Soc., XIX, p. 490): In the *M. turgidus* there are no cingula on the fourth [upper] premolar. It is wider than long, and the external face is a little flattened. The tubercles are conic; the external has a small one at the anterior base, and a rudiment at the posterior base, and there is a low one on the posterior side at the middle. The second true molar is wider than the first. The tubercles are all round in section. Besides those already mentioned, there is a rudiment of a posterior inner on the first, which is represented by a cingulum on the second. The latter has basal cingula all around except on the inner side; the same are visible on the first true molar in a rudimental condition. Enamel nearly smooth.

The inferior molars are of robust proportions. Their sizes are, commencing with the largest: P–m. IV; M. II; M. I; M. III. The last molar is only half as large as the penultimate [areally]. It has two anterior and an external lateral tubercles, and a heel. On the penultimate molar, there are two anterior tubercles with a trace of anterior inner; also a broad flat heel, with a low tubercle on the external side. The constitution of the first true molar is identical. The fourth premolar has a rudimental heel consisting of a low tubercle only. The principal cusp is conic and is over the middle of the transverse diameter, and a little behind the middle of the antero-posterior diameter. No cingula. Enamel nearly smooth. [Measurements follow.]

A wide range in individual variation is included under the species *M. turgidus* as here limited. The size of the molars, their width, the relative reduction of m_3^3, vary about 20 to 25 per cent, as shown in the table of measurements. The hypocones range from distinct, although small, cusps, to wholly absent. The paraconids are always present but vary in distinctness. The third upper premolar is sometimes transversely symmetrical, the inner cusp directly internal; sometimes it is strongly asymmetric, the inner cusp antero-internal. These variations, although inconstant, are partly associated together into varieties or mutants.   Specimens from the upper level of the Torrejon are prevailingly of large size, p^3 oblique, cingula and hypocones on m^{1-2} distinct, $m^3_{\overline{2}}$ less reduced.   But these characters, some or all of them, also appear in certain specimens from the lower level and are shown in varying degree.   The type of the species is a rather small individual, but No. 3153 has much smaller teeth.   The type of *M. zittelianus* has teeth of nearly the same size as typical *M. turgidus* and is in no way distinguishable, as shown by Matthew in 1897.

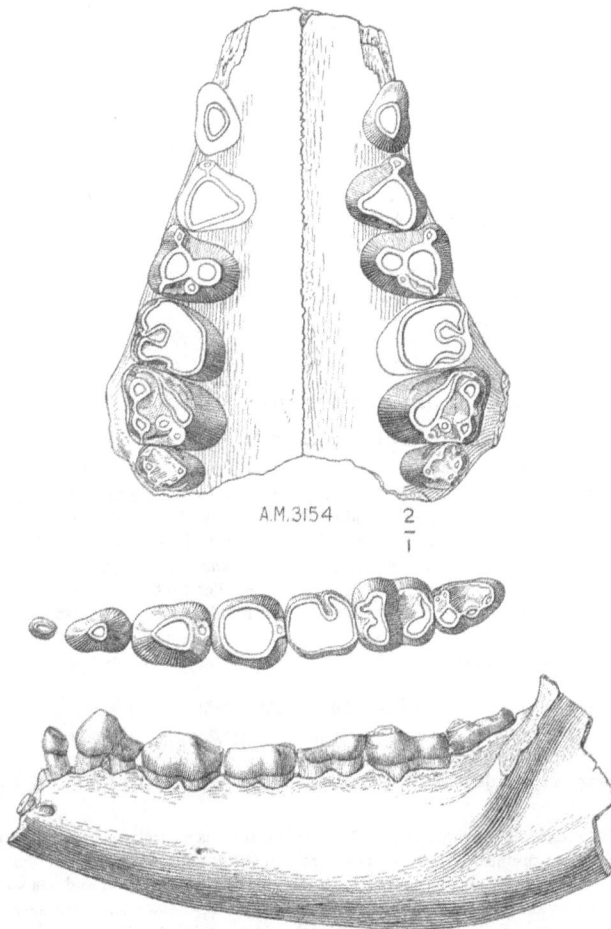

FIG. 50.   *Mioclaenus turgidus*, reconstructed palate and teeth, crown views; lower jaw, side view, A. M. No. 3154. Twice natural size.

	No. 3135 Type	Nos. 3163 16628	No. 3153	No. 3154	No. 15965	No. 16620	No. 3136
p^1–m^3.			39.0	39.8			
p^{1-4}.			23.9	24.9			
*diameters, p^2.			4.8 × 3.3	5.4 × 4.0	5.8 × 4.2		
" p^3.		6.0 × 7.1	5.5 × 6.0	6.3 × 7.0	6.8 × 6.8	6.1 × 7.0	
" p^4.	5.8 × 7.4	6.1 × 7.5	5.4 × 6.8	6.3 × 8.1	6.2 × 8.4	6.8 × 7.9	
m^{1-3}.	e 15.8	16.2	15.3	15.5	17.2	14.9	17.0
diameters, m^1.	5.8 × 7.3	5.7 × 7.1	5.7 × 6.6	5.5 × 7.2	6.5 × 7.8	5.7 × 8.3	5.9 × 8.6
" m^2.	5.6 × 9.3	5.9 × 9.0	5.2 × 7.8	6.1 × 9.7	6.0 × 9.5	5.7 × 9.5	6.1 × 10.2
" m^3.		4.1 × 5.5	3.3 × 5.4	3.7 × 5.9	3.5 × 5.3	3.5 × 5.3	4.2 × 6.2
p_1–m_3.		41.0		42.0			e 47.0
p_{1-4}.		22.7		22.8			e 25.5
diameters, p_1.				2.1 × 1.7			
" p_2.		5.9 × 3.6		5.9 × 3.6			6.4 × 3.8
" p_3.		7.0 × 4.8		7.0 × 5.3	7.5 × 5.2	7.2 × 4.9	7.8 × 5.3
" p_4.	6.7 × 5.5	7.5 × 5.4		6.8 × 5.8	7.2 × 5.8	6.4 × 5.1	7.7 × 6.8
m_{1-3}.	18.0	18.3		19.5	20.7		20.7
diameters, m_1. . . .	6.1 × 5.6	6.2 × 5.4		6.3 × 5.5	6.5 × 6.2	6.3 × 5.8	6.9 × 6.0
" m_2. . . .	6.3 × 6.1	6.1 × 5.9		6.8 × 7.2	6.7 × 6.5	6.7 × 6.4	6.9 × 6.7
" m_3. . . .	5.7 × 4.2	6.6 × 4.8		6.2 × 4.4	6.1 × 4.6		6.7 × 4.7

* Diameters = length × width.

Mioclaenus lydekkerianus Cope

Mioclaenus lydekkerianus Cope, 1888, Trans. Amer. Phil. Soc., XVI, N.S., p. 328.

Type: A. M. Cope Coll. No. 3201, part of right lower jaw with m_{1-2} and broken m_3.
Horizon and Locality: Paleocene, San Juan basin, probably from the Torrejon horizon.
Author's Diagnosis: This species is characterized by the presence on the inferior molar teeth of a distinct curved ledge in front of the two principal anterior cusps, which terminates in a more or less distinct fifth cusp at its internal extremity. The ledge with its curved anterior edge is unusual in this genus, though more or less developed in some of them, especially the *M. opisthacus*. The present species is much larger than the latter, and exceeds the *M. subtrigonus* also, but is smaller than the *M. protogonioides*.

The heel is rather wide and the edges elevated and with only an external cusp-like elevation, giving a basin-like surface. The last molar is not enlarged or reduced, but is narrower than the m. ii. The crowns are without cingula, and the enamel is obsoletely plicate. No premolars preserved.

Ellipsodon Scott, 1892

Scott, W. B., 1892, Proc. Acad. Nat. Sci. Phila., XLIV, p. 298

Type: Mioclaenus inaequidens Cope, from the Paleocene, Torrejon formation, San Juan basin, New Mexico.
Author's Diagnosis: The systematic position of this form is entirely obscure; it agrees with *Tricentes* in having but three upper premolars, but differs entirely in the construction of the molars. The premolars are relatively broad and massive, almost as in *Mioclaenus;* p^4 is especially broad and has a very large deuterocone. The molars are oval in shape and have no hypocone; m^2 is the largest of the series, m^3 very greatly reduced and forms a mere oval-shaped rudiment without recognizable elements. Inferior dentition unknown. One species: *E. (Tricentes) inaequidens* Cope. Puerco.

Matthew in 1897 retained *M. inaequidens* and the allied species *acolytus, lemuroides* and *turgidunculus* under *Mioclaenus*, regarding them as rather nearly allied to the type species of that genus, *M. turgidus*, despite some obvious differences in proportions of the teeth.   Further study and a large series of additional specimens have not materially changed that conclusion.   The details of construction and characteristic form of the cusps throughout all the upper and lower teeth prove the relationship beyond doubt to anyone who studies the original specimens.[110]   It may, however, be advisable to revive *Ellipsodon* for the benefit of the group of small species, of which *acolytus* differs rather widely from *Mioclaenus turgidus*, although the other species are partly intermediate.

There is a sufficient resemblance in the teeth of these small species to those of *Hyopsodus* to have caused Cope to refer *M. acolytus* at first to that genus, and to subsequent students it has suggested relationship.   This evidence—not very conclusive by itself—is strongly supported by the marked agreement in the limb and foot bones, in particular the astragalus of *Ellipsodon aequidens* (see below).   This astragalus is certainly not primate, insectivore or artiodactyl in type, and it may be considered to dispose of the suggestions that have been made of reference to any of these orders.   It is of a very primitive condylarthran or creodont type, and the teeth quite definitely refer it to the former order.

The evidence of teeth and skeleton places *Ellipsodon* nearer to *Hyopsodus* than are any of the other mioclaenids, and so far as *E. acolytus* is concerned it might well be regarded as directly ancestral, the dentition being but little removed from that of *H. simplex*, the oldest and most primitive species of *Hyopsodus*.   On the other hand, the close relationship of the genus to *Mioclaenus* is obvious and certain, despite Winge's dictum; and the relationship to *Protoselene* and *Oxyacodon*, although not quite so close, appears to be equally beyond reasonable doubt.

Key to Species of Ellipsodon

A. Last lower molars moderately or much reduced.
    1. Molar-premolar series = 21 mm.   Premolars smaller and more pointed. M_3
       much reduced.. *E. priscus*
    2. Molar-premolar series = 29 mm.   Premolars larger. M_3 moderately reduced. *E. lemuroides*
    3. Molar-premolar series = 23 mm. (approximate).   Last lower molar oval with
       flattened crown... *E. inaequidens*
B. Last lower molar unreduced or little reduced.
    1. P_1-m_3 = 22 mm. (approximate). M^3 oval, somewhat reduced, p^4 subquadrate. *E. acolytus*
    2. P_1-m_3 = 25 mm. (estimated). M^3 trigonal, unreduced, p^4 subtrigonal...... *E. aequidens*

Ellipsodon inaequidens (Cope)

Tricentes inaequidens Cope, 1884, Paleont. Bull. No. 37, Jan. 2, and Proc. Amer. Phil. Soc., XXI, p. 317.

Type: A. M. Cope Coll. No. 3095, upper jaws and part of skull, p^{2-3} and m^{1-3} l., p^4 and m^{2-3} r.

Horizon and Locality: Middle Paleocene, Torrejon formation, Gallegos Canyon, San Juan basin, New Mexico.

Author's Description: Besides its inferior size, other characters distinguish this species.   The simplicity of the superior molars is seen in no other, and the very reduced size of the third superior molar is not found in any of its allies.   This is correlated with an oblique reduction of the maxillary bone behind, which gives the second true molar an oblique external border instead of the longitudinal one seen in the other species.   The external cusps of the molars are conic, and are not in contact at the base.   The internal cusp

---

[110] Winge, however, does not appear to regard these small species as nearly related to *M. turgidus*.— W.D.M.

is also conic, and is larger than the external. The internal cusp of the fourth premolar is large. It is probable that the third premolar supports an internal cusp, as the crown base is as wide as long. The premolars are spaced in this species, as in the last [*Tricentes crassicollidens*], but the diastema is shorter than in the *T. crassicollidens*, not exceeding the premolar interspaces. The external cingulum is quite weak. The canine alveolus is large. The incisors are wanting, but the premaxillary region is wide. The inferior dentition is unknown. [Measurements follow.]

Scott in 1892 [111] separated this species from *Tricentes*, to which it is not very nearly related, making it the type of *Ellipsodon* as above. Matthew placed it under *Mioclaenus*, but as a separate group which is here accorded generic rank.

It is rather singular, in view of the relative abundance of small mioclaenids in the 1896, 1912 and 1913 collections, that only one additional specimen was found in them referable to this species. It still remains as left by Matthew in 1897, known only from the specimens in the Cope Collection: the type No. 3095, a palate with badly worn teeth partly buried in hard matrix; No. 3096, upper teeth partly buried in flinty matrix; a third specimen, No. 3296, incorrectly referred by Cope to "*Mioclaenus minimus*" and figured by him in 1888; and a fourth specimen of doubtful reference, No. 3299, a fragment of the lower jaw with p_4–m_2. There is nothing therefore to add to the 1897 description, which is as follows:

Revised Diagnosis: The front [upper] teeth are very vaguely indicated and might be interpreted in at least two ways. Either the first premolar is minute or absent, as Prof. Cope believed, and the canine moderately large, or the first premolar is spaced (displaced probably) and the canine unknown. The latter view brings the front teeth into harmony with the other Mioclaeni. The upper molars and last two premolars are quite of the type of *M. lemuroides* but more simple, the molars trituberculate with no internal cingulum and obsolete external one, intermediates minute or absent, no hypocone. Third and fourth premolar with strong internal cusp, others apparently simple. Last molar minute, transversely oval. Lower molars very simple, paraconid absent, protoconid and metaconid approximated, opposite, blunted; small hypoconid on heel, basin very much depressed internally, no entoconid. Last molar much reduced, oval, cusps low and flattened, basin shallow and nearly as high as the cusps.

Ellipsodon acolytus (Cope)

Hyopsodus acolytus Cope, 1882, Paleont. Bull. No. 35, Nov. 11, and Proc. Amer. Phil. Soc., XX, p. 462; 1884, Tertiary Vertebrata, p. 238, pl. xxiii *d*, figs. 5, 6; (*Mioclaenus*) 1888, Trans. Amer. Phil. Soc., XVI, N.S., p. 335; Matthew, 1897, Bull. Amer. Mus. Nat. Hist., IX, p. 317, fig. 18.

Mioclaenus acolytus Osborn, 1902, Bull. Amer. Mus. Nat. Hist., XVI, p. 170, fig. 1A.

Mioclaenus minimus Cope, 1882, Paleont. Bull. No. 35, Nov. 11, and Proc. Amer. Phil. Soc., XX, p. 468. (Not *M. minimus* Cope, 1888.)

Type: A. M. Cope Coll. No. 3208, upper and lower jaw, $p_{\overline{3}}^2$ to $m_{\overline{3}}^2$ l. Type of *Mioclaenus minimus:* No. 3294, a lower jaw with p_3–m_2 l.

Horizon and Locality: Paleocene of San Juan basin, probably Torrejon formation, Middle Paleocene.

Author's Description: This the least species of the genus [*Hyopsodus*] is also the oldest. . . . The species differs from those hitherto described in other characters than the minute size. One of these is the absence of posterior interior cusp, the heels of the first and second true inferior molars being bounded by a ridge only at this point, as in most of the species of *Pelycodus*. The last inferior molar is not smaller than the second, nor longer. The anterior cusps of all the molars are robust, so that on the first and second true molars they are separated by a shallow notch only. There is a rudiment of the anterior inner cusp on the

---

[111] Scott, W. B., 1892, Proc. Acad. Nat. Sci. Phila., XLIV, p. 298.

first true molar but none on the second and third.   The posterior external is obtuse and has a triangular section on all the molars; a crest is continued from the heel of the third molar on the inner side of the crown half way to the anterior inner cusp.

The *Microsyops* [*Haplomylus*] *spierianus* differs from this species in its smaller size (true molars .008) and in the presence of posterior internal cusps of the true molars.

Type Description of *Mioclaenus minimus:* This is one of the least mammalia of the Puerco fauna, exceeding by a little the *Hyopsodus acolytus*. It is represented by parts of two mandibles, which display all the true molars. As there are no premolars preserved, its reference to the genus *Mioclaenus* is provisional only, but its true molars have the peculiar characteristics of those of the *M. turgidus*.

The two anterior cusps of the true molars are higher than the heel, and they are united together to a point above the level of the heel.   The section of both those of the M. ii is round; that of the external one of the first [molar] is crescentic; of the inner cusp, round.   The heel is wide, and supports a cusp at the posterior external angle.   It is bounded posteriorly, and on the inner side by a raised ridge, which gives with the cusp, on wearing a comma-shaped surface.   A transverse ridge closely appressed to the anterior cusps connects them anteriorly.   In one of the specimens there is a cingulum on the external side of the second inferior molar; on the other specimen it is wanting.   Enamel smooth.

The mandibular ramus is rather deep and compressed and displays an external ridge on the anterior border of the coronoid, which is not continued downwards.   [Measurements follow.]

Matthew in 1897 identified with this species the type specimen of *Mioclaenus minimus,* but referred to *M. inaequidens* the " *M. minimus,*" No. 3296.   Many additional specimens are in the collection of 1913, mostly upper and lower jaws, and a few fragments of the skeleton associated with one of them.   The dentition can now be more fully described and compared with that of other species of *Ellipsodon.*

Revised Description: Dentition $\frac{?\cdot1\cdot4\cdot3}{?\cdot1\cdot4\cdot3}$.   Incisors, canine and first premolar small, with round, slender roots, set much as in *Hyopsodus.*   Number of incisors unknown.   Lower canine a little larger than i_3 and about as large as p_1^1, but its crown not preserved.   The upper canine doubtfully indicated as of considerably larger size than the lower.   Premolars

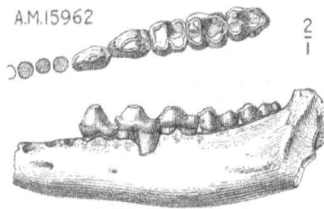

FIG. 51. *Ellipsodon acolytus,* lower jaw, A. M. No. 15962, crown and side views.   Twice natural size.

simple, compressed and moderately inflated, the first quite small and one-rooted, the others of nearly uniform length and height but progressively thicker transversely.   P² two-rooted, with a rudimentary thickening on the inner face at the base of the crown, foreshadowing the inner cusp and root of the following premolars.   P³⁻⁴ three-rooted; a rudimentary inner cusp on p³, a large subcrescentic one on p⁴.   A weak external cingulum on p³, external and lateral cingula on p⁴ and small stylar cusps at its external angles.   Upper molars rounded-trigonal, with subcrescentic protocones, small conules, conical para- and metacones, external, anterior and posterior cingula, but no hypocones.   M² slightly larger than m¹, m³ much smaller, with metacone reduced and set inward, no metaconule.

The first lower premolar is small, one-rooted, with moderately compressed, pointed, lanceolate crown. The following premolars are two-rooted, with simple crowns, progressively thicker transversely, heel very rudimentary on p_2, distinct on p_3, broad and somewhat basined on p_4. An inner crest and very rudimentary cusp on p_4. The lower molars are simple, quadrate and subequal. The trigonid is slightly higher than the talonid, of about the same width, and the protoconid and metaconid subequal, well separated, paraconid vestigial or wholly absent. The heel is basined, with entoconid distinct, hypoconulid imperfectly separated from it and submedian in position, hypoconid external, somewhat higher than the inner cusps. The third molar is about as long as the first and second, but of less width, the heel narrowing posteriorly to a median hypoconulid somewhat larger and more distinct than on m_{1-2}.

The lower jaw is rather thick, but not deep; it decreases anteriorly to a slender symphysis, much as in *Hyopsodus*.

The foregoing description of the teeth is based chiefly upon No. 15949, palate and lower jaws from the upper level of the Torrejon, found in Torrejon Arroyo by the Museum expedition of 1912. A few fragments of the skeleton are associated with this specimen. Numerous upper and lower jaws or parts of jaws, some with the upper and lower dentition associated, the majority isolated, represent *E. acolytus* in both the upper and lower levels of the Torrejon. We are unable to detect any constant differences between the specimens from the upper and lower levels.

Ellipsodon lemuroides (Matthew)

Mioclaenus lemuroides Matthew, 1897, Bull. Amer. Mus. Nat. Hist., IX, p. 314, figs. 15, 16.

Type: No. 2421, lower jaws with p_2–m_3 l., p_4–m_2 r.

Horizon and Locality: Middle Paleocene, Torrejon formation, head of Rio Torrejon, A. M. Exped. 1896.

Author's Diagnosis (op. cit., p. 315): There were three small incisors; the canine was somewhat larger, about the size of the first premolar; the first premolar was one-rooted, the second is two-rooted, larger, recurved, and minute-heeled. The third and fourth are still larger, stouter, recurved and moderately inflated, the heel of the fourth larger than the others. The molars are very simple, broad and short and considerably inflated, with more or less oval outline, the cusps low, no paraconid, no entoconid, and a simple shallow basin in the heel. Length [in meters] from i_1–m_3 = .0343; p_{2-4} = .0135; m_{1-3} = .0129.

A.M.16636 $\frac{2}{1}$

Fɪɢ. 52. *Ellipsodon lemuroides,* upper teeth, A. M. No. 16636, crown view. Twice natural size.

The shape of the jaw is that of *Euprotogonia* and of *M. turgidus;* the teeth are much less inflated than in *M. turgidus,* but more so than in the preceding species [*M. turgidunculus,* now referred to *Oxyacodon*]. . . . The third upper premolar is subtriangular with a well-separated internal cusp; the fourth is wider transversely, more nearly oval, the internal cusp almost as large as the protocone. The first and second molars are slightly larger than p^4, with rudimentary hypocone and minute intermediates. The third molar is but two-thirds the diameter of the second. All the teeth have low rounded cusps, and external cingula obsolete except on m^1 and m^2. Length, p^3–m^3 = .019; m^{1-3} = .011.

Many additional specimens of the upper and lower jaws are in the 1913 collection. They confirm the association of upper teeth with the type jaw, but add little to the characters of the dentition, although some are rather better preserved.

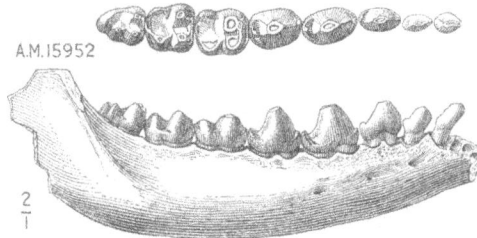

FIG. 53.   *Ellipsodon lemuroides*, lower jaw, A. M. No. 15952, crown and side views.   Twice natural size.

? Ellipsodon aequidens, new species

Indrodon malaris Osborn and Earle, 1895, Bull. Amer. Mus. Nat. Hist., VII, p. 16, figs. 2, 3 (exclusive of lower molar tooth).
Supposed Primate, Osborn, 1902, *ibid.*, p. 170, figs. 1B–C, 2.

Type: No. 823, upper and lower jaw, p^4–m^3, damaged m_{1-2}, parts of vertebral column, limb and foot bones.

Horizon and Locality: Middle Paleocene, Torrejon formation, Chaco Canyon, San Juan basin, A. M. Exped. 1892.

Diagnosis: Molar teeth slightly larger than in *Ellipsodon acolytus*, m^3 unreduced, p^4 somewhat smaller and more trigonal in outline, the inner cusp smaller and more compressed.   Lower jaw deeper, longer posteriorly, skeleton considerably larger.

The teeth of the type and only specimen of this species are badly worn and damaged, so that their structure is not exactly determinable.   The species appears to be allied to *E. acolytus*, but is clearly distinct from this and from any other described species.   The skeleton bones are about a third larger lineally than in *E. acolytus*, No. 15949, and decidedly more robust.

No. 16039, a lower jaw fragment with m_{1-3} well-preserved, is possibly referable to this species.   If so, it is distinguished by a well-developed entoconid, complete absence of paraconid and the molars, and the trigonid and talonid cusps are somewhat more separated, the lateral valleys between them being broader.

Ellipsodon priscus, new species

Type: No. 16403, lower jaws with p_2–m_3 l., p_3–m_3 r.

Paratypes: No. 16401, lower jaw fragments, m_{1-3} l., m_{2-3} r.; No. 16530, parts of upper and lower jaw, p^{3-4} l. and most of m^{1-2} r., m_{1-3} and root of p_4 r., partly buried in matrix.

Horizon and Locality: Type and No. 16401 from the lower level of the Puerco, 3 miles east of Kimbetoh; No. 16530 from the upper level of the Puerco, near Ojo Alamo.   All from Lower Paleocene of the San Juan basin, A.M. Exped. 1913.

Diagnosis: Size of *Ellipsodon acolytus,* but m_3 reduced as in *E. lemuroides,* and m_{1-2} very short and wide. Paraconids small but distinct on all three molars. Premolars more pointed, less blade-like than in the Torrejon species. Inner cusp of p^4 conical. Closely allied to *E. inaequidens* of the Torrejon and of about the same size. M_3 equally reduced in size, but retaining the cusp construction of the preceding molars, the heel much narrower than the trigonid and deeply basined, the trigonid cusps opposite instead of alternate, and

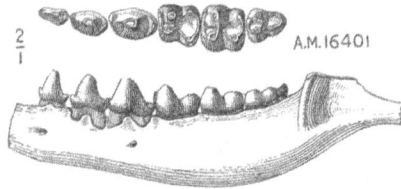

FIG. 54. *Ellipsodon priscus,* type lower jaw, A. M. No. 16401, two premolars from A. M. No. 16403, crown and side views. Twice natural size.

the crown not flattened. Premolars with smaller and less compressed main cusps and larger heel than in *E. acolytus* or *lemuroides,* m_3 more reduced than in the latter, much more than in the former species.

This species is readily recognized as an *Ellipsodon* by the characteristic molars, but it is distinct from any of the Torrejon species and mainly synthetic in its characters, combining the small size of *acolytus* with the reduced m_3 of *lemuroides,* and the more primitive premolars that apparently distinguished *E. aequidens.* The round cusps of the molars, reduced m_3^3, etc., readily distinguish it from *Oxyacodon.*

Protoselene Matthew, 1897

Matthew, W. D., 1897, Bull. Amer. Mus. Nat. Hist., IX, Nov. 16, p. 317

Type: Mioclaenus opisthacus Cope, 1882, from the Paleocene, Torrejon formation, San Juan basin, New Mexico.

Author's Diagnosis: Premolars not much inflated, lower ones trenchant. Fourth upper premolar with strong internal cusp (deuterocone) and distinct postero-external cusp (tritocone). Third premolar nearly simple. Last lower premolar with strong heel, anterior ones simpler. Entoconid well developed on lower molars.

Protoselene opisthacus (Cope)

Mioclaenus opisthacus Cope, 1882, Amer. Nat., XVI, Sept. 28, p. 833.
Hemithlaenus [= *Mioclaenus*] *baldwini* Cope, 1882, Amer. Nat., XVI, p. 833.

Type: A. M. Cope Coll. No. 3275, p_4–m_3 l., m_{1-3} r., also premolars.

Horizon and Locality: Paleocene of San Juan basin, undoubtedly Torrejon formation, Middle Paleocene.

Author's Diagnosis: The species of this genus brought thus far from the Puerco formation have no internal cusp, but a ridge on the internal side of the heel of the inferior true molar teeth [i.e. no distinct entoconid]. The *M. brachystomus* of the Wasatch [= *Diacodexis chacensis*] has such a cusp. The present species from the Puerco also possesses this cusp [i.e. entoconid distinct]. It differs from the *M. brachystomus* in its much larger size and more robust premolars. The latter are, however, less robust than in *M. turgidus* and have an oval anteroposterior section. The fourth has a small heel, but no anterior basal lobe. The true molars are of subequal size and not smaller than the premolars. No anterior inner nor posterior median cusps. [Measurements follow.]

Type Description of Mioclaenus baldwini: The description of the last species ["*M.*" *opisthacus*] applies to this one in many respects, including the posterior inner lobe of the inferior true molars, but the size is less, and the last inferior molar is materially smaller. There is also a well defined anterior internal cusp on the second true molar. The ramus becomes quite slender anteriorly. [Measurements follow.]

A number of additional specimens referred to this species illustrate the upper and lower cheek teeth and some characteristic parts of the skeleton.

The incisors are not preserved, but appear to have been much as in *Mioclaenus*. The canine is small, semi-procumbent, about the same size as p_1. P_1 is one-rooted, rather small; p_{2-4} are two-rooted, moderately compressed, elongate, little inflated, p_3 with very rudimentary, p_4 with fairly well-developed anterior basal cusp and heel. The molars have a small submedian paraconid, large, rounded subequal paraconid and metaconid, large hypoconid and entoconid only a little lower than the trigonid cusps, and distinct hypoconulid, of large size on m_3. The heel is considerably "basined." There are no cingula and the molars are subequal in size; p_3 and p_4 are approximately as long as m_1.

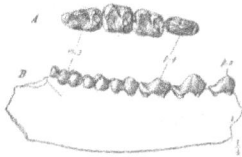

Fig. 55. *Protoselene opisthacus,* lower jaws, A. M. No. 3278, crown view; A. M. No. 2435, side view. (From Matthew, 1897.) Natural size.

The upper molars are tritubercular, subequal, m^3 a little smaller than m^1 or m^2. They have a very small but distinct hypocone on m^{1-2}, small mesostyle on m^{1-3}, external, anterior and posterior cingula on all three molars. P^4 has large protocone, rudimentary metacone on the crested posterior slope of the paracone, small parastyle, anterior and posterior cingula and obsolete external cingulum.

The humerus is moderately long and not especially wide at its distal end; the deltoid crest ends abruptly but is comparatively low; the distal end is only moderately prominent. The radius has an oval head, a shaft somewhat smaller than that of the ulna, a low anterior crest. The ulnar shaft is oval, not exceptionally wide, narrowing toward the distal end. The femur has a longish shaft, a narrow trochlea, a third trochanter high up on the shaft.

Oxyacodon Osborn and Earle, 1895

Osborn, H. F., and Earle, Charles, 1895, Bull. Amer. Mus. Nat. Hist., VII, p. 25; Matthew, 1897, *ibid.*, IX, p. 292

Type: Oxyacodon apiculatus Osborn and Earle, 1895, from the Puerco formation, New Mexico.

Author's Diagnosis: Fourth lower premolar strongly compressed laterally, with only a very minute talon, no deuteroconid. Crowns of inferior true molars high with sharp cusps, trigonid not elevated above talonid, paraconid reduced. Hypoconulid of last molar high and sharp. . . . The true molars in this genus resemble somewhat those of *Anisonchus*, but the structure of the last premolar is widely different. The general structure of the teeth differs decidedly from that seen in *Chriacus* or *Tricentes* and appears to be more of the insectivorous type.

Matthew in 1897 figured and referred to this genus *Anisonchus agapetillus* of Cope, of which a more complete lower jaw was figured, and placed it as *incertae sedis* without reference to any order and with the following diagnosis:

Lower premolars laterally compressed, high and trenchant with minute talons, deuteroconid on p_4 minute or absent. Molars short and wide with high angular cusps, trigonid somewhat elevated above talon, paraconid reduced. Hypoconulid on m_3 high and sharp. Premolars not crowded, the anterior ones not reduced in size. Upper teeth unknown. It may belong to the Anisonchinae, but the molars do not show the closed basin of the heel peculiar to that group, and are wide instead of compressed.[112]

Various additional specimens, jaws or fragments of jaws of this genus have been found, and a few doubtfully associated fragments of the skeleton. The upper teeth are known only in *Oxyacodon turgidunculus*. There is nothing to contradict the view that it is really what it seems to be, an intermediate between the Mioclaenidae, Anisonchidae and Oxyclaenidae. The astragalus described below, if it actually belongs, would indicate fundamentally near relations to *Loxolophus*, although the teeth dispose one to refer it to Mioclaenidae. But the Mioclaenidae and the small Periptychidae are not very distant from *Loxolophus* in the little that is known of their skeleton characters—nearer indeed than they are to *Periptychus* or *Ectoconus*. In fact, the Puerco " Amblypoda " and " Creodonta " are none of them very distantly related to each other, and synthetic genera are to be expected.

Four species are now referred to *Oxyacodon*, as follows:

> *Anisonchus agapetillus* Cope, 1884
> *Mioclaenus turgidunculus* Cope, 1888
> *Oxyacodon apiculatus* Osborn and Earle, 1895
> *Oxyacodon priscilla*, new species

Revised Diagnosis: Canines smaller, lower premolars simple, somewhat inflated, high and moderately compressed. Upper premolars with massive rounded cusps, p^3 simple, p^4 with large inner cusp. Upper molars tritubercular with subangular cusps and strong cingula; last upper molar unreduced, partly transverse. Lower molars rather short and wide, with subangulate cusps; trigonid higher than talonid, tricuspid but with paraconid reduced to a varying degree; talonid deeply basined but not closed internally; hypoconulid of m_3 high and prominent.

Distinguished from *Ellipsodon* and *Mioclaenus* by the unreduced last molar, more acute and angular cusps and trigonal rather than ovate outline of the molars, distinct paraconid, etc.; from *Protoselene* by the shorter lower molars, more compact, with higher and more acute cusps, lack of mesostyle or hypocone and other distinctions of pattern in the upper molars. The premolars resemble closely those of *Ellipsodon*, the molars differ widely both in relative proportions and in the character of the cusps. The teeth are much more like those of *Mixoclaenus* (*Coriphagus*) but the last molar is unreduced, with high, prominent hypoconulid, p^4 has a round-conical inner cusp, in place of the smaller, less symmetrical and more crest-like inner cusp of *Ellipsodon*, the molars are less angulate, the inner half less compressed antero-posteriorly, external cingula complete and external angles less extended; m^3 is unreduced with well-developed metacone.

[112] Matthew, W. D., 1897, Bull. Amer. Mus. Nat. Hist., IX, p. 292, fig. 10.

Oxyacodon apiculatus Osborn and Earle

Oxyacodon apiculatus Osborn and Earle, 1895, Bull. Amer. Mus. Nat. Hist., VII, p. 25, fig. 6.

Type: A. M. No. 816, part of lower jaw, with left p_4–m_2.

Paratype: No. 806, part of lower jaw, with right m_{1-3}.

Horizon and Locality: Lower Paleocene, Puerco formation, Coal Creek Canyon, San Juan basin, New Mexico. A. M. Exped. 1892.

Author's Description: Last lower premolar higher than the first true molar, and the crown of same as long antero-posteriorly as the latter. Hypoconulid of m_3 well constricted off; very sharp and curved forwards. The last lower premolar is flattened with sharp anterior and posterior cutting edges; there is only a very slight enlargement behind. This tooth differs from that of *Protochriacus* in being more flattened and trenchant. The second true molar is high and narrow with four principal cusps inclined forward; these cusps are also less connected than in the typical genera of the Chriacidae. The structure of the talon of the last lower true molar is peculiar, in arising from the height of the hypoconulid, which is unusually sharp and pointed. The jaw is deep, and was probably short. This character relates this genus to the Primates.

Matthew in 1897 distinguished the species by the larger size, more compressed premolars and absence of deuteroconid.

The type of the species is somewhat crushed, as are the specimens referred to it in 1897. Better material shows that the premolars are not especially compressed and that there is a rudimentary deuteroconid (metaconid) on p_4, as in the smaller species. There is somewhat more of a basin to the heel, and both molars and premolars are relatively more elongate. The jaw is of moderate depth, not short anteriorly but somewhat slender and elongate, the first two premolars spaced, the canine small, p_1 one-rooted, symphysis loosely sutured. It is the largest of the three species, $m_{1-3} = 13$ mm.

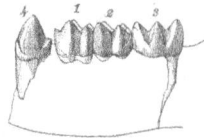

FIG. 56. *Oxyacodon apiculatus*, lower jaw, A. M. No. 816, side view. (From Osborn and Earle, 1895.) One and a half times natural size.

All the specimens found in 1913 are from the upper horizon of the true Puerco.

A few fragments of the skeleton are preserved in No. 16368, but not such as to give any important data as to the affinities. The femur is creodont-like, with moderate neck and head, great trochanter scarcely overtopping the head, lesser trochanter postero-internal, third trochanter with process not distinct but the crest very marked. Tibia has the cnemial crest not very prominent, but the compressed ridge that supports it very well-developed.

Oxyacodon priscilla, new species

Type: A. M. Cope Coll. No. 3547a, right lower jaw with p_2–m_3. Figured by Matthew in 1897 under name of *Oxyacodon agapetillus*.

Diagnosis: Size intermediate between *O. apiculatus* and *agapetillus*; $m_{1-3} = 11.2$ mm. Anterior basal cusp of p_4 distinct, otherwise very like *agapetillus*. Eight specimens besides the type are referred to this species, all lower jaw fragments; seven are from the lower level of the Puerco, the other two unrecorded.

Oxyacodon agapetillus (Cope)

Anisonchus agapetillus Cope, 1884, Paleont. Bull. No. 37, Jan. 2, and Proc. Amer. Phil. Soc., XXI, p. 320; 1885, Trans. Amer. Phil. Soc., XVI, N.S., p. 305; (*Oxyacodon*) Matthew, 1897, Bull. Amer. Mus. Nat. Hist., IX, p. 292, fig. 10.

Cotypes: " Parts of six mandibular rami."
Lectotype: A. M. No. 3557, two fragments of the lower jaw, m_{1-2} r., m_2 l.
Horizon and Locality: Paleocene, probably Puerco formation, San Juan basin, no exact locality recorded.

Author's Description: The inferior molars have the anterior inner cusp [paraconid] moderately well developed, as in *Anisonchus gillianus*. The crowns of the true molars consist of two Vs; of which the posterior base of the posterior one [talonid], is rendered irregular by the presence of a small posterior median tubercle [hypoconulid]. Of the anterior pair of cusps, the external [protoconid] is a little the more elevated, and the internal [metaconid] is more elevated than any of the posterior ones [talonid cusps]. The internal posterior [entoconid] as well as the external posterior [hypoconid] cusp has a V-shaped section, because its anterior border is continued as an oblique ridge to the base of the anterior internal cusp [metaconid]. Internal cingula none; a slight one on the external base of the large anterior external cusp [protoconid]. The heel of the third true molar is well developed, and rises into an acute cusp. That of the fourth premolar is short and flat. The anterior cusp of the same is basal and rudimental. This tooth is not enlarged as is usually the case in the *Periptychidae*, and it first here differs from these animals, and agrees with the unguiculate types in that its lateral faces are unequally convex.

Revised Diagnosis, Matthew, 1897: Premolars moderately compressed, minute deuteroconid on p_4. Dimensions: p_2-m_2, .0103 [meters].

Revised Diagnosis: $m_{1-3} = 9.5$ mm. Anterior basal cusp of p_4 rudimentary or absent.

Four fragments of lower jaws are referred to this species, two of them from the lower level of the Puerco, the others unrecorded.

Oxyacodon turgidunculus (Cope)

Mioclaenus turgidunculus Cope, 1888, Trans. Amer. Phil. Soc., XVI, N.S., p. 334 (type specimen only); Matthew, 1897, Bull. Amer. Mus. Nat. Hist., IX, p. 313, fig. 14.

Type: A. M. Cope Coll. No. 3291, upper jaw fragment, left p^4-m^2.
Horizon and Locality: Puerco formation, San Juan basin, New Mexico. No exact locality recorded.

Revised Diagnosis: Size of *Oxyacodon apiculatus* but premolars decidedly more robust, molars broader, lower crowned, with cusps less acute, paraconids on lower molars small, tending to become connate with metaconids. Upper molars trigonal, simple cusp construction with no distinct hypocones, upper premolars somewhat inflated, simple in construction, sub-oval trigonal.

The upper teeth approach those of *Ellipsodon* in their simple character, low cusps and general construction but they are less rounded, m^3 less reduced. The lower teeth differ chiefly in retention of a small paraconid. It appears very probable that this species is the ancestor of *E. lemuroides*. It is the only one of the species of *Oxyacodon* in which the upper teeth are known, and the marked resemblances in the premolars to those of *Ellipsodon* and *Mioclaenus* afford sufficient reason for referring the genus to the Mioclaeninae. The resemblance in the lower teeth to those of *O. agapetillus* and through this species to *priscilla* and *apiculatus* warrants the reference to *Oxyacodon*.

The principal specimens referred are:

A. M. No. 3291, type, upper jaw, p^4–m^2 l.

A. M. No. 3212, figured, upper and lower jaws, m^{1-3} l., p_1–m_1 r., p_{3-4} l.

A. M. No. 3474a, lower jaw, m_{2-3} l.

A. M. No. 16402, lower jaw fragment, m_{1-2} l., and astragalus.

A. M. No. 16400, lower jaws, p_{3-4}, m_{2-3} r., p_{3-4} l., upper premolar.

A. M. No. 16404, lower jaw, p_3–m_3 r., three upper premolars and one upper molar.

A. M. No. 16399, lower jaw fragment, m_{1-2} r.

A. M. No. 16489, upper jaws, p^4–m^3 l., m^{2-3} r.

An anterior upper premolar, probably p^2, is of rounded trigonal form, no trace of inner cusp but a faint encircling basal cingulum on the two inner sides, and three subequal roots. P^3 (?) has a large rounded inner cusp about equal to the protocone, which has a faint anterior and a distinct posterior crest, and basal cingula running around the external angles. P^4 (?) is quite similar but the internal cusp somewhat larger, the basal cingula raised into small definite cusps. All the premolar cusps are low, robust, moderately inflated and with smooth enamel. The inner root on p^{3-4} and on the molars is much larger than the outer roots. M^1 is subtrigonal, the paracone and metacone low, round, well-separated, protocone crescentic, conules not distinguishable, heavy external cingulum but no stylar cusps, distinct anterior and posterior cingula but no hypocone or protostyle, no internal cingulum. M^2 is quite similar but metacone somewhat smaller than paracone. On m^3 the metacone is quite small and the postero-external angle of the tooth reduced, the whole tooth materially smaller than m^2.

The four lower premolars are much like those of *Ellipsodon lemuroides* but smaller and somewhat less inflated. All have distinct posterior heels but no anterior basal cusps or cingula; the third and fourth are subequal, the first and second much smaller. The molar trigonids are very short but as wide as the talonids, protoconid and metaconid subequal, paraconid small and internal in position, heel almost as high as trigonid, hypoconid strong, conical entoconid and hypoconulid imperfectly separated, enclosing a deep contracted basin. Heel of m_3 reduced in width, scarcely longer than on m_{1-2} but much narrower, the whole tooth somewhat smaller than m_2.

4. ORDER INSECTIVORA

CLASSIFICATION OF THE INSECTIVORA

To this order are referred four groups of Paleocene placentals, all of small size and for the most part very imperfectly known. The Insectivora are generally recognized as an order of wider scope than other placental orders; their existing representatives, while retaining many primitive characters, belong to four or five distinct groups, the relationships of which are rather remote. Their Tertiary fossil record is scanty and does little toward clearing up the exact derivation of the several groups, but adds to the complexity of the problem by introducing certain extinct groups of more or less equivocal relationship.

Repeated attempts have been made to split up the Insectivora into two or more orders which might be thought to correspond more nearly to the scope of the better-defined orders, but none of these is altogether satisfactory.

Gill in 1872 divided the order into Zalambdodonta and Dilambdodonta, a distinction based upon the characters of the molar teeth. The antiquity of this distinction has been confirmed by all the palaeontological evidence, which shows the two types of molar—the one a pre-tritubercular stage, the other tritubercular or derivable therefrom—as distinct throughout the Tertiary and even in the later Cretaceous. But the evidence in regard to the zalambdodonts is so scanty that no great weight can be placed upon it. Leche in 1883 after a careful study of the comparative anatomy of Insectivora, together with such fossil evidence as was available, distinguished the Tupaiidae and Macroscelididae as a separate order Menotyphla, leaving the remaining families in the Insectivora. Broom has in more recent years distinguished the Chrysochloridae as a separate order, basing his conclusions primarily upon the study of the organ of Jacobson in modern mammals. To classify mammals upon the evidence of one organ, concerning the evolution and characters of which there is no fossil evidence, seems rather rash; nevertheless, Broom's insight into comparative anatomy and phylogeny is vindicated by the study of the skeleton in general, which indicates that the Chrysochloridae must be very distantly related to any other family, and that the similarity in molar characters to other zalambdodont insectivores is more nominal than real. Carlsson, who has made a rather careful comparative study of the anatomy of the Tupaiidae and Macroscelididae, concludes that the latter group shows very real relationship to the Erinaceidae and ought not to be separated in a distinct order. And yet they have also a considerable approach to the tupaiids, confirmed by the recent discovery in the Oligocene of Mongolia of two genera which seem to be largely intermediate.

We find further, in the older Tertiary, Paleocene and Eocene, two extinct families, Plesiadapidae and Pantolestidae, the former intermediate between Tupaiidae and primitive Lemuroidea so far as can be judged from the skull and the known parts of the skeleton, the latter with teeth suggesting creodont relationships but skull and skeleton of insectivore affinities. A third extinct family, Leptictidae, has generally been cited as related both to Erinaceidae and Tupaiidae, but without any conclusive proof that the relationships to either family are due to real affinity.

Present evidence on the whole would rather suggest that the Insectivora, if divided at all, should be divided into not two but six orders, as follows:

CHRYSOCHLOROIDEA: Chrysochloridae, Necrolestidae
CENTETOIDEA: Centetidae, Apternodontidae, Solenodontidae
SORICOIDEA: Soricidae, Talpidae, Myogalidae
ERINACEOIDEA: Erinaceidae, Leptictidae, ?Nesophontidae
MENOTYPHLA: Tupaiidae, Macroscelididae, Plesiadapidae, ?Mixodectidae
PANTOLESTOIDEA: Pantolestidae

It is furthermore difficult to place in any of these groups the Cretaceous mammal skulls from Mongolia, which might necessitate a seventh ordinal group. It would be difficult to combine any two of these groups without providing plausible grounds for reducing the others to similar rank. The zalambdodont dentition suggests relationship of the first two, yet Broom would separate the Chrysochlorids from all the rest. Leche regards the soricoids as related not to erinaceids but to centetids, and would distinguish the Menotyphla only as a separate order; yet Carlsson and Winge insist upon erinaceid relations in the Macroscelididae, which cannot be so distant from Tupaiidae in view of the Oligocene intermediate genera. Stehlin regards the Plesiadapidae as Primates, but their tupaiid affinities are difficult to set aside; they in turn are apparently related to the Mixodectidae, which can hardly be classed as Primates whatever their real position.   The pantolestoids show no especial affinities to the other groups.

It appears on the whole most convenient to retain the order Insectivora to cover all these diversely specialized and distantly related groups, with the several groups as suborders.

Many authors have regarded the Insectivora as representing the primary stock of the placental mammals.   To the extent that this is true it affords a logical reason for the

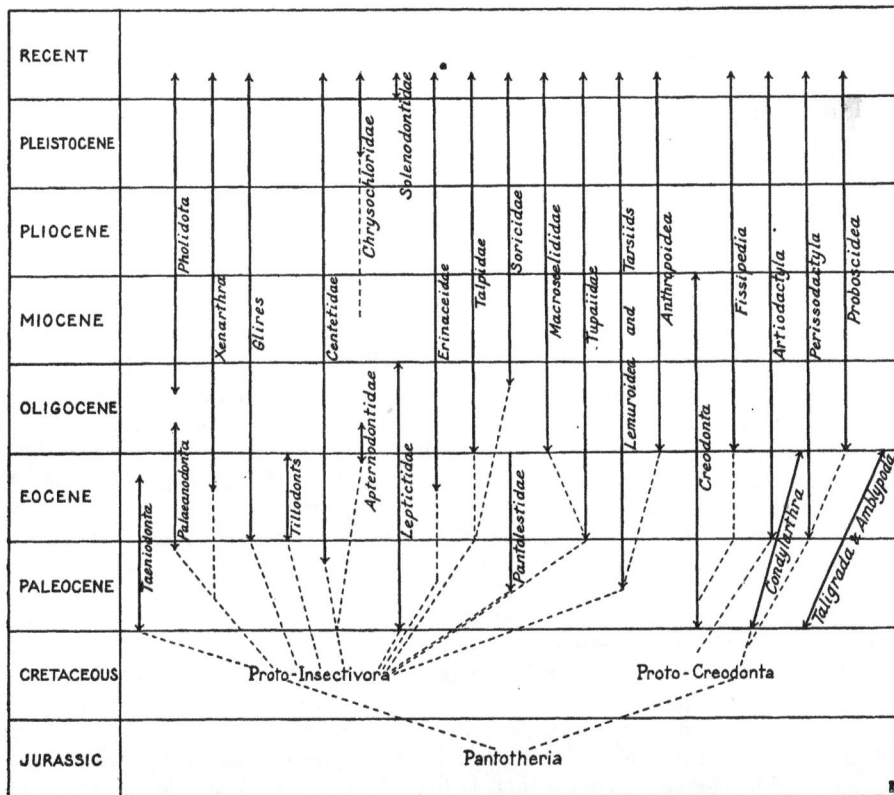

FIG. 57.   The Origins and Geological Range of the Mammalia.

deep-seated divergences of the several surviving groups.   It has, however, been pointed out by Cope and Wortman that the early Tertiary Creodonta have strong claims to be regarded as the central ancestral group of the placentals, and in fact their much greater nearness to the primitive stock, due to their geological antiquity, makes them in many respects better representatives of such an ancestral group than are any of the modern Insectivora, much specialized in one way or another so as to obscure or lose the primitive characters.   It is not at all clear that the earliest Tertiary Insectivora are as primitive a stock as the earliest creodonts.   They show more specialization and more fundamental diversity in the dentition, and they clearly are not so primitive in tarsal construction. Matthew pointed out in 1909 that the tarsus of all insectivores shows distinctive and characteristic specializations in the astragalus not seen in the early creodonts, and that the insectivore tarsus, while it could specialize readily into that of rodents and edentates, could not be ancestral to that of creodonts, from which the condylarth, amblypod and probably the perissodactyl and artiodactyl tarsi appear to be derived. From this and other evidence it would appear that Insectivora are structurally ancestral as an order to a part, but not to the whole, of the placentals, the remainder being in a similar way structurally derived from the creodonts. To what precise degree these relations accord with the Mesozoic origin of the placental orders will not be proved until we know far more about Mezosoic mammals.

The four families represented in the Puerco-Torrejon fauna are:

1. Leptictidae: *Prodiacodon, Acmeodon*
2. Mixodectidae: *Mixodectes, Indrodon*
3. Pantolestidae: *Pentacodon*
[4. Palaeoryctidae: *Palaeoryctes*][113]

LEPTICTIDAE Cope, 1882

Cope, E. D., 1882, Proc. Amer. Phil. Soc., XX, p. 156

The typical genus, *Leptictis*, was briefly described by Leidy in 1868 [114] and more fully described and figured in 1869. It was based upon a well-preserved skull from the White River Oligocene. The nearly related genus *Ictops* was described by Leidy in 1868; [115] the two genera, he states, "belong to a peculiar family related with that of the hedgehogs." [116] In the description of the *Leptictis* skull he says: " It apparently belongs to an animal of the insectivorous order, but exhibits sufficient resemblance to the skulls of the Opossums to lead to the suspicion that it may possibly pertain to a member of the Marsupialia. Among the Carnivora it exhibits more affinity to the canine family than any others, and appears more nearly related to the Viverrine than the Musteline family " (*loc. cit.*, p. 345). On a later page (p. 358) he places both genera under the family Erinacidae.

Cope in 1895 included the Leptictidae provisionally in the creodont division of his Bunotheria, referring to it besides the typical group a number of genera of true creodonts with primitive dentition, later grouped as Proviverridae and subsequently distributed in various families. He recognized nevertheless the insectivore affinities of the typical group. Schlosser in 1888 referred the group definitely to the Insectivora, where it has been retained by all subsequent authors. He observes with regard to the affinities of the family:

Ictopsidae (Leptictidae Cope partim). . . . Alle M haben gleiche Grösse, und unterscheiden sich die Ictopsiden folglich ganz wesentlich von den Erinaceiden und *Gymnura*, denen sie sonst im Schädelbau sehr ähnlich sehen; sie schliessen sich eher an die Tupajiden—*Cladobates*—an. Bei diesen ist jedoch der Schädel

---

[113] *Palaeoryctes* was regarded by Dr. Matthew as belonging to the Centetidae (= Tenrecidae). It was inadvertently omitted in this general discussion of the Insectivora and in the systematic consideration of genera and species. See p. 362 of the Addendum.—THE EDITORS.

[114] Leidy, J., 1868, Proc. Acad. Nat. Sci. Phila., p. 315.

[115] *Op. cit.*, p. 316.

[116] Leidy, J., 1869, "The Extinct Mammalian Fauna of Dakota and Nebraska . . .," p. 345.

mehr modernisirt, die Zähne, wenigsten die M eher noch ursprünglicher und die Pr in beiden Kiefern noch sehr viel einfacher. Auch scheint der Kiefer der Ictopsiden viel plumper zu sein als bei den Tupajiden.

Der Schädel der Ictopsiden erscheint noch ziemlich flach; die Gesichtspartie hat, abgesehen von Leptictis, noch eine nicht unbeträchtliche Länge. Der Jochbogen ist zwar nicht sehr massiv, aber gleichwohl sehr gut entwickelt, ähnlich wie bei Erinaceus.

Wahrscheinlich haben wir es hier mit einem ganz selbständigen, in der Gegenwart aber vollständig erloschenen Formenkreis zu thun, der mit den Erinaceiden und Tupajiden aus einer gemeinsamen Stammform hervorgegangen ist.[117]

Douglass in 1905 [118] described a number of skulls and parts of the skeleton of *Ictops*, but does not discuss the affinities of the family beyond citing the opinions of Leidy and Cope.

Matthew in 1909 [119] defined the family as follows:

Insectivora allied to the Hedgehogs and Tree-Shrews but with tritubercular molars. Limbs not fossorial.

Dentition $\frac{2 \cdot 1 \cdot 4 \cdot 3}{3 \cdot 1 \cdot 4 \cdot 3}$. Molars and premolars with high, sharp cusps. Fourth premolar molariform. Replacement of teeth retarded, taking place only after the animal has attained full adult dimensions. Face elongate, basicranial region short, tympanic chamber partially formed by alisphenoid and basisphenoid processes. [There is, however, a true tympanic bulla in *Ictops*, loosely coössified to the walls of the tympanic chamber.]

Vertebral formula not known. Lumbar zygapophyses simple, flat. Tail reduced.

Humerus with moderately prominent deltoid and supinator crests and an entepicondylar foramen. Radius and ulna separate, subequal. Femur with strong hooked internal trochanter, rudimentary third trochanter, narrow and elongate patellar trochlea. Tibia and fibula completely fused in distal half; no fibulo-calcanear facet. Astragalus with short, wide trochlea, shallow but sharply defined at edges, distinct neck and wide convex head, with slight contact with cuboid. Cuboid narrow, navicular with strong, stout, inferior hook. Cuneiform separate. Digits of pes five, the laterals reduced, hallux not opposed. [A key to the genera follows.]

The above definition was based primarily upon a series of skulls and skeletons of *Ictops* and related genera from the White River Oligocene, and in some particulars it does not apply to the genera of the Eocene and Paleocene, now known from more extensive and complete material.

In 1918 Matthew revised the Lower Eocene Leptictidae, showing that *Palaeictops* was the same as *Diacodon* Cope, and described a species from the Torrejon as representing a new subgenus *Palaeolestes* of *Diacodon*, since re-named *Prodiacodon* as the name *Palaeolestes* was found to be preoccupied. Additional material both of typical *Diacodon* from the Huerfano of Colorado and of *Prodiacodon* from the Torrejon shows that the skeletal characters are distinct enough to make a full genus out of *Prodiacodon*. We are now able to compare three geological stages—Paleocene, Eocene and Oligocene—in the leptictid phylum, from well-preserved and associated material, as follows:

3. Middle Oligocene: Skulls and principal parts of skeleton of *Ictops*, skulls of *Mesodectes* and *Leptictis*.

2. Lower Eocene (Huerfano and Upper Wind River): Skull and considerable part of skeleton of *Diacodon*.

1. Middle Paleocene (Torrejon): Palate, jaws and part of skeleton of *Prodiacodon*.

[117] Schlosser, Max, 1888, Die Affen, Lemuren, . . . p. 140.
[118] Douglass, Earl, 1905, Mem. Carnegie Mus., II, pp. 210–23.
[119] Matthew, W. D., 1909, Mem. Amer. Mus. Nat. Hist., IX, p. 534.

The genera *Parictops* of the Wind River, *Myrmecoboides* of the Fort Union, *Acmeodon* of the Torrejon, *Xenacodon* and *Leptacodon* of the Upper Paleocene, are known only from the jaws; but of the three principal genera, *Prodiacodon*, *Diacodon*, *Ictops*, we are able to compare skull and skeleton characters as well as teeth. It is remarkable that the molar and premolar pattern in these three genera endures with very little change from Paleocene to Middle Oligocene but there are considerable progressive changes in the skull and skeleton, as also in the front teeth.

Oligocene Leptictidae: Two upper incisors. Cranium more massive with paired supratemporal crests.

1. P³ with distinct parastyle and metacone. Tibia and fibula solidly united in distal third. Lateral digits reduced.. *Ictops*.
2. P³ with metacone, no parastyle.. *Mesodectes*.
3. P³ with neither metacone nor parastyle.. *Leptictis*.

Lower Eocene Leptictidae: Three upper incisors. Canine small. Cranium thinner with median sagittal crest. Tibia and fibula solidly coössified at distal ends. Lateral digits unreduced.

4. Teeth as in *Ictops*.. *Diacodon*.
5. P₂₋₃ broad-bladed, with accessory cusps.. *Parictops*.

Paleocene Leptictidae:

6. Teeth as in *Ictops*. Canines larger, incisors smaller. Cranium small, with median sagittal crest. Tibia and fibula separate. Lateral digits unreduced............. *Prodiacodon*.
7. Protoconids of lower molars reduced... *Myrmecoboides*.
8. Paraconids well developed on molars, metaconid of p₄ small......................... *Leptacodon*.
9. Fourth premolar more simple, non-molariform, no metaconid......................... *Acmeodon*.
10. Fourth premolar with large metaconid, minute paraconid and very short heel; p₁ two-rooted.. *Xenacodon*.

Of the five Paleocene genera *Prodiacodon* appears to be the ancestral stock of all the later Leptictidae; the other four, known only from the dentition, are aberrant variants from a more primitive stage in the evolution of the family. They are sharply distinct generically, but show nevertheless the leptictine stamp in the details of their molar and premolar construction, so that, except perhaps for *Acmeodon*, there can be no reasonable doubt of their relationship. They serve, therefore, to confirm the antiquity of the leptictid phylum.

The precise affinities of the family are not very clear. It is an extinct phylum of very ancient origin—probably pre-Tertiary—but whether derived from the ancestral stock of the Lipotyphla alone, or, as is perhaps more probable, from the common ancestral stock of Lipotyphla and Menotyphla, it is difficult to say. A thorough and judicious comparison of the skull and skeleton of the three best-known genera with the various existing Insectivora (*sensu lato*) might decide this problem, but this is beyond the scope of the present memoir.

Prodiacodon Matthew, 1929

Matthew, W. D., 1929, Jour. Mammalogy, X, p. 171

Palaeolestes Matthew, 1918, Bull. Amer. Mus. Nat. Hist., XXXVIII, p. 576.

Type: Diacodon Palaeolestes puercensis from the Torrejon of New Mexico.

Author's Diagnosis: The characters indicated as subgeneric in the original description are: " Canine larger, incisors reduced." Complete preparation of the paratype specimen mentioned in the above description and comparison with a partial skeleton of *Diacodon* from the Huerfano basin in Colorado enable us to distinguish adequate generic characters as follows:

Revised Diagnosis: Dentition, $\frac{2 \cdot 1 \cdot 4 \cdot 3}{7 \cdot 1 \cdot 4 \cdot 3}$; teeth as in *Ictops* except that the premolars are less blade-like and relatively smaller, the external stylar cusps of the molars larger, hypocones smaller, paraconids less reduced; canines less reduced, incisors presumably $\frac{3}{3}$.   Skull much less massive posteriorly, with a faint median sagittal crest, no paired postorbital crests.   Limbs and feet of moderate length; humerus with flattened abruptly-ending deltoid crest; tibia and fibula discrete.

The separate fibula affords a distinction from *Diacodon* and *Ictops;* the deltoid crest of the humerus is like that in *Diacodon;* in *Ictops* it is not flattened, and disappears more gradually at the middle of the shaft.   The proportions of the limb and foot bones to the skull are like those of *Ictops;* in *Diacodon* the limbs and feet are relatively small.   In the cranial portion of the skull *Prodiacodon* resembles *Diacodon* and differs from the Oligocene genera.

Prodiacodon puercensis (Matthew, 1918)

Diacodon (Palaeolestes) puercensis Matthew, 1918, Bull. Amer. Mus. Nat. Hist., XXXVIII, pp. 576–9, figs. 6–9.

Type: No. 16011, fragmentary skeleton from the Torrejon of New Mexico.

Paratype: No. 16748, palate, lower jaws, and part of skeleton from the same horizon and locality.

Horizon and Locality: Middle Paleocene, Torrejon formation, San Juan basin, New Mexico.

Revised Diagnosis: See the diagnosis for the genus, above.

The original description is based chiefly upon the type, the better-preserved paratype specimen not being prepared for study.   The following additional data are now determinable:

The upper canine is known only from an incomplete alveolus, but was evidently considerably larger than p[1].   The premolars are less blade-like than in *Ictops*, somewhat smaller relatively to the molars.   P[1] is one-rooted; in *Ictops* it is two-rooted.   P[3] has the inner cusp relatively larger than in *Ictops* and internal instead of postero-internal; the postero-external cusp (metacone) is somewhat more distinct.   P[4] has a distinct parastyle, vestigial in *Ictops;* the inner cusp (protocone) is more compressed, and the hypocone is represented only by a low and narrow cingulum.   The molars have also more compressed protocones, with very rudimentary hypocones.

Affinities of *Myrmecoboides* Gidley, 1915 [120]

A lower jaw from the Fort Union of Montana was described and figured by Gidley as a marsupial and a probable relative of *Myrmecobius.*

It is unfortunate that Gidley failed to make any comparisons with the genera of Leptictidae, which the specimen closely resembles in dentition and to which in my opinion it is unquestionably nearly related.   The author has, I believe, misinterpreted the dental formula.   The tooth which he regards as a fourth milk molar and considers evidently as

[120] Gidley, J. W., 1915, Proc. U. S. Nat. Mus., 48, pp. 395–402, pl. xxiii.

persistent in this genus, as in the marsupials generally, appears to be p₄. If it were a milk molar it would be rash to assert in the present instance that it had no premolar succeeding it. But the photograph shows quite clearly that it is less worn than the tooth behind it and less completely emerged from the jaw; this shows that it came into use later than the tooth behind it, and not earlier, as would be the case were it a milk molar. Both the photograph and the description show that it has the typical construction of p₄ in the Leptictidae. There is no doubt at all as to the premolar replacement in this family, and it occurs very late, after the animal is adult, as Matthew has elsewhere pointed out.[121] The tooth next in front of it is a milk molar, dp₃, again typically leptictid; the canine and p₁ and p₂ are permanent teeth.

The molars are wholly leptictid in pattern, the only difference from the described genera being the small protoconid, " with inner cusps of trigonid (paraconid and metaconid) as high or higher than main outer cusp (protoconid)." This condition is closely approached, if not equalled, in certain species of *Diacodon*, but in general the protoconid is distinctly higher and the paraconid more reduced in this family.

Gidley thinks that the very complex structure of the " dp₄" is against its being a premolar; also that " it has the same number of cusps as the molars, and these, with the exception of the paraconid, have the same general form, proportions, and arrangement. The crown is proportionately narrower, the talonid is relatively smaller, and the large paraconid is directed well forward, making up the whole anterior portion of the trigonid, and is quite distinct. In this respect the tooth differs markedly from the true molars."

But every detail of the above citation (*op. cit.*, p. 397) applies accurately to p₄ of the Leptictidae, which is positively known to be a replacement tooth.[122] Comparison of Gidley's figures with the photographs of *Prodiacodon* and figures of *Diacodon huerfanensis* will show that there is a close relationship, although his genus may perhaps be retained as valid.

Acmeodon Matthew and Granger, 1921

Matthew, W. D., and Granger, W., 1921, Amer. Mus. Novitates No. 13, p. 3

Type: Acmeodon secans from the Middle Paleocene, Torrejon formation, of New Mexico. Based upon the lower jaw.

Authors' Diagnosis: Dentition $\overline{\tfrac{?.1.3.3}{}}$. Molars of leptictid type but trigonid not so high and paraconid better developed than in *Diacodon* or *Palaeolestes*. P₄ of peculiar pattern, the principal cusp (protoconid) much compressed and crested, with strong accessory cusps (paraconid, protostylid) on the anterior and posterior edges and a somewhat weaker posterointernal cusp (metaconid) connected by a prominent crest with the apex of the protoconid; also a well-developed basined talonid with acute postero-external and posterointernal cusps (hypoconid, entoconid). P₃ large, simple, high, acute and compressed, with anterior postero-external and posterointernal crests, and a small, low, simple, acute heel-cusp. P₂ much smaller, simple, with anterior and posterior crests and a minute heel-cusp. Canine rather small, oval in cross-section at base. At least one small incisor is present.

Acmeodon secans Matthew and Granger

Acmeodon secans Matthew and Granger, 1921, Amer. Mus. Novitates No. 13, p. 3.

Type: No. 16599, a part of the lower jaw with left p₂–m₂ and the root of the canine preserved.

[121] Matthew, W. D., 1918, Bull. Amer. Nat. Hist., XXXVIII, p. 573.

[122] Skulls of *Ictops* in the American Museum collection show the permanent upper and lower premolars preformed in the jaws, with the milk predecessors still in position.—w.d.m.

Paratype: No. 16600, right lower jaw with p_4–m_1, the root of m_2 and alveoli of m_3.

Horizon and Locality: Both are from the upper level of the Torrejon, the type from Torrejon Arroyo, the paratype from the Escavada wash.

Revised Diagnosis: Size of *Prodiacodon puercensis*, but with shorter and deeper jaw, somewhat larger canine, the premolars compressed and knife-like and not at all recurved, p_1 absent, the others of very peculiar construction, suggestive of *Parictops* and *Leptictis*. There is a very faint rudimentary anterior basal cusp on p_2 and p_3; on p_4 it is distinct, as in *Prodiacodon*. The protostylid appears as a faint rudiment on p_3 about half-way down the postero-external crest; on p_4 it is a strong cusp connected by a crest with the apex of the protoconid. There is no trace of this cusp on p_4 of any other leptictid, but it is prominent on p_2 and p_3 of *Parictops*. The distinct separate metaconid internal to the protoconid in other Leptictidae is here apparently represented by the inner posterior crest of p_4, terminating in a distinct cusp somewhat lower than the paraconid. The heel of p_4 is much smaller and less separated than in *Prodiacodon* or any of the typical leptictid genera. A rather obscure obsolete cingulum at the external base of the flattened three-cusped, blade-like trigonid of this tooth. The third premolar (p_3) is very like that of an undescribed species of *Leptictis* in the Kansas University Museum; it is much higher than p_4 and much larger than p_2, although equally simple in structure.

The first molar is not so typically leptictid as in *Prodiacodon*, and in constructive details approaches some of the Oxyclaenidae. The trigonid is high, triangular, capped by three trigonal cusps, of which the protoconid is a little the largest, the other two subequal. The heel is relatively small and the entoconid low as compared with other leptictid genera.

Acmeodon would seem to be a primitive but aberrant leptictid. P_4 is less molariform than in the more typical genera; its unusual construction is suggestive of p_{2-3} in *Parictops*, which have the blade-like crest with anterior and posterior cusps flanking the main central cusp, although the postero-internal crest and cusp are absent in the *Parictops* teeth, and p_4 is normally molariform. The high, simple p_3 of *Acmeodon* is another peculiar feature which is paralleled in the Leptictidae, as indicated above, and the construction of the molars, while not typically leptictid, comes nearer to this group than to the Oxyclaenidae.

MEASUREMENTS

A. M. No. 16599, type:	p_2–m_2 = 19.3 mm.
A. M. No. 16600, paratype:	m_1–m_3 = 10.6 mm.

MIXODECTIDAE Cope, 1883

Cope, E. D., 1883, Proc. Acad. Nat. Sci., Phila., XXXV, p. 80

This family was based by Cope upon *Mixodectes* and *Cynodontomys*, to which he added *Microsyops* in 1885. In Tertiary Vertebrata (p. 239) he observes under Prosimiae:

The suborder may be differentiated from the Mesodonta by the possession of an opposable hallux of the posterior foot. This character is, however, not yet demonstrated in the genera of the American Eocene, which I provisionally give to it, nor is the absence of the character known to belong to any of the genera of Mesodonta excepting *Pelycodus*.[123] It is, however, very probable that the other genera referred to the

[123] Cope's belief that the hallux was not opposable in *Pelycodus* was an error due to association of certain creodont skeletal parts with the teeth of a *Pelycodus* from the New Mexican Wasatch. The hallux is in fact like that of *Notharctus*, wholly lemuroid in character.—W.D.M.

Mesodonta agree with *Pelycodus*.   It is also possible that some of the genera here referred to the Prosimiae agree with *Pelycodus*.

In the uncertainty which exists as to the reference of the genus *Cynodontomys* and its immediate allies, I compare the genera of the Eocene lemuroids as follows.   I premise by observing that the genus *Chiromys* clearly represents a primary division of the Bunotheria, which occupies a position between the Prosimiae and the Tillodonta.   The rodent-like incisors with permanent pulps are those of the Tillodonta, but the opposable hallux of the posterior foot is not found as yet in that suborder.   The suborder has been named by Gill [Footnote: "Arrangement of the Families of Mammals," 1872, p. 54] the superfamily Daubentonioidea.

We can distinguish three families among our Eocene forms of lemuroids, in the dental characters, as follows:

Inferior premolars, four.. *Adapidae.*
Inferior premolars, three....................................... *Mixodectidae.*
Premolars, two, with internal lobes in the upper jaw................ *Anaptomorphidae.*

Cope distinguishes the mixodectid genera thus:

a. Last premolar without inner tubercle.
      A very large incisor; canine smaller; first premolar, only one-rooted... *Mixodectes.*
aa. Last premolar with internal tubercle.
      A very large ? canine; first premolar only one-rooted............. *Microsyops.*
      A very large ? canine; first and second premolars each one-rooted... *Cynodontomys.*

Matthew in 1897[124] transferred *Mixodectes* provisionally to the Rodentia in the following terms:

The discovery of some skeleton fragments in good association with a lower jaw of *Mixodectes pungens* makes it probable that this genus should be removed from the Primates and placed as an extremely primitive Rodent.   *Microsyops* may perhaps go with it, but this is extremely doubtful, as the type of its lower molars is much more primitive and persistently so, and in several other respects different from *Mixodectes*.

The principal evidence for the transfer lay in the characters of the astragalus, which is " quite unlike that of the contemporary Primates, and still more different from *Chiromys*."

Osborn in 1902[125] placed the Mixodectidae, including *Microsyops*, *Cynodontomys* and also *Indrodon* (referred by Cope to the Anaptomorphidae), in a primitive suborder of Rodentia, the *Proglires*.   He added a new genus, *Olbodotes*, to the family, based upon a lower jaw which is closely allied to that of *Mixodectes*, the distinctions made resting upon an interpretation of the tooth formula which I believe to be incorrect.   He defined the suborder as " distinguished by the presence of rooted incisors, and canine teeth, and by the absence of any considerable diastemata and of antero-posterior motion of the jaw."   The family definition was as follows:

Median lower incisors close to symphysis, enlarged and elongating (unlike Tillodontia, in which second incisor is enlarged), lateral incisors early reduced; canines persistent (unlike Rodentia); no diastemata (unlike Rodentia), first and second premolars rapidly reduced; third premolar slowly reduced, fourth premolar progressively molariform (as in Tillodontia and Rodentia); lower molars with narrow, slightly elevated trigonid, but early reduced paraconid; talonid broad, hypoconulid small, except in third lower molar; superior molars tritubercular.   A feature of the jaw is the sharp definition of a ridge descending from the coronoid and defining the masseteric insertion anteriorly.

Wortman in 1903[126] questioned the association of the astragalus described by Matthew with the jaw of *Mixodectes* and criticized the evidences for rodent relationship presented by

[124] Matthew, W. D., 1897, Bull. Amer. Mus. Nat. Hist., IX, pp. 265, 267.
[125] Osborn, H. F., 1902, Bull. Amer. Mus. Nat. Hist., XVI, p. 203.
[126] Wortman, J. L., 1903, Amer. Jour. Sci. (4), XVI, p. 364.

Osborn as being equally characteristic of *Chiromys*, which is an undoubted primate. He further showed that in the method of complication of p₄ the rodents differed from most mammals, and that the " Microsyopsidae " showed the normal method of complication as in primates and other orders of mammals unlike the rodents.

Matthew in 1909 [127] discussed the affinities of the family. He stated more exactly the evidence upon which the astragalus and other fragments were regarded as in " good association " with the jaw of *Mixodectes*, summarized the arguments of previous writers and remarked:

It is curious that Osborn, Wortman and the present writer have overlooked the possibility of insectivore relationship of the Mixodectidae, which may after all be the true solution of this problem. Decisive evidence indeed is lacking, but in one or another of the genera there are several indications that point towards the Insectivora [indications cited in detail].

It is by no means certain that the Torrejon *Mixodectes*, *Olbodotes* and *Indrodon* belong to the same family as the Wasatch, Wind River and Bridger group (*Cynodontomys* and *Microsyops*). The teeth differ very considerably in the form and position of the cusps, and the fourth premolar in the older group is of entirely different and rather specialized type (compare the Insectivore *Pentacodon*....) The present evidence, however, seems to favor referring both groups to the Insectivora.

In 1915 Matthew [128] revised the species of the Eocene genera *Cynodontomys* and *Microsyops*, retaining them provisionally in the Mixodectidae, but pointing out important differences in the dentition which make their reference to it " open to serious doubt."

Stehlin in 1916,[129] in discussion of the relationships of various North American genera to the Plesiadapidae, distinguishes the Mixodectidae and Microsyopidae as distinct families, but is disposed to regard them, especially the latter, as more probably chiromyoid primates than Insectivora.

In 1918 Matthew,[130] while accepting Stehlin's separation of the two families as distinct, continued to regard the Mixodectidae as probably of insectivore affinities.

The additional material now at hand affords a better acquaintance with the upper dentition of *Mixodectes* and confirms its relationship to *Indrodon*. It tends to confirm the view expressed by Matthew and Stehlin that the group is distinct from the Microsyopidae and Plesiadapidae, and has no close affinities to any known Eocene genera. It does not add much to the evidence for deciding its ordinal affinity. The upper molars in *Mixodectes* and *Indrodon* are of peculiar type, with a sharp crested mesostyle, and prominent cingular hypocone, the protocone, paracone and metacone sharp, high, crescentic, with deep basin between them. They resemble those of *Chriacus* in the sharp offsetting of the hypocone, but differ in presence of mesostyle and absence of any trace of internal cingulum. In the last two features they resemble the molars of *Protoselene*, but differ in the hypocone and in proportions. The fourth premolar is unreduced, simple, the principal cusp rather high, round-conical, with a peculiar backward pitch or recurved form more marked in the lower premolar, and with sharp, heavy heels internal on p⁴, posterior on p₄. This type of tooth is widely different from the creodont type, nearer to Condylarthra, Primates or Insectivora. The anterior premolars are much reduced. As in Plesiadapidae, a lower incisor is enlarged, and two upper front teeth, the other anterior teeth tending to disappear.

[127] Matthew, W. D., 1909, Mem. Amer. Mus. Nat. Hist., IX, pp. 547-9.
[128] Matthew, W. D., 1915, Bull. Amer. Mus. Nat. Hist., XXXIV, p. 466.
[129] Stehlin, H. G., 1916, Abh. schweiz. pal. Ges., XLI, p. 1503.
[130] Matthew, W. D., 1918, Bull. Amer. Mus. Nat. Hist., XXXVIII, p. 568.

But it is far from clear whether their enlarged teeth are the same in the two families, or whether the enlarged teeth of the Microsyopidae accord with either.

Mixodectes Cope, 1883

Cope, E. D., 1883, Proc. Amer. Phil. Soc., XX, March 16, p. 559, and Paleont. Bull. No. 36

Type: Mixodectes pungens from the Middle Paleocene, Torrejon formation of New Mexico.

Synonyms: Indrodon Cope, 1883; type: *I. malaris.*

Olbodotes Osborn, 1902; type: *O. copei* (= *M. pungens* Cope).

Author's Diagnosis: The position of this genus is uncertain, but may be near to *Cynodontomys* Cope, which I have provisionally placed among the Prosimiae. It is only known from mandibles, which have presumably the following dental formula. I. 0; C. 1; P–m. 4; M. 3. An uncertainty exists as to the proper names of the anterior teeth, which cannot be decided until the discovery of the superior series. For instance the formula may be I. 1; C. 1; P–m. 3.

The supposed canine is a large tooth, issuing from the ramus at the symphysis like a rodent incisor, and has an oval section, with long diameter parallel to the symphysis. The crown is lost from all the specimens. The second tooth is similar in form to the first, but is much smaller. It is situated posterior and external to the first. The next tooth is still smaller and is one-rooted. The third and fourth premolars have simple conic crowns, and more or less developed heels without cusps. The true molars are in general like those of *Pelycodus; i.e.,* with an anterior smaller [trigonid], and a posterior triangle or V [talonid]. The supplementary anterior inner cusp [pa^d] is quite small, while the principal anterior inner [me^d] is elevated. The posterior inner [en^d] is much more elevated than in the species of *Pelycodus.* Last inferior molar with a fifth lobe.

The upper dentition of *Mixodectes* was unknown to Cope and, save for a rather doubtfully associated upper molar, was unknown to Osborn in 1902. The expeditions of 1912 and 1913 secured a number of upper and lower jaws, one of which, No. 16012, provided the positive association of upper and lower teeth of the same individual which confirmed the tentative correlation of *Mixodectes* and *Indrodon* as the lower and upper dentition of closely related if not identical genera. The new material enables us to figure and describe the upper cheek teeth of *Mixodectes*, and from their close relationship to those of *Indrodon* it is safe to infer similarity in the upper front teeth. A number of specimens referable to *Indrodon malaris* shows that the lower dentition of this species agrees generically with *Mixodectes.*

Type Description of *Indrodon* Cope, 1883: Family Anaptomorphidae, suborder perhaps Lemuroidea, as indicated by the dentition only. It differs from *Anaptomorphus* in three points: first, there are three incisors; second, the first superior premolar has no internal lobe; third, there is a distinct posterior internal tubercle on the first and second superior true molars. [Free rendering of Cope.]

The superior dental formula of Indrodon is I., ^{3}; C., ^{1}; P–m., ^{2}; M., ^{3}. The canine is compressed and acute; the third premolar is compressed conic, and has two roots. The fourth premolar has but one external cusp. The external cusps of the true molars are conic and acute, and are connected with the internal cusp by ridges which form a V. Posterior inner cusp distinct on Ms. I and II, a part of the posterior cingulum. Intermediate tubercles present, small. The superior incisors are well developed and display no tendency toward the rodent type. A portion of lower jaw adheres to the skull, and may belong to the same animal. It supports the last two molars. These have two anterior, opposite, approximated cusps. The heel of the penultimate molar is rather large, and has a raised edge, which develops two tubercles at the angles.[131]

The type actually consists, as far as one can judge, of the front half of the skull and the lower jaws, buried in very hard matrix. The bone, one may suspect, is partly or wholly

[131] Amer. Nat., XVIII, Dec. 29, p. 60.

absorbed into the concretionary matrix and is no longer distinguishable.   The teeth, however, are well preserved except so far as they have been damaged by surface weathering or by unskillful attempts to remove the matrix.   The upper left teeth had been partly cleared for Cope's description but considerable matrix still remained on them.

In 1888 Cope published a figure of the type dentition but gave no further description.

In 1895 Osborn and Earle referred to *Indrodon* a specimen giving certain characters of the skeleton, but the specimen was subsequently shown to be erroneously referred to this genus and was provisionally placed by Osborn in 1902 in *Mioclaenus*, *Indrodon* being removed to the Mixodectidae.

Whatever be the ordinal position of *Indrodon*, it is not to be doubted that it is a close relative of *Mixodectes*.   The construction and form and proportions of the upper molars are precisely alike in the two genera, and quite distinct from any other known form.   The specimen referred by Osborn to *Indrodon* (A. M. No. 833) differs very much in the p⁴, but this tooth is probably wrongly associated (see p. 224).   The lower jaw herein referred to *Indrodon* is closely related to *Mixodectes*.

The molars are, on the other hand, very different from any of the " anaptomorphid " (tarsiid) genera.   They approach more nearly to some of the modern lemurs.   From *Mixodectes* it differs in smaller size, also in a somewhat simpler construction and less specialization of the premolars.   These are minor distinctions to which we are unable to accord generic value.

Osborn interprets the teeth as i?, c^1, p^2, m^3, regarding the third tooth instead of the fourth as a canine.   The maxillo-premaxillary suture is not distinguishable, but this interpretation appears more probable than Cope's.

Cope's view of the relationships of *Indrodon* is thus expressed (1883, p. 61):

The discovery of this type in the Puerco formation is a fact of interest.   In the shortening of its dental series it is the most specialized genus of the epoch, while the forms of its true molars are like those of the simpler Creodonta, and more specialized than those of *Anaptomorphus* and the lemurs generally.   In the simplicity of its premolars, however, it maintains the general character of the Puerco fauna, and is more primitive than the forms just named.   Its nearest ally of the Puerco yet known is *Chriacus*.

Mixodectes pungens Cope

Mixodectes pungens Cope, 1883, Proc. Amer. Phil. Soc., XX, March 16, p. 559, and Paleont. Bull. No. 36.
Olbodotes copei Osborn, 1902, Bull. Amer. Mus. Nat. Hist., XVI, p. 206, fig. 29.

Type: A. M. Cope Coll. No. 3081, lower jaw with p₃–m₃ r. and roots of anterior teeth. Type of *O. copei:* A. M. No. 2385, lower jaw, p₄, broken m₂₋₃ and roots of alveoli of remaining teeth of the left side, also an upper molar somewhat doubtfully associated. From the Torrejon formation, Rio Torrejon, San Juan basin, expedition of 1896.

Horizon and Locality: Paleocene of San Juan basin, undoubtedly from the Torrejon formation, Middle Paleocene.

Author's Description: Char. specif.   The mandible of the *Mixodectis pungens* is about the size of that of the mink.   Its inferior outline is straight to below the second premolar, whence it rises upwards and forwards like that of a rodent.   The anterior masseteric ridge is very prominent, but terminates below the middle of the ramus.   Inferior masseteric ridge much less pronounced.   The inferior part of the ramus is robust below the base of the coronoid process, but there is no indication of recurvature of the edge.   Mental foramina two; one below the front of the first true molar, and one below the second premolar.

The oval base of the canine [i.e. incisor] is not flattened on either side; that of the second tooth [canine] is flattened on the inner side. There is a great difference between the sizes of the last three premolars. The fourth is twice as large as the third, and the second, judging from the space and size of its alveolus, was much smaller than the third, and the crown was probably a simple acute cone. The crown of the third is of that form, with the addition of a short heel. The long axis of the base of the crown is diagonal to that of the jaw. The fourth premolar has a relatively larger heel than the third, but it is shorter than the diameter of the base of the cusp. Its posterior edge is elevated. The cusps of the anterior pair of the true molars are elevated, but the interior is the most so. The supplementary one is not exactly in the line of the interior border of the crown. Each of the inner cusps are connected with the base of the external by a ridge, which together form a V. The posterior base is nearly surrounded by a raised edge, which rises into cusps at the posterior lateral angles. Of these the internal is the more prominent. The edge connecting these cusps is slightly convex backwards, and evidently bears a part in mastication. The lateral borders of the last molar are somewhat expanded, and the fifth lobe is very short. No cingula on any of the teeth. [Measurements follow.]

The upper teeth of this species are found associated with the lower jaw in No. 2385, an upper molar, somewhat doubtful, and No. 16012, parts of both upper and both lower jaws and a few skull fragments. No. 16038 consists of a fragment of the upper jaw with p⁴, and three skull fragments. Another upper jaw fragment has p³⁻⁴ well preserved. The remainder of the identified specimens are lower jaws or parts of jaws which add very little to the data on which the foregoing descriptions are based.

The upper teeth are identical with those described under *Mixodectes crassiusculus*, except for somewhat smaller size and less robust proportions, most distinctly seen in the premolars. The lower teeth are a little narrower, and p_4 less massive, although variable in size, and the jaw is more slender anteriorly with more sloping chin. It is, however, doubtful whether *M. pungens* and *M. crassiusculus* are fully separable species.

Mixodectes crassiusculus Cope

Mixodectes crassiusculus Cope, 1883, Proc. Amer. Phil. Soc., XX, March 16, p. 560, and Paleont. Bull. No. 36.

Type: A. M. Cope Coll. No. 3085, lower jaw with right m_2.

Horizon and Locality: Paleocene of San Juan basin, undoubtedly Torrejon formation, Upper Paleocene.

Author's Description: The species differs from the last [*M. pungens*] in its greater size, and in the relatively greater length of the last inferior molar. The length of the posterior four molars of the *M. pungens* equals that of the three true molars of the *M. crassiusculus;* and the last true molar of the latter is half as long again as the penultimate, while in *M. pungens* it exceeds it but little.

The best preserved true molar is the second. Its most elevated cusps are the anterior and posterior inner, of which the anterior is subconic and more elevated. The anterior external cusp is crescentic in section, and sends crests to the supplementary, anterior, inner [paraconid] and the posterior anterior inner [metaconid], both of which descend inwards. The posterior crest reaches the posterior base of the anterior inner cusp.

The posterior external cusp [hypoconid] is an elevated angle, sending crests forward and backwards. The former reaches the base of the anterior external cusp [protoconid] (not reaching the inner), while the latter passes round the posterior edge of the crown. As in *M. pungens*, it is convex posteriorly, and rises to the posterior internal cusp [entoconid]. In both species its appearance indicates that it performs an important masticatory function in connection with the superior molar. No cingula. [Measurements follow.]

Only the third and fourth upper premolars are known. In advance of the third is a short diastema with a rather small alveolus indicated in front of it, probably for a p² like that of *Indrodon malaris*. P³ is a rather small tooth, two-rooted (except in one specimen doubtfully referred to *Mixodectes crassiusculus*), with simple high-pointed oval crown, and no interior cusp or root, the basal cingulum faint externally and absent internally. P⁴ is much larger and has a high conical stout outer cusp, an external basal cingulum ending in minute cusps at the external angles, a large high inner cusp almost equalling the outer cusp, conical except for well-defined crests, its apex with the external basal angles. The outer cusp has a slight backward pitch, while the inner cusp is vertical, and this, with their height and robustness and the slight development of the external angles and absence of accessory cusps, gives a rather characteristic form to the tooth so that it is readily recognized.

The upper molars are equally characteristic. Their construction is very well shown in Osborn's figure No. 29a of m¹ of " *Olbodotes copei* " (= *Mixodectes crassiusculus*). They are tritubercular but with a well-developed hypocone crested and sharply separated from the protocone. The paracone and metacone are sharp, markedly crescentic, with the external stylar cusps well developed, the mesostyle being a prominent, sharp transverse crest. The protocone is widely crescentic, with prominent crests sweeping from its apex downward and outward to the external basal angles. They are slightly interrupted by the small and obscure conules, which are reduced to low crescents, the outer wing continuous with the protocone crescents, the inner wing in each extending across the basin to the inner base of the paracone and metacone respectively. There is an anterior basal cingulum along the middle third of the anterior side of each tooth. The last upper molar differs from the others only in the reduction of the hypocone to a short postero-internal basal cingulum and the absence of metastyle and obsolescence of the posterior wing of the metacone.

The above description applies equally to the teeth of *Indrodon malaris*. There is no distinction save in size.

Mixodectes malaris (Cope)

Indrodon malaris Cope, 1883, Amer. Nat., XVIII, p. 60; 1884, Paleont. Bull. No. 37, Jan. 2, and Proc. Amer. Phil. Soc., XXI, p. 318; 1888, Trans. Amer. Phil. Soc., XVI, N.S., p. 337, fig. 10; Osborn, 1902, Bull. Amer. Mus. Nat. Hist., XVI, p. 208. (Not Osborn and Earle, 1895, *ibid.*, VII, p. 16.)

Type: A. M. Cope Coll. No. 3080, a " skull " in hard concretion, with the left upper teeth showing.

Horizon and Locality: From the Paleocene of New Mexico, presumably Torrejon formation.

Author's Description (p. 61) : *Char. specif.* The first and third superior incisors are a little larger than the second. Canine preceded and followed by diastemata, each of which is 1.5 times as long as the long diameter of the crown. Premolars separated from each other and from the first true molar by interspaces half as long as the diastema. Neither tooth has any basal tubercles, but the posterior has a weak external cingulum, which is stronger posteriorly. The internal cusp of the same tooth is anterior, is acute and elevated. The superior true molars have a strong external cingulum, which rises into a small tubercle opposite the space between the external principal cusps. Of the latter, the anterior is a little more conic than the posterior, and both are well within the external border. On the last molar, the posterior external cusp is continuous with the external intermediate tubercle, and forms a cutting edge within the posterior margin of the crown. The posterior inner tubercle is rather large, and projects further inwards than the apex of the anterior V on the second true molar, but not so far as in the species of *Anisonchus* and *Haploconus*.

The surface of the cranium is too much obscured by cracks and films of matrix to permit a view of the sutures and foramina. The face is wide, as the posterior part of the maxillary and malar bone are expanded outwards. I have not yet been able to ascertain the condition of the orbit posteriorly. The mandibular ramus is rather slender. [Measurements follow.] The skull is about the size of that of the *Bassaris astuta*.

The status of *Indrodon* has been much confused by erroneous references of material to *I. malaris*. Nos. 823 and 829, referred in 1895 to this species, consisting of upper and lower jaws with badly worn teeth, hind limb bones, etc. (No. 823), and a lower jaw (No. 829), associated with a molar tooth, are certainly not *Indrodon;* the teeth agree with those of *Ellipsodon*, the skeleton fragments are the very primitive type now known to belong to the smaller Mioclaenidae, the Oxyclaenidae and perhaps other Paleocene groups. None of these has anything to do with *Indrodon*. On the other hand No. 833, referred by Osborn and Earle in 1895 and by Osborn in 1902 to *I. malaris*, agrees exactly with that species as far as the molars are concerned, while the premolar is widely different and like that of some small Oxyclaenidae. The premolar is upon a separate fragment of matrix, and although it is like the molars in wear and preservation, it seems probable that it does not belong to the same animal. If it does, the specimen could represent only an otherwise unknown genus of Mixodectidae not closely allied to *Indrodon* or *Mixodectes*. Excluding this anomalous premolar, No. 833 illustrates the molar structure of *I. malaris* much better than the type, where the teeth were all " obscured by films of matrix," as Cope phrases it, until the matrix was removed by one of us.

The principal specimens obtained by the later expeditions are lower jaws, one of them, No. 17064, nearly complete with well-preserved teeth, from the northern margin of the basin.

?PANTOLESTIDAE

No typical Pantolestidae have been found in the Paleocene, but the following problematic genus has been provisionally referred to the family.

Pentacodon Scott, 1892

Scott, W. B., 1892, Proc. Acad. Nat. Sci. Phila., XLIV, p. 296

Type: Chriacus inversus Cope, 1888, from the Paleocene of New Mexico, presumably Torrejon formation, Upper Paleocene.

Author's Diagnosis: Upper teeth unknown. Anterior lower premolars very small and simple in construction, $p_{\overline{4}}$ is large, with large heel and very distinct deuteroconid. [*] The molars increase in size from the first to the third, which is very large proportionately; its talon is remarkable for the entire absence of the entoconid; the hypoconid and hypoconulid form very acute cusps upon the external and posterior borders of the talon respectively, the valley of which thus opens inward without obstruction.

The two jaw fragments upon which the above diagnosis was based belong to two different animals; the fragment containing the molars is *Deltatherium fundaminis*, and the type is restricted to the fragment containing the premolars. This invalidates the greater portion of Professor Scott's description that follows * in the quotation above and only the first sentence has any application. Although this error was discovered by Matthew in 1896 when revising the Paleocene faunas, it has not been specifically stated in print and the only allusions to *Pentacodon* are as " incertae sedis " in a list of the Torrejon fauna in 1897 [132] and a provisional reference of the genus to the Pantolestidae in 1909,[133] in the following terms:

[132] Matthew, W. D., 1897, Bull. Amer. Mus. Nat. Hist., IX, p. 264.
[133] Matthew, W. D., 1909, Mem. Amer. Mus. Nat. Hist., IX, p. 523.

To this family may also be provisionally referred the Basal Eocene genus *Pentacodon* which shows the highly diagnostic Insectivore feature of the position of the mental foramen beneath m_1. *Pentacodon* is known only from the lower jaw, and the teeth are of peculiar form and proportions, the molars reduced, tritubercular, with large heels, the fourth premolar enlarged, its principal cusp twinned and pointing backward, the third premolar quite small, the second a little larger, both simple and trenchant.

Eight specimens from the Torrejon are referable to *Pentacodon*, one of them with part of the upper jaw, the rest lower jaws. The singular proportions of the teeth are very characteristic, the very large robust p_4 in contrast with the small blade-like premolars in front of it, and the molars reduced regularly in size from first to third. The construction of the molars, however, makes it very difficult to decide the affinities of the genus. They are most like those of the Pantolestidae, but in neither *Pantolestes* nor *Palaeosinopa* is there any suggestion of the peculiar relative proportions of the *Pentacodon* teeth. The front teeth are unknown, but the jaw is so rapidly reduced in depth and mass towards the symphysis that they were evidently small; there was no enlarged front tooth as in the Mixodectidae, nor a large canine as in the Oxyclaenidae. The upper jaw shows the infraorbital foramen of moderately large size, placed far forward, in advance of p^3, while the anterior border of the orbital rim is above the anterior angle of m^1; the lachrymal foramen (as far as one can judge from the damaged specimen) lay within the rim of a fairly well-defined orbital border; a broken surface showing a cross-section of lachrymal and maxillary bones seems to indicate that the lachrymal had a considerable extension upon the face, but its suture with the maxillary is obliterated. The suture with the superior branch of the jugal reaches almost up to the lachrymal foramen. These are the normal primitive conditions seen in the inadaptive creodont families (modified in the Miacidae and fissipede Carnivora by the pushing forward of the orbit). There is but little approach to the peculiar conditions of *Pantolestes* with its very large round infra-orbital foramen situated close in front of the orbit. In *Palaeosinopa*, however, the conditions are intermediate; the infraorbital foramen is a little larger and more circular than in *Pentacodon* and is situated much farther backward, but is not so close and by no means so large as in *Pantolestes*.

P^4 is the largest of the upper teeth series; it is a robust, massive tooth with a heavy internal cusp, but too much worn to determine its construction. It is considerably pinched in between the inner and outer cusps, in contrast with the oval outline of *Mioclaenus*. The molars are very wide transversely, little extended at the external angles, the inner half subquadrate and separated from the outer half by a peculiar pinching in of the outline at the sides. The outer cusps are conical, the protocone somewhat crescentic, the hypocone minute, and the cingula are limited to the anterior and posterior sides of the inner half of the crown. No internal or external cingulum, although the outer cusps are set somewhat inward so as to make an obscure external shelf. No stylar cusps. M^1 is considerably larger than m^2, although scarcely any wider transversely. M^3 was evidently smaller, and quite narrow anteroposteriorly, to judge from its alveoli.

P_2 and p_3 are rather small teeth of usual simple cutting type, subequal in size, with very rudimentary heels. P_4 is much larger and of very characteristic form. The protoconid is large, robust, conical and moderately high, pitched backward and with a prominent but much smaller inner cusp (metaconid) on its inner flank, slightly posteriorly placed. The heel is large, high, the hypoconid forming a low median crest internal to which lies a very shallow basin, bounded by a low marginal internal crest rising postero-medially to a high point (hypoconulid); the entoconid is rudimentary. The molars decrease in size from first to third; they have a moderately high three-cusped, equilateral trigonid, the paraconid smaller and lower than the other two cusps; and a deeply-basined heel slightly narrower and materially shorter than the trigonid, with the three heel cusps only moderately distinct on its margin.

The lower jaw is deep under the molars, thickened directly under p_4, rapidly decreasing forward in depth and thickness. The posterior mental foramen under m_1 as in *Pantolestes* and *Palaeosinopa*, but is not so large; the anterior mental foramen is under p_2.

The upper and lower molars are more like those of *Pantolestes* than any other genus, but the resemblance is not convincing, as the construction is little removed from the primitive type of the Oxyclaenidae, and there is nothing in the premolars to support it. The exceptional relative sizes of the molars suggests *Coriphagus*, but there is nothing else in this genus to confirm the likeness. The peculiar position of the mental foramen is characteristic of the Pantolestidae and some other Insectivora; and the conditions of the infra-orbital foramen, while in the main primitive creodont, show some apparent approach to the specialized conditions of *Pantolestes*, *Palaeosinopa* being intermediate. The enlargement of p_4^4 and reduction of the teeth in advance of it suggest comparison with *Mixodectes*, but the construction of the tooth is wholly different, there is no enlarged incisor or canine, and no particular resemblance in molar construction.

The genus is therefore left as referred by Matthew in 1909, provisionally placed in the Pantolestidae.

Pentacodon inversus (Cope)

Chriacus inversus Cope, 1888, Trans. Amer. Phil. Soc., XVI, N.S., p. 342 (in part); (*Pentacodon*)˙ Scott, 1892, Proc. Acad. Nat. Sci. Phila., XLIV, p. 297 (in part).

Type: A. M. Cope Coll. No. 3129, part of lower jaw, p_{2-4} and broken m_1. With this fragment Cope associated with doubt a second portion of a lower jaw with m_{2-3}, which belongs to *Deltatherium fundaminis*. The greater part of Scott's description of *Pentacodon* is based upon this erroneously referred fragment. It was recognized by Matthew in 1896 and separately catalogued as No. 3130.

Horizon and Locality: Torrejon formation, San Juan basin, New Mexico.

Author's Description: The first premolar [p_4] has a well-developed interior cusp, and a large heel. The latter has an internal vertical, and an external oblique side, terminating in a cutting edge, the internal curving round the posterior border to meet the external. There is no anterior basal cusp, nor any cingulum. The second and third premolars are compressed, and have cutting edges before and behind. The second [p_3] has the heel slightly transverse, and a mere trace of anterior basal lobe. The third [p_2] is larger than the second and is more compressed. Externally its face is regularly convex, but internally its convexity is a vertical median rib, and in front of this the face is concave, thus maintaining the acuteness of the anterior cutting edge on wear [i.e. anterior crest thinned out towards margin and concave on inner face]. The heel is small and compressed; anterior lobe none. The enamel of all the teeth is smooth. No cingula. [Measurements follow.]

To this species are referred, besides the type:

No. 17038, lower jaw, p_4–m_2 l., well preserved.

No. 17073, jaw fragment, p_4–m_1.

No. 3385, parts of lower jaws, p_4–m_2 l.

No. 3386, upper jaw, p^4–m^2 l., parts of lower jaw, m_3 l.

Revised Diagnosis: P_{2-3} quite small, subequal, simple, moderately compressed, anterior edge bladed, posterior edge, especially in p_3 more robust, with a weak postero-internal subsidiary crest and a minute heel. P_4 nearly twice as large lineally, with very distinct metaconid, large, stout, conical protoconid, very rudimentary and minute basal paraconid. Heel large, broad-based, projecting cusp-like upward and backward to a point, the hypoconid constituting a low, broad median crest, and an inner marginal crest enclosing a shallow basin. The principal cusps lean backwards in a very characteristic manner; they are not recurved in any degree. The surface of the protoconid is wrinkled, otherwise the enamel is nearly smooth. No cingula.

The molars are nearly twice as long as wide, divided into subequal moieties of trigonid and talonid, the height of the trigonid nearly twice that of the heel; m_1 a little larger than m_2, which is slightly larger than m_3. Protoconid and metaconid on all the molars of equal size, well separated, the metaconid slightly farther back, paraconid smaller and lower, the three forming an equilateral triangle on m_1, the transverse diameter of the trigonid somewhat greater relatively on m_2 and m_3. The heel cusps are more distinct in this species, the hypoconid highest, set somewhat inward, the hypoconulid and entoconid fully marginal, the three cusps forming angles of a continuous crest that encloses a rather deep basin shallowed by wear. No cingula except a slight one around the antero-external portion of the base. The enamel is nearly smooth.

The depth and robustness of the jaw are moderate and quite variable, but it is most robust under p_4, deepest under the molars and becomes slender and small anteriorly.

The upper jaw and teeth have already been described in some detail in the generic discussion.

MEASUREMENTS (IN MM.)

	No. 3129 Type	No. 17038	No. 3385	No. 3386
Upper jaw, p^4–m^3 (estimated)				20.8
Upper molars (estimated)				13.8
Diameters, P^4, ap. $\times$ tr.				6.9 $\times$ 9.4
" M^1, "				6.0 $\times$ 8.0
" M^2, "				4.7 $\times$ 7.3
Lower jaw, p_3–m_3 (estimated)		23.8	23.5	
" p_3–m_2		19.3		
" p_{2-4}	13.6	(e) 13.8		
" p_4–m_3			20.3	
Lower molars, m_{1-3}			13.9	
Diameters, P_2, ap. $\times$ tr.	3.7 $\times$ 1.8			
" P_3, "	3.5 $\times$ 2.1	3.2 $\times$ 2.0		
" P_4, "	6.9 $\times$ 3.5	6.5 $\times$ 3.5	6.3 $\times$ 3.8	
" M_1, "		4.9 $\times$ 3.4	5.2 $\times$ 3.6	
" M_2, "		4.2 $\times$ 3.2	4.2 $\times$ 3.2	
" M_3, "				4.5 $\times$ 3.0
Lower jaw, depth below p_2		8.3	.	
" " m_3		11.5	9.4	
" thickness below p_4	5.1	5.8	5.7	

Pentacodon occultus, new species

Type: No. 16592, lower jaw, p_4–m_3 and alveoli of p_{2-3}.

Horizon and Locality: Upper Paleocene, Torrejon formation, upper beds, Alamito wash, San Juan basin, expedition of 1913.

Diagnosis: Nearly one-third larger than *Pentacodon inversus;* teeth more robust. P_4 much wider at base posteriorly, the protoconid with distinct angles separating a rugose outer face from nearly smooth posterior and antero-internal faces; metaconid less distinct; heel more broadly buttressed at external base; an obsolescent cingulum connecting metaconid with the very rudimentary paraconid. The heel cusps of the molars are more obscure, the marginal crest more continuous around the basin. Jaw relatively deep and more massive than in *P. inversus.*

No other specimens than the type have been found.

MEASUREMENTS (IN MM.)

Lower teeth, p–m, length	29.8
Lower molars, m_{1-3}, length	16.2
Diameters, P_4 ap. × tr.	9.5 × 5.3
" M_1 "	6.6 × 4.8
" M_2 "	5.3 × 4.0
" M_3 "	4.3 × 3.2
Depth of jaw at p_3	10.2
" " " " m_3	14.0
Thickness of jaw beneath p_4	8.9

5. ORDER TAENIODONTA

Synonyms: Ganodonta Wortman, 1896. Stylinodontia Marsh, 1897.

TAXONOMIC POSITION AND AFFINITIES

This order was based by Cope in 1876 upon *Calamodon* and *Ectoganus* of the Lower Eocene. The original diagnosis was as follows:

The feet are armed with compressed claws. The dental characters are seen first in the supposed superior incisors. Unfortunately, they have not yet been found in place in the cranium, but their association with a rodent type of inferior incisors, which have been found in place in the mandible, confines us to the alternative choice between superior incisors and canines. From the small size, or absence, of inferior canines, a similar character may be inferred for the superior canines.

These superior incisors present two bands of enamel, an anterior and a posterior. They are compressed in form, the sides presenting a surface of dentine or cementum. Attrition produces a truncate or slightly concave extremity. The inferior incisors are rodent-like.

. . . The general structure of *Calamodon* affords some points of approximation to the *Edentata,* which indicate that the *Taeniodonta* partially fill the interval between that order and the *Insectivora,* presented by the existing fauna.[124]

In 1885 [125] Cope added the following features of distinction:

[124] Cope, E. D., 1876, Proc. Acad. Nat. Sci. Phila., XXVIII, p. 39.
[125] Cope, E. D., 1885, Tertiary Vertebrata, p. 187.

The *Taeniodonta* agree with the *Tillodontia* in the possession of a pair of inferior incisors of rodent character, but . . . the inferior canines . . . in *Calamodon* are of large size, and though not as long-rooted as the second incisors, grow from persistent pulps.   They have two enamel faces, the anterior and posterior, the former like the corresponding face of the rodent incisors. . . . The external incisors, wanting in *Tillotherium*, are here largely developed, and though not growing from persistent pulps, have but one, an external band-like enamel face. . . . The superior dentition of the *Taeniodonta* is unknown.

. . . The great reduction in the extent of the enamel investment is an interesting approximation to the *Edentata*, where this substance is altogether wanting. . . . In addition, there are a heavy cementum investiture and undivided roots in the genus *Calamodon*, features essentially characteristic of the *Edentata*.

Psittacotherium, first described in 1882 and referred to the Tillodontia, was subsequently transferred by Cope to the taeniodonts.   "*Hemiganus*" *otariidens* (1885), *Conoryctes* (1882), and *Onychodectes* (1888) were regarded in 1888 by Professor Cope as a series connecting the taeniodonts with generalized creodonts, but in accord with his taxonomic principles were retained in the latter order.

In 1896 and 1897 [136] Wortman referred all these genera along with the little-known *Stylinodon* Marsh to the Ganodonta as a new suborder of Edentata, grouping them under the two families Stylinodontidae and Conoryctidae.   The former he considered as directly ancestral to the ground-sloths, the latter as related to the armadillos.   He declares that the stylinodont series is " as complete and perfect a phylum as has ever been deciphered within the whole range of paleontology," that the evidence for its edentate affinities is " overwhelmingly conclusive," and cites the following features:

(1) In the skull there is great similarity in form; the muzzle is short, the sagittal crest is low, and the occipital plane slopes forward as in *Mylodon*, *Megatherium* and *Megalonyx*.   (2) The lower jaw is short, deep and robust, with a greatly enlarged coronoid, a prominent angle, and a position of the condyle high above the tooth line.   (3) The incisors are reduced to a single pair in the lower jaw of *Calamodon*, and are probably completely absent in *Stylinodon*.   (4) The posterior portion of the tooth line below passes well behind the anterior border of the coronoid.   (5) The canines in all are enlarged, and in *Calamodon* and *Stylinodon* grew from persistent pulps, as in *Megalonyx*.   (6) All the molars and premolars in *Stylinodon* are greatly elongated, of persistent growth, and the enamel is confined to narrow vertical bands.   (7) There is a thick deposit of cementum on the dentine in those situations in which the enamel disappears.   (8) The cervical vertebrae strongly resemble those of the Gravigrada.   (9) There were well-developed calvicles present.   (10) The humerus bears a striking resemblance in all of its essential features to those of *Mylodon*, *Megalonyx* and *Megatherium*.   (11) The ulna and radius are also similar.   (12) The manus is almost identical with that of the Ground Sloths.   (13) The humerus and ulna and radius have no medullary cavities; and (14) the femur has all the characteristic features of the Gravigrada.   (15) The lumbar vertebral formula was the same as in the Edentata.   (16) The pelvis is decidedly Edentate; and (17) the caudals bear a striking resemblance to those of the Ground Sloths.

If this astonishing array of similarities is accidental and does not indicate genetic affinity, then all that can be said is that palaeontological evidence is worthless in the determination of the various successive steps in the descent of a group or species.   I hold that, in view of all the evidence above set forth, the proposition that the one has descended from the other may now be regarded as a positively demonstrated fact.

Wortman rejects Cope's name of Taeniodonta as erroneously defined and includes the Ganodonta in the order Edentata, although he admits that this makes it impossible to frame an ordinal definition, save for " an early disposition to the loss of the enamel from the crowns of the teeth, as well as the loss of incisors from both jaws."   The Ganodonta he defines (p. 108) as

Primitive Edentates, characterized in the earlier forms by rooted teeth with divided fangs, with their crowns having a more or less complete enamel investment, in the later forms by the teeth becoming hypso-

[136] Wortman, J. L., 1896, Bull. Amer. Mus. Nat. Hist., VIII, p. 260; *ibid.*, IX, p. 103.

dont, rootless, of persistent growth, and by limitation of the enamel to vertical bands in progressive decrease. They are further characterized by the presence of incisors in both jaws, by a typical molar and premolar dentition, by a trituberculate molar crown, which disappears early in life through wear, leaving the dentine exposed.

I further add that the vertebral articulations are not complex and hence nomarthrous.

The two families are distinguished (p. 109) by the position of p_{2-3}, longitudinal in Conoryctidae (but oblique in *Conoryctes*), transverse in Stylinodontidae. *Conoryctes* is distinguished from *Onychodectes* by the reduction of the upper premolars to three. The stylinodont genera are more adequately defined.

Marsh in 1897 [137] described and figured for the first time parts of the skeleton of *Stylinodon* and proposed the name Stylinodontia.[138]

Subsequent opinion as to the affinities of this group has ranged from uncritical acceptance to total rejection of Wortman's views, but in general the more conservative views expressed by Cope have been followed, the taeniodonts being more or less provisionally annexed to the Edentata as a primitive group whose exact affinities are not clear. Scott [139] has wholly rejected their edentate affinities, but gives no extended discussion of his reasons. Winge [140] likewise rejects any such relationship and refers the group to the Insectivora as a progressively specialized series, the most primitive stages nearly related to the Leptictidae. In this relationship he associates them with the Tillotheriidae and Periptychidae, regarding the three families as derived from Leptictidae, which in turn are derived from Galeopithecidae!

In his volume on the Edentata of Lagoa Santa,[141] Winge argues strongly against Wortman's view. He declares that far from being " almost identical with that of the ground-sloths " the manus of *Psittacotherium* has not a single one of the peculiarities that are characteristic of them; there is no trace of the shortening of the inner digits and lengthening of the two outer metapodials or of the peculiar characters of the outer side of the manus. There is some general resemblance in the hypsodont rootless teeth at first glance between the latest Ganodonta and some of the ground-sloths, especially *Megalonyx*, but closer study shows the greatest dissimilarity. The large front teeth in the jaws of the Ganodonta are canines; with the ground-sloths it is a cheek tooth (premolar). (Even if their upper teeth should be canines, the lower cannot be, Winge states. Wortman, he says, seems not to have noted this, but reckons these teeth as canines in the ground-sloths both in upper and lower jaw.) Ganodonta are developed in clear manner from mammals with unreduced dentition, have had stronger and stronger teeth, jaws and musculature; megatheres (i.e. ground-sloths) show that they are derived from animals with vestigial dentition and in an unusual manner have acquired heavy jaws and teeth; ganodonts lack a whole series of the characteristics of ground-sloths; they have not lost the enamel, they have not the vestigial premaxillary, they lack the " tudformede " lower jaw, the jugal is not free but is unusually heavy and lacks the peculiar form of the ground-sloths; the coronoid process is notably strong, the angular small, the lower jaw articulation has not the peculiar ground-sloth

---

[137] Marsh, O. C., 1897, Amer. Jour. Sci. (4), III, p. 137.

[138] To replace the supposedly preoccupied term Ganodonta, according to Wortman, 1897, p. 107; Marsh makes no explanation of his proposal of the term.—w.d.m.

[139] Scott, W. B., 1905, Report Princeton Expeditions to Patagonia, V, Edentata, p. 361.

[140] Winge, H., 1923, Pattedyr-Slægter, I, pp. 128, 133.

[141] Winge, H., 1915, Jordfundne og nulevende Gumlere, p. 300.

form but is entirely a hinge joint. Many other peculiarities are lacking in the ganodonts, as in the form of the infra-orbital canal or the mental foramen, the anterior border of the coronoid process, etc. In short, in spite of some superficial resemblance, there is no real relationship between ganodonts and ground-sloths. But even if all Wortman's points of similarity were sound, they would not aid in explaining the derivation of the ground-sloths, for the likenesses are only between the ganodonts and the most specialized ground-sloths, particularly *Megalonyx*, and Wortman fails to show how *Megalonyx* could be regarded as the stem form for the other edentates. There is no way leading from generalized mammals through *Megalonyx* into the other Edentata.[142]

Scott has discussed the matter only very briefly, but has pointed out that the enlarged teeth in the Ganodonta are canines, while those of the ground-sloths are premolars, that the Miocene ground-sloths of Patagonia, instead of being intermediate between the Pleistocene genera and the Ganodonta, are less like the specialized ganodonts, and that edentates do not make their first appearance in South America in the Santa Cruz Miocene as Wortman supposed but are recorded from the older Tertiaries of that country as well. He omits to note, however, that the remains are very imperfect and provide little or nothing of evidence as to the skull, teeth or skeleton structure of the early Tertiary Xenarthra, so that comparison of these with the Ganodonta is not practicable. I do not agree with Scott's statement that the Miocene ground-sloths are less like the specialized ganodonts than is *Megalonyx*.

In his Pattedyr-Slægter (vol. I, pp. 128, 133) Winge refers the " Ganodonta " to the Insectivora, family Stylinodontidae, with three subfamilies, Onychodectinae, Conoryctinae and Stylinodontinae. This arrangement, in view of the rather wide difference between *Onychodectes* and *Conoryctes*, appears more logical than Wortman's, and is here adopted, except that the Taeniodonta are retained as a distinct order, instead of expanding the already over-large order Insectivora to accommodate them, and that *Psittacotherium* and *Wortmania* are placed in a distinct subfamily from the Eocene stylinodonts.

Nomenclature

Cope's name Taeniodonta was proposed in 1876 to include the only genera of this group then known and was subsequently expanded to cover other related forms discovered later. The facts that he misinterpreted the enlarged front teeth and failed to include in the group the primitive genera which Wortman considered ancestral to the more specialized forms afford no sufficient reason for replacing the name by a new one. The name Taeniodonta is therefore preferred to Ganodonta.

For the family name, however, Stylinodontidae Marsh must be used as the earliest published family name, and based upon the earliest described genus. Marsh referred the family to the Tillodontia. In 1876 he described the lower jaw of *Dryptodon* (= ?*Calamodon*) from the Lower Eocene of New Mexico, but he added nothing to his description of the very fragmentary type of *Stylinodon* until after Wortman's first article in 1896 on the " Ganodonta." Marsh then [143] described and figured considerable parts of the skeleton of *Stylinodon* and removed it, along with *Dryptodon*, to Stylinodontia, a new suborder of Edentata, without reference either to the Ganodonta or Taeniodonta, of which it is a synonym. The name appears only in the title and is not defined.

[142] Free translation of Winge's argument.
[143] Marsh, O. C., 1897, Amer. Jour. Sci. (4), III, pp. 137–46.

DISCUSSION OF AFFINITIES

General opinion has not sustained Wortman's view that the stylinodonts were the direct ancestors of the ground-sloths, but they have been regarded as having some degree of connection with the Edentata, except by Scott and Winge. Little has been added to the evidence so carefully studied and described by Wortman, but the discovery of *Metacheiromys* Wortman 1903, Osborn 1904, and *Palaeanodon* Matthew 1918, has provided very important indirect evidence bearing on the relationships of the taeniodonts.

The arguments of Scott and Winge seem to be quite conclusive against any direct ancestral relation to the ground-sloths. The enlarged teeth are not the same; few if any of the special *gravigrade* characters in skull or feet are to be found in the taeniodonts; and the characters of the xenarthral dentition are best interpreted (following Winge) as the result of loss of the front teeth and progressive reduction of the cheek teeth to simple small enamel-less pegs of the type preserved in armadillos, followed by a re-integration and re-complication of these teeth in gravigrades and glyptodonts, the lost enamel being functionally replaced by the hardened and inflected outer layer of dentine.

That the armadillos most nearly represent the ancestral Xenarthra is generally accepted, as also the derivation of the order from proto-Insectivora, presumably of later Mesozoic age. It is beyond the scope of this discussion to present the argument for these views. The fossil record of the Xenarthra is an exceptionally full one as far back as the Santa Cruz Miocene. At this epoch the several groups of Gravigrada are very close to a common ancestry, but both these and the glyptodonts are still wholly separate from the Loricata. The earlier history of the Xenarthra is very little known. Scutes of armadillos and glyptodonts are known from the *Colpodon* and *Pyrotherium* beds, and of armadillos from the *Notostylops* beds; but such few fragments of the skeleton of any of the edentates as have been described from pre-Santa Cruzian horizons are too doubtful and fragmentary to throw any light on the skull or skeletal characters.[144] Nevertheless the relative simplicity of the teeth and less diversely specialized characters of skull and feet show a marked approach in the Santa Cruz glyptodonts and gravigrades to a generalized armadillo.

Between the rather late Miocene Santa Cruz stage and the supposed ancestral group of palaeanodonts in the early Eocene there is a wide gap, and the real gap is even wider, for it is evident that the metacheiromyids are an aberrant phylum and that any ancestral stock common to them and the Xenarthra must date back to the Paleocene. But *Palaeanodon* of the Lower Eocene is very nearly what we might expect of an ancestor of the Xenarthra at the beginning of the Eocene. The cheek teeth are simple, oval, peg-like, the skull is armadilloid in various details, the feet show already a number of features in proportions and articulations peculiar to the more primitive Xenarthra or common to them and the Pholidota. There is no evidence of a bony carapace, but that, as Scott has pointed out, is not to be expected in a common ancestor of the Xenarthra, the ossifications, if present, probably being incomplete and buried in the skin as they are in *Glyptotherium*. The evidence for relationship of *Palaeanodon* to the edentates was more fully discussed by

[144] The reader is referred to a recent paper by Simpson, in which the jaw of an armadillo from the *Notostylops* beds is described (Simpson, G. G., 1932, Amer. Mus. Novitates No. 567). It might be well to note in this connection that much of the skull, jaws, skeleton and carapace of this armadillo from the Casamayor formation (= *Notostylops* beds) is preserved. The specimen tends to substantiate a derivation from palaeanodont-like forms and to contradict any special relationship to the taeniodonts.—THE EDITORS.

Matthew in 1918; it will be sufficient to point out here that it is not far removed from primitive Insectivora but is in nearly all respects what the theoretical common ancestor of the edentates should be at the beginning of the Tertiary. Winge [145] has taken exception to this statement, his chief stated objections referring to the unreduced enamel-covered canines and the union of tibia and fibula. These characters constitute a real objection to direct descent of the Xenarthra from *Palaeanodon*, which, as I pointed out, is on other grounds improbable. They do not make any less probable the derivation of Xenarthra from early Paleocene or late Cretaceous Palaeanodonta. The Paleocene ancestor of *Palaeanodon* might be expected to have tibia and fibula separate, just as in the nearly related Leptictidae *Diacodon* of the Lower Eocene has these bones coössified distally, while in *Prodiacodon* (*Palaeolestes*) of the Paleocene they are separate. But the peculiar character of the articulation of tibia and fibula in the ground-sloths points to derivation from the same type of tibio-tarsal articulations that is shown in *Prodiacodon*, *Onychodectes*, *Palaeanodon* and the proto-insectivoran group generally.

Onychodectes shows the proto-insectivoran construction very little modified. It does, however, display the beginnings of certain taeniodont features. The inner half of the upper molars is partly converted into a high semi-circular crescent, composed of protocone and the two conules, the same construction that is carried further in *Wortmania* and *Psittacotherium*, while the lower molars correspondingly have a reduced low paraconid and a tendency to robust, blunt-pointed cusps arranged in two transverse pairs. The premolars tend to be short, blunt-pointed and simple, but are not crowded or obliquely set. The skull has in the cranial region something of the same peculiar elongate form seen in taeniodonts—not in palaeanodonts or edentates. The feet have the proto-insectivore astragalus and short, flattened phalanges as in proto-Insectivora, not so much shortened as in taeniodonts or palaeanodonts, and the distal metapodial joints are somewhat " squared " as in proto-Insectivora but have not assumed the peculiar articulations of *Metacheiromys* and the armadillos. The claws are compressed, but not notably enlarged as in the stylinodonts.

In this genus we have a connecting link between the taeniodonts and proto-Insectivora, much as *Palaeanodon* is a connecting link between Xenarthra and proto-Insectivora.

The characters of *Conoryctes* afford another connecting link, serving to explain the already highly specialized characters of *Wortmania* (although the latter is the older genus geologically). The premolars are crowded and obliquely set, approaching the stout, short, transversely-set premolars of *Wortmania*. The canines are considerably enlarged; the upper incisors are reduced to two, both of large size, but the lower incisors, three in number, are quite small. The skull is shortened in the facial region, the zygoma is much heavier than in *Onychodectes*, and the overlap of squamosal and jugal much longer. The little that is known of the limbs and feet suggests that they had already assumed some taeniodont peculiarities. *Conoryctes* is evidently only partly intermediate, in other features off the line of descent, and geologically it is of Torrejon age, later than *Wortmania* of the Puerco, so that it is quite out of the genetic series; nor, in view of the wide differences between it and *Onychodectes*, does it seem probable that it is directly descended from the Puerco *Onychodectes*.

Wortmania of the Puerco is very clearly of the stylinodont series, and quite near to the somewhat better-known *Psittacotherium*. The incisors are specialized in the same

[145] Winge, H., 1923, Pattedyr-Slægter, I, pp. 343–4.

fashion; the canines are robust but less flattened and elongate than in *Psittacotherium*, the cheek teeth are similar, except for retaining one more upper premolar and probably less specialized molars; the skull, limb bones, and what is known of the fore foot are in all respects like the Torrejon genus but a little more primitive throughout.

Psittacotherium of the Torrejon is uniformly more specialized than *Wortmania* and has the taeniodont characters well advanced, although the canines still have roots, and the cheek teeth are short and still show something of the more primitive construction of the crown. The fore foot is highly specialized, with very short phalanges, large compressed claws, three stout metapodials and the first and fifth much reduced or vestigial. (The early stage of this seen in *Onychodectes* shows the first digit much reduced and quite slender, the fifth considerably reduced but somewhat stouter, but metapodials long and slender.) These proportions in reduction of the lateral digits are also seen in some armadillos and in the metacheiromyids. In the primitive rodents—also of proto-insectivore derivation—the manus has the first digit reduced and short, the fifth unreduced, and in the pes the digits are all unreduced. The leptictiderinaceid insectivores also have the lateral digits unreduced except for Mc. I.

The hind limbs of *Psittacotherium* are imperfectly known; the femur is relatively large, stout and massive, with a good deal of general resemblance to the Pleistocene ground-sloths, not so much to the earlier types. The tibia and fibula were separate. The pes has somewhat the same proportions as the manus except that the claws are not compressed, the first and second row of phalanges are short and flattened with very limited movement, the first digit is fair-sized but much smaller than the second, and is not opposable or even divergent, the distal metapodial facets are like those of the manus.

The feet in *Calamodon* and *Stylinodon* are very little known but appear to carry somewhat further the specialization of *Psittacotherium*. In none of the genera do we find developed the peculiar distal metapodial articulations characteristic of the palaeanodonts and armadillos. The cheek teeth have become elongate and rootless, the enlarged canines are fully gliriform and rootless, the incisors progressively reduced and the anterior lower premolars become sub-gliriform. The remaining premolars and molars become finally a series of rootless round peg-like teeth with a strip of enamel on each side, which may in the Upper Eocene disappear wholly, leaving only dentine as in Edentata.

A review of the characters of the taeniodonts as now known shows that they constitute a fairly good anatomical sequence from *Onychodectes* to *Stylinodon*, not a genetic series, but serving to show approximately the successive stages in the special adaptation of the group. It parallels the xenarthral specializations, and is derived from a common source among the proto-Insectivora. While the palaeanodonts appear to represent approximately the ancestral source of the Xenarthra and Pholidota, the taeniodonts appear to be a side branch of parallel adaptation to the Gravigrada, but a much earlier specialization. They attain in the Eocene a corresponding extreme of specialization in many points to that attained by the gravigrades at the end of the Tertiary. They wholly lack many of the characteristic peculiarities of the gravigrades, some of which are conditioned by the latter group having passed through a prolonged armadilloid adaptation, others due to various other differences in their adaptational history.

It appears better therefore to retain the taeniodonts as a separate order, related to the palaeanodonts, Xenarthra and Pholidota, and perhaps included with them in a super-order

Edentata, but if so as a marginal and outlying group. They do not appear to have any especial relationship to the Tillodontia, other than a common descent from ?proto-insectivore stock; the enlarged teeth are not the same, the pattern of molars and premolars is quite different, the characters of the feet have little in common.[146]

STYLINODONTIDAE Marsh, 1875

Marsh, O. C., 1875, Amer. Jour. Sci. (3), IX, p. 221; separata Feb. 18, 1875

Synonyms: Calamodontidae Cope, 1876, Proc. Acad. Nat. Sci. Phila., XXVIII, p. 39. Ectoganidae Cope, 1876, *ibid.* Hemiganidae Cope (as subfamily), 1888, Trans. Amer. Phil. Soc., XVI, N.S., p. 310. *Including:* Conoryctidae Wortman, 1896; Onychodectinae Winge, 1923.

Marsh based this family on the single genus *Stylinodon* of the Middle Eocene, referring it to the order Tillodontia and distinguishing it from the Tillotheriidae by the hypsodont rootless cheek teeth.   He made no suggestion at this time as to the affinities of the group except that he suggests ungulate affinities for *Tillotherium.* In 1876 [147] he added to the family *Dryptodon* (= ?*Calamodon*) of the Lower Eocene and described the lower jaw, interpreting the enlarged front tooth as the third incisor.

Cope in 1876 [148] defined the order Taeniodonta with two families, Ectoganidae and Calamodontidae, diagnosing the order as cited above (p. 637) and the families as follows:

Two families represented this suborder in the Eocene period in New Mexico. The first, or *Ectoganidae,* possesses molar teeth with several roots; in the *Calamodontidae,* each molar has a simple conic fang. . . . The general structure of *Calamodon* affords some points of approximation to the *Edentata,* which indicate that the *Taeniodonta* partially fill the interval between that order and the *Insectivora,* presented by the existing fauna.

In 1877 [149] Cope stated more explicitly his reasons for this view as to the affinities of the Taeniodonta:

The great reduction in the extent of the enamel investment is an interesting approximation to the *Edentata,* where this substance is altogether wanting. . . . In addition, there are a heavy cementum investiture and undivided roots in the genus *Calamodon,* features essentially characteristic of the *Edentata.* Thus we have in the *Taeniodonta* the first hint as to the relations of the *Edentata* in early Tertiary time.

Marsh in 1877 [150] suggested the derivation of the Edentata from the Tillodontia, in which he included the Stylinodontidae, through the genus *Moropus,* which he regarded as an edentate, and apparently believed that the South American Edentata migrated to that country from North America in post-Miocene time.

Wortman in 1896 [151] characterized the " Ganodonta " (= Stylinodontidae of this paper) as " Primitive Edentates, characterized in the earlier forms by rooted teeth with divided fangs having a more or less complete enamel investment, in the later forms by the

[146] It is not certain that the tillodonts are derived from the proto-insectivore stock. The characters of the skeleton have never been fully described or adequately analyzed, either in *Esthonyx* or the more specialized members of the order.—w.d.m.

[147] Marsh, O. C., 1876, Amer. Jour. Sci. (3), XII, p. 403.

[148] Cope, E. D., 1876, Proc. Acad. Nat. Sci. Phila., XXVIII, p. 39.

[149] Cope, E. D., 1877, Rept. U. S. Geog. Surv. W. of 100th Mer., IV, p. 157. [This was expanded in 1885. *See* footnote 135.—THE EDITORS.]

[150] Marsh, O. C., 1877, Amer. Jour. Sci. (3), XIV, pp. 337–78.

[151] Wortman, J. L., 1896, Bull. Amer. Mus. Nat. Hist., VIII, p. 260.

teeth becoming hypsodont, rootless, of persistent growth, and by limitation of the enamel to vertical bands in progressive decrease.   They are further characterized by the presence of incisors in both jaws, by a typical molar and premolar dentition, by a trituberculate molar crown, which disappears early in life through wear, leaving the dentine exposed "; and the Stylinodontidae (= Psittacotheriinae + Stylinodontinae of this paper) " by having a remarkably short, deep, and heavy lower jaw with an enormously developed coronoid process reaching even with, or in advance of, the posterior termination of the tooth-line.   The fore foot is short, with remarkably abbreviated, deeply excavated first and second phalanges (unknown in *Hemiganus*), together with a powerful, highly compressed, deep claw; to this should be added a highly characteristic shortening of the facial portion of the skull."

Winge's diagnosis of the family includes the following features: (1) nasal region short, not tubular and elongate; (2) external stylar cusps of upper molars degenerate; (3) canines becoming notably stout and specialized.   The first character associates them with Galeopithecidae, Leptictidae, Tillotheriidae and Periptychidae, the second excludes Galeopithecidae and the third is peculiar to the Stylinodontidae.

Revised Diagnosis of Stylinodontidae: A progressively specialized series of placental mammals of the Paleocene and Eocene derived from proto-insectivoran stock, with diprotodont dentition, incisors reduced in number, the canines becoming enlarged and gliriform, the cheek teeth at first tritubercular, becoming hypsodont and peg-like, enamel becoming reduced to longitudinal strips or caps, muzzle and jaws at first slender, becoming very short and deep.   No postorbital process; mesocranial region elongate; the cranial region very long and narrow with high but thickened sagittal ridge.   Limbs and feet at first slender, but becoming stout, massive, with large claws, compressed on fore feet, rather broad on hind feet, intermediate phalanges progressively short, toes 5–5, the pollex and hallux reduced, the metapodials in the more specialized types short and stout, hinge-jointed, developing other peculiarities much like some edentates.   Ulna and radius separate, the ulnar shaft somewhat heavier, the olecranon massive and long, the distal cotylus of the radius single; lunare, centrale and trapezoid unusually wide, scaphoid small and thin; tibia and fibula separate, the astragalus almost wholly beneath the tibia, and the tibio-astragalar facet progressively flattened with reduction of the trochlear grooves.   Pelvis with progressively broadened ilium, wide and deep obturator foramen, the pubis and the symphyseal process of the ischium elongate and general form intermediate between proto-insectivoran and gravigrade construction.   Tail long and heavy.

Four subfamilies:

1. ONYCHODECTINAE.   Dentition unreduced; canines of moderate size, incisors small, premolars short, high, simple, moderately compressed, somewhat inflated but not crowded or oblique, molars brachydont tritubercular, the upper with protocone and conules somewhat elevated in a strong crescent, the lower with small paraconid and basined heel. Muzzle long, jaws slender, sagittal ridge long but not very high; zygomatic arch very slender, almost thread-like.   Limbs and feet comparatively long and slender; metapodials quite long, claws small.

2. CONORYCTINAE.   Two enlarged upper incisors, lower incisors small, canines somewhat enlarged, premolars obliquely set and crowded, $p\frac{4}{4}$ sub-molariform, inner crescent of upper molars high and dominant, lower molars with no trigonid and tending to paired

cusps on trigonid and heel. Muzzle moderately short and wide, sagittal ridge long and moderately elevated, zygomatic arch moderate, jaws of moderate depth but short and heavy anteriorly. Skeleton almost unknown. Phalanges appear to be nearer to (4) than to (2) in proportions.

3. PSITTACOTHERIINAE. One enlarged upper and lower incisor, canines much enlarged, subgliriform, rooted, premolars short-crowned, transverse, p^4 simple, upper molars consisting of a stout inflated inner crescent and a connate pair of outer cusps, lower of an anterior and a posterior pair of connate cusps, the premolars of a large outer cusp and small inner heel. Crowns little elongated, but roots connate and enamel reduced. Jaws very massive and short, especially deep in front, coronoid process long and wide, moderately recurved, angular process wide, flat, not inflected. Muzzle extremely short, cranial region long with high and thickened sagittal ridge. Limbs short and massive, feet short, stout, five-toed, but the two lateral digits much reduced in the fore foot, metapodials very short, with hinged distal ends, phalanges extremely short, flattened, with very restricted movement fore-and-aft, claws very large, high and compressed on manus, not compressed in pes.

4. STYLINODONTINAE. Incisors $?\frac{1}{0}$, canines greatly enlarged and rootless, gliriform, cheek teeth hypsodont and rootless, with a strip of enamel at front and back. Pattern of unworn molar crowns as in Psittacotheriinae except that paracone and metacone are wholly fused into a single outer cusp. Jaw extremely deep and short, limb bones and manus extremely short and stout, digits five in manus but the fifth (also first?) much reduced.

GENERA, SPECIES AND GEOLOGICAL RANGE OF THE STYLINODONTIDAE

| | PALEOCENE | | | | | EOCENE | | | | | | | |
| | LR. | MID. | UPR. | | | LOWER | | | | MID. | | UPR. | |
	Puerco	Torrejon	Ft. Union	Tiffany	Clark Fork	Sand Coulee	Gray Bull	Lysite	Lost Cabin	L. Bridger	U. Bridger / L. Washakie	L. Uinta	U. Uinta
ONYCHODECTINAE													
Onychodectes tisonensis	×	–	–	–	–	–	–	–	–	–	–	–	–
Onychodectes rarus	×	–	–	–	–	–	–	–	–	–	–	–	–
CONORYCTINAE													
Conoryctes comma	–	×	–	–	–	–	–	–	–	–	–	–	–
PSITTACOTHERIINAE													
Wortmania otariidens	×	–	–	–	–	–	–	–	–	–	–	–	–
Psittacotherium multifragum	–	×	×	–	–	–	–	–	–	–	–	–	–
Psittacotherium aspasiae	–	×	–	–	–	–	–	–	–	–	–	–	–
Calamodon arcamoenus	–	–	–	–	–	–	×	–	–	–	–	–	–
Calamodon simplex	–	–	–	–	–	–	×	×	–	–	–	–	–
"Dryptodon" crassus	–	–	–	–	–	–	×	–	–	–	–	–	–
"Ectoganus" gliriformis	–	–	–	–	–	–	×	–	–	–	–	–	–
STYLINODONTINAE													
Stylinodon cylindrifer	–	–	–	–	–	–	–	–	–	×	–	–	–
Stylinodon mirus	–	–	–	–	–	–	–	–	–	–	×	–	–
Stylinodon sp.	–	–	–	–	–	–	–	–	–	–	–	×	–

ONYCHODECTINAE Winge, 1923

Winge, H., 1923, Pattedyr-Slægter, I, pp. 128, 133

Onychodectes Cope, 1888

Cope, E. D., 1888, Trans. Amer. Phil. Soc., XVI, N.S., p. 317; Osborn, H. F., and Earle, Charles, 1895, Bull. Amer. Mus. Nat. Hist., VII, p. 40; Wortman, J. L., 1897, *ibid.*, IX, p. 97; Winge, H., 1917, Vidensk. Medd. fra Dansk naturh. Foren., LXVIII, p. 105

Type: Onychodectes tisonensis Cope, from the Puerco of New Mexico.

Author's Diagnosis: Superior molars tritubercular, the external cusps distinct; the internal with the intermediate confounded in a prismatic form with flat grinding surface, and whose internal angle rises claw-like to an elevation equal to that of the external cusps, and without cingula or appendicular cusps. First premolar [i.e. p⁴] with but one external and one internal cusps.   Inferior molars seven, the true molars with five cusps, the anterior triangle distinct.   Last inferior molar with a heel; canine large.

Caudal vertebrae robust.   Ilium rather slender, flat-triangular in section, and with a small anterior-inferior spine.   Scapula with coracoid hook, and abruptly rising spine.   Astragalus with unequal trochlear ridges, the internal the lower.   Internal face oblique, but less so than in the species of Mioclaenus where it is known [this refers to *Claenodon ferox*], and not produced farther posteriorly than the external face, which is vertical.   Head depressed, convex, and without angles.   Cuboid with a small external distal facet.

The genus is intermediate in character of teeth between Conoryctes and Mioclaenus.   The molars are those of the former as to the internal portion of the crown.   The external cusps are more those of Mioclaenus, and there is but one external cusp of the first premolar [p⁴] while there are two in Conoryctes.   It is in the remarkable table-like form of the interior part of the crown and the hoof-like production of the internal angle, that Onychodectes differs from Mioclaenus.

Cope fully recognized the affinities of the genus to *Conoryctes* and *Hemiganus*, but referred all three to the Creodonta as annectant types leading into the Taeniodonta.

Osborn and Earle in 1895, in figuring a skull of *O. tisonensis*, refer the genus to the Tillodontia.

Wortman in 1897 redescribed the genus, referring it to the Ganodonta (= Taeniodonta), family Conoryctidae.

Winge, 1917, refers the genus to the Stylinodontidae and to the order Insectivora, along with the Leptictidae, Tillotheriidae and Periptychidae.   He regards it as the first stage in the stylinodont series, and as in general a primitive type of dentition, but conformant to his theories of molar evolution he considers the absence of external stylar cusps and the simple construction of the fourth premolar as specialized characters.

The expedition of 1913 was fortunate in securing considerably better-preserved specimens of this genus than had been previously known.   The skull and feet especially can now be more fully described and the affinities of the genus better estimated.   The views of Cope and Wortman appear to be confirmed as to the relationship of this genus to *Conoryctes* and "*Hemiganus*" (*Wortmania*).   It cannot, however, be regarded as a creodont, as placed by Cope, nor does it appear to have any near affinities to the Tillodontia.   On the other hand, it is clear that there is much to be said for Winge's view that it is related to the primitive Insectivora, although not to the families with which he associates it.

Onychodectes tisonensis [152] Cope

Onychodectes tisonensis Cope, 1888, Trans. Amer. Phil. Soc., XVI, N.S., p. 318 and pl. V, figs. 8, 9; Osborn and Earle, 1895, Bull. Amer. Mus. Nat. Hist., VII, p. 40, fig. 12; Wortman, 1897, *ibid.*, IX, p. 97, figs. 31–4.

Type: A. M. Cope Coll. No. 3405, upper and lower jaws, astragalus, cuboid, parts of scapula and pelvis.

Horizon and Locality: Puerco formation, San Juan basin, New Mexico.

Author's Description: It is characteristic of the superior molars that the external cusps have a lenticular section, and not triangular or a round one as in the species of Mioclaenus and Chriacus.[153]   The external cusp of the first premolar [i.e. p⁴] is large and elevated, and has the same fore and aft lenticular section with obtuse cutting edges.   The internal table of the crown is of parabolic outline and its edges are right angles.   The sides ascend perpendicularly to the alveolar border without the least trace of cingulum or other irregularity.   The crown has a weak external cingulum, which does not support any cusps.   The posterior of the external cusps of the third molar is well developed, and nearly in longitudinal line with the anterior.

The anterior triangle of the second inferior molar has a broadly rounded external apex, and it is a little elevated above the heel.   The latter has two internal marginal cusps, but its summit is so worn that the form of the surface cannot be further determined.   No cingula.   The manner of mastication is such as to wear the crown obliquely from within outwards in conformity with the form of the inner table of the superior molars.   The anterior triangle fits, as usual, between two adjacent superior molars, and the claw-shaped internal border of the superior molar worked, scoop-like across the heel, the inferior molar moving from without inwards.   The motion was the same as in Conoryctes.   In neither genera do I possess the glenoid surface for the mandibular condyle, but it is highly probable from the evident lateral movement of the lower jaw that neither genus possessed a preglenoid crest as is found in Mioclaenus [i.e. *Triisodon*, etc.].

The mandibular ramus is slender and moderately stout [i.e. thick but not deep], especially so at the anterior base of the coronoid process.   It follows that the anterior border of the masseteric fossa is well marked, but there is no distinct inferior border.   The angle is prominent, straight, and compressed; apex lost.   The dental foramen is below the middle of the base of the coronoid; and in line with the alveolar border.   The symphyseal surface is smooth.   The fourth premolar [i.e. p₁] is close to the canine, and to the third [p₂], and has one root.   Premolars all closely contiguous.   From the appearance of the alveoli the last inferior molar is of reduced size.   Enamel everywhere smooth.

Osborn and Earle (1895, p. 41) describe the cranium (A. M. No. 785, expedition of 1892):

The skull is about as large as that of a small *Didelphys*.   It is much lengthened between the glenoid facet and the last molar.   The cranium is long and narrow, and there is no depression between the cranial and facial portions.[154]   There is a very faintly developed sagittal crest, which extends as far forward as the posterior boundary of the orbit.   The nasals are narrow and elongate, and the anterior nares are terminal in position.   The palate is long and narrow, and the palatines and pterygoids form very narrow posterior nares quite different from that of the Lemuroidea.

The upper teeth are mostly broken off.   The fangs of the anterior teeth indicate that there is a well-developed incisor shortly in front of the canine; the latter tooth is laterally compressed, and the first premolar is small and single-rooted.   The second and third premolars are double-rooted; the fourth premolar is three-rooted.   It is evidently nearly as large as the molars.   There is no preglenoid ridge.   The angular region of the lower jaw is partly preserved, showing that the condyle is obliquely transverse; the coronoid

[152] The specific name is derived from the Pueblo Indian name for the great Colorado River, the Tison. See Amer. Nat., XXII, p. 45.—w.d.m.

[153] This is an effect of crushing in the type specimen.   The external cusps in uncrushed specimens are round, both in the molars and p⁴.—w.d.m.

[154] *Id est*, the basifacial and basicranial axes are in the same plane—true of most primitive mammals and of nearly all the Paleocene genera.—w.d.m.

is rather broad and the posterior border of the angle extends backwards. The inferior premolars are not spaced, and the posterior members of this series are robust.

Wortman's redescription in 1897 (p. 98) adds to a brief description of the skull and jaws the following points not noted or not clearly explained in the above:

The dental formula is somewhat in doubt, owing to our lack of knowledge of the incisors; that there was one pair at least in the upper jaw is certain, and I think it more than likely that there were two. The same uncertainty prevails with reference to the lower jaw. It may be provisionally written $I_{\frac{2}{2}}(?)$, $C_{\frac{1}{1}}$, $Pm_{\frac{4}{4}}$, $M_{\frac{3}{3}}$.

. . . The superior premolars are all simple, with the exception of the fourth, which has a strong external and internal cusp. The unworn crowns of the upper molars show three principal cusps, with a weak external cingulum. The two outer cusps, paracone and metacone, are subequal, and their summits are slightly inclined inwards, giving them a somewhat claw-like appearance; hence the name *Onychodectes* (claw biter). The internal cusp or protocone is large and lunate, having upon the limbs of the crescent faint intermediate cusps. The inner face of the crown is remarkably deep, and the enamel has unusual vertical extent.

The premolars of the inferior series are simple, laterally compressed cones, except the third, which has a faint posterior heel, and the fourth, which has a broader and better developed heel. The pattern of the molars is of the imperfect tuberculo-sectorial type, the anterior cusp of the trigon being poorly developed. The two principal cusps of the trigon are subequal and are placed in such a manner as to occupy a transverse position. The heel is broad and lunate, and in the unworn condition carries three principal cusps and a small though distinct cusp at its anterior inner termination. The great development of this element in one specimen has led Osborn and Earle to establish a second species, *O. rarus*.

. . . The head of the bone [humerus] displays that characteristic pyriform articular surface so constant in all the edentates; in the arrangements of the tuberosities and the deltoid crests it resembles the corresponding parts of the Armadillo.

A number of metapodials of both the fore and hind feet show that these bones are moderately short and stout, while the phalanges are more elongated and slender than those of *Hemiganus* [*Wortmania*]. The claw has essentially the same shape, but is relatively much smaller. The astragalus has a well-grooved trochlear surface, and can be readily distinguished from its cotemporaries by the absence of the astragalar foramen.

The principal specimens referred to this species are:

No. 3405, A. M. Cope Coll., type specimen, upper and lower jaws, astragalus, cuboid and parts of scapula and pelvis.

No. 785, skull and lower jaw.

No. 16528, skull, lower jaws, parts of fore and hind feet, vertebrae, parts of several limb bones.

No. 16410, parts of lower jaws, with vertebrae, limb bones and various other fragments.

Revised Description

DENTITION. $i_{\frac{3?}{3}}$ $c_{\frac{1}{1}}$ $p_{\frac{4}{4}}$ $m_{\frac{3}{3}}$. The first and second upper and the first and third lower incisors much reduced, i^1 perhaps absent, i^3 and i_2 somewhat enlarged. Canines of moderate subequal size and laniary form, considerably blunted by wear in all our specimens. Premolars rather high and pointed, only moderately compressed, less trenchant than with most creodonts, regularly graded in size and complexity from first to fourth, not crowded or obliquely set, but slightly spaced, without any considerable diastemata. First upper and lower premolar one-rooted, with simple pointed crown, oval in cross-section. P^2 has two roots widely separated, moderately compressed oval crown and slight posterior heel. P^3 is three-rooted, with trigonal base, the crown consisting of a large round-oval cusp, small inner basal cusp, and obsolete encircling cingulum. P^4 is more nearly molariform in proportions, the inner cusp is large and semicircular, but not so large as in the molars;

the outer cusp is still single; but with a faint metacone budding off its posterior flank; the wings of the inner crescent are continued down to the basal outer angles of the tooth, but there are no definite stylar cusps and the outer cingulum is obsolete. The upper molars are tritubercular, with large rounded crescentic inner half composed of protocone and conules, no trace of hypocone or protostyle, the paracone and metacone sub-marginal, low, rounded cusps, well separated, with the external cingulum distinct and at the external angles expanded into rudimentary stylar cusps or crests. The third molar is smaller than m^{1-2}, the metacone and postero-external portion reduced to a varying degree in different specimens.

The second, third and fourth lower premolars are two-rooted with progressively larger heels, the principal cusp high, oval in cross-section, somewhat inflated, no accessory cusps, the heel quite rudimentary on p_2, small and simple on p_3, larger with a small inner cusp on p_4. The lower molars are tritubercular, with trigonid and talonid subequal in size, the talonid a little lower, the three trigonid cusps all well developed and well separated but the paraconid lower than the others, the heel somewhat basined with well-developed hypoconid, hypoconulid and entoconid, the first a little larger than the others. The third molar is somewhat smaller than m_{1-2}, and the hypoconulid projects somewhat more posteriorly, the heel being narrower.

SKULL. The skull is long, narrow, strikingly insectivore-like in its proportions and in many characteristic points of construction. The long muzzle, terminal nares, relatively large premaxilla, its superior branch well extended backward, the long nasal bones, the absence of any postorbital process, the long mesocranial region, the low sagittal crest and ridge-like building up of the cranium beneath it, the moderate-sized lachrymal, the very much reduced zygomatic arch, the zygomatic branches of both jugal and squamosal very slender and meeting only for a short distance towards their tips, the lack of transverse expansion of the glenoid facets—these and other characters are distinctive of the Insectivora, and most distinctive of the early Tertiary Insectivora and of the most primitive living types. The nasals are longer and more expanded posteriorly than in any known Insectivora, in this respect approaching marsupials. The lachrymal foramen and tubercle are on the margin of the orbit, and the lachrymal has a moderate expansion on the face above the tubercle, nearly reaching the broad posterior expansions of the nasals (cf. Didelphys, etc.). The anterior border of the lachrymal is not distinguishable, and the outlines of the frontals cannot be clearly made out; anteriorly their relations to nasals and lachrymals seem to be much as in Didelphys, but posteriorly the parietal contact is not certainly recognizable. The superior plate of the squamosal is rather small, and the construction of the glenoid region much as in creodonts except for the reduced zygomatic process, small postglenoid and short, rounded glenoid fossa. As in many creodonts this is limited mediad by a small process of the squamosal, which in Centetes is much higher and plastered against the alisphenoid. There is a small postglenoid foramen and a rather shallow and short otic notch separating the postglenoid process from the stout post-tympanic process, which lies postero-external to the mastoid and sutured against it. The conditions approach those of Deltatherium so far as they can be determined, and indicate similar relations and proportions but with a decided approach towards insectivore relations. The alisphenoid must have been small (unlike marsupials).

The under side of the skull is largely buried in flinty matrix in No. 785, and much obscured by crushing and shattering in No. 16528.   It is evident, however, that the palate was fully ossified without vacuities, and with anterior palatine foramina small and restricted backward.   The posterior border of the palate was much as in *Centetes*, with a rather weak post-palatine ridge, and a long narrow open post-nareal canal behind it, walled on each side by high parallel crests composed of the posterior wings of the palatines and the pterygoid plates.   Whether and to what extent the alisphenoids extended forward to take part in these posterior nareal crests is not clear.   They are nearest to *Centetes* and other Insectivora in proportions.   In creodonts they are equally prominent, often closed in for a considerable distance back of the palate.   In *Didelphys* they are short and low.

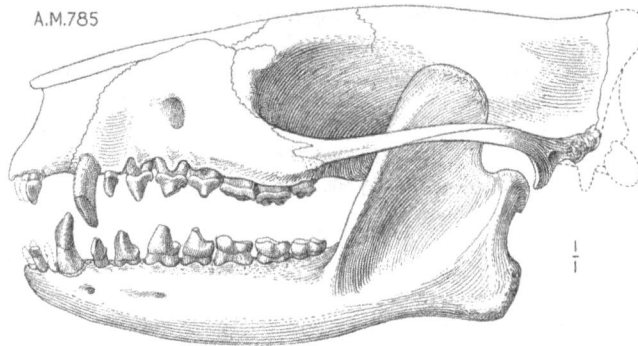

FIG. 58.   *Onychodectes tisonensis*, skull and lower jaw, A. M. No. 785, side view.   Natural size.

The jugal in No. 16528 appears to be considerably reduced in its anterior branch, as well as in the antero-inferior branch; a long and very slender, nearly straight posterior branch extends well backward but not to the glenoid region, and has a narrow and weak suture for the squamosal process in the posterior third of its length.   This process is preserved in No. 16528; it has an equally slender, almost vestigial character.

In neither of the skulls is the occiput preserved, and little can be seen of the otic region.   Apparently, however, the petrosal prominence was rather large and distinct, pyriform in shape, with no suggestion of any bulla or false bulla covering it, lying directly opposite the otic notch, and having the *fenestra ovalis* and *fenestra rotunda* in the same position as in creodonts.

The lower jaw is preserved more or less complete in many specimens.   It is of moderate depth, rather long, somewhat suggestive of Insectivora in the comparatively weak and shallow though rather long symphysis, the small canines and lack of specialization in the cheek teeth, the coronoid process broad and flat, not recurved, the angle flat and rather wide, the condyle transverse but short, trihedral in form, the transverse diameter about one and a third times the sagittal.

On the whole this skull and jaws of *Onychodectes* show a number of distinctively insectivore characters as compared with the more primitive creodonts, with which it has much in common.   It also has a good deal in common with the opossum and other primitive marsupials, but these are merely primitive characters.   The skull shows clearly its placental relationships in the reduced jugal, the small alisphenoid, the character of the posterior nareal channel and of the bony palate, as well as in the dentition. The distinctively insectivore characters have been summarized on a preceding page. It is to be observed that the relationship to the more primitive creodonts is far closer than to marsupials. Comparison with the *Loxolophus* skull and jaws figured and described in a preceding section of this memoir will show the degree of this relationship.

Taeniodont characters are very rudimentary in *Onychodectes*. They appear in the tendency to specialization of the incisors, which indicates that in this order it is the *third* upper and the *second* lower incisor that are enlarged in the more specialized genera, the others being lost. They appear more clearly in the construction of the upper molars with their high, massive, inner crescent, very evidently related to the molar construction of *Conoryctes*, which in turn is related in the molars to *Wortmania* and *Psittacotherium*. A third taeniodont feature is seen in the elongate narrow cranium, already built up to a noticeable degree beneath a low, inconspicuous sagittal crest. This character, also seen in *Conoryctes*, is carried much further in *Wortmania* and *Psittacotherium*.

The lack of post-canine diastemata and short, high premolars may be regarded as the foundation upon which the peculiar taeniodont specializations of the premolars were based.

SKELETON. The skeleton of *Onychodectes* is large in comparison with the skull, somewhat in contrast to most of the creodonts, in which the skull is relatively large. The proportions of skull and skeleton are roughly comparable to those of *Tamandua*, the tail being extremely long and heavy, limbs of moderate length, but longer than in *Tamandua*, comparable in proportions with the creodonts, feet five-toed with short phalanges, small uncompressed claws, metapodials moderately long, especially those of the pes.

A few vertebrae, badly crushed and incomplete, are preserved in Nos. 3405, 3476 and 16410, and a number of caudals in No. 16528. Part of a cervical centrum indicates a very short neck, as in Taligrada. The dorsals, to judge from a few fragments, are much smaller than the lumbars, about as in the more primitive Carnivora and Creodonta. A part of the sacrum and some fragments of the pelvis are preserved, but show no unusual characters. The sacro-iliac suture is strong and close, the number of sacral vertebrae unknown. The caudals are unusually large, the proximal caudals being about the size of the lumbars, while the middle caudals are considerably longer and the posterior caudals are long and slender. They carry throughout much more the proportions and construction of the Creodonta than of any of the later Edentata, in which the caudals, while numerous, are all short and of rather uniform construction. Two large chevron bones are preserved in No. 16528, along with a number of caudals from the middle part of the tail. They are coössified proximally and extend downward into plates elongate antero-posteriorly, enclosing between them a moderately wide channel for the caudal artery. The vertebrae in this region have lost the neural arch but retain considerable anterior and posterior transverse processes, in addition to an anterior sub-median spine and a widely separated pair of posterior processes (?metapophyses) on the dorsal side. The reduction of the neural arch has gone further than in *Tritemnodon*, of the transverse processes not so far. The poor preservation of the vertebrae in all these specimens and some doubt in Nos. 3405 and 16410 as to whether all of them belong to *Onychodectes* make it inadvisable to describe them in detail.

Fore Limb and Foot. The humerus is best preserved in No. 16410. It is of about the same size as in *Goniacodon* or *Deltatherium*, and very much resembles that of creodonts generally in its proportions and construction. The head, however, is more extended downward on the posterior border, making an approach to the "pyriform" construction which Wortman described as characteristic and compared with the Edentata (apparently from a badly crushed upper part of humerus in No. 3576a). The bone has, however, very little suggestion of edentate characters. It is not at all like that of *Palaeanodon*. The greater tuberosity is somewhat less, and the lesser tuberosity somewhat more developed than in most Creodonta; the deltoid crest is prominent, moderately broadened in its upper part, extending about halfway down the shaft and narrowing to an abruptly ending crest about the middle of the shaft. The supinator crest is prominent, as much so as in *Goniacodon*, less than in *Deltatherium*, the entepicondylar bridge is present, and the process prominent, the supra-trochlear fossae are shallow, with no foramen, the inner crest of the distal trochlea is prominent both in front and behind, the external crest is prominent behind, and the radial surface is slightly flattened.

The ulna is complete in No. 16410, parts of it in other specimens. It is as long as the humerus and quite creodont-like in aspect, comparing closely with *Goniacodon* in size and proportions. The shaft is flattened, moderately wide, deeply grooved on the antero-internal side; the olecranon is high, somewhat overhanging forward but not notably compressed; the distal end of the shaft is somewhat narrowed, with a prominent oblique crest on its postero-internal and posterior face; this end is also slightly widened at the radial facet, and is strongly oblique with a small convex knob for pisiform and cuneiform. The proximal facet for the radius is nearly flat, the sigmoid fossa of moderate depth with prominent marginal crests above and below for the humeral facet. There is little or nothing in this bone to distinguish it from the Creodonta or to suggest edentate or even taeniodont affinities.

The radius is not complete in any of our specimens, but parts are preserved, especially in Nos. 3576a, 16410 and 16528. The head is oval, with a moderate supporting ledge on the external side of the humeral facet, a shallowly convex facet for the ulna, deeper but not so wide as is usual in creodonts. The upper part of the shaft is round, the bicipital tubercle obscure; distally the shaft broadens with a prominent anterior or antero-internal ridge comparable with that of *Palaeanodon* but not so conspicuous, continuing down somewhat obliquely to the internal side of the distal facet, which is round-oval, simply concave, without separation of scaphoid and lunar portions of the facet. The shaft has a considerable anterior convexity. The radius differs from that of Creodonta chiefly in the form of the head, in the prominence of the anterior crest on the distal portion of the shaft and in the relative enlargement of the distal end.

The manus is more distinctive of insectivore relationships. Most of the right manus is preserved in No. 16528, but the cuneiform, trapezoid and trapezium in the carpus, and about half the phalanges are lacking.

The carpus has much the same construction and proportions as in Creodonta, so far as preserved, but the lunar is perhaps somewhat wider, the scaphoid rather small and the centrale unusually wide, extending under more than a third of the lunar, which rests about equally upon the centrale, magnum and unciform. The unciform has the usual form, with a wide facet distal-internal chiefly occupied by Mc. III, a proximal-internal facet of about equal width for the lunar, and the usual proximal and distal facets for cuneiform and

Mc. IV and V. The magnum is rather small dorsally, but not so small as it usually is in creodonts, extends beneath the lunar in the usual manner, and barely touches the unciform at the dorsal surface, but has a wider contact below. The trapezoid is not preserved, but must have been comparatively wide and shallow, and the trapezium was evidently small.

There are five metacarpals, all complete with phalanges but the first and fifth much reduced in size, about half as long as the others, and Mc. I about half the diameter, while Mc. V maintains nearly the same diameter of shaft.

Mc. I has the usual convex head for the trapezium but with very little dorso-palmar convexity. It has no facet for Mc. II, and hence was apparently divergent but not opposable, unlike the creodonts, in which some degree of opposability (dorso-palmad movement on the carpus) exists. This does not of course preclude possible movement of the digit, including the trapezium, in this direction. The distal end of Mc. I is moderately expanded, with a hinge-joint faintly convex from side to side. All the other metacarpals and metatarsals show this type of distal end, permitting moderate flexure and extension of the phalanges but very little lateral movement. In creodonts the distal metapodial joints are decidedly more convex laterally as well as dorso-palmad, approaching the ball-ends of

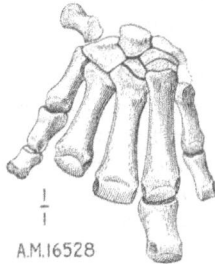

A.M.16528

Fig. 59. *Onychodectes tisonensis*, fore foot, A. M. No. 16528, anterior view. Natural size.

the Primates. In taligrades and condylarths the joint tends to become similarly limited in lateral movement.

Mc. II has a comparatively wide head with the usual saddle-shaped articulation (concave laterally, convex dorso-palmad) for the trapezium, relatively wider than in creodonts, a small overlap on Mc. III enabling it to attain a narrow contact on the magnum, a large nearly flat facet internad for the trapezium. The distal end is somewhat asymmetric, considerably broadened, somewhat "squared" or hinged as to the facet.

Mc. III is a little larger and of about the same length, symmetrical distally, the proximal end having a wide overlap on Mc. IV, with an unusually broad unciform contact facing proximal-internal, and narrowing toward the palmar surface. The magnum contact is moderately wide and of the usual saddle-shaped form.

Mc. IV is about as large as Mc. II, but not so long, but the successive overlaps of the third and second make the distal end symmetrical with the distal end of Mc. II. The proximal end has a convex facet for the unciform and a narrower facet for Mc. III obscurely separated from the unciform but facing partly entad. On the external face Mc. IV overhangs the fifth metacarpal with a rather large concave facet.

Mc. V is short and stout, the shaft nearly as large as that of Mc. IV, but the distal end much smaller, very little expanded, and the length about half that of Mc. IV.   The convex head fits proximally against the unciform and cuneiform about equally, and on the internal side against Mc. IV; the process on the external side of the head was apparently of good size, but is broken off.

The phalanges are remarkably short, wide, flattened, the length varying from one and a half to one and a quarter times the width of the proximal end.   They are all widest proximally, narrowing in the shaft, and not expanded distally.   The first and second series have about the same proportions, the proximal facets in the first series are faintly concave in either direction, the distal facets hinge-joint with a rather limited movement of flexion and extension but not so limited dorsad as in the larger taligrades.   The second series of phalanges is very like the first in proportions and form of facets, but their proximal facets are flatter and their distal facets more convex dorso-palmad and comparatively narrow.   The ungual phalanges are small claws, neither compressed nor flattened, and without any distal fissure.

The above-described characters of the fore limb and foot are in many respects suggestive of insectivore relationship, in some of taeniodont affinities, but for the most part are merely primitive characters common to the earliest creodonts, insectivores and other eutherian mammals.   The humerus makes a slight approach towards insectivores and taeniodonts in the tendency to a posterior-facing, sub-pyriform head, to a subequal development of the inner and outer tuberosities, and the relative shallowness and breadth of the distal end.   The radius and ulna have no very distinctive features, save for the tendency to expansion of the distal half of the shaft of the radius and cresting of its antero-internal face.   The manus is distinctive in the marked reduction of the lateral digits which follows the relative proportions seen in *Psittacotherium*, in the relative width of the lunar and wide shallow centrale, again foreshadowing the taeniodonts, and in the shortening of the phalanges, seen also to a varying extent in the insectivores.   The relative shortening of the first row of phalanges, which are scarcely any longer than the second series, is quite as in the manus of *Psittacotherium*.   The hinge-joints between the metapodials and phalanges compare with insectivore conditions and differ from the more rounded joints of creodonts.   There is no suggestion of the peculiar armadillo-like form of these and the inter-phalangeal joints seen in *Palaeanodon* and *Metacheiromys*.   The claws are quite small in *Onychodectes*, and show nothing of the enlargement and compression characteristic of the stylinodont manus.   On the whole, it may be regarded as a very early stage in the evolution of the taeniodont foot, much more primitive than the contemporary *Wortmania*, but affording a valuable illustration of the origin of the peculiarities that characterize the more specialized genera.   Its relations to *Conoryctes* are not clear, as the skeleton of that genus is practically unknown.

Hind Limb and Foot.   The hind limb bones are very incompletely known.   Parts of the femur and tibia are preserved in No. 3405, and fragments of these bones in other specimens.   The hind limb was of heavier proportions than the fore limb, but not much longer.   The femur has a broad, shallow digital fossa, a moderately wide third trochanter set high on the shaft, an equally prominent internally-placed second trochanter, a much flattened shaft (exaggerated by crushing to an unknown extent).   The patella (No. 3576a) is subcircular and exceptionally flat.   The tibia has a moderately long shaft with low cnemial crest.

Parts of the hind foot are preserved in Nos. 16528, 3576a, and 3405.   It is five-toed, the lateral digits considerably reduced as in the fore foot, the astragalus of insectivore type, the tarsus not wholly serial, as there is a distinct astragalo-cuboid facet, the phalanges short and flattened and the unguals small and like those of the manus, but the first row of phalanges is longer than the second. Cope in 1888 described and figured the rather badly crushed astragalus of the type (No. 3405). The uncrushed bone in No. 16528 shows more clearly the symmetric character of the trochlear crests, the shallow but rather wide trochlea between them and entire absence of astragalar foramen. The neck is distinct and the head a flattened oval. It compares in these features with *Prodiacodon*, *Pantolestes*, and Insectivora in general, but the neck is much shorter than in *Prodiacodon*, the trochlea wider, shallower and shorter, the head more obliquely set. The proportions are nearer to those of *Pantolestes*, but the trochlear crests are sharper, the external one is more extended at its distal end, the fibular facet is little excavated, the internal crest is less extended towards the distal end, the neck and head are wider and more flattened, the sustentacular facet is more clearly separate from the navicular facet, the astragalo-calcanear facet is not so concave and of less width.

The calcaneum has a rather short tuber, deep but not wide, the end not much expanded. The astragalo-calcanear facet is oblique, rather narrow and shallow, faces more upward than in most creodonts, but is otherwise similar. The peroneal tubercle is prominent but

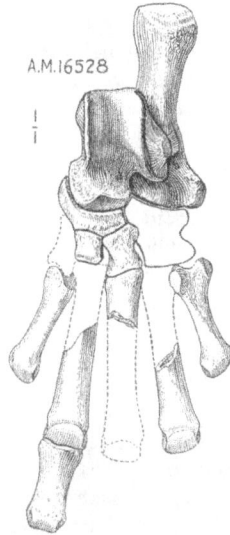

FIG. 60. *Onychodectes tisonensis*, hind foot, A. M. No. 16528, anterior view. Natural size.

not more than in many creodonts, and the cuboid and sustentacular facets are similar to this group, the process beneath the cuboid facet being unusually prominent. The navicular is rather small, shallow and not very wide, the inferior hook has a distinct neck, the internal hook is reduced. The facets for ecto- and meso-cuneiform are at a distinct angle, about 120°, the ectocuneiform facet facing partly ectad, the mesocuneiform partly entad. The entocuneiform facet appears to be greatly reduced. The ectocuneiform has about twice the height and nearly twice the width of the mesocuneiform; it has a deep, narrow, oblique facet facing proximo-ectad for the cuboid, the usual distal facet for Mt. III, and on the distal half of the inner side a dorsal and a palmar facet for Mt. II, facing entad. The mesocuneiform is narrow, with oblique dorsal surface, with deep and narrow facets for navicular and second metatarsal, both of them narrowing toward the plantar surface. On the inner face a reduced and somewhat obscurely facetted surface for the trapezium indicates the considerable reduction of this bone.

The metatarsals are from a half longer to twice as long as the metacarpals, but very little larger in the shafts or distal ends; but the first and fifth are somewhat less reduced in proportion to the rest. The first metatarsal has somewhat the proportions of Mc. I, about three-fifths the length of the second, and much more slender in shaft and smaller in distal end. The proximal end is not much enlarged, narrow and deep, with the trapezium facet strongly concave from side to side, moderately convex from front to back. The second metatarsal is much larger and more robust, the head deep and rather narrow, with somewhat oblique and slightly concave mesocuneiform facet, two rounded facets at the dorsal and plantar ends of the ectocuneiform contact, and an internal facet for the trapezium. The distal end is somewhat "squared" or hinge-jointed in comparison with most creodonts, rather less so than in the manus. The third metatarsal is known only from the proximal end. It appears to be slightly larger than the second, and has no overlap on the fourth, but a somewhat overhanging concave facet for its head, confined to the dorsal part of the head. The fourth metatarsal is distinctly longer than the second, but less robust in the shaft, its distal end more oblique than that of Mt. II. It has a deeply concave facet on the outer side of the head for Mt. V, permitting a certain amount of pivoting. The fifth metatarsal is scarcely over half the length of Mt. IV, its shaft much flattened so that it is wider than that of the fourth although much less robust. The proximal end carries a large facet for Mt. IV, facing proximad-entad and considerably dorsad, also a very small narrow cuboid facet and a large external process. The distal end is very little expanded, strongly oblique, moderately convex both ways.

The phalanges are of much the same character as those of the fore foot, but the proximal series are longer. The unguals are small and pointed, not distinguishable from those of the fore foot.

The hind foot shows clearly its insectivore relationship in the astragalus with its broad, shallow trochlea bounded by two sub-symmetric crests, with no trace of astragalar foramen, the sub-vertical fibular facet, and other indications of the position of the astragalus beneath the tibia instead of wedged in between tibia and fibula as is the primary position in creodonts, condylarths and taligrades. The short, wide, flattened phalanges are also suggestive of insectivore relationships, although this specialization is carried further than in the most primitive living insectivores.[155] The reduction of the lateral digits is not especially characteristic of insectivores and is not seen in creodonts or taligrades, but does occur in some condylarths. In the present case, being associated with shortening of the phalanges and retention of claws, it points towards Taeniodonta, like the similar but more marked reduction of the lateral digits of the manus.

[155] It compares with *Erinaceus*, but there are reasons to believe that this is partly a specialized character in *Erinaceus*, although Wortman regarded it as a primitive feature of the Insectivora. It is paralleled in taligrades, condylarths, perissodactyls, etc., but in all these associated with broadening of claws into hoofs. —W.D.M.

There are no significant points of approach towards the edentates, save for the general proto-insectivore characters already noted.

Onychodectes rarus Osborn and Earle

Onychodectes rarus Osborn and Earle, 1895, Bull. Amer. Mus. Nat. Hist., VII, p. 42, fig. 13.

Type: A. M. No. 824, lower jaw, m_{1-2} left. [M_{2-3} according to Osborn and Earle.—The Editors.]

Horizon and Locality: Puerco formation, Coal Creek Canyon, San Juan basin, New Mexico.

Authors' Description: A prominent external cusp on each lower true molar, placed between the outer lobes.

This new species is established upon a jaw fragment which contains two of the lower true molars (No. 824). The most striking character is the very prominent cusp which is placed upon the external side just in front of the posterior lobe. The trigonid is well raised above the talonid. The paraconid is well developed; the protoconid is relatively robust and placed at the apex of the triangle and at an equal distance between the para- and metaconids. The talonid is broad and deep and extends into a basin on the inner side. The external interlobular cusp of the second molar is smaller than that of the first; it arises from the base of the hypoconid, and is placed just opposite the convexity of the latter.

The principal referred specimen is No. 16405, maxilla and lower jaws, p^3–m^3 l., c_1–m_3 of both sides.

Revised Description: This species was recognized by Osborn and Earle from the fragmentary type. Our additional material serves to validate and extend its characters. The first and second molars are wider than in *O. tisonensis*, and have a small but distinct extra cusp on the outer flank of the talonid. The third molar does not show this cusp, but has a somewhat wider trigonid than in Cope's species. The fourth lower premolar is broadened at the base, especially posteriorly, and has a distinct anterior accessory cusp in addition to the double heel cusp. The remaining lower teeth show no constant differences from *O. tisonensis*. The upper molars are somewhat broader than in *O. tisonensis*, the fourth premolar of somewhat larger size and the external cingula of p^4–m^3 are relatively prominent. The third premolar has a minute postero-external basal cusp and a small internal cingulum.

CONORYCTINAE Winge, 1923

Winge, H., 1923, Pattèdyr-Slægter, I, pp. 128, 133

Conoryctes Cope, 1881

Cope, E. D., 1881, Amer. Nat., XV, Sept. 22, p. 829

Type: Conoryctes comma Cope, Upper Paleocene, Torrejon formation, San Juan basin, New Mexico.

Synonym: Hexodon Cope, 1884, Amer. Nat., XVIII, July 17, p. 794; type: *H. molestus.*

Author's Diagnosis: Char. gen.—Allied to *Esthonyx.* Inferior canines not rodent-like, with conic crowns. Molars 3-3, the first one-rooted, the second two-rooted, the third with an anterior conic cusp and a posterior grinding heel. True molars consisting of two lobes, of sub-cylindric section, separated by deep vertical grooves. Enamel developed on internal and external faces of crowns. . . . This genus differs from *Esthonyx* in the form of the fourth premolar. In the latter the anterior lobe is compressed and trenchant.

Type Diagnosis of Hexodon: Family Periptychidae. Three premolars. Fourth superior premolar like molars; inferior premolars without internal ledge. . . . In *Hexodon* Cope the [periptychid] type is most developed in the direction of the dental prehension. With the shortening of the jaws comes the loss of a premolar.

Conoryctes comma Cope

Conoryctes comma Cope, 1881, Amer. Nat., XV, p. 829.
Hexodon molestus Cope, 1884, Amer. Nat., XVIII, p. 795, fig. 3.

Type: A. M. Cope Coll. No. 3395, lower jaw with p_4–m_2, l., canine. Type of *H. molestus:* No. 3396, upper and lower jaws.

Horizon and Locality: Paleocene, Torrejon formation, San Juan basin, New Mexico.

Author's Description: The external faces of the molars are much more exposed than the internal and are somewhat contracted inwards. In the unworn crown there is a distinct anterior inner cusp, which is soon confounded on attrition. The heel of the last premolar has a crescentic section, the internal horn the narrower. The anterior lobe is a robust cone. The base of the second (third) premolar is oblique to the axis of the ramus outwards and forwards. It is possible that there is a minute first premolar filling the short space between the second and the canine. No cingula; enamel obscurely plicate; ramus robust. [Measurements follow.]

Type Description of *Hexodon molestus* (p. 796): It differs from the *E.* [*Ectoconus*] *ditrigonus* in the short, rounded incisive region and closely-placed incisor teeth, the small posterior superior molar, and the more robust and more vertical canine teeth. It is about the size of the red fox, but much more robust. It is one of the few species of the family which is armed with large canine teeth, and evidently stood pre-eminent in its powers of offence and defence. In the typical specimen the teeth are all worn by the mastication of hard or tough substances, so that the structure of the crowns of the true molars is not entirely known.

[Footnote] This species is represented by a specimen which is referred by me to the *Conoryctes comma*, in the Vol. III of the Report of the U. S. Geological Survey of the Terrs., p. [198], and are represented in Figs. 1–5, Pl. xxiiie of the same. Better specimens of the *C. comma* show that the canine (or ?incisor) teeth are of very different character from those of this animal.

As may be seen by the above diagnoses, Cope at first supposed *"Hexodon"* to be related to *Periptychus*. In 1888 [156] he recognizes the true relationships in a revised description:

The position of the genus is doubtful, owing to the absence of the ungual phalanges. It is probably Creodont, rather than Condylarthrous, for two reasons; one is the close resemblance of the dentition to those of Onychodectes and Hemiganus [= *Wortmania*], between which it takes a natural position. The other is, that it displays no resemblance to any of the Condylarthra in the details of its structure.

He describes the character of the upper molars as follows:

The superior true molars, and the first premolar, have two external conical cusps, and an internal triangular table, whose inner angle is produced downwards to a line with the apices of the internal cusps. The inferior premolars and the first premolar have the anterior part much elevated above the posterior. The former consists of a large external and a small internal cusp joined to near their summits, except on the first premolar which has but one anterior cusp, which is simple acute cone. A rudimental fifth cusp is present on the true molars.

The species description of *Conoryctes comma* (*op. cit.*, p. 317) gives the construction of the lower teeth:

This animal was about the size of a wolverine, of which species one is reminded by its robust characters. It has an elevated sagittal crest and a strong inion. In a series of teeth which are but little worn the following characters may be discerned. The crown of the inferior canine has a flat inner face, beyond which the anterior surface extends inwards, forming a rib-like border. The enamel on the internal and posterior faces extends but a short way from the apex, and is thin, while on the convex anteroexternal face, it extends

---

[156] Cope, E. D., 1888, Trans. Amer. Phil. Soc., XVI, N.S., p. 316.

below the usual position. It thus approaches the condition seen in Hemiganus. A similar state of affairs is seen in the molars, where the enamel is extended much further on the external face of the inferior and the internal face of the superior molars than is usual, approaching the genera mentioned and also the Taeniodonta in this respect. There are no cingula on any of the molars of either series excepting on the external side of the superiors; and there it sends out a process or cusp between the two external cusps. The crowns of the inferior molars are notched at the junction of the anterior and posterior parts. The notch is the section of a vertical groove from the base of the crown on the external side, and of a very short superficial one of the internal side. The fifth cusp is median, and about opposite the rim of the heel in elevation. The grooves of the first premolar are similar to those of the true molars. There is no anterior basal cusp. The heel is large and has a raised border on the posterior and inner sides, and an external median lateral conic cusp. This when worn joins the curved crest, forming a comma-shaped figure.

Wortman in 1897 [157] redescribes this species and gives a restoration of the skull, based chiefly upon the type of *Hexodon molestus* but in part upon No. 3398.

Some additional fragments pieced on to Nos. 3396 and 3398 add a little to the known characters of *Conoryctes*. No. 15939, skull and jaws, is somewhat more complete than either of the preceding specimens. In all these specimens the teeth are heavily worn. No. 16029 shows unworn m_{1-2} and other isolated teeth show the pattern of the unworn upper molars. Very little is known of the skeleton. In No. 3396 a few fragments are preserved.

Revised Description: Dental formula $i\frac{2}{2}$ $c\frac{1}{1}$ $p\frac{3}{4}$ $m\frac{3}{3}$. Teeth brachydont, tritubercular, pattern modified, between *Onychodectes* and *Wortmania*. Canines stout, upper incisors somewhat enlarged, premolars crowded, oblique, $p\frac{4}{4}$ molariform, molars more oval than in *Onychodectes*, the upper series with inner crescents higher, mesostyles distinct, the lower with vestigial paraconid. Skull broader and more massive than in *Onychodectes*, with shorter face and more elongate cranium, high built-out sagittal ridge narrowing posteriorly into a sharp crest, zygomatic arch moderately heavy, nasals long, broad posteriorly, postorbital crests midway between fronto-nasal and fronto-parietal sutures. Jaws moderately heavy, of about equal depth and massiveness beneath premolars and molars.

The upper incisors are two in number on each side, with simple pointed crowns that wear to a somewhat spatulate edge like the lower canines but much smaller. The enamel extends farther down on the antero-external face of the tooth. The first upper incisor was figured as the second by Wortman (1897, fig. 36). They are in fact materially larger than he supposed. The second upper incisor is quite like the first, except for somewhat larger size, and wears both in front and behind, giving a blunted end similar to the upper canine wearing surface. The upper canine is a stout tusk, with heavy root, the enamel limited to the upper third of the tooth, and extending farthest down on the external side. It wears against the lower canine to a flat vertical transverse surface in front, and also has a more rounded postero-internal surface of wear, these with the external enamel surface making a spatulate tip.

There is a short diastema in front of the canine but none behind it. The first premolar is absent. The second premolar is a small pointed tooth much like the incisors but smaller than either of them, and with a double root, closely connate. The third premolar is a much larger tooth but smaller than p⁴, and is three-rooted with round-oval crown slightly broader than long, stout massive principal cusp somewhat inflated, and rudimentary internal cusp. The fourth premolar is sub-molariform, with protocone and metacone

[157] Wortman, J. L., 1897, Bull. Amer. Mus. Nat. Hist., IX, pp. 101–2.

distinct, round-conical, the protocone larger and still sub-central in position, a large inner crescent, which is very like that of the molars. There seems also to be a postero-median cusp in the position of the metaconule. The first and second molars have the protocone and metacone low, conical, subequal, well separated, a smaller mesostyle between and external to them, and the protocone forms a wide, high, rounded crescent much larger and higher than the outer cusps, with a shallow concave upper surface pitching outwards and its slightly crested inner border continued as a crest to the base of the external angles joining the external cingulum which continues around to the mesostyle. Whether any trace exists of the conules is not known; they are not indicated in the moderately worn teeth. These two teeth are of about the same size as p⁴, the inner crescent larger but the external side not so wide antero-posteriorly. The third upper molar is smaller than the others, about the size of p³, more nearly circular in outline. The metacone is reduced and the mesostyle vestigial.

In the lower jaw there are two incisors on each ramus, one nearly as large as the upper incisors, the other much smaller. They appear to be of similar type to the upper incisors. The lower canine is nearly as stout as the upper, the root more curved and somewhat more slender, the enamel similarly limited to the outer face of the tooth, the heavy wear on the posterior face of the tooth making a flat surface, the antero-internal face worn to a less degree and its surface rounded. There are four lower premolars, all two-rooted, fair-sized, and the first three set more or less obliquely in the jaw.

The molars are seen unworn in No. 16029; in all other specimens they are more or less heavily worn. They are considerably longer than wide, the trigonid somewhat shorter than talonid, and slightly higher, composed of two principal low, rounded cusps, and a minute median basal paraconid. The talonid is deeply basined, with a strong hypoconid at the postero-external angle, a small cusp directly in front of it and a high posterior and internal crest crowned with four or five small cusps. The crown as a whole is considerably elevated, without basal cingula.

The skull is much larger than that of *Onychodectes*, nearly twice as long, and at least twice as wide, and differs in its shorter face, wider muzzle and interorbital space, heavier zygomata and much more robust proportions throughout. The premaxillae have somewhat the same width and backward extension but are much stouter, carrying the two considerably enlarged incisors. The nasals are long, narrow, moderately expanded backward, reaching a point about opposite to where the postorbital processes would be if present, a little in advance of the union of the very weak postorbital ridges. There are no postorbital processes, however, and the lachrymals extend back to the point where they should come, thus excluding the frontals from the orbital rim. Little else is preserved of the lachrymals; they extend out anteriorly to an uncertain amount on the face. The maxilla is a large and stout bone, excluded apparently from the orbital rim and having a rather short contact with the frontals as far as can be judged. The infra-orbital foramen is above the posterior part of p³ and is rather distinctly doubled. The frontals are comparatively short, extending forward on each side of the posterior expansion of the nasals to a contact with the maxillaries, limited laterally by the lachrymal expansion, and posteriorly overlapped by the parietals as far forward as the postorbital constriction. A low sagittal ridge on their posterior portion divides toward the anterior half of the frontals into the faint, rapidly divergent postorbital crests, which fade away completely before reaching the posterior ends of the lachrymals. The parietals are long and narrow, the sagittal ridge narrowing up into a crest of considerable height toward their posterior part, then expanding rather suddenly into the occipital crest, which is quite high, and projects somewhat backward; but very little is shown of the occiput in any of our specimens.

The jugal, as in *Onychodectes*, has the upper and lower branches well separated, the lower branch wider and more massive, the upper branch extending farther along the inferior and anterior border of the orbit, apparently joining the lachrymal and so excluding the maxilla from the orbital rim. The posterior branch of the jugal is much heavier than in *Onychodectes*, reaches almost to the glenoid facet, and the zygomatic process of the squa-

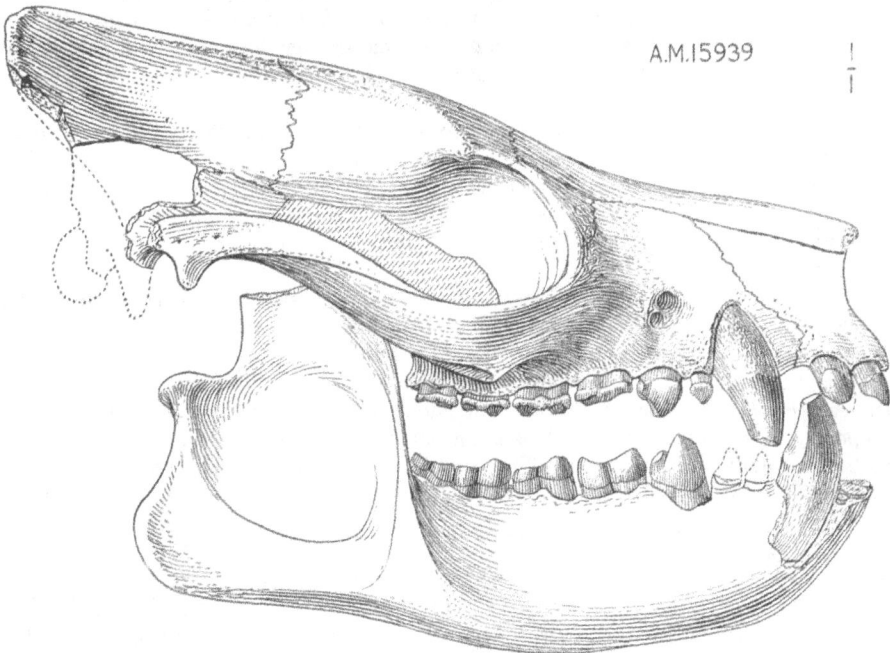

A.M.15939

FIG. 61. *Conoryctes comma*, skull and jaws, A. M. No. 15939, side view. Natural size.

mosal extends well forward, closely coössified with the jugal and making with it a zygomatic arch decidedly stronger and distinctly wider, and shorter than in *Onychodectes*.

The glenoid region shows a flat, moderately wide glenoid facet, with a low postglenoid process and quite small postglenoid foramen. Internal to the glenoid facet rises a low crest or process, with the suture between alisphenoid and squamosal running obliquely across it, so that the anterior part, with the foramen ovale on its inner side, is in the alisphenoid, the posterior part divided between squamosal and alisphenoid, the posterior lacerate foramen on the inner side of this part of the crest lying postero-externally to the foramen ovale and quite close to it. Postero-externally to this lies the suture between the squamosal and petrosal bones. This crest between alisphenoid and squamosal is seen in *Onychodectes* and in *Deltatherium*, but in the latter genus at least the foramen ovale and foramen lacerum posterius are considerably farther apart.   In armadillos and such modern Insectivora as have been compared the conditions differ too much for profitable comparison.

The lower jaw is of moderate depth and massiveness, about as deep anteriorly as posteriorly, the coronoid process arising on the outer side of the jaw just opposite the back of m_3, of moderate height and width, but not complete in any of our specimens.   The condyle is transverse, wide but not strongly convex, and less heavily braced than in most Creodonta, only slightly above the level of the tooth row.   The angle is flat, rather wide, projecting backward as a broad plate somewhat behind the vertical line from the back of the condyle.   The suture is close and deep, not very long antero-posteriorly.

A large part of the humerus is preserved in No. 3396; a part of it was figured by Cope (Tertiary Vertebrata, pl. xxiiie).   It is of the same size as the humerus of *Onychodectes* and very much like it in all respects except that the broadening of the deltoid crest extends to its lower end.

The distal half of the radius (figured by Cope as the tibia) shows the same notable expansion toward the distal end as in *Onychodectes*, with a lower anterior crest; the distal facet is simple, strongly oblique, shallower and less rounded than in *Psittacotherium*.

Part of the shaft and distal half of the tibia is preserved and appears to compare in size and proportions with that of *Onychodectes*, but there is nothing very characteristic in the parts shown.

It would appear therefore that *Conoryctes* had a skeleton no larger than that of *Onychodectes* and quite similar in proportions of the limb bones, although the skull is nearly twice as large, and more specialized in various particulars.

From the above data it is evident that *Conoryctes* is largely intermediate anatomically between *Onychodectes* and *Wortmania*, but later geologically than either genus, and con-temporary with *Psittacotherium*.   It is not at all clear that it is a direct or closely approximate descendant of *Onychodectes*.   The specialization of the incisors appears to differ in the three Paleocene groups, as follows:

PUERCO	TORREJON	I^1	I^2	I^3	I_1	I_2	I_3
Onychodectes		vestigial	vestigial	sl. enlarged	vestigial	sl. enlarged	vestigial
	Conoryctes	absent	mod. enlarged	mod. enlarged	absent	sl. enlarged	vestigial
Wortmania		absent	absent	much enlarged	absent	enlarged, specialized	absent
	Psittacotherium	absent	absent	much enlarged	absent	enlarged, specialized	absent

The premolars are unreduced in *Onychodectes*, one upper premolar is lost in *Conoryctes* and in *Wortmania*, two in *Psittacotherium*.   The lower premolars are unreduced in any of the genera; their heels are posterior in *Onychodectes*, oblique in *Conoryctes*, internal in *Wortmania* and *Psittacotherium*.   The pattern shows many points of relationship both in premolars and molars, and so also with the skull characters.   The long and strongly built-up sagittal ridge on the cranium narrows to a true sagittal crest posteriorly; the character of the occiput is not well shown in this or in any of the taeniodont genera.

PSITTACOTHERIINAE, new subfamily

Psittacotherium Cope, 1882

Cope, E. D., 1882, Amer. Nat., XVI, Jan. 25, p. 156

Type: Psittocotherium multifragum Cope, from the Middle Paleocene, Torrejon formation, of New Mexico.

Synonym: Hemiganus Cope, 1882, Amer. Nat., XVI, Sept. 28, p. 831; type: *Hemiganus vultuosus.*

Author's Diagnosis: Owing to the absence of the superior dental series it is not possible to be sure which tooth is the canine. The inferior dental formula may be therefore written, I.$_2$; C.$_1$; Pm.$_3$; M.$_3$; or I.$_3$; C.$_0$; Pm.$_3$; M.$_3$; or I.$_3$; C.$_1$; Pm.$_2$; M.$_3$. The first and second incisors are large and rodent-like, growing from persistent pulps; the second are the larger. The third, or canines, are small and probably not gliriform. There is no diastema. The first premolar (or canine) has a compressed crown with two cusps placed transversely to the jaw axis, and has a complete enamel sheath, and probably two roots. The succeeding tooth is also transverse, and is two-rooted, judging from the alveolus. The first and second true molars are rooted, and the crown consists of two transverse separated crests, each partially divided into two tubercles. On wearing, the grinding surface of each assumes the form of a letter B with the convexities anterior. The last inferior molar is injured. The rami are short, and the symphysis deep and recurved.

The above diagnosis is copied without change in Paleontogical Bulletin No. 34.

Type Diagnosis of Hemiganus: Probably a taeniodont, and allied to *Calamodon*, but the absence of the canine teeth renders the determination incomplete. The incisors, while of the form of those of *Calamodon*, had a limited period of growth, and the root displays a contracted base. The enamel also extends but a short distance on the anterior face of the tooth. The probable first inferior incisors are quite small, but are generally like the second or large ones. The superior molars have but a single conic root, but in some of them a fissure of the external side marks the usual place of division. The crowns are narrow and transverse to the axis of the jaw.

Psittacotherium multifragum Cope

Psittacotherium multifragum Cope, 1882, Amer. Nat., XVI, Jan. 25, p. 157; Proc. Amer. Phil. Soc., XX, p. 191.
Hemiganus vultuosus Cope, 1882, Amer. Nat., XVI, p. 831.

Type: A. M. Cope Coll. No. 3413, lower jaws.

Horizon and Locality: Paleocene of San Juan basin, undoubtedly from the Torrejon formation, Middle Paleocene.

Author's Description: The base of the coronoid process is opposite the junction of the second and third true molars. The ramus is deep and moderately stout. The enamel of the first incisor does not extend below the alveolar border, at the internal and external faces, and does not reach it at the sides. It has a few wrinkles on the anterior face. The anterior enamel face of the second incisor is thrown into shallow longitudinal grooves with more or less numerous irregularities from the low dividing ridges. There is a deeper groove on each side of the tooth, and there are about a dozen ridges between these on the anterior face. Both cusps of the first premolar are conic, and the external is the larger. The second true molar is a little smaller than the first. The enamel of the premolars and molars is smooth, and there are no cingula. [Measurements follow.]

Type Description of " *Hemiganus vultuosus* ": Large incisors strongly curved, robust, wearing with a strong posterior shoulder. Shaft with the dentine finely and sharply ridged. Inferior apex compressed; front regularly rounded. Enamel ? ridged or smooth. Superior molar with narrowed transverse crowns, and roots covered with a thin layer of cementum. There are one, perhaps two external cusps, but the crowns are all much worn. One crown, perhaps inferior, is subround with a notch, as in *Calamodon* sp. Enamel short, with equal base, smooth. [Measurements follow.]

Revised description, Wortman, 1897 [158]:

In the skull [figured], the facial portion is seen to be short and deep, the sagittal crest low and inconspicuous, and there is but a faint indication of postorbital processes upon the frontals. The anterior root of the zygoma is situated well forward; it has a considerable vertical depth and projects outwards, downwards, and backwards. In front of and below the zygomatic root is a shallow fossa, at the upper extremity of which is the anterior opening of the infraorbital canal, which is double. . . . the main canal is below and the smaller one above. Both, however, are placed unusually high upon the face. No evidence of a distinct lachrymal foramen is to be seen. . . .

The dental formula is not completely known, the discrepancy being in the number of superior molars and premolars. It is certain, however, that there were five, and possibly there were six teeth behind the canine. In the lower jaw there are nine teeth upon either side. The dental formula can therefore be written: $I.\frac{1}{1}$, $C.\frac{1}{1}$, $Pm.\frac{?^{(?)}}{4}$, $M.\frac{?^{(?)}}{3}$. . . . The upper incisors are strong, curved teeth, deeply implanted in the premaxillary bones, with the anterior face covered with a thick layer of enamel, and the posterior portion having the dentine exposed. It results from this arrangement of the dentinal tissues that the tooth wears to a chisel point, as in the typical rodent incisor. They did not, however, grow from persistent pulps, although some specimens show that the dentinal pulps were active throughout a large part of the animal's life.

The incisors of the lower jaw [figured] are relatively smaller. In the younger specimens the entire crown is covered with enamel, but owing to its thinness and small extent upon the posterior surface, it is soon worn away, leaving an external enamel covering only; the tooth then wears into the typical chisel point. They likewise were not of persistent growth.

The canines are large, powerful, curved teeth, being deeply imbedded in the maxillary bones; the anterior surface is covered with a thick layer of enamel, but the posterior surface is devoid of any enamel. The canines did not grow from persistent pulps, but the cavities remained open and the pulps were active long after the animal was adult.

It frequently happens that the crowns of the molars and premolars are so much worn that the average specimen does not give one any clue to their pattern, but there are, fortunately, a few specimens in the collection from which a tolerable idea of the crown pattern can be had. This applies, however, only to the lower teeth, that of the upper teeth being totally unknown.

In one specimen (No. 3413) [159] of a lower jaw [figured] the second premolar and the first and second molars are preserved in place, and the crowns are sufficiently unworn to permit of a determination of their structure. The second premolar displays two conical cusps placed at right angles to the long axis of the jaw; of these the external or labial cusp is the larger, and has its apex bent slightly inward giving to it a distinct hook-shaped appearance; the internal or lingual cusp is smaller and stands vertical. There are in the collection a number of loose teeth of this pattern, and it is more than probable that they represent both superior and inferior premolars. It is, moreover, highly improbable that any of the premolars, with the possible exception of the fourth, reached a further stage of complication than that just described, *viz.*: the bicuspid stage.

The true molars of the lower jaw present a crown pattern identical with that of *Hemiganus* already described. In the younger specimens, in both molars and premolars, there is more or less evidence of a division of the root into fangs, but as the animal approached maturity the crowns of the teeth were rapidly worn away, and the fangs of the root completely disappeared. The appearance of the molars and premolars at this stage may perhaps best be likened to a row of pegs deeply planted in the jaw. There was a strong tendency to the formation of a prismatic or hypsodont dentition. In the upper teeth, however, evidence of a more or less divided fang persisted even in the oldest individuals.

The description of the skeleton material which follows (*op. cit.*, pp. 76–88) is too extended for citation. He compares it with "*Hemiganus*" (*Wortmania*) and shows that the two are very nearly related. On the other hand, he shows a near relationship to *Calamodon*, the chief differences being that in *Calamodon* the canines (are said to) grow from persistent pulps, and the premolars are enlarged and show a limitation of the enamel

[158] Wortman, J. L., 1897, Bull. Amer. Mus. Nat. Hist., IX, p. 72.
[159] This is the type specimen.—THE EDITORS.

to anterior and posterior vertical bands as in the canines of the older form; also the cheek teeth are hypsodont.

Revised Description, Matthew: *Psittacotherium* is chiefly known from Wortman's detailed description and figures of the specimens studied by him in 1896–7. Some additional parts have been pieced on to the skull and fore foot which he figured, and a number of other specimens have been referred to the genus. These serve to show that the cranium carried a longer and higher sagittal crest than his restoration allows; that the premaxilla was extended backwards to a point directly above the infra-orbital foramen; they confirm the course assigned by him (on somewhat doubtful evidence) to the maxillo-premaxillary suture; they show the presence of a large and characteristic centrale in the carpus; and they show in part the characters of the hind limb and foot, which were unknown.

It is not altogether certain that Wortman was right in regarding all the specimens of this genus as belonging to a single species. Cope described three distinguished chiefly by size, but not comparable because the smaller species are based on young individuals with unworn teeth. The adult specimens conform as far as comparisons have been made, but it may prove that the differences cited by Cope in describing *P. aspasiae* and *megalodus* are not wholly matters of age.

Psittacotherium is one of the largest animals of the Paleocene, exceeding all but *Pantolambda cavirictus* and *Triisodon conidens*. *Eoconodon* has a larger skull, but the skeleton is smaller.

The principal specimens are:

No. 754, large part of skull and lower jaw.

No. 2453, lower jaw, upper teeth, ulna, radius, fore foot.

No. 16560, fragmentary hind limb and part of pes.

No. 16661, parts of upper and lower jaws.

No. 16662, lower jaw.

[No. 2456 was erroneously listed by Wortman in 1897, Bull. Amer. Mus. Nat. Hist., IX, p. 72.]

SKULL. The additional fragments of the skull in No. 754 show that the premaxillary was extended backward to a point just above the infra-orbital foramen and that the cranial region is more extended backward than Wortman supposed. These features reduce the resemblance to the ground-sloths and increase the resemblance to *Conoryctes* and *Onychodectes*.

The nasals are exceptionally long, broad posteriorly as in other taeniodonts, ending posteriorly in a wedge between the frontals, which extends back almost to the beginning of the sagittal crest. The frontals in turn extend backward apparently to the middle of the cranial region, but the fronto-parietal suture is not clearly shown; anteriorly they expand, embracing the nasals and extending laterally almost to the border of the orbits. The postorbital ridges are low and situated on the frontals close to and parallel with the fronto-nasal suture. Anteriorly the frontal terminates in a wedge between nasals and maxillary at a point about opposite the middle of the orbit, but there is no postorbital process and the frontal does not actually touch the orbital rim. It is possible that a small vestigial postorbital bone is present, but apparently it is the maxillary that lies external to the frontal at this point and makes the orbital rim. The lachrymal appears to be very much reduced or vestigial, not expanded on the face, and the tubercle and foramen lie within the orbital rim. The maxillary is a large and massive bone enclosing the enlarged canine and the series of cheek teeth behind it; superiorly it lies against the premaxillary and nasal and posteriorly against the frontal; it forms almost the whole of the inner and anterior and part of the inferior border of the orbit. The superior branch of the jugal is thick and short, extending up along the inferior rim of the orbit; the inferior branch is likewise thick and very short; it is sutured to the postero-external end of the maxillary a little above the tooth-row. There is no evidence as to the length of the posterior branch or other characters of the zygomatic arch. Wortman has restored it as comparatively broad and heavy and like that of *Conoryctes* except for greater massiveness. This is very probably correct enough.

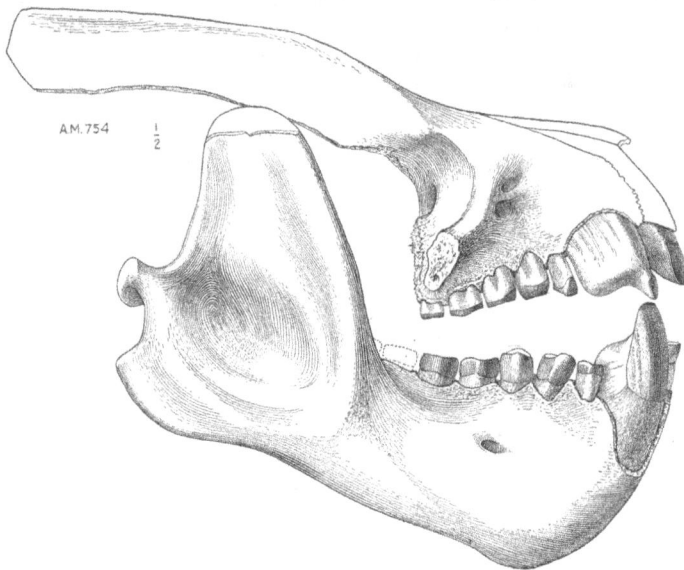

Fig. 62. *Psittacotherium multifragum*, part of skull and lower jaw, A. M. No. 754, side view. One-half natural size

The premaxillary is imperfectly known. The superior branch extends posteriorly in a long and rather narrow wedge between maxilla and nasal to a point above the fourth premolar, somewhat in advance of the infra-orbital foramen. The anterior branch, containing certainly one large incisor, perhaps two as in *Conoryctes*, is massive, but its front border is not shown in any specimen.

The teeth are preserved in a number of specimens collected in 1913, in addition to those described by Wortman. No. 16661 shows the teeth only moderately worn, so that something of their cusp construction can be seen. In No. 16662 the sockets or roots of i^3-m^2 are in place in the upper jaw. Other specimens show a number of loose teeth. No. 754 has been more completely pieced together in connection with the present restudy, with the result of adding a number of teeth more or less complete, as well as fragments of the skull and lower jaws. It is certain that there was but one lower incisor, and that the number of premolars was two in the upper, four in the lower jaw. There were three molars in each jaw. The dental formula is therefore $i_1^{1?}$, c_1^1, p_4^2, m_3^3, as against i_2^2, c_1^1, p_4^3, m_3^3 in *Conoryctes*. In *Wortmania* the number of incisors is the same, but there was one more upper premolar. In both genera there were four premolars in the lower jaw.

The canines are greatly enlarged, sub-gliriform, the enamel being limited to the front of the tooth, and the crown and root greatly elongated. They are not rootless as in *Calamodon*, but have lost the creodont-like form of *Wortmania* and *Conoryctes*. The enamel wears to a sharp edge, but on account of the transverse curvature it is spatulate rather than chisel-shaped. The upper and lower canines are very much alike, but the lower is somewhat less curved and more compressed, with longer crown and root, the anterior face more regularly and convexly curved, lacking an obscure longitudinal groove that lies on the outer side of the anterior enamelled face of the upper canine. It wears also to a more acute edge in front than does the upper canine, which is sometimes blunt and stubby through wear.

The upper incisor is much like the canines, but smaller and more compressed, with long curved root and crown, the enamel limited to the anterior face. The lower incisor has a shorter root, not curved, and the crown in all our specimens is worn down so that the enamel is lost.

The cheek teeth, five above and seven below, are partly preserved in No. 754, but heavily worn and fragmentary, most of them out of the sockets so that they cannot all be identified. Parts of all the upper teeth of the right side are present and have been fitted in the sockets, but only p^3 and m^3 are at all complete. The premolars have a large but short outer root and a small inner root, closely connate. The molars, also short-rooted, have a large inner and two small outer roots, likewise connate but not so much as in the premolars. The crowns are so heavily worn as to leave no trace of the cusp composition and little of the enamel is left at the sides except the heavy inner crescent on m^3 and heavy antero-external crescent on p^3. A narrow strip is also left on the external side of m^1; beyond this the teeth are insufficiently preserved to show their form.

The upper teeth are preserved better in other specimens, particularly in No. 15938, which shows p^4–m^3, p_4 of both sides, and an upper incisor. These teeth are only moderately worn, and although loose they are identified with reasonable certainty. Other isolated unworn upper teeth show the pattern more exactly. The lower teeth are mostly in position in No. 2453, and i_1, p_1 and p_4 in No. 754. Unworn lower teeth are present in the types of *P. multifragum* and *aspasiae* and enabled Cope to identify them in part and Wortman more completely. The upper cheek teeth have not hitherto been identified in *Psittacotherium* or any of the later genera.

P^3 is the smallest tooth of the series. It is preserved although not in place in No. 754, and the alveolus in No. 16662 shows that it is crowded in between p^4 and the *outer* angle of the canine. It is one-rooted, with simple oval crown, traces of a small connate inner root, the pattern of the crown not preserved.

P^4 has two subequal connate roots, the outer one somewhat larger. The crown consists of large outer and inner cusps, conical and blunt-tipped, the outer somewhat higher and much wider, a little flattened on the external face and pitched somewhat backward. The inner cusp is opposite the posterior half of the outer cusp, is round-conical and well separated. A very rudimentary postero-external cusp is obscurely indicated behind the outer cusp. This tooth is curiously like p^3 of *Wortmania*.

M¹ has two small outer roots and a large inner root, all closely connate and transversely compressed. The crown consists of two small connate external cusps, the metacone lower and smaller than the paracone, and a large protocone which, with the conules, forms a wide massive crescent. The cusps are massive and the crown considerably inflated.

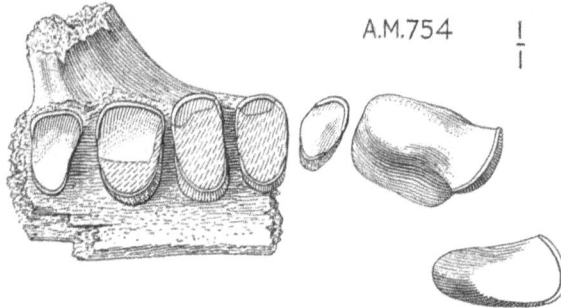

A.M.754

FIG. 63. *Psittacotherium multifragum*, upper jaw, A. M. No. 754, occlusal view with well-worn teeth. Natural size.

M² is slightly larger than m¹, the metacone more closely connate and reduced, the root beneath it smaller, more closely connate and postero-internal in position.

M³ is smaller than the other molars, the metacone and the root beneath it have practically disappeared, and only the outer cone and inner crescent are left with a crenulated ridge swinging around the posterior border between paracone and metaconule representing

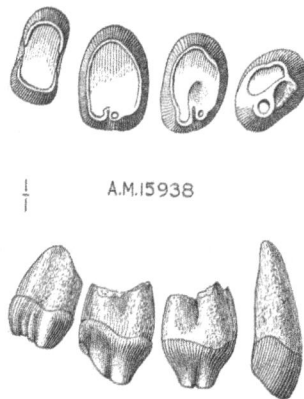

A.M.15938

FIG. 64. *Psittacotherium multifragum*, upper teeth, moderately worn, A. M. No. 15938, crown and external views. Natural size.

the remains of the metacone. The entire construction of m³ is much like that of *Conoryctes*, but the preceding molars differ in the partial union of the outer cusps into a higher cone, the greater inflation of the crown beneath the cusps and more oval, less trigonal form of the tooth. The premolars are reduced to two instead of three as in *Conoryctes*. The upper molars of *Wortmania* are imperfectly known, but there were probably three premolars as in *Conoryctes*. The upper teeth of *Calamodon*, as far as known, appear to have a similar cusp pattern except that the metaconids are more reduced, but the form is decidedly different, the crown and root making a straight-sided cylinder in place of the much shorter, convexly-inflated crown of *Psittacotherium* cheek teeth. The roots are more closely connate, the enamel of the teeth shows a marked tendency to be limited to anterior and posterior bands, while that of *Psittacotherium* is so limited only on the incisors and canines. There is also a considerable cement investment on the base of the cheek teeth of *Calamodon*, of which no trace is seen in *Psittacotherium*. *Stylinodon* is similar to *Calamodon* except for its longer and more completely rootless hypsodont teeth. *Ectoganus* is probably a synonym

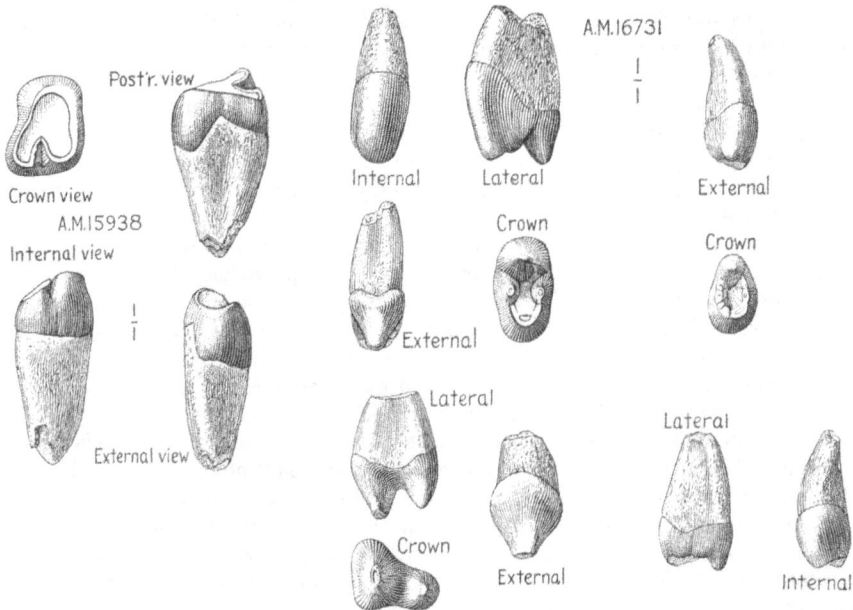

Fig. 65. *Psittacotherium multifragum*, views of unworn or little-worn upper teeth, A. M. No. 16731; fourth lower premolar, A. M. No. 15938. Natural size.

of *Calamodon*, over which it has priority (*see* page 235), but is too incompletely known for any profitable comparisons.

The lower cheek teeth have been described and figured by Cope and Wortman; both worn and unworn stages are known. There are four premolars. All of them are single-rooted, the first is smaller and may have also a simple crown, but the second and third have two large rounded cusps, a larger and higher external, a smaller and lower internal one. The fourth premolar is larger than the preceding or following teeth, and sub-molariform, with large subequal external and internal cusps (pr^d, me^d) and a transverse talonid crest on the postero-internal border connected by a rising ridge with the summit of the antero-external cusp (protoconid). The molars are somewhat smaller and shorter than p_4, with distinct anterior and posterior roots and the crown composed of two equal conical trigonid cusps (protoconid and metaconid), the talonid of nearly the same size as trigonid but considerably lower and separated from it by a distinct transverse valley. The talonid is also bicuspid, but the outer cusp (hypoconid) is larger and more central in position. Low transverse commissures connect the anterior and posterior pairs of cusps, which are low round-conical in form. The talonid of m_3 is a little narrower than the trigonid, otherwise the cusp composition of the three molars is much the same.

As compared with *Conoryctes* the lower cheek teeth are more crowded and shortened, but p_1 is of good size as in *Wortmania*, whereas it is either vestigial or absent in *Conoryctes*. The second cusp on p_{2-3} is internal instead of postero-internal, and p_4 has a bicuspid trigonid of more molariform type but a much smaller trigonid. The molars are more nearly equal in length and width, and their cusp construction simplified; the paraconid has wholly disappeared while the entoconid is a simple conical cusp instead of the serrate curved crest of *Conoryctes*.

In *Calamodon* the lower cheek teeth are much more hypsodont and cylindrical in form, the crowns of the molars sub-quadrate with talonid and trigonid of equal height, the two anterior premolars, p_{1-2}, are enlarged and sub-gliriform, p_1 being somewhat larger and more like the canine, and both having an oblique set, the inner narrower portion of the tooth postero-internal.

In *Wortmania* the number of lower premolars is four, as in *Psittacotherium*, and the inner cusp in internal but quite rudimentary on p_2, p_3 and p_4; p_4 has a heel of the same type as in *Psittacotherium* but of smaller size. The pattern of the molars appears to be similar.

SKELETON. The ulna, radius and manus have been described in considerable detail by Wortman; additional specimens show the characters of the humerus and of the hind limb and part of the pes. Wortman has also described the pelvis, lumbar and caudal vertebrae, but it is doubtful whether they really belong to *Psittacotherium*.

The skeleton bones are of somewhat smaller size than those of *Pantolambda cavirictus*. The humerus is known only from the proximal and distal ends preserved in No. 3962 along with a metacarpal bone. The head of the humerus faces more proximad than in *Onychodectes*, much as in *Pantolambda*, but the internal tuberosity is much larger and more prominent than in the latter, as large as the external although lower set. From the front of the external tuberosity a heavy crest runs down the anterior face of the bone, towards what was presumably a high deltoid crest, as in primitive placentals generally, but the shaft of the bone is not known. The distal end of the humerus is expanded broadly with prominent epicondyles and large entepicondylar foramen. The supracondylar fossae are very shallow and the inner trochlear crest very little developed, differing in these points from *Pantolambda* and other taligrades, creodonts and condylarths, and making a marked approach toward the peculiar construction seen in the Miocene ground-sloths and other Edentata. The humerus of *Onychodectes* conforms to most of these features, although much less massively proportioned, but the inner trochlear crest is fairly well developed. In *Stylinodon* the humerus appears from Marsh's figure to be much of the same peculiar type, but the distal articulation more oblique and the internal tuberosity absent (possibly broken off; Marsh does not allude to the fact that the absence of this tuberosity would be in strange contrast to the edentates, with which he compares *Stylinodon*).

The ulna and radius are short and massive, the ulna larger with broad flattened shaft, stout and wide and long olecranon (incomplete in the specimen described by Wortman), the distal end strongly oblique, with the cuneiform facet facing as much inward as distad, and nearly flat. In *Pantolambda* the distal facet is but little oblique, the proportions of the ulna not very different otherwise. The radius has a round-oval head with moderate convexity on the ulnar facet, a stout shaft convex anteriorly, heavy rounded distal end, differing from *Pantolambda* in the single undivided concave facet for scaphoid and lunar. In *Dissacus saurognathus* the radius and ulna are considerably longer, more slender, the scaphoid and lunar facets obscurely distinguished.

The manus has been described in considerable detail by Wortman, but additional piecing up of the specimen which he discovered in 1896 and described in 1897 has enabled us to extend and modify his description. Both lunars and centralia are present, and a small bone which may be the right trapezium, also a number of fragments of lateral metacarpals and phalanges which were not known to him.

The foot is probably pentadactyl, with the first digit much reduced, very short metapodials and phalanges, the unguals large, high, much compressed claws. The carpus has many primitive features, but differs primarily from that of the creodonts in the relatively large size of the lunar and centrale and reduction of the supposed trapezium.

The cuneiform and unciform are imperfectly preserved; the outer borders of both are missing. The cuneiform has the usual two proximal facets for ulna and pisiform, both shallow concave and at a slight angle with each other; distally it rests with a shallow concave facet on the unciform, which is of the same height but probably greater width, having the usual proximal-internal facet for the lunar and a much wider proximal-external facet for the cuneiform. The distal facet of the unciform faces externally to an unusual degree, covering the heads of the fourth and presumably the small fifth metacarpals, and the internal facets for the third metacarpal and magnum face partly distad, the magnum facet being quite narrow and in a plane with that for Mc. III.

The lunar is of large size, wide and deep, with the customary wedge form. The distal-external facets for cuneiform and unciform are flat, indistinctly separated, of about equal width but the unciform facet deeper. The magnum facet is rather narrow at the dorsal surface, but widens out palmad and is excavated beneath into a rounded pocket. The distal-internal face of the wedge is made by the centrale facet, equal in width to the combined facets of cuneiform and unciform, and somewhat saddle-shaped. The much shorter inner face of the lunar is occupied by the somewhat concave scaphoid facet. The proximal facet is rather strongly convex both ways, and occupies most of the distal facet of the radius.

The scaphoid is not preserved, but it must have been small and quite shallow, the distal surface taken up by a wide flat facet for the centrale.

The centrale is exceptionally large and wide, extending obliquely beneath scaphoid and lunar, narrowing internad, with a wide, slightly concave distal facet for the trapezoid. Its scaphoid facet has a somewhat rugose surface indicating a tendency to coössify the two bones.

The magnum is incompletely preserved, but shows the usual primitive characters of small size, especially dorsad, narrow lunar facet widening below and extending beneath the lunar into a rather large rounded knob-like keel. It abuts externally against the unciform and internally against the trapezoid or the overlap of Mc. II, while the centrale overlaps its proximal-internal surface.

The trapezoid is represented only by uncharacteristic fragments, but appears to have been exceptionally large and comparatively wide. The supposed trapezium on the other hand is greatly reduced, almost vestigial, but retains proximal, distal and a small external facet, presumably for the scaphoid, Mc. I, and trapezoid respectively.

The metacarpals were five in number, but only Mc. III and IV are complete. The distal end of Mc. V is preserved. Mc. II of the left side is represented only by fragments but is preserved complete in two other specimens. The fourth metacarpal is longer than the third, but considerably more slender; the second is shorter than the third and overlaps

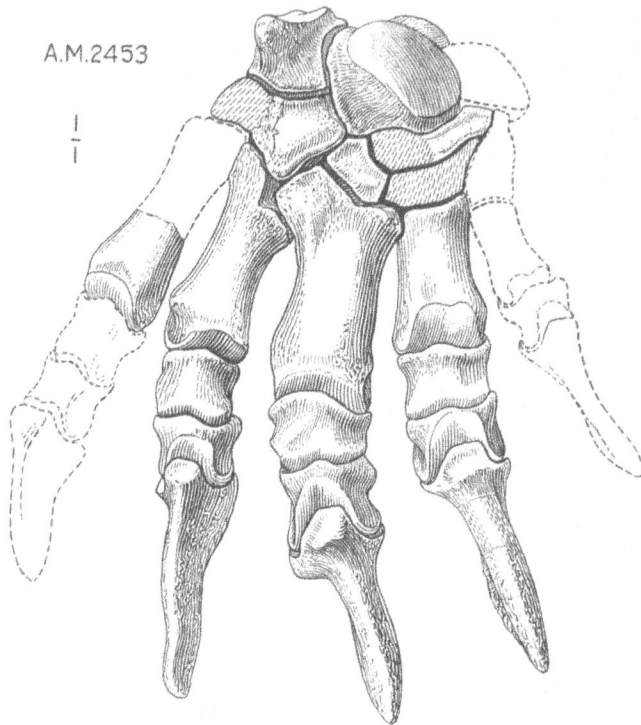

Fig. 66. *Psittacotherium multifragum*, fore foot, A. M. No. 2453, dorsal view. Natural size.

it proximally, increasing the reduction in length of the inner digits seen also in many Edentata.

Mc. I must have been vestigial, to judge from the facet on the supposed trapezium, but the identification of this bone is in doubt, and of the metacarpal no fragments have been recognized. Mc. II is about half as wide as its length. The proximal end has the usual facet for the trapezoid, and on the external side overlaps the head of Mc. III, but not to any marked extent. The distal facet is somewhat concave from side to side, and strongly convex dorso-palmad, with a short median keel on the palmar face only, and the side-to-side concavity flattened out at the margins dorsad into low marginal trochleae.

The third metacarpal is about a fifth longer than Mc. II, but no more robust. Proximally it has a heavy overlap over the head of Mc. IV; the distal end is similar to that of Mc. II. The fourth metacarpal is as long as the third, but much more slender and the head more strongly oblique, overlapped on the inner side by the head of Mc. III, and projecting proximally on the outer side below the unciform, so that its convex unciform facet faces as much internad as proximad. On the outer side is a concave facet for the head of Mc. V, the borders being broken off so that its full size is not seen. The distal end is a hinge facet with a slight median keel on the plantar face, but not concave from side. The distal end of the fifth metacarpal is preserved; the facet is strongly oblique, not much convex dorso-plantad, permitting of but limited movement in this direction, the internal side wider and flattened, the external side obscurely defined and the median keel distal.

The phalanges are extremely short, the first and second series of about equal length dorsally, but on the plantar side the second is longer than the first, extending beneath its distal end. They have two shallow trochlear grooves proximally and corresponding lateral keels distally. The second series has the distal keels very convex, permitting a wide range of movement in the unguals. The ungual phalanges are large, very high, and much compressed, with stout and prominent superior and sub-ungual processes, the upper border of the phalanx being much curved, the tips incomplete so that their exact length is not known. They are much like the claw phalanges of the Miocene ground-sloths and of *Megalonyx*, but higher and more compressed.

The fore limb and foot show many features of resemblance to one or another of the edentate groups. Some are probably due to similar adaptation, others probably indicate some degree of real affinity. In the humerus the strong development of the inner tuberosity and the position of the tuberosities relative to the head, the shallow supratrochlear fossae and lack of crest on the external margin of the trochlea suggest affinity. The radius and ulna have the characters of various other primitive mammals; where they differ from *Pantolambda* they conform to edentate characters, but this does not carry much weight as to special affinity. In the manus there are many specialized features, departing widely from the primitive insectivore characters and still more from those of the creodont-condylarth-taligrade group, and these are at least in part quite characteristic of some or all of the Edentata. The peculiar proportions of the lunar, the very broad centrale and trapezoid, the well-developed facet between lunar and cuneiform, the slight overlap of Mc. II on Mc. III and wide overhang of Mc. III on Mc. IV, obliquity of head of Mc. IV, the longer outer metacarpals and short, stout inner metacarpals, the extremely short phalanges and the character of their articulations, the gravigrade-like unguals, are not wholly explained as " convergent," but at least must be regarded as parallel adaptations in closely related stocks, and in some degree due to actual community of descent. In the edentates the scaphoid and centrale have been coössified, but the form of the conjoined bone shows very evident traces in some of them (especially some armadillos) of the peculiar size and proportions of the centrale in *Psittacotherium*. In others (ground-sloths, etc.) the centrale process of the scaphoid has been progressively reduced. The peculiar proportions of the metacarpals are characteristic of many edentates, lost in some armadillos, etc. The high compressed claws are characteristic of the more primitive ground-sloths, lost in the specialized genera of the Pleistocene but retained in *Megalonyx*.[160]

[160] Scott's dictum, that the Miocene ground-sloths are less like the taeniodonts than are their Pleis-

The characters of *Onychodectes* enable us to interpret the significance of those of *Psittacotherium*, and show that most of the peculiarities discussed above are specializations. *Onychodectes* of the Puerco has already acquired the exceptionally wide centrale, the heavy overlap of Mc. III on IV and of IV on V, with slight overlap of Mc. II on III; its metacarpals are normal in length, but the phalanges are somewhat shortened, the unguals small but somewhat compressed. In the contemporary *Wortmania*, a much more nearly direct ancestor of *Psittacotherium*, the specialization of the manus had gone much further; as far as one may judge from the two bones of the manus that are known, it was nearly as specialized as *Psittacotherium*.

Wortman has referred to *Psittacotherium* a pelvis with posterior dorsal, lumbar, and anterior caudal vertebrae associated. We have not succeeded in finding any of the fundamental differences which he states separate it from *Pantolambda*. The characters of the pelvis are not unlike those of *Pantolambda bathmodon;* the form of the ilium, the elongate ischium, separate and long pubis, are present in *Pantolambda*, and, as far as appears, it may belong to *P. cavirictus*. Wortman attributes only three lumbars to it, but there appears to be no evidence that all the lumbars were present. He states that " the pelvis is so characteristic that it requires but a passing glance on the part of an anatomist at all familiar with the osteology of the Edentata to demonstrate its marked similarity to the corresponding parts of these forms; this is most strikingly apparent in the direction of the gluteal surfaces of the ilia, the lengthened pubic bones, the character of the obturator foramen, and the deeply impressed, roughened surface for the attachment of the sacrum." But with all due respect for Wortman's great anatomical knowledge, I am unable to find any conclusive reasons for referring this specimen to *Psittacotherium* rather than *Pantolambda cavirictus*, nor any especial likeness to Edentata aside from the retention of primitive characters (e.g. the second and third of the characters mentioned) or those associated with the size and general similarity of proportions in this part of the skeleton (e.g. first and fourth characters). I regard this specimen as indeterminate.

The femur is incompletely preserved in No. 16560. It has a very short, wide, flattened shaft, a moderately rounded head set on a wide neck, stout and prominent external trochanter, almost as high as the head and nearly as bulky, broad and shallow digital fossa, small internally placed and low-set lesser trochanter and a mere trace of the third trochanter set high up on the shaft. The bone is badly crushed and of its distal half only fragments are preserved. The condyles are somewhat like those of the larger Miocene ground-sloths, but a little less convex from side to side, the epicondylar processes less prominent laterally, and the depth of the distal end appears to have been greater; but of the distance between the condyles and the character of the patellar trochlea nothing is known. The bone has the same general proportions as in the larger taligrades, but differs in the vestigial third trochanter, well developed in taligrades and set farther down on the shaft, in the less transverse convexity of the condyles, probably in other characters. Comparison with *Wortmania* is difficult on account of the distortion by crushing to which most of the apparent differences must be ascribed. The shaft appears to be straighter in *Psittacotherium*, the third tro-

tocene descendants, is certainly not true of this or of various other peculiarities of the manus, and only partly true of the skull and dentition. In fact, most of the comparisons that we have been able to make show significant points of resemblance in the Miocene ground-sloths that have been lost in the more specialized genera.—W.D.M.

chanter has the same position but is quite vestigial, the digital fossa has perhaps less depth, the bone as a whole is more massive, as well as of considerably larger size.

The tibia is best preserved in No. 15938. It is short, massive, heavy, and comparatively straight in the shaft, much flattened obliquely toward the proximal end, which is missing. The astragalar trochlea is strongly concave antero-posteriorly and faces distally with but little obliquity, showing that the body of the astragalus lay directly beneath the tibia and is not to any extent wedged in between tibia and fibula; the keels are not deeply excavated, the internal malleolus has its facet at right angles to the principal astragalar facet. These are the characteristic relations of primitive Insectivora and related groups (rodents, edentates, taeniodonts). Another feature of interest is that, although the upper part of the tibial shaft is much flattened, the cnemial crest is vestigial, reduced to a short obscure ridge about half-way down the shaft.

As compared with *Wortmania* the tibia is quite evidently related. It is a little longer and much heavier in the shaft. The astragalar trochlea is very similar in all details except for somewhat more grooved inner trochlea. The tibia of the Taligrada as shown in *Ectoconus*, *Periptychus* and *Pantolambda* has about the same relative length, but is widely different in the much more flattened middle portion of the shaft, with the cnemial crest high and prominent, extending over half-way down; the distal end has a very oblique trochlea, almost ungrooved, and not concave but flat antero-posteriorly, the astragalus being wedged in between tibia and fibula, instead of lying under the tibia as it does in the stylinodonts. In *Onychodectes* the tibia is quite slender, but the distal end has much more the relations of *Wortmania* and *Psittacotherium*.

Part of the left pes is preserved in No. 16560, consisting of the navicular, cuneiforms and the two inner digits. The navicular is deeply excavated proximally, showing that the astragalus had a very convex head; it is extended in a stout, hook-like process beneath the astragalar head. The cuneiforms have the usual relations, the second being as wide as the third but only three-fifths as long, while the first (entocuneiform) is large, rather deep and narrow, overlapping the inner side of the navicular and extending down beyond the mesocuneiform with a convex facet for the second metatarsal, and a deep but narrow oval facet for Mt. I, somewhat concave dorso-plantar and convex transversely.

The second metatarsal is short and stout but somewhat longer than the third metatarsal and nearly as robust. Proximally it has the usual narrow and deep facet for the mesocuneiform, somewhat concave transversely, with very little dorso-plantar curvature, and on each side of this facet are the smaller facets for the first and third cuneiforms. Distally it has an oblique hinge-facet with median keel on the plantar side. The first metatarsal is smaller, about three-fourths as long, and more unsymmetrical, the distal end very oblique, with a similar hinge-joint.

The first and second phalanges are very short, somewhat wider than their length, notably flatter than those of the fore foot, and their facets are not so much keeled. The ungual phalanx is large, moderately curved, but not high and compressed like the unguals of the fore foot, the width about equal to the height, superior surface round-convex, inferior surface flat transversely. They are not unlike the unguals of *Tillotherium* but shorter, stouter, the under side more plane, the upper side more regularly rounded without the distal flattening characteristic of *Tillotherium* unguals. There is no indication of a hood.

Save in general proportions these foot bones have little resemblance to those of *Panto-*

lambda. They have much more of the characters of the fore foot bones of *Psittacotherium*, but are less specialized. The uncompressed unguals differ notably. There is much more suggestion of the armadillos and the Miocene glyptodonts, but not enough to prove more than a common descent from proto-insectivoran stock.

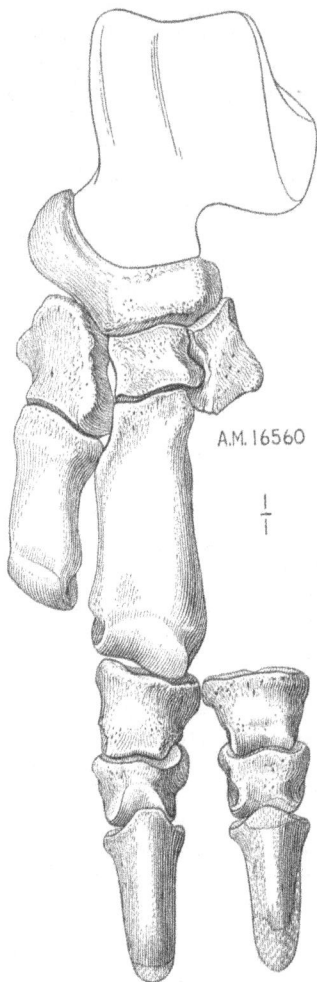

Fig. 67. *Psittacotherium multifragum*, part of hind foot, A. M. No. 16560. Natural size.

AFFINITIES OF *Psittacotherium*

The foregoing review of the genus *Psittacotherium* as now known enables us to evaluate its characters more satisfactorily. It is evident that its relationships lie somewhere between the extreme views expressed by Wortman on the one hand and by Winge and

Scott on the other. The numerous points of approach to the Edentata, or to some of them, seen in characters of the teeth, skull and skeleton, are undeniable. They are not confined to general or formal resemblances, but extend to innumerable minor details in teeth, skull, limbs and feet. Yet there are great and fundamental differences that exclude them apparently from direct ancestry and indicate that the resemblances must be interpreted as parallel but distinct specializations from a common primitive stock among the proto-Insectivora of the late Cretaceous.

The characters of the cheek teeth may well suggest relationship, if not ancestry, to sloths, but would exclude ancestry to armadillos, and the same is true of some characters of the fore foot. On the other hand, the characters of the hind limb and foot, while suggesting relationship to the armadillo group, have no particular suggestion of gravigrade affinity, and the enlargement and specialization of $i_{\frac{3}{2}}$, $c_{\frac{1}{1}}$, is in contrast with Xenarthra, where, except in *Peltophilus*, the incisors are reduced or absent, and the enlarged teeth, if any, are premolars. It should be noted, however, that in Stylinodontinae the lower premolars become sub-gliriform, and it is conceivable that they might replace the anterior gliriform teeth in the lower jaw, thus bringing about a pair of enlarged teeth consisting of c/p_1 or p_2, which would be a *possible* interpretation of the teeth in Megalonychidae.

Affinities to *Wortmania* are evidently rather close, the most striking difference being in the canines, which are more nearly of the normal form in the Puerco genus. The dental formula is nearly the same, but in *Wortmania* there are three upper premolars, the posterior lower premolars are more oblique and p_4 is not enlarged; the cranial region is less elongate than in *Psittacotherium*, the jaw has less depth anteriorly, the coronoid process and angular process are relatively smaller, the limb bones less massive in the shafts and what is known of the fore foot is very similar to *Psittacotherium*.

Affinities to *Calamodon* are not so close as to *Wortmania*, the Eocene genus being much more specialized in the hypsodont columnar cheek teeth, rootless anterior teeth, enlarged and sub-gliriform p_{1-2}, along with the greater specialization in various details. The pattern of the cheek teeth is much the same, but the upper molars have only a single outer cusp and their roots are more connate.

Conoryctes is a survival of an earlier stage in the phyletic specialization already showing nearly all the characteristics of the group, but less advanced than *Wortmania*. Its skeleton unfortunately is almost unknown; what there is of it is of primitive construction approaching *Onychodectes* rather than *Psittacotherium*.

Psittacotherium aspasiae Cope

Psittacotherium aspasiae Cope, 1882, Paleont. Bull. No. 34, Feb. 20, and Proc. Amer. Phil. Soc., XX, p. 192.

Type: A. M. Cope Coll. No. 3416, lower jaw fragments with left m_3 and alveoli.

Horizon and Locality: Paleocene of San Juan basin, New Mexico. Presumably Torrejon formation.

Author's Description: The most obvious difference from the *P. multifragum* is its inferior size, which can be readily perceived from the measurements given. The posterior crest of the molars appears to have less transverse extent than in the larger species. This crest in the last inferior molar has a curved crenate edge, with a small conic tubercle at its external extremity. The anterior crest consists of two conic tubercles, whose apices converge, but whose bases are closely appressed, and only distinguished by a superficial fissure. The valley between the crests is uninterrupted. The preceding molar is larger, and its posterior crest is like that of the lost molar. The apex of the anterior crest is broken off.

The ramus deepens rapidly forwards, and contains the enormous alveolies for the incisors. The coronoid process leaves the alveolar border at the line separating the last two molars, or, in the smaller specimen, a little anterior to this point, and is quite prominent. The masseteric fossa is well marked, but shallows gradually anteriorly and inferiorly.

Wortmania Hay, 1899

Hay, O. P., 1899, Science (2), IX, p. 593

Type: Hemiganus otariidens Cope, from the Lower Paleocene, Puerco formation, of New Mexico.

Synonym: Hemiganus Wortman, 1897, Bull. Amer. Mus. Nat. Hist., IX, p. 67. (Not *Hemiganus* Cope.)

Diagnosis: Cope in 1885 [161] gave the following amended diagnosis of *Hemiganus*, based upon *H. otariidens:*

One of the most remarkable of the Eocene Mammalia yet discovered. The claws are large and compressed like those of a prehensile-footed carnivore. The astragalotibial articulation is nearly flat. The femur is very robust, and has a low third trochanter, as in Bunotheria generally. The vertebrae of the neck are short and wide. The jaws have a very large and wide coronoid process, as in Calamodon, and the horizontal rami are very robust. The molar teeth of the lower jaw have but one root. Only one true molar (the first) is preserved, and it has the crown worn. Its outline is sub-round, with a notch on the internal side. There are probably four premolars, and their crowns are short, obtuse cones, with a low heel-like expansion at the inner side of the posterior base. They resemble very nearly the teeth of some of the eared seals. There is a robust canine tooth in the upper jaw, which is not separated from the premolars by a diastema. There is at least one superior incisor, but the exact number is unknown. There is a large tooth on each side of the symphysis of the lower jaw, but in the specimens it is not in place. It has enamel on the anterior face only, and its apex is worn transversely. The wear descending passes to one side of the middle line. It evidently has a median position, and may therefore be an incisor. Its form reminds one of that of the second inferior incisor of Calamodon, but the enamel-face is much shorter.

Should the large inferior teeth be canines, the mandibular dentition will greatly resemble that of the seals, as does that of the maxillary bone. The absence of postorbital angles resembles the condition in the Phocidae. The wide vertical coronoid process and the flat vertical angle are as in Calamodon. The sagittal crest is elevated, and the brain-case very small.

Hemiganus may for the present be referred to the Creodonta where it will stand quite alone, and next to the Taeniodonta.

In 1888, Cope [162] gave a detailed description with figures of the species *"Hemiganus" otariidens*. He compares the genus with *Calamodon* and places it in the Creodonta next to the Taeniodonta. Wortman in 1897 [163] removes the genus to the "Ganodonta," excludes the type species from it as being a synonym of *Psittacotherium*, and gives a revised description of *"H." otariidens*, comparing it with the edentates in various points and showing very clearly its near relationship to *Psittacotherium*.

Hay in 1899 [164] proposed the generic name *Wortmania* for *"Hemiganus" otariidens*, pointing out that Wortman's transfer of the type species of *Hemiganus* to *Psittacotherium* left *H. otariidens* without a name.

[161] Cope, E. D., 1885, Amer. Nat., XIX, p. 492.

[162] Cope, E. D., 1888, Trans. Amer. Phil. Soc., XVI, N.S., p. 311, pls. IV, V, figs. 1–7.

[163] Wortman, J. L., 1897, Bull. Amer. Mus. Nat. Hist., IX, p. 67.

[164] Hay, O. P., 1899, Science (2), IX, p. 593.

Revised Diagnosis: Dentition $i_1^1 c_1^1 p_4^3 m_3^3$?.   Proportions and pattern of teeth much as in *Psittacotherium*, but incisors and canines less enlarged, not so long-rooted or compressed, retaining more of the creodont character.   Upper premolars less reduced, p_4 less molariform; pattern of upper molars unknown, lower molars like those of *Psittacotherium*, but with shorter crowns.   Jaw deep and short, but not so deep and massive anteriorly as in *Psittacotherium*.   Skull and skeleton similar to the Torrejon genus but less massive and less specialized in various details.

The type and only described species is *Wortmania otariidens* (Cope), known principally from a single individual, No. 3394, in which the upper and lower jaws, part of the cranium, most of the radius, ulna, femur, tibia and three bones of the manus are preserved.   This specimen was studied and described in considerable detail by Cope and subsequently restudied by Wortman in connection with *Psittacotherium*.

Wortmania otariidens (Cope)

Hemiganus otariidens Cope, 1885, Amer. Nat., XIX, p. 492.

Type: A. M. No. 3394, parts of skull, lower jaw, femur, metapodial, ungual phalanx.

Horizon and Locality: Lower Paleocene, Puerco formation, San Juan basin, New Mexico.

Author's Diagnosis: Species . . . may be characterized as follows: Enamel of teeth everywhere smooth.   Posterior true molars smaller than the anterior.   [Measurements follow.]

In 1888 [165] Cope described the species in detail as follows:

Only one individual of this species has been found, but it is represented by many parts of the skeleton. It was a plantigrade beast of about the size of a black bear, of robust proportions, and with a wide head with an exceedingly short thick muzzle, armed with some formidable teeth in front.   These, with its sharp claws, made it the most formidable animal yet known of the Puerco fauna, excepting its larger and more powerful congener, the *H. vultuosus*.

The nares are well roofed by the nasal bones, which border the premaxillaries above to the line of the front of the second superior incisor, by a wide sutural surface.   The superior process of the premaxillary bone is short, not extending posterior to the vertical line of the posterior face of the superior canine tooth. A small foramen, perhaps the infraorbital, issues above the second tooth posterior to the canine.   Exterior to the third tooth that follows the canine, the external face of the maxillary bone spreads outwards as though forming the malar process, and that this is the case is rendered probable by its smooth superior surface, which is the inferior orbital border.   Just anterior to the orbital border, a large foramen from the maxillary antrum perforates the maxillary bone.   The two teeth in the maxillary bone are injured, but the anterior has a conical crown and a single root, while the crown and base of the second are widened a little transversely.   I can find no superior true molars in the collection.

The mandibular rami are remarkable for the shortness of the dentary portion, and the elevation and width of the coronoid process.   The condyle is elevated above the alveolar border of the lower jaw, when the inferior border of the ramus is horizontal.   The ramus increases in depth anteriorly, as in Taeniodonta, to accommodate the large anterior teeth.   The inferior border is straight and compressed, and the posterior border is gently concave to a short rectangular angle, which does not extend posteriorly to the line of the base of the condyle.   It is therefore much less prominent than in Creodonta generally, resembling in this respect the Taeniodonta.   There are four alveoli for single-rooted molars, and apparently another one in front of the anterior one of the four.   This would give seven molars, the first true molar having the form of a premolar; but the distribution of the teeth is not quite certain.   As already described, the heel of these premolariform teeth is partly internal.   The first true molar may be one of these simple teeth.   The second has two roots, and the crown is about as wide as long.   The crown consists of an anterior portion,

---

[165] Cope, E. D., 1888, Trans. Amer. Phil. Soc., XVI, N.S., p. 311.

which is slightly elevated above a posterior heel.   The superior face of the crown is worn by mastication so that its construction is not evident, but there is no trace of a division between fourth and fifth tubercles, so that I suspect that the latter did not exist.   It is not probable that there were well-marked cusps on the heel.

The parietal region of the skull is very much compressed, and the sides slope regularly upwards to the elevated sagittal crest.   The temporal ridge is an oblique angular line of the surface, and the frontal region is flat.   No other parts of the skull are preserved.   [Measurements of skull follow.]

Only cervical vertebrae are preserved.   These have small anteroposterior diameter, and their transverse exceeds their vertical diameter.   In general they resemble those of Periptychus.   The atlas is peculiar in the small anteroposterior diameter of the paradiapapohysis, whose base is perforated by an anteroposterior canal.   It sends upwards a vertical keel to opposite the middle of the facet for the axis.   The axis has a cylindric and rather slender odontoid process whose superior extremity is obliquely beveled on a curve.   Its articular surface is continuous with the large atlantal facets laterally and inferiorly.   The longitudinal axis of the cervical centra is oblique to the horizontal, showing that the head was elevated above the body.   The floor of the neural canal is pierced by a foramen of considerable size on each side.   A posterior (?seventh) cervical has a greater anteroposterior diameter than the two which precede it, and the vertical diameter is relatively greater.   The posterior articular face of all three is slightly concave.   [Measurements of vertebrae follow.]

The anterior limb is represented by parts of both ulnae, part of one radius, and a metacarpal of the pollex.   More than half of one ulna is preserved.   The olecranon appears to be short and terminating in an acute apex at the basal border; but it may have been broken off.   The humeral cotylus is oblique, extending backwards and outwards, and inwards and forwards.   The posterior border is elevated into a ridge, which is convex forwards.   The anterior marginal ridge is limited to the external part, and it extends outwards, downwards, and then backwards, overhanging the internal face of the ulna.   The radial facet is flat, and slopes gently, and not steeply inwards, it is bordered on the outer side by a low ridge, external to which is a longitudinal groove.   Both do not extend far distad, and the superior edge of the shaft of the ulna is narrow and convex.   The inferior edge is similar except below the humeral cotylus, where it is transversely flattened, the inferior face turning upwards on the inner side posteriorly.   The internal side of the shaft of the ulna is concave, and the external side is convex.   The head of the radius is a transverse oval, with subequal broadly rounded extremities.   The superior border is openly shallowly excavated, while the inferior is obliquely beveled for the ulnar face.   A short tuberosity projects longitudinally from the middle of the ulnar facet.   A metacarpal, supposed to be that of the pollex, is quite short and robust and has a proximal excavation of the internal side for the trapezium.   This concave facet extends half its length.   The distal end is a subround convex facet which presents outwards.   It has neither median keel nor groove.   It indicates a robust digit.   [Measurements of anterior limb follow.]

The femur lacks the distal extremity so that it is not possible to determine its exact length.   Its proximal portion is robust, and about as large as that of a fully grown pig.   That the animal is not fully grown is shown by the fact that the epiphysis of the head is not united, although it is preserved.   The projection of the great trochanter is about equal to that of the head, and is robust, and truncate both proximally and externally, and incloses a considerable trochanteric fossa.   The head has a large fossa for the round ligament which is near the neck, from which it is separated by a low border.   The little trochanter is quite prominent, and is reverted, but it is not connected with the great trochanter by a ridge.   The third trochanter is well developed, and has a wide external surface, whose anterior edge is recurved forwards.   Its upper portion overlaps the line of the inferior edge of the little trochanter, being higher up than in Pachyaena.   The middle of the shaft is somewhat depressed, and its margins are rounded.

The proximal part of the left tibia, and the distal part of the right,[166] give the characters of that element.   The proximal part is laterally crushed.   It is evident, however, that the crest is large and obtuse at the apex, and that the spines are low ridges.   The externo-posterior border forms a roughened ridge for 35 mm. below the internal femoral surface, and ceases rather abruptly below.   At or near the middle the shaft is normally somewhat compressed, and slender.   The malleolus is very prominent, and terminates in an apex in its internal plane.   The astragalar surface is but little oblique; the fibular articular surface is large.

---

[166] As Wortman subsequently determined, both halves belong to the left tibia.—W.D.M.

One of the metatarsals [167] shows that the foot was short, since it is neither the first nor the fifth. Its proximal face is concave in the transverse direction, but nearly straight anteroposteriorly. The arc of the phalangeal face is less than half a circle, and is slightly concave in transverse section. It is divided medially at its inferior fourth by a short, narrow, and low trochlear keel. Inferior border of phalangeal face prominent and openly emarginate. An ungual phalange is preserved, but whether of the anterior or posterior foot I do not know. Its apex is lost. It is strongly compressed, and has a narrowly rounded superior border. The phalangeal cotylus is deeply excavated, and is rather narrow, and has a weak median keel. The superior process overhangs the phalangeal surface rather further than the inferior. The tuberosity for the flexor tendon is a longitudinal oval, with surface transversely convex, which gradually ascends to the narrow but flat inferior surface of what remains of the phalange. The large nutritive foramen enters above its middle. [Measurements of posterior limb follow.]

Wortman [168] redescribed the same specimen in 1897, comparing it with the armadillos and other edentates and correcting certain errors:

In the skull [figured], of which a considerable part is preserved, the face is short, the sagittal crest is long and not very prominent, the lower jaw is short, deep and robust, with a greatly enlarged coronoid and a pronounced angle. The tooth-line passes to the inside and slightly behind the root of the coronoid, so that the last molar is partly concealed in a side view; the condyle is situated unusually high above the tooth-line.

The complete dental formula cannot be stated at present, but it is certain that there were four premolars and three molars in the lower jaw. There was also a pair of canines and one or two pairs of incisors. In the upper jaw there was at least one pair of incisors and very probably two; there was also a pair of large and powerful canines; the upper molar dentition is unknown.

In structure several isolated incisors show a long tapering root closed at its extremity, and having the enamel limited to the anterior face of the crown. The lower canines also exhibit a long tapering root closed at the base, and having the enamel limited to the anterior of face of the crown. Just how much this limitation of the enamel is due to wear, however, is not easy to say; it is more than probable that in a perfectly unworn young tooth, the entire crown is covered with enamel, but upon the posterior surface it is very thin and is soon worn away. The superior canine shows complete investment of the crown with enamel, although the covering upon the posterior portion of the crown is very thin. Of the premolars, two of the inferior ones (3d and 4th) are in place, together with the first molar. The third premolar consists of a principal cone with a slight cingulum, situated internal and posterior to the principal element of the crown, while in the fourth the internal cusp is larger and more posteriorly situated. It is highly probable that this posterior cusp is the incipient heel, the greater development of which would produce the posterior part of the true molar.

There are three lower molars preserved, the crowns of which are so much worn as to obscure considerably the pattern of the grinding surface; it can be stated, however, that it is composed of the usual four cusps which go to make up the quadratubercular crown. The four cusps were apparently fused into two transverse crests, the posterior of which is much the lower. The anterior cusp of the trigon is persistently absent, and there is much evidence of the fact that the molars did not pass through the typical 'tuberculo-sectorial' stage to reach the quadratubercular form. The superior molars are entirely unknown. All the molars of the lower jaw had well-developed roots with divided fangs.

It is a fact worthy of note, to which Cope has called attention, that the superior surface of the premaxillary is marked by a suture throughout its entire extent. This would indicate that the snout was, in some degree at least, tubular. The only similar condition that I have met with in the mammalia is that of the Armadillo, in which the nasals cover in the premaxillae throughout their entire extent above.[169]

A number of cervical vertebrae are represented by their centra, which are remarkable for their great transverse diameter in comparison with their antero-posterior dimension, as in the living Armadillos. The arches are not preserved, so it is impossible to determine their characters.

Of the fore limb the proximal parts of both ulnae, a nearly complete radius, a lunar, the metapodial of the second digit, and a terminal phalanx, are represented in the specimen. The proximal end of the ulna

[167] Metacarpals.—W.D.M.
[168] Wortman, J. L., 1897, Bull. Amer. Mus. Nat. Hist., IX, pp. 68–71.
[169] This condition is rather characteristic of the Insectivora.—W.D.M.

shows a marked resemblance to that of the Gravigrada, especially *Mylodon robustus*, the olecranon being relatively short and the sigmoid portion for articulation with the humerus wide. The radius is short and robust, with a well-excavated head and an ulnar articulation which permitted free pronation and supination of the manus. It increases in diameter distally, and in its ridges and surfaces resembles both that of *Mylodon* and *Megalonyx*.

The lunar [figured] is free, presenting a convex proximal facet for the radius, a posterior extended portion, a cup-shaped facet for the head of the magnum, and lateral facets for the scaphoid and cuneiform; its proportions and relations to the surrounding bones of the carpus are very similar to those of the corresponding bone of *Mylodon robustus*.

The metapodial of the index [figured] or second digit [Footnote: Cope has determined this metapodial to be a metatarsal, but there can be little doubt that it not only belongs to the fore foot but that it is the second metacarpal.] is remarkable for its brevity; it shows a fore and aft grooved articular facet for the trapezoid, and facets upon either side for the articulation of the first [170] and third metapodials. The metapodial keel is but faintly indicated, and is confined wholly to the palmar aspect of the distal facet.

The single ungual phalanx [figured] is proportionally very large, exceeding the second metapodial in length; it is compressed and high, with a marked dorsal curvature. The articular surface for the second phalanx is deeply excavated, having a median vertical ridge, and a backwardly prolonged overhanging upper portion. The whole facet describes almost a semicircle. The subungual process is large, and is perforated by a considerable foramen.

Of the hind limb only the proximal two-thirds of a femur and a complete tibia are known. The femur is a short, stout bone with a short, globular, sessile head, which does not rise above the great trochanter; the whole bone is markedly flattened from before backwards; there is a strong lesser trochanter and a weak third trochanter. The general shape of the bone recalls at a glance the femur of the edentates, especially the Ground Sloths.

The tibia [Footnote: Cope makes out that this bone, of which there are two pieces, pertains to both sides, the head to the left and the distal half to the right side. When the matrix was removed, and the two ends fitted together, they were found to make a complete bone of the left side.] in comparison with the femur, is short and small. The proximal portion is crushed so as to obscure the form of the head of the bone, but the distal end is well preserved. The distal trochlea for articulation with the astragalus is not very well grooved, but yet the grooves are better developed than in any other Puerco mammal of corresponding age. The internal malleolus is well developed. The whole character of this part of the tibia resembles the corresponding part of this bone in the Armadillo.

Revised Description: Although there is little additional material of this interesting genus, the single specimen found by the Museum party of 1913 serves to place beyond question the character and sequence of the lower premolars, and a more careful comparison of the loose teeth of the type with the dentition of *Psittacotherium* enables us to identify them with reasonable certainty, and brings out a closer comparison with *Psittacotherium* than was apparent to Wortman.

The dentition appears to be $i_1^1 c_1^1 p_4^3 m_3^3$, as in *Psittacotherium*, except for one more upper premolar, and each of the teeth is characteristically like the corresponding tooth in the Torrejon genus in construction, but most of them more brachydont and the canines decidedly more primitive.

There is a single enlarged upper incisor, which is worn to a spatulate tip as in *Psittacotherium*, and the root may have been expanded antero-posteriorly but is broken so that this point cannot be determined. The upper canine is enlarged, robust, worn apparently in somewhat the same fashion, but crown and root are more round-oval in cross-section, less elongate and retaining much more of the primitive creodont-like proportions. P^2 is single-rooted and simple crowned. P^3 has a partly divided root, the principal cusp round,

[170] The inner facet is really for the trapezium.—W.D.M.

with the postero-internal side of the crown expanded at the base, but no distinct inner cusp. Whether a cingulum is present is not certain. P⁴ has a large inner cusp and larger outer cusp, both round-conical toward the tip, the outline of the crown rounded, subtrigonal, and apparently the tooth had three roots. The molars have two small outer roots and a large inner root, are nearly twice as wide as their antero-posterior diameter, and the third, if present, must have been much reduced. The crowns are too much worn to determine anything of their pattern.

The lower incisor, like that of *Psittacotherium*, has a much compressed root and small crown. The lower canine is large, robust, strongly recurved, heavily worn on the posterior face, but retaining much of the proportions of creodont canines. The lower premolars are all much alike, having large, stout, rounded and incurved principal cusp, with a heel or buttress on the postero-internal side which in p₄ makes a more definite transverse basal crest limited to the inner half of the posterior side. The first premolar has the principal cusp somewhat smaller than it is in the others, but is not otherwise different. All of them appear to be one-rooted. The lower molars, as far as one can judge from the single heavily worn tooth preserved in the type, are of similar pattern to those of *Psittacotherium* and *Conoryctes*.

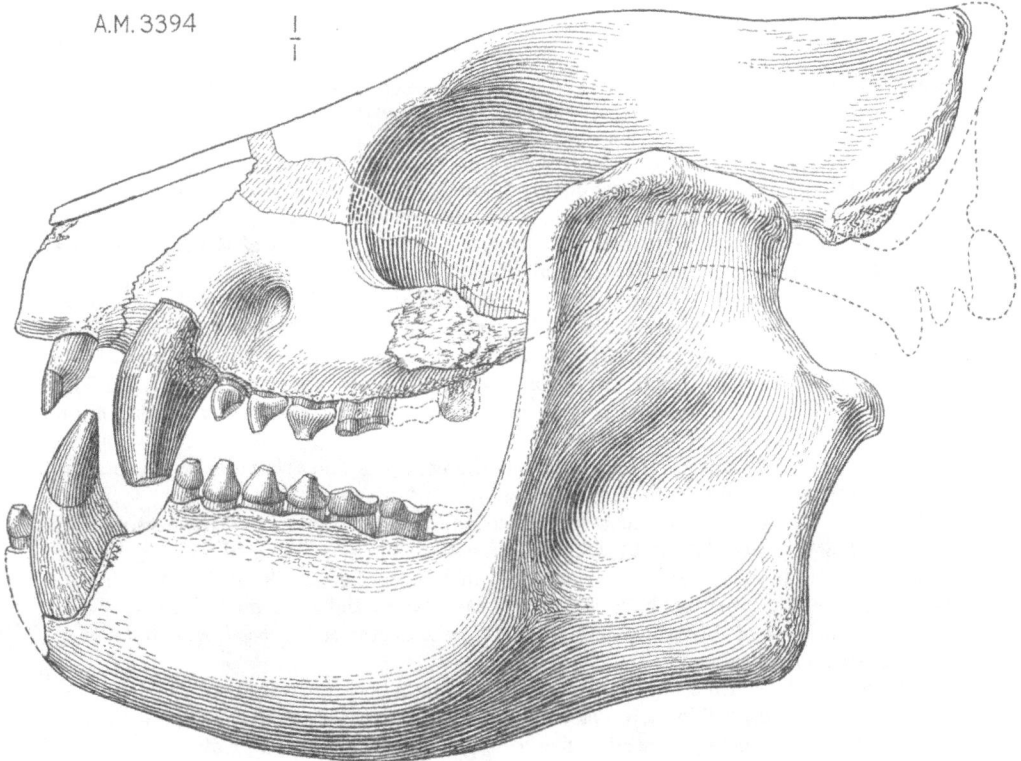

A.M. 3394

FIG. 68.   *Wortmania otariidens*, skull and jaws, A. M. No. 3394, type, side view.   Natural size.

The skull has a short face and a long narrow cranial region with the sagittal crest obscure but the cranium beneath it built up in a ridge in the same manner as in *Onychodectes, Conoryctes* and *Psittacotherium*.   The nasals extend back to the postorbital crests, as in the last genus; the frontals are very short, the parietals extending forward medially almost to the junction of the postorbital crests; the parietals are nearly flat, sloping out at an angle of about 45° from the sagittal crest; their posterior portion is concave, sweeping up towards the occipital crest, which appears to have been wide and prominent.

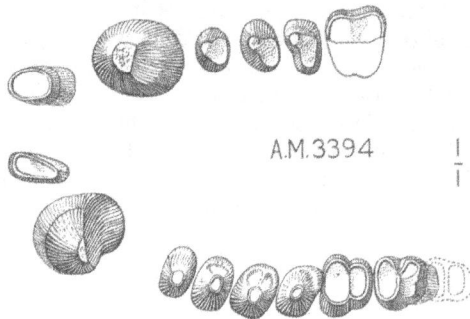

Fɪɢ. 69.   *Wortmania otariidens*, upper and lower teeth, crown views, from the type specimen, A. M. No. 3394, supplemented from other specimens.   Natural size.

The premaxillary extends backward between maxilla and nasals as a stout wedge, ending at a point above the post-canine diastema.   There is a short diastema behind the canine; the upper cheek teeth are a little spaced, but without diastema.   The infra-orbital foramen is above the last premolar.   The zygomatic arch and base of the skull and the occiput are unknown save for a few uncharacteristic fragments.

The lower jaw is short, massive and deep anteriorly, somewhat shallowed posteriorly, the symphyseal suture deep and massive, but rather short, the coronoid process high and very wide, not recurved, the condyle considerably above the level of the tooth row, moderately transverse, the angle flat, moderately deep, but not extended backward.   The jaw is intermediate in characters between those of *Conoryctes* and *Psittacotherium*, but more suggestive of the latter.

A number of cervical vertebrae show that the neck was quite short, as much so as in *Ectoconus*. but less than in *Periptychus*.

The ulna and radius are much like those of *Psittacotherium*, but less massive, the olecranon is comparatively short, while in *Psittacotherium* it is long and wide; the shaft of the ulna is reduced towards its distal end; in neither bone is the distal end known. The manus is represented by the lunar, second metacarpal and an ungual phalanx, which are quite of the type of the corresponding bones in the manus of *Psittacotherium* but somewhat smaller and less specialized.

The proximal two-thirds of the femur is known, and is a massively proportioned bone with wide shallow digital fossa, short rounded head and the great trochanter high and wide, rising slightly above the head.   The lesser trochanter is internal, rather prominent, and the third trochanter well developed, situated high up on the shaft.   The shaft is considerably broadened, especially toward the distal part, and the axis of the shaft is not nearly so straight as it is in *Psittacotherium*, the bones being otherwise much alike in construction.

The tibia is correspondingly stout, with the upper part of the shaft much flattened, the cnemial crest long and strong, but not so prominent or abruptly ending as in the taligrades. The distal end is characteristic in that the astragalar trochlea, while shallow, faces almost distad, the astragalus lying beneath the tibia instead of being wedged between the tibia and fibula, and the internal malleolus prominent, its astragalar facet almost at a right angle to the plane of the trochlea. The pes is wholly unknown.

AFFINITIES OF *Wortmania*

The construction and proportions of the teeth show throughout a near relationship to *Psittacotherium*, and this is confirmed throughout by all that is known of the skull and skeleton. The differences are not much greater than those that separate the Puerco and Torrejon species of *Periptychus* or *Anisonchus*, and are all of them points of approach to the more primitive genera *Conoryctes* and *Onychodectes*, and through these to the generalized proto-Insectivora and Creodonta. In the teeth the less reduction of the upper and less specialization of the lower premolars, the shorter crowns and less compressed or elongate roots of the incisors and canines, the less enlargement of i^3 and c_1^1, more primitive pattern of p_4, may be noted. The jaw is less enlarged and deepened anteriorly, the cranium has the same characteristic high narrow form built up beneath the low and inconspicuous sagittal crest, the relations of the bones of the face are the same throughout, the neck is short, the limbs stout and massive, the olecranon is less developed, the femoral shaft less straight, the construction of the manus appears to be similar, as compared with *Psittacotherium*. All the evidence points to *Wortmania* as being the direct ancestor of *Psittacotherium* and its position in the same subfamily of Taeniodonta, as clearly shown. The affinities to the Edentata are considered elsewhere.

ALLOTHERIA [171]

6. ORDER MULTITUBERCULATA

DISCUSSION

The question of priority as between the names Multituberculata Cope and Allotheria Marsh is not easily settled. Marsh's name is the older one, but is open to three objections:

1. It was proposed without any definition that would distinguish it from the Marsupialia, the order to which its included types had been previously referred by all writers. Marsh's definition is as follows:

> (1) Teeth much below the normal number.
> (2) Canine teeth wanting.
> (3) Premolar and molar teeth specialized.
> (4) Angle of lower jaw distinctly inflected.
> (5) Mylohyoid groove wanting.[172]

[171] Marsh, 1880 (as an order); Granger and Simpson, 1929, p. 674 (as a distinct subclass).
[172] Marsh, O. C., 1880, Amer. Jour. Sci. (3), XX, p. 239.

The above list is intended to distinguish the Allotheria from the Pantotheria (trituberculate mammals), and Marsh adds:

> These characters alone do not indeed separate the *Plagiaulacidæ* from some of the Marsupials, and future discoveries may prove them to belong in that group, where they would then represent a well marked suborder.

He gives no hint of how this " well marked sub-order " is to be distinguished from the diprotodont marsupials—in which group, it must be remembered, the Plagiaucidae were placed by all previous authors. In 1887 Marsh repeated the same list of characters, adding: " These characters alone do not separate the *Plagiaulacidae* and *Microlestidae* from some of the Marsupials, and the facts now known seem to prove that they belong in that group, where they represent, at least, a well-marked sub-order." [173]   Again there is no hint of what *Marsh* regarded as the subordinal characters, although in his description of the genera *Ctenacodon* and *Allodon* he gives certain characters that *in the light of our present knowledge* would be regarded as of subordinal or higher value (character of the angle, high number of premolars, etc.).   The importance of these characters was first pointed out by Broom in 1914.

Cope in 1884 proposed the name Multituberculata, with the following definition:

> The extinct marsupials belong to three types as distinguished by the form of their superior molar teeth. These are trituberculate, quadrituberculate or multituberculate.   To the first division belong the carnivorous types or Sarcophaga of Owen; to the second the Kangaroos and the wombats, to which Owen's name of Poëphaga may be applied.   The third division is entirely extinct, and is characterized by having at least three longitudinal series of tubercles in its superior molar teeth.   To this suborder I apply the name of Multituberculata.[174]

The above does constitute a valid definition, as it distinguishes the group from the other marsupials, to which order its members had been previously referred.[175]   It is necessary to be thus specific in view of Marsh's subsequent statements [176] that he had " defined " the order Allotheria in 1880 and that Cope had not defined the Multituberculata when first published in 1884.   Neither statement is true.   When an author characterizes a " new " order by characters that all apply to that part of the old order to which its members had previously been referred, he does *not* define the order.   A definition must necessarily delimit or distinguish the group in some way from the order to which it had previously been referred.   Cope's definition, scanty as it is, does distinguish the Multituberculata from any known marsupials; Marsh's characterization of the Allotheria not only does not distinguish them from marsupials, but he adds that they do not separate them from that order save as a suborder, which again he does *not* define.

2. The names of the primary groups of mammals end in -theria, and although there are exceptions it is not a usual termination for ordinal names.

3. The name Multituberculata has been very generally used and if possible should be retained.

But in fact both names are needed and can be used appropriately by reserving the

[173] Marsh, O. C., 1887, Amer. Jour. Sci. (3), XXXIII, p. 345.

[174] Cope, E. D., 1884, Amer. Nat., XVIII, p. 687.

[175] It would, if applied exactly, exclude certain multituberculate genera (*Bolodon*, etc.) the upper teeth of which were not then known to be associated with plagiaulacoid lower teeth.   But it is valid and distinctive in the light of what was then known of the group.—w.d.m.

[176] Marsh, O. C., 1891, Proc. Acad. Nat. Sci. Phila., XLIII, p. 239.

term Allotheria for the super-ordinal or sub-class division that includes the order Multituberculata.

The Multituberculata were generally regarded as allied to the diprotodont marsupials until 1888, when Poulton's study of the young *Ornithorhynchus* showed teeth of a character suggestively like those of certain multituberculates.   Cope, as a result, concluded that they were more probably related to the monotremes, and this viewpoint was more or less generally accepted until 1909, when Gidley described [177] a fairly complete skull and jaws with parts of the skeleton of *Ptilodus* from the Fort Union beds of Montana.   From the characters of this specimen Gidley concluded that the group belonged with the diprotodont marsupials.   Broom in 1905 had restudied the classic skull of *Tritylodon* from the South African Triassic, concluding that its affinities lay rather with the monotremes than with higher forms, and in 1910 he discussed the relations of the multituberculates in the light of Gidley's contribution and concluded that they had no very near affinities with the living monotremes, marsupials or eutherians.

In 1914 Broom described the skull of *Polymastodon* (= *Taeniolabis*) from the Puerco, and again discussed the ordinal affinities, maintaining that they were an early offshoot of the common ancestral stock of all mammalia, and that the monotremes were degenerate descendants from Middle Jurassic multituberculates.   This last view was, however, partly based upon a misinterpretation of the pelvis of *Ptilodus*, which Broom supposed to be a shoulder girdle.

In 1929 Granger and Simpson described a partial skeleton of *Eucosmodon* in the course of a complete review of the known Tertiary multituberculate remains.   In the meantime much had been added to our knowledge of Mesozoic multituberculates by Simpson in a series of contributions from 1925 to 1929, and in the above-cited paper the evidence as to their affinities is discussed, with the conclusion that they should be placed in a separate subclass, to which the name Allotheria Marsh is appropriately applied.

Classification: The following arrangement of the Paleocene genera and species is taken from Granger and Simpson, 1929: [178]

I. Premolars $\frac{1}{1}$, last lower premolar reduced, not trenchant..........................Taeniolabididae.
   A. Lower molars narrower, cusps not more than 6:5, 4:2.............................*Catopsalis.*
   B. Lower molars broader, cusps not less than 7:6, 5:4..............................*Taeniolabis.*
II. Premolars $\frac{4-3}{2-1}$, last lower premolar enlarged, trenchant, serrated......................Ptilodontidae.
   A. Lower incisor not gliriform, P₃ present.
      1. P⁴ with outer tubercles numerous...*Ptilodus.*
      2. P⁴ with outer tubercles 1 or 2 in number.................................*Ectypodus.*
   B. Lower incisors gliriform, P₃ absent, P⁴ with 3 or 4 outer tubercles.................*Eucosmodon.*

TAENIOLABIDIDAE Granger and Simpson, 1929

Granger, W., and Simpson, G. G., 1929, Bull. Amer. Mus. Nat. Hist., LVI, p. 603

Taeniolabis Cope, 1882

Cope, E. D., 1882, Amer. Nat., XVI, p. 604

Type: Taeniolabis sulcatus Cope from Middle Paleocene of New Mexico.
Synonyms: Polymastodon Cope, 1882; type: *P. taöensis.*
          Catopsalis Cope, 1882; type: *C. pollux.*[179]

[177] Gidley, J. W., 1909, Proc. U. S. Nat. Mus., XXXVI, pp. 611–26, pl. LXX.
[178] Granger, W., and Simpson, G. G., 1929, Bull. Amer. Mus. Nat. Hist., LVI, p. 603.
[179] Cope, E. D., 1882, Amer. Nat., XVI, July 28, p. 684.

Author's Diagnosis: Char. gen. This genus is established on a tooth whose position is on the arc of the alveolar line which connects the molar and middle incisor regions. It is probably either the third incisor of the superior or inferior series, or the canine of the inferior series. In either case it differs from the corresponding tooth of any known genera of *Tillodonta* or *Taeniodonta*. The long diameter of the root being placed antero-posteriorly, that of the crown makes with it an angle of 30°.

Section of the crown oval; the grinding surface scalpriform in the manner of a rodent incisor; but beveled on side of the long diameter instead of on the end as in that order. Enamel consisting of a wide band on the external side of the tooth, which embraces more of the circumference near the apex than elsewhere. Apex grooved behind.

If this be an inferior canine tooth it differs from that of the *Tillodonta* in its large size and incisor-like form. It most resembles the external or third inferior incisor of *Calamodon*. From this it differs in the scalpriform wear, and the oval instead of triangular section, and in the absence of cementum layer.

Type Diagnosis of Polymastodon: Char. gen. Known only from the inferior dentition [actually the superior dentition]. Supposed formula: I.1; C.0; P.–m.0; M.2. The first true molar is large, exceeding the second, and supports three longitudinal series of tubercles. Function of the molars grinding.

In this genus the molar part of the dentition assumes the exclusive control of mastication, having already displayed a predominance in *Catopsalis*. The molars are similar in their general character to those of *Ptilodus* and *Catopsalis*, but the three rows of tubercles distinguish them from both.

In the same paper the supposedly new species *Catopsalis pollux*, another synonym, was characterized as to the astragalus, etc., as follows:

It considerably resembles that of a kangaroo; the reduced navicular facet and the large cuboid facet indicate the predominant development of the external digits, and the reduction of those of the inner side of the foot. Caudal vertebrae indicate a large tail.

The astragalus here ascribed to *Catopsalis* and subsequently figured by Cope in Tertiary Vertebrata (pl. xxiiic, fig. 6) under the name of *Polymastodon* is, however, so much like that of *Ectoconus ditrigonus* that Matthew in 1897 questioned its association with " *Polymastodon.*" The discovery of the characters of the pes in the allied genus *Ptilodus*, with a totally different kind of astragalus, agreeing fairly well with such primitive marsupials as *Didelphys* and wholly unlike any placental astragalus, confirms the dissociation of the *Ectoconus*-like astragalus from *Taeniolabis*. As for its resemblance to that of the kangaroo, it can hardly be said to extend beyond the character of the superior surface of the bone. The kangaroo has the shallow-bodied marsupial astragalus, devoid of distinct neck and head, lacking also the characteristic differentiated sustentacular and ectal calcaneal facets that characterize all or almost all placentals. *Ptilodus* was typically marsupialoid in all these features, and so probably was *Taeniolabis*.

In 1885 (Amer. Nat., XIX, p. 493) Cope recognized the identity of *Polymastodon* and *Taeniolabis* but retained the former name for the genus.

The characters of this genus, now known from a fairly complete jaw (A. M. No. 16310), from a fragmentary skull with lower jaws (A. M. No. 3036, type of *T. taöenis*) and from palates with teeth, various lower jaws more or less complete and many upper and lower teeth, have been described in considerable detail by Broom, 1914, and Granger and Simpson, 1929. It is not necessary to repeat these descriptions. It is the largest and most specialized of the Multituberculata. As far as known all the specimens found are from a single layer in the Puerco formation about forty feet above the base; nor has the genus been discovered in any other Paleocene formations. This curiously limited occurrence is suggestive of the incomplete character of our record of the Puerco fauna.

Three species, the first indeterminate, are admitted by Granger and Simpson: *Taeniolabis sulcatus*, *T. taöensis*, *T. triserialis*.

Taeniolabis sulcatus Cope

Taeniolabis sulcatus Cope, 1882, Amer. Nat., XVI, June 22, p. 604; (*T. scalper*) 1884, Tertiary Vertebrata, p. 193, pl. xxiiie, fig. 7; Granger and Simpson, 1929, Bull. Amer. Mus. Nat. Hist., LVI, p. 615, fig. 7.

Type: A. M. Cope Coll. No. 3038, an upper incisor.

Horizon and Locality: Puerco formation, San Juan basin, New Mexico.

Author's Description of *T. sulcatus* (p. 605): *Char. specif.* The enamel band does not cover the entire width of the external face, but leaves exposed a part of the dental surface anterior and posterior to it except at the apex. At the latter point there are seven coarse shallow grooves of the enamel surface; the posterior of these split up below, and become narrowed, while the anterior run out at the more curved anterior edge of the enamel band. The posterior apical groove has a flat bottom. At the front of the apex the enamel is involute to the inner side for a short distance. The inner face of the tooth displays five facet-like bands of the dentinal surface, which soon disappear inferiorly.

FIG. 70. *Taeniolabis sulcatus*, type right upper incisor, A. M. No. 3038, external view. (From Granger and Simpson, 1929.) Natural size.

In reading the above description it must be kept in mind that Cope regarded the tooth as facing antero-externally—not anteriorly—as being the canine or third incisor in a jaw like that of *Calamodon*. He had no suspicion of its plagiaulacid affinities.

This type is regarded as specifically indeterminate, and no additional material has been referred to it.

Taeniolabis taöensis (Cope)

Polymastodon taöensis Cope, 1882, Amer. Nat., XVI, p. 684.
Catopsalis pollux Cope, 1882, Amer. Nat., XVI, p. 685.
Polymastodon latimolis Cope, 1885, Amer. Nat., XIX, p. 385.
Polymastodon attenuatus Cope, 1885, Amer. Nat., XIX, p. 494.
Polymastodon selenodus Osborn and Earle, 1895, Bull. Amer. Mus. Nat. Hist., VII, pp. 12, 15.

Type: A. M. Cope Coll. No. 3036, parts of right and left maxillae and fragments of skull; also lower jaws described as *Catopsalis pollux* but subsequently found by Cope to belong to the same individual as the upper jaws. The upper jaws were wrongly identified in the original description as lower jaws. Type of *Catopsalis pollux:* A. M. No. 3036, lower jaws of the above individual. Type of *Polymastodon latimolis:* A. M. No. 3045, lower jaw with m$_{1-2}$. Type of *P. attenuatus:* A. M. No. 3046, lower jaw with p$_4$-m$_2$. Type of *P. selenodus:* A. M. No. 749, lower jaw with p$_4$-m$_1$.

Horizon and Locality: Puerco formation, San Juan basin, New Mexico. Probably all the above are from the upper level of the true Puerco, to which the genus is limited as far as known.

Type Description of *Polymastodon taöensis: Char. specif.* The first true molar is two-fifths of itself longer than the second molar, and viewed from above, it has an oval outline, a little narrowed anteriorly and with rounded extremities. Its tubercles are small and closely packed together, so that those of the middle row have a subquadrate outline. There are eight tubercles in the internal row, twelve in the external and nine in the median. There are no basal cingula. The second and last true molar has a pyriform outline when viewed from above, the posterior extremity being the narrow one. The contraction of the outline is regular on each side, and the posterior extremity is rounded. There are seven tubercles in the external

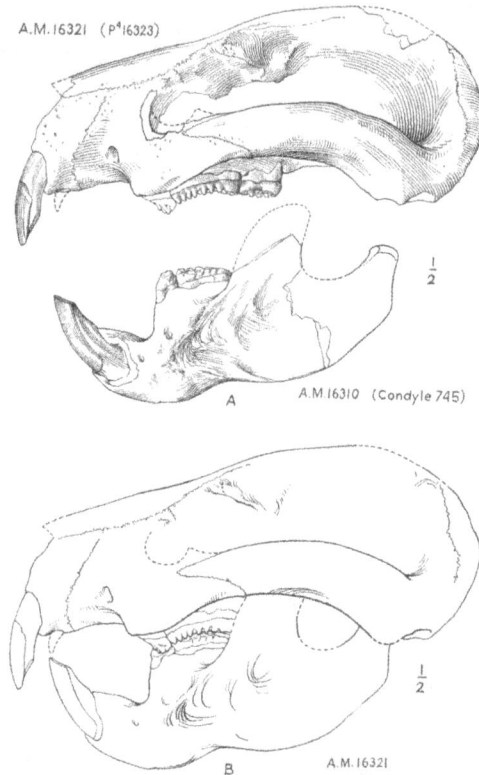

Fig. 71.   *Taeniolabis taöensis*, skull and jaw, composite, A. M. Nos. 16310, 16321 and 745. Side views: above, with jaws separated; below, teeth in occlusion. (From Granger and Simpson, 1929.)   One-half natural size.

row, five in the middle row and only two in the internal, since the middle row forms the internal edge of more than half the length of the crown. No cingula. [Measurements follow.]

Type Description of *Catopsalis pollux:* The size of this species exceeded that of *Macropus giganteus* and still more that of the *Catopsalis foliatus*. The ramus has the form of that of a rodent, being vertically narrowed at the diastema, and deep at the molar region. The inferior face widens and becomes flat posteriorly, and is more oblique than in the *C. foliatus*, from the greater downward extension of the external or masseteric edge. The interior edge on the contrary, ascends a little from the anterior inferior border, enclosing the large internal pterygoid fossa. The inferior plane commences below the anterior part of the first true molar. The symphysis is short, and was not probably strongly united, as is indicated by the few rugosities of its surface.   The coronoid process rises from a point opposite the posterior extremity of the first true molar.

The incisor is relatively large, and is more curved than that of a kangaroo, having the general form of that of a rodent.   The acumination or bevel of the posterior face is less rapid than that of a rodent, and is perfectly gradual.   The enamel band covers the antero-external face as far as exposed, which is below the anterior part of the diastema, and is gently convex in transverse section.   It does not cover the entire

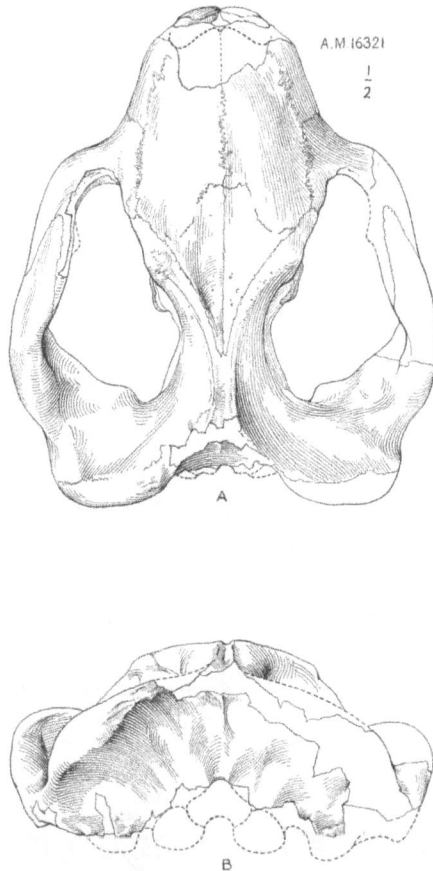

FIG. 72.   *Taeniolabis taōensis*, skull, A. M. No. 16321, superior view above, occipital view below.   (From Granger and Simpson, 1929.)   One-half natural size.

external face, as its width is equal, while the antero-posterior diameter of the tooth increases below.   The posterior face is convex and is not much narrowed.   The internal face is slightly concave, and the enamel is recurved so as to form a band on its anterior part, thus differing from most rodents.   The enamel surface is delicately obsoletely line-ridged.   The length of the diastema is equal to that of the combined P.–m.IV and M.1.   The fourth premolar is a simple tooth with a triangular transverse section, the obtuse apex of the triangle looking forward.   This edge is continued downwards by reason of the exposure of the anterior root, and is not acute.   The first true molar is an elongate-oval, with six tubercles on each side.   These are so closely placed that their outlines are angular, and they are only separated by fissures. No cingula. The second true molar is three-fifths the length of the first, and is broadly rounded posteriorly. It supports four tubercles on the internal, and five on the external sides, and a raised edge connecting the sides posteriorly. The tubercles are appressed as in the first molar. No cingula. [Measurements follow.]

Type Description of Polymastodon latimolis: This marsupial equals the *P. taöensis* in size, and is therefore larger than either the *P. fissidens* or the *P. foliatus* [*Catopsalis fissidens* and *C. foliatus*]. It

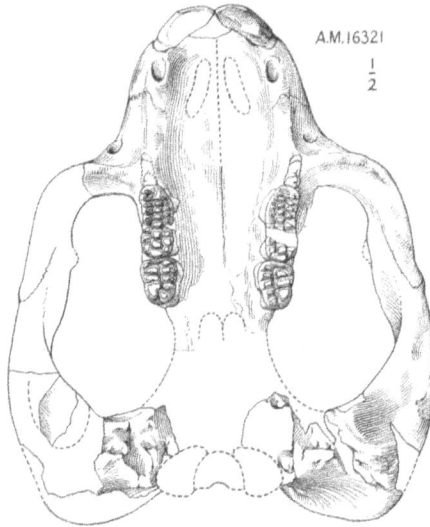

FIG. 73. *Taeniolabis taöensis*, skull, A. M. No. 16321, palatal view. (From Granger and Simpson, 1929.) One-half natural size.

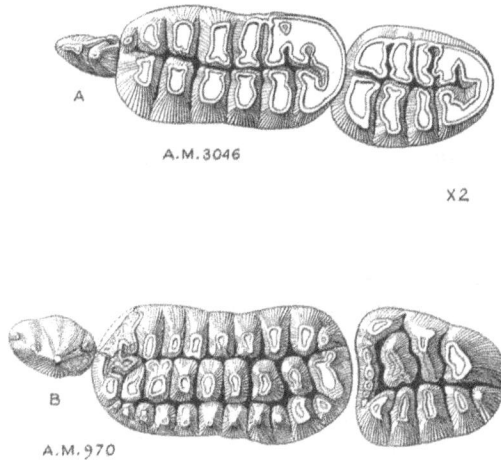

FIG. 74. *Taeniolabis taöensis*, *A*, lower cheek teeth, right side, A. M. No. 3046; *B*, upper cheek teeth, left side, A. M. No. 970, crown views. (From Granger and Simpson, 1929.) Twice natural size.

differs especially from both the *P. taöensis* and the *P. foliatus* in the great shortness of the first true inferior molar, which is only one-half longer than the second or last true molar. The latter is as wide as long in the type, and a little narrower in a second specimen. It supports four tubercles on the inner side; outer side worn. The first true molar appears to have five tubercles on the inner side, although the anterior edge is injured. In *P. taöensis* there are six or seven. The fourth premolar is two-rooted. The enamel of the last inferior molar is faintly longitudinally wrinkled. The coronoid process rises opposite the middle of the second true molar. [Measurements follow.] Besides the shortness of the second true molar, the width of the same tooth and of the last true molar distinguish this species from the *P. taöensis*. The inflection of the angle of the ramus of the lower jaw is as well marked as in other species of the genus.

Type Description of *Polymastodon attenuatus:* This form is represented by a mandibular ramus with entire dentition, of one individual, and by a superior incisor with portions of inferior molars of a second. The specific character is seen in the very compressed incisors and general lightness of structure of the ramus, in which it is quite different from the species of similar size, the *P. taöensis* and *P. latimolis*. The tubercles and proportions of the true molars are as in *P. taöensis*. The apex of the fourth premolar is transversely fissured. The superior incisor is much more compressed than in that of *P. taöensis*, and is more rapidly acuminate in its form, to the subacute apex. There are no facets of the internal side as in that species. The enamel covers almost the entire external face, and is marked by rather coarse parallel grooves. A groove runs along the concave edge of the crown, forming the edge of the enamel excepting for its distal half, where the enamel crosses it, and covers the internal side for its distal fourth. The inferior incisor is also much compressed so that the enamel is presented externally rather than anteriorly, and its cutting edge is nearly anteroposterior and not transverse, as in *P. taöensis*. Its surface is obsoletely grooved. Length of superior incisor .25; diameters do. at middle; anteroposterior .013; transverse .006. Length of inferior true molars .032; depth of ramus at middle M.1. .034.

Type Description of *Polymastodon selenodus:* The type (No. 749) lower molar is widely distinct from the above [*P. fissidens*] in the *crescentic* form of its molar cusps.

The anterior cusps are distinct, the posterior are low and irregular; there are 7 in the outer and 6 in the inner row; the anterior border is convex, the posterior is flattened, giving a sub-crescentic section, which reminds us strongly of the cusps of *Meniscoëssus conquistus* of the Laramie. The fourth premolar is very small.

Taeniolabis triserialis Granger and Simpson

Taeniolabis triserialis Granger and Simpson, 1929, Bull. Amer. Mus. Nat. Hist., LVI, p. 619, fig. 9.

Type: A. M. No. 725, part of left lower jaw with broken m₁ and complete m₂. A. M. Exped., 1892.

Paratypes: A. M. No. 748, right lower jaw nearly complete, with i and m₂; A. M. No. 727, fragment of right lower jaw with p₄ and m₁. Both of same origin as type.

Horizon and Locality: Upper level of Puerco, San Juan basin, New Mexico. Type and paratypes from Coal Creek Canyon.

Diagnosis: Lower molars generally more robust, cusp number tending to increase. Molars with numerous accessory cuspules and especially a row passing around the posterior end of the tooth and then for a short distance forward outside the external main cusp row, developed to a greater or less degree on both m₁ and m₂.

Catopsalis Cope, 1882

Cope, E. D., 1882, Amer. Nat., XVI, p. 416; Granger, W., and Simpson, G. G., 1929, Bull. Amer. Mus. Nat. Hist., LVI, p. 621

This genus, although generally regarded as a synonym of *Taeniolabis*, appears to be really distinct, and limited to the Torrejon formation as *Taeniolabis* is to the Puerco. It is revalidated by Granger and Simpson to include *C. foliatus* and *C. fissidens* Cope, as well as the recently described *C. calgariensis* of the Paskapoo. It is known only from lower jaws, of smaller size than in *Taeniolabis*, with narrower teeth and fewer cusps on the molars.

Type: Catopsalis foliatus Cope, from the Paleocene of New Mexico.

Author's Diagnosis: Family . . . Plagiaulacidae . . . two inferior true molars . . . fourth premolar trenchant. . . . One large premolar, which presents anteriorly. Fourth premolar with a cutting edge anteriorly, and a free posterior cusp; molars with numerous cusps.

In 1885 [180] Cope described the genus more at length as follows:

This genus is known from part of a mandibular ramus with a few other bones associated.[181] The jaw is broken off in front of the fourth premolar, and the fracture displays the shaft of a large incisor tooth. It is impossible to state how many premolars there are. The fourth is of large size, and is exceedingly compressed. The alveolar border descends abruptly from its posterior root, having the outline of the diastema of the jaw in various rodents, where, however, it is edentulous. The result of this form is, that the crown is presented forwards in an acute edge. The inferior two-thirds of this edge is broken off, so that it is not possible to state whether it is grooved or serrate. The superior part is neither, and rises into a cusp posteriorly. The two molar teeth are very peculiar, and the first is much larger than the second. The arrangement of the cusps is alternating on opposite sides of a median groove. The grooves are deep, and resemble the impression of a simply pinnate leaf with alternating leaflets.

The coronoid process rises opposite the second molar. The inferior face of the posterior part of the ramus is flat, owing to both internal and external inflections. Both are well marked, the latter bounding the masseteric fossa, which is open in front, and without foramen. The internal inflection bounds a deep fossa, like that seen in *Hypsiprymnus* and *Macropus* to terminate in the dental foramen.

Catopsalis foliatus Cope

Catopsalis foliatus Cope, 1882, Amer. Nat., XVI, April 24, p. 416; (*Polymastodon*) Cope, 1884, Amer. Nat., XVIII, p. 690, fig. 5; 1885, Tertiary Vertebrata, pl. xxiiid, fig. 2 (same figure as 1884 *supra*); (*Catopsalis*) Granger and Simpson, 1929, Bull. Amer. Mus. Nat. Hist., LVI, p. 621, figs. 10, 11a.

Type: A. M. Cope Coll. No. 3035, part of lower jaw with p_4–m_2 and root of incisor.

Horizon and Locality: " Puerco " (probably Torrejon) formation, San Juan basin, New Mexico. One additional specimen of this species is known and is certainly from the Torrejon.

Author's Diagnosis: Char. specif. The mandibular ramus which represents this animal, is robust and deep. The alveolar line rises from behind forwards, as in *Elephantidae* and various rodents, and then suddenly descends. The inner side of the ramus is concave, while the external side, anterior to the masseteric fossa is convex. The incisive alveolus is thus thrown inside the line of the molars in front. There is a large fossa exposed by weathering, below and behind the last molar, which is identical with that seen in *Hypsiprymnus* and *Macropus*, and indicates a large dental foramen. Below the middle of the fourth premolar tooth, the incisor tooth is quite large, suggesting whether it had not a persistent growth, as in the rodentia.

The posterior cusp of the fourth premolar is triangular in profile, the anterior edge descending steeply. It is uncertain whether the edge of the crown rises again, forming another lobe. The apex of the cusp is conic. The first true molar is of large size and remarkable form. The crown viewed from above is a long oval. It has a deep median longitudinal groove, which sends out branch grooves alternately, and at right angles to the edge. The spaces between the grooves form block-shaped tubercles, four on the inner and five on the outer sides, whose transverse diameter generally exceeds their anteroposterior. The median groove is open at its anterior extremity; the posterior is closed by an elevated convex margin. The apices of the lobes are obtuse where not distinctly worn. The last (second) true molar is much shorter, and a little

---

[180] Tertiary Vertebrata, p. 170.

[181] Probably some unrecognizable fragments. There is no record to identify them in the Cope Collection, if they are still preserved.—w.d.m.

wider than the first, and has the same character of surface.   There are two large tubercles on the inner side, and four smaller on the external side.   The posterior end of the crown is narrower than the anterior.   The anterior base of the coronoid process is opposite the posterior extremity of the first true molar tooth.   The jaw with its dentition, in its present condition, has a curious resemblance to that of a tubercular-toothed *Mastodon*, with the order of size of the molars reversed.   [Measurements follow.]

The above description is of interest as being the first really detailed description of the construction of multituberculate molars.

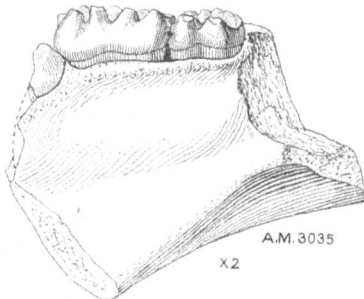

FIG. 75.   *Catopsalis foliatus*, type lower jaw, A. M. No. 3035, internal view, right side.   (From Granger and Simpson, 1929.)   Twice natural size.

In his later papers Cope regards this species as congeneric with " *Polymastodon* " *taöensis*.   Osborn and Earle in 1895 also make it a species of *Polymastodon*.   Matthew in 1897 listed it as distinct generically but gave no reasons for separating it.   Granger and Simpson in 1929 showed that the genus is clearly distinct from *Taeniolabis*, and refigured the type.   It should be noted that the premolar, although represented in their figure as complete, is more or less broken and incomplete, its actual size uncertain and possibly considerably larger than represented.

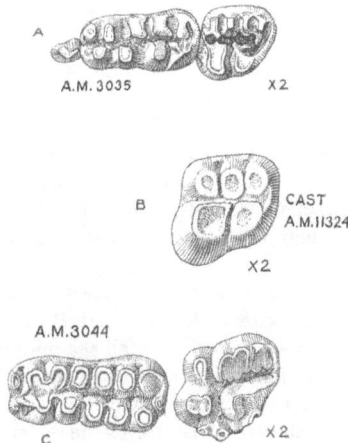

FIG. 76.   A, *Catopsalis foliatus*, right p$_4$-m$_2$ of type, A. M. No. 3035; B, *Catopsalis calgariensis*, right m$_2$ of plasto-type, A. M. No. 11324 (cast); C, *Catopsalis fissidens*, right m$_{1-2}$ of type, A. M. No. 3044, crown views.   (From Granger and Simpson, 1929.)   Twice natural size.

Revised Diagnosis:[182] Cusp formula: M_1, 5:4; M_2, 4:2.   Molar lengths of type: M_1, 10.7 mm.; M_2, 6.6 mm.   No internal accessory cuspules on M_2.

Catopsalis fissidens Cope

Catopsalis fissidens Cope, 1884, Paleont. Bull. No. 37, Jan. 2, and Proc. Amer. Phil. Soc.,
    XXI, p. 322; (*Polymastodon*) Cope, 1884, Amer. Nat., XVIII, p. 689; (*Catopsalis*)
    Granger and Simpson, 1929, Bull. Amer. Mus. Nat. Hist., LVI, p. 623, fig. 11 c.

Type: A. M. Cope Coll. No. 3044, lower jaw with m_{1-2}, and alveolus of p_4.

Horizon and Locality: Middle Paleocene, Torrejon formation, San Juan basin, New
Mexico.

Author's Description: In size this species is intermediate between the small *C. foliatus* and the
large *C. pollux*.   The first molar is the longer and narrower, and the second the shorter and wider, as in the
known species.   The first molar differs from that of both the latter, in having the tubercles of one side separated nearly to the base.   These tubercles are conic, and not flattened as in *C. foliatus* and *C. pollux*, and
the two rows are separated by a distinct valley, as in the first named.   There are five tubercles on one side,
and four on the other side of the crown, and in addition, two small cusps at the anterior extremity of each
row, and another at the posterior extremity of one of the rows.   These additional cusplets are not present
in the other species.

The last molar is relatively wider than in the other species.   Its crown is a good deal worn, but there
are probably more than two rows of tubercles, as there are some appendicular rows on one side of the crown
at least.   [Measurements follow.]

Cope states definitely that this specimen is from the " Upper Puerco," i.e. Torrejon
formation.   This is confirmed by the original label record of the specimen, which was
dated, as was Cope's custom, with the time the specimen was collected.   This was 11-26-'83;
evidently Baldwin discovered a rich pocket on that date in Gallegos Canyon, for a large
series of specimens of various Torrejon species bears that date, sometimes with, sometimes
without, the locality.   It was this shipment of new material which is described in the
" Second Addition to the Puerco Fauna."   It is of course possible that Baldwin obtained
this specimen from the Puerco at an earlier or later date at some other locality (it is not
exposed in this canyon) and that it was accidentally mislabeled, but it is very unlikely.
The matrix is of a character more commonly found in the Torrejon, but it does occasionally
occur in the Puerco.

The specimen consists of two molars in place upon a jaw which is wholly buried in
hard concretion and so far destroyed by concretionary processes that very little of it remains, and it is quite impossible to develop any considerable part of the surface by removal
of the concretion.   Cope evidently attempted this, and succeeded in showing in front of
m_1 what appears to be a small matrix-filled cavity that may be the alveolus for a small p_4
or for one root of it.

The species is considerably smaller than any of the species of *Taeniolabis*.   The molars
differ from that genus and agree with *Catopsalis* and *Ptilodus* in their manner of wear,
forming a deep longitudinal valley or groove through the middle of each tooth.   In *Taeniolabis* the molars wear to a flat or shallowly warped surface, indicating a lateral play of the
molars in grinding, while the deep furrowed wear of the molars of *Ptilodus* and of all the
genera with large knife-like premolars must be due to the restriction of the jaws to a directly

[182] Granger, W., and Simpson, G. G., 1929, Bull. Amer. Mus. Nat. Hist., LVI, p. 622.

fore-and-aft motion caused by these teeth. It is not certain to what extent the premolars were reduced in *Catopsalis;* in *C. foliatus* p_4 is broken and *may* have been much larger than it seems, and in *C. fissidens* the size and character of p_4 is still more doubtful. Both species agree apparently with *Ptilodus* and its allies in this manner of wear of the teeth and are distinct from *Taeniolabis* in various details of form and construction of the molars.

Granger and Simpson remark, however, that this apparent difference in manner of wear of the teeth cannot be regarded as certainly shown in absence of more material. Their specific diagnosis is as follows: [183]

Cusp formula: M_1 6:5; M_2 apparently 3:2. Molar lengths of type: M_1 14.0 mm.; M_2 9.2 mm. Accessory row of tubercles on M_2 internal to the second main row.

PTILODONTIDAE Simpson, 1927

Simpson, G. G., 1927, Amer. Mus. Novitates No. 267, pp. 1–2; 1928, Catalogue of Mesozoic Mammalia in the Geologic Department of the British Museum, p. 52

Chirogidae Cope, 1887, Amer. Nat., XXI, p. 567. Cimolodontidae Marsh, 1889, Amer. Jour. Sci. (3), XXXVIII, p. 84. Neoplagiaulacidae Ameghino, 1902, Ann. Mus. Nac. Buenos Aires (3), I, p. 11.

As a family name Neoplagiaulacidae Ameghino, 1902, has priority, as also Cimolodontidae Marsh, if the genera on which these names are based are shown to be valid genera and of the same family as *Ptilodus*. However, *Ptilodus* is certainly the " leading genus " of the family in the sense of being the best known and most important, although Ameghino's name was based on the first described genus of the family, and Cope's and Marsh's names were the first family names attached to the group. Cope's name is certainly based on an invalid generic name; Marsh's may be so.

Definition: Dental formula $\frac{2 \cdot 0 \cdot 3 - 4 \cdot 2}{1 \cdot 0 \cdot 1 - 2 \cdot 2}$. Enlarged incisors rooted, with extra-alveolar portion almost completely enameled or with restricted enamel band. P_3, when present, small, one-rooted, fitting into a notch in P_4. P_4 always very large, laterally compressed, trenchant, with serrate edge and curving ridges and grooves on sides. Anterior upper premolars not opposed by any lower teeth, grasping, with three to six conical cusps. Only the last upper premolar shearing in function, enlarged. M^1 with three cusp rows, the inner usually incomplete anteriorly; M^2 with three cusp rows, the outer always incomplete posteriorly. Molar cusps more or less definitely crescentic. Wear generally not reducing molar surfaces to planes but accentuating the longitudinal ridges and grooves. First molars much larger than second. Skull notably triangular in outline as seen from above. Animals of relatively small size.[184]

One genus of this family occurs in the Puerco and two in the Torrejon.

Ptilodus Cope, 1881

Cope, E. D., 1881, Amer. Nat., XV, Nov. 12, p. 921

Type: Ptilodus mediaevus Cope, from Middle Paleocene of New Mexico.

Synonym: Chirox Cope, 1884, Proc. Amer. Phil. Soc., XXI, p. 321; type: *C. plicatus.*

Author's Diagnosis: The characteristic obliquely ridged cutting tooth well known in *Plagiaulax.* It presents the following differences from those of *Plagiaulax* and *Ctenacodon*, which I regard for the present as generic.

[183] Granger, W., and Simpson, G. G., 1929, Bull. Amer. Mus. Nat. Hist., LVI, p. 623.
[184] Granger, W., and Simpson, G. G., 1929, Bull. Amer. Mus. Nat. Hist., LVI, p. 625.

Char. gen. Cutting edge convex and continuous with the anterior edge of the crown, and serrate from the union of ridges which ascend on each side. Ridges curved backwards, all reaching the edge excepting above the posterior root of the tooth, where they are discontinued, leaving a smooth edge. In *Plagiaulax* the ridges are continued to the posterior edge of the crown, and in *Ctenacodon* the ridges do not extend on the sides of the crown. In *Hypsiprymnus* the ridges are vertical.

In 1882 new material enabled Cope to revise his diagnosis of the genus as follows:

One large cutting premolar. Inferior premolars with several tubercles. Large premolar without posterior cusp; edge directed upwards; sides ridged.[185]

In 1884 [186] Cope described the upper teeth of *Ptilodus* as a new genus, *Chirox*, with the following diagnosis:

Char. gen. These are known from three superior molars; viz: the last premolar, and the second and third true molars [actually the second, third and fourth upper premolars]. The fourth premolar has two external, and one internal cusps, and the true molars have four cusps each. The cusps are of peculiar form. The second true molar resembles a convex body which has been divided by two cuts at right angles to each other, from which the quarters thus produced have spread away from each other subequally. The external faces of the cusps are convex. The apices are acute. The last superior molar is larger anteroposteriorly than transversely. The fourth premolar (supposed) is two-rooted.

These molar teeth remind one of the inferior molars of *Ptilodus*, though they differ much from them. The genus is probably nearer to *Catopsalis*, and belongs to the Marsupial order. The presence of only two series of cusps in the superior molars, distinguishes it from these genera, which have presumably three series of such cusps. Lemoine has shown this to be the case in *Neoplagiaulax*.

Although Cope does not refer to the " Allodontia " described and figured by Marsh from the Jurassic of Wyoming, his recognition of the plagiaulacid affinities of *Chirox* must have been based upon comparison with Marsh's published figures.

This genus is best known from the skull, jaws and partial skeleton described by Gidley in 1909 from the Silberling quarry in the Fort Union of Montana. The Torrejon species are nearly related. The following diagnosis cited from Granger and Simpson, 1929, is chiefly based upon the Gidley specimen:

Dental formula $\frac{2 \cdot 0 \cdot 4 \cdot 2}{1 \cdot 0 \cdot 2 \cdot 2}$. Lower incisors long and slender, pointed, not gliriform. P_3 present. P^1 with three cusps, P^2 with four. P^3 narrower than P^2 and with 4–6 cusps. P^4 with two complete longitudinal rows of cusps and an incipient anteroexternal third row. Upper molars with the inner row of M^1 incomplete. Cusps moderately crescentic.

Ptilodus mediaevus Cope

Ptilodus mediaevus Cope, 1881, Amer. Nat., XV, Oct. 28, p. 921; 1884, *ibid.*, XVIII, p. 694,
    fig. 8; 1885, Tertiary Vertebrata, p. 173, pl. xxiii*d*, fig. 1; Gidley, 1909, Proc. U. S.
    Nat. Mus., XXXVI, pp. 612–13; Granger and Simpson, 1929, Bull. Amer. Mus. Nat.
    Hist., LVI, pp. 629–31, figs. 14c, 15.
Chirox plicatus Cope, 1884, Paleont. Bull. No. 37, Jan. 2, and Proc. Amer. Phil. Soc.,
    XXI, p. 322.
Ptilodus plicatus Gidley, 1909, Proc. U. S. Nat. Mus., XXXVI, p. 614; = *Ptilodus
    mediaevus* Granger and Simpson, 1929, Bull. Amer. Mus. Nat. Hist., LVI, p. 629.

Type: A. M. Cope Coll. No. 3019, a premolar tooth, p_4. Type of *Chirox plicatus:* A. M.
Cope Coll. No. 3032, upper jaw fragment, p^{2-4}.

[185] Cope, E. D., 1882, Amer. Nat., XVI, May 20, p. 520.
[186] Cope, E. D., 1884, Paleont. Bull. No. 37, Jan. 2, and Proc. Amer. Phil. Soc., XXI, p. 321.

Horizon and Locality: Middle Paleocene, Torrejon formation, San Juan basin, New Mexico. Found with type of *Triisodon quivirensis.*

Author's Diagnosis (p. 922): *Char. specif.* The tooth is much larger than that of any of the *Plagiaulacidae* yet known, exceeding the corresponding one of the kangaroo-rat of Australia. There are twelve ridges on the side of the crown, extending from the base. They are crowded anteriorly and become more widely spaced posteriorly. The anterior margin is acute from near the base; the latter projects a little beyond the root. The most elevated point of the crown is between the roots. Ridges fine, enamel smooth. Length of base of sculptured part of crown, .0062; elevation of do., .0047; thickness of do. at base, .0025.

In 1884 Cope figured the lower jaw of *P. mediaevus* in the American Naturalist. The same figure appears in Tertiary Vertebrata, 1885 (pl. xxiiid, fig. 1), and is a composite of two specimens.

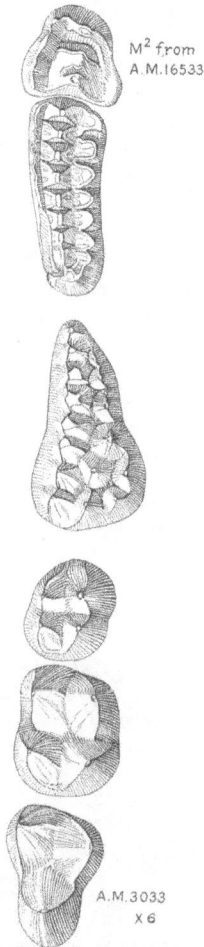

Fig. 77. *Ptilodus mediaevus,* right upper cheek teeth, A. M. Nos. 16533 and 3033, crown views. (From Granger and Simpson, 1929.) Six times natural size.

Type Description of *Chirox plicatus*: The external cusps of the fourth premolar [p²] are flattened on the external side, and lean a little inwards. The internal cusp (probably homologically the anterior) is opposite the anterior external, and has a convex internal face. Its apex is acute and compressed; the apices of the external cusps are trihedral and acute.

The cusps of the second true molars [p³] are more widely separated transversely than anteroposteriorly; that is, the longitudinal fissure is wider than the transverse. The apices are all acute, the internal trihedral, the external more compressed.

The transverse diameter of the last true molar [p⁴] is smaller than that of the second true molar, while the longitudinal is nearly the same. The crown projects convexly posterior to the posterior pair, and there is a small tubercle at the anterior base of the external anterior cusp.

None of the teeth preserved display cingula. The bases of the crown are smooth, but the cusps are sharply and finely parallel-grooved on their external faces. [Measurements follow.]

To this species Granger and Simpson refer the palate with finely preserved teeth described by Cope under the name of *Chirox plicatus*, as well as the upper jaw which served as type of that species; their diagnosis is as follows:

Length P_4 8.5–9.0 mm., 12 serrations on edge. Length M_1 (referred specimen, slightly smaller than type) 3.7 mm., cusp formula 6:4, anteroexternal cusp minute. P³ with four cusps, P⁴ with eight cusps in inner main row, six in outer main row, and two anteroexternal accessory cusps.

The species characters are described by them in detail and need not be repeated here.

Ptilodus trovessartianus Cope

Ptilodus trovessartianus Cope, 1882, Amer. Nat., XVI, July 28, p. 684; 1885, Tertiary Vertebrata, p. 737, pl. xxvf, fig. 19; Amer. Nat., XIX, p. 493; Osborn, 1893, Bull. Amer. Mus. Nat. Hist., V, p. 315, fig. 1; Gidley, 1909, Proc. U. S. Nat. Mus., XXXVI, p. 614; Granger and Simpson, 1929, Bull. Amer. Mus. Nat. Hist., LVI, p. 630, figs. 13A, 14A.

Type: A. M. Cope Coll. No. 3025, three lower premolars, one on fragment of jaw; incisors, etc.

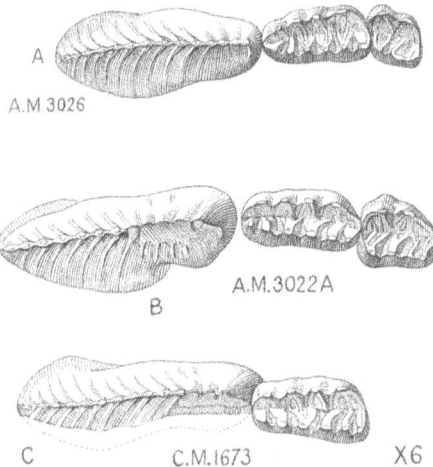

FIG. 78.   *A, Ptilodus trovessartianus,* left p₄–m₂, A. M. No. 3026; *B, Ptilodus ?montanus,* left p₄–m₂, A. M. No. 3022A; *C, Ptilodus montanus,* left p₄–m₁ of type, Carnegie Museum No. 1673, crown views. (From Granger and Simpson, 1929.)   Six times natural size.

Horizon and Locality: Paleocene of San Juan basin, undoubtedly from the Torrejon formation, Upper Paleocene.

Author's Description: This species is represented by three of the characteristic fourth inferior premolars, one of which stands on a part of the ramus, giving its depth. These differ from those of the *P. mediaevus* in their uniformly smaller size, and in their strongly serrate posterior edge. The number of lateral edges [misprint for ridges] is 12, as in *P. mediaevus*. [Measurements follow.]

In 1885 in a note in the American Naturalist (XIX, p. 493), Cope referred to this species two well-preserved mandibles containing both molars and premolars. This specimen, No. 3026, is recorded as from Gallegos Canyon, and " middle horizon of the Puerco," probably the lower fossiliferous horizon of the Torrejon. It was figured by Osborn in 1893 in his review of Marsh's Cretaceous Mammalia.

No additional material of importance has been referred to this species in recent years. Gidley considered it is poorly characterized except for its small size. Granger and Simpson regard the species as well distinguished, with the following diagnosis:

Length of P_4 5.8–5.9 mm., 13 or 14 serrations. Length M_1 (referred specimen) 3.1 mm., cusp formula 6:4, anteroexternal cusp stronger than in *Pt. mediaevus*. Length M_2 1.7 mm. Cusp formula 3:2, M_2 smaller relative to M_1 than in other known species, and M_1 relatively larger and more slender.

Eucosmodon Matthew and Granger, 1921

Matthew, W. D. and Granger, W., Amer. Mus. Novitates No. 13, p. 1

Type: Neoplagiaulax americanus Cope.

Author's Diagnosis: Agrees with the true *Neoplagiaulax*, of the Cernaysian of France, in absence of P^3 (present in *Ptilodus*), but differs from both of these genera and from *Ectypodus* in the large, compressed, fully scalpriform incisor, rootless or nearly so. The species are of considerably larger size than those of the three above-named genera.

Revised Diagnosis, Granger and Simpson, 1929: P_3 absent, P_4 typically lower relative to length than in other ptilodontids. Lower incisors strongly compressed laterally, with enamel limited to a narrow longitudinal band. Species largest of this family.[187]

This genus is distinguished from all other Plagiaulacidae by the peculiar compressed and thoroughly gliriform incisors. Four species are known, and one subspecies, from the Puerco-Torrejon and Wasatch formations. They are the largest of the Ptilodontidae. Granger and Simpson remark that the long geological range, beyond that of any other Paleocene mammal, indicates that a better knowledge would probably result in showing that more than one genus is included in this assemblage.[188]

To *Eucosmodon*, sp. indet., is referred the important specimen, A. M. No. 16024, from the lower beds of the true Puerco, consisting of lower jaws associated with many skeletal fragments. This was described and figured in 1929 (Granger & Simpson, *op. cit.*) and its bearing upon the affinities of the Multituberculata discussed. Simpson and Elftman in 1929 (Amer. Mus. Novitates No. 333) reconstructed the musculature as indicated in this and other specimens and discussed the probable habits and environment. It has considerable analogy but no very close relationship to the more primitive diprotodonts, and seems to have been a forest-dwelling type, with some indications of arboreal adaptation.

[187] Granger, W., and Simpson, G. G., 1929, Bull. Amer. Mus. Nat. Hist., LVI, p. 635.

[188] A recent paper by Jepsen (1930, Proc. Amer. Phil. Soc., LXIX, p. 121), in part substantiates this observation.—THE EDITORS.

Eucosmodon americanus (Cope)

Neoplagiaulax americanus Cope, 1885, Amer. Nat., XIX, p. 493; (*Eucosmodon*) Matthew
and Granger, 1921, Amer. Mus. Novitates No. 13, p. 1; Granger and Simpson, 1929,
Bull. Amer. Mus. Nat. Hist., LVI, p. 648, fig. 27.

Type: A. M. Cope Coll. No. 3028, lower jaws, and p_4 l., p_4 r., m_{1-2} r., parts of other
teeth.

Horizon and Locality: Lower Paleocene, Puerco formation, San Juan basin, New Mexico.

Author's Description: Size a little exceeding that of the *Ptilodus mediaevus*, and many times larger
than the *N. eocænus* Lemoine. The large fourth premolar is less elevated than in the two species mentioned.
Its cutting edge is obtusely serrate, and the lateral keels though fine, as in *P. mediaevus*, are only seven in
number instead of twelve. The posterior base wears into a little truncation. The molars are much as in
the species named. The tubercles are coarse and number four on each side on the first, and two on each side
on the second. The incisor is much compressed, and the enamel band is perfectly smooth. The coronoid
process rises opposite the second tubercle of the first true molar. [Measurements follow.]

Revised Diagnosis, Granger & Simpson, 1929: P_4 with not less than 12 serrations, 10.6 mm. long
in type. Maximum height of incisor 6.0 mm., width 2.7 mm.; ratio height: width ca. 2.22. Length M_1
5.6 mm., width 2.6 mm., cusp formula 5:4.

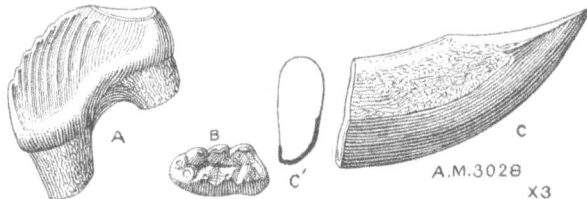

Fig. 79. *Eucosmodon americanus*, cotypes, A. M. No. 3028. *A*, left p_4, external view; *B*, right m_1, crown view;
C, right lower incisor, external view; C^1, transverse section of lower incisor. (From Granger and Simpson, 1929.) All
three times natural size.

To this species Granger and Simpson refer a premolar and incisor found by the expedition of 1913. They remark that the pelvis and hind limbs mentioned above may belong
to this form, which they regard as a distinct subspecies, *Eucosmodon americanus primus*, and
diagnose as follows:

Incisor smaller than in *E. americanus*, max. height 4.2 mm., width 1.9 mm., ratio ca. 2.21. P_4 shorter
than in *E. americanus* and with only nine serrations.

Eucosmodon molestus (Cope)

Neoplagiaulax molestus Cope, 1886, Amer. Nat., XX, April 24, p. 451; 1888, Trans. Amer.
Phil. Soc., XVI, N.S., p. 307 and pl. V, fig. 10 [not fig. 11 of same plate]; (*Eucosmodon*)
Granger and Simpson, 1929, Bull. Amer. Mus. Nat. Hist., LVI, p. 650, figs. 29, 30.

Type: A. M. Cope Coll. No. 3029, a lower premolar, with skeletal fragments.
Neotype: No. 17063, parts of lower jaw with incisor and p_4.
Horizon and Locality: Paleocene, north side of San Juan River, New Mexico. Probably
Middle Paleocene, the neotype certainly so.

Author's Description: Established on an entire inferior fourth premolar. The length of the base
of this tooth is one third greater than that of the corresponding tooth of the *N. americanus*, and there are
fifteen keel-crests on the side of the crown, while there are but seven in the N. americanus, and thus not so

elevated as in our species of Ptilodus.   The irregularity in the outline of the base of the crown is less than in the other species, and the diameter of the roots is subequal.   The anterior base of the crown is not excavated for the fourth premolar as in the species of Ptilodus.   [Measurements follow.]

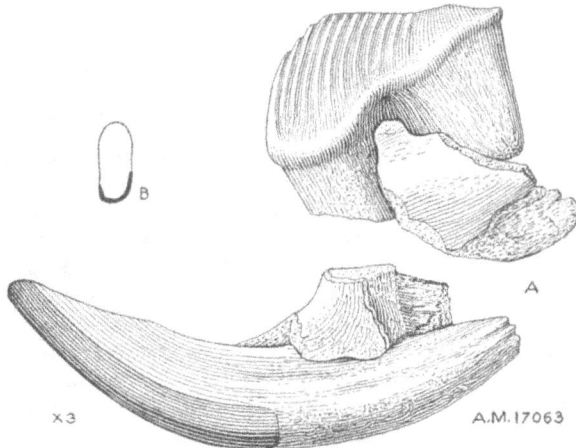

FIG. 80.   *Eucosmodon molestus*, neotype, A. M. No. 17063.   *A*, parts of left lower jaw with incisor and p₄; *B*, transverse section of incisor.   (From Granger and Simpson, 1929.)   Three times natural size.

Cope referred to this species in 1888 a lower jaw without teeth from the Puerco, A. M. No. 3030, but Granger and Simpson regard the specimen as indeterminate as to genus; they point out that it is considerably too large to belong to *Eucosmodon molestus* and comes from the Puerco, whereas the type and neotype are from the Torrejon.

Eucosmodon teilhardi Granger and Simpson

Eucosmodon teilhardi Granger and Simpson, 1929, Bull. Amer. Mus. Nat. Hist., LVI, p. 651, figs. 31, 32, 33.

Type: A. M. No. 16024, lower jaws with roots of teeth, associated with various skeleton fragments.

Paratypes: A. M. No. 2375, lower jaw with broken p₄ and part of incisor; A. M. No. 16023, lower jaw with incisor complete.

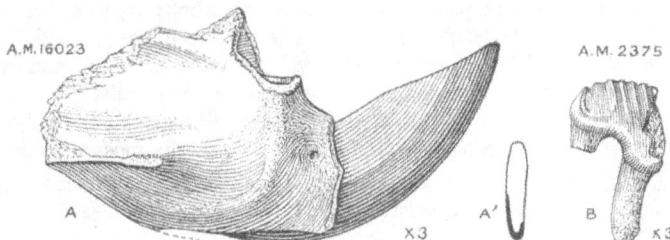

FIG. 81.   *Eucosmodon teilhardi*, paratypes.   *A*, part of right lower jaw, A. M. No. 16023, external view; *A*¹, transverse section of incisor; *B*, right p₄, A. M. No. 2375, external view.   (From Granger and Simpson, 1929.)   Three times natural size.

Horizon and Locality: Torrejon formation, San Juan basin, New Mexico.

Author's Diagnosis: Incisor, max. height 5.2 mm., width 1.6 mm., ratio 3.25.   Length P₄ at alveoli
5.5 mm.   P₄ relatively short and high.

This species is quite distinct from the two preceding and its generic reference is somewhat doubtful.

PLAGIAULACIDAE Gill, 1872

Gill, T., 1872, Smithsonian Misc. Coll. No. 230, p. 27

The family Plagiaulacidae was proposed by Gill in 1872 to contain *Plagiaulax* Falconer, 1857; it was placed as of uncertain affinities among the marsupials, but was not defined or characterized.   Marsh in 1879 and 1880,[189] in describing *Ctenacodon*, associated it with *Plagiaulax* in the family Plagiaulacidae, to which he added *Allodon* in 1881.[190]

Cope in 1882 [191] specified definite characters by which the Plagiaulacidae could be distinguished from the Macropodidae, referring them to the order Marsupialia.   He points out that Marsh's list of characters of the "Order" Allotheria does not in any respect separate it from the Marsupialia, to which the single known family, the Plagiaulacidae, had been referred by all previous authors.   He gives a key to the genera, which was corrected two months later as a result of better knowledge of the dentition of *Ptilodus* and *Thylacoleo* of the Australian Pleistocene added to the family.

In 1884 Cope published in the American Naturalist (June 17, p. 687) a review of the "Tertiary Marsupialia," in which he named and characterized the Multituberculata as a suborder, with three families distinguished as follows:

Fourth premolars (and probably others) more simple than first true molars...........Polymastodontidae
Fourth superior premolar (at least) like true molars....................................Tritylodontidae
Fourth premolars (and often others) developed into flat cutting blades..................Plagiaulacidae

Catopsalis is here made a synonym of *Polymastodon*, the jaw of *C. foliatus* and jaws and skeleton parts of *P. taöensis* are figured, and the view is advanced that the family "is probably the ancestral type of the kangaroos. . . . On the other hand, the Polymastodontidae may well have derived their origin from the Tritylodontidae."

The Plagiaulacidae in turn are regarded as probably ancestral to *Thylacoleo*.   Seven genera are admitted, *Plagiaulax*, *Ctenacodon* and *Plioprion* from the Jurassic, *Meniscoëssus* from the Cretaceous, and *Ptilodus*, *Neoplagiaulax* and *Liotomus* from the Eocene.   *Ptilodus mediaevus* is figured and distinguished from *Neoplagiaulax* and other genera.   A phylogeny of the genera is presented, which derives *Thylacoleo* directly from the Eocene plagiaulacids but, disagreeing with the text, places *Polymastodon* on a wholly separate line of descent from the kangaroos, although *Tritylodon* is the ultimate ancestral type for both.

Cope's discussion of the group in Tertiary Vertebrata, although published later, was written considerably earlier than the above article in the Naturalist.

Osborn and Earle in 1895 [192] and Matthew in 1897 [193] regarded the Polymastodontinae as a subfamily of Plagiaulacidae, an arrangement accepted by most subsequent writers.

[189] Marsh, O. C., 1879, Amer. Jour. Sci. (3), XVIII, p. 397; 1880, *ibid.*, XX, p. 239.   Marsh apparently intended to propose the family as new, as far as one may judge from the wording of his papers.—w. d. m.

[190] 1881, *ibid.*, XXI, p. 511.

[191] Cope, E. D., 1882, Amer. Nat., XVI, p. 416.

[192] Bull. Amer. Mus. Nat. Hist., VII, p. 11.

[193] *Ibid.*, IX, p. 263.

METATHERIA

7. ORDER MARSUPIALIA

DIDELPHIIDAE Gray, 1821

Gray, J. E., 1821, London Med. Repos., XV, p. 308 (Didelphiidae)

The true opossums are represented in the Tiffany fauna by well-preserved and nearly complete upper and lower jaws described [194] under the name of *Peradectes elegans*. Another species of *Peradectes* occurs in the Bridger,[195] and specimens of doubtful reference have been found in the Lower Eocene formations. Among the trituberculate mammals of the Lance (" Laramie ") fauna occur various fragments of jaws and isolated teeth of minute species, some of which have been named and figured by Marsh under the names of *Pediomys*, *Batodon* and *Telacodon*. These probably belong either to the Didelphiidae or to the closely allied Cimolestidae, so far as one can judge from Marsh's figures and from the characters of similar material in the American Museum Lance collection. The lower teeth show the characteristic didelphid position of the hypoconulid. The correlation of the upper and lower teeth of these minute species from the Lance, and the synonymy of the genera, are quite uncertain. *Pediomys* is founded on a last upper molar, with which Marsh subsequently correlated a jaw fragment with the last lower molar; *Batodon* is founded on a lower jaw fragment with the last two molars, *Telacodon* on a jaw fragment with premolars.[196]

Batodon and other small related genera have been recorded from the Fort Union and from the Paskapoo, but the reference is very doubtful.

From the Torrejon no true marsupials are recorded, and in the entire collection studied by us for the present revision there is apparently not a single tooth referable even provisionally to this order. The same may be said of the Puerco collection, with a single exception— a fragment of the lower jaw with m_{1-3} of a species about the size of *Marmosa*. It is distinct generically from *Peradectes* and about twice as large lineally as *P. elegans*. The molars do not agree closely with any corresponding teeth in the American Museum Lance or Paskapoo collections; they are distinct generically from the lower molar referred by Marsh to *Pediomys*, and also from *Batodon* if Marsh's figure be accurate. With *Telacodon* no comparisons are possible. It appeared best therefore to give a new generic name to the Puerco opossum.

Thylacodon Matthew and Granger, 1921

Matthew, W. D., and Granger, W., 1921, Amer. Mus. Novitates No. 13, p. 2

Type: Thylacodon pusillus from the Lower Paleocene, Puerco formation of New Mexico.

Author's Description: Molar teeth of didelphid type, rather narrow as a whole, the trigonid relatively high, with reduced paraconid. Metaconid and protoconid high, well separated, acute, the protoconid considerably the higher. Talonid deeply basined with acute marginal cusps, the entoconid internal, hypoconulid well developed on all molars and nearly postero-internal in position (a character-

[194] Matthew, W. D., and Granger, W., 1921, Amer. Mus. Novitates No. 13, p. 2.

[195] Matthew, W. D., 1909, Mem. Amer. Mus. Nat. Hist., IX, p. 339, pl. L, fig. 9.

[196] In this connection the reader is referred to: Simpson, G. G., 1928, Amer. Mus. Novitates No. 307. Simpson, G. G., 1929, "American Mesozoic Mammalia," Mem. Peabody Mus., Yale Univ., III, Pt. I.— THE EDITORS.

istic didelphid construction), hypoconid postero-external. The high bicuspid trigonid and reduced paraconid serve to distinguish this from other didelphid genera; the tooth is also unusually narrow and the hypoconulid and entoconid more distinct than usual.

Thylacodon pusillus Matthew and Granger

Thylacodon pusillus Matthew & Granger, 1921, Amer. Mus. Novitates No. 13, p. 2.

Type: No. 16414, a fragment of lower jaw with m_{2-3}, the heel of m_1 and part of alveolus of m_4.

Horizon and Locality: Upper level of the Puerco formation near Ojo Alamo, San Juan basin, New Mexico. A. M. Exped., 1913.

Diagnosis: Size of the larger species of *Marmosa*, about a third larger than *M. chapmani*.

The type and only specimen is a fragment of the right ramus of the lower jaw, with the second and third molars, the heel of m_1 and part of the alveolus of m_4. The teeth are un-

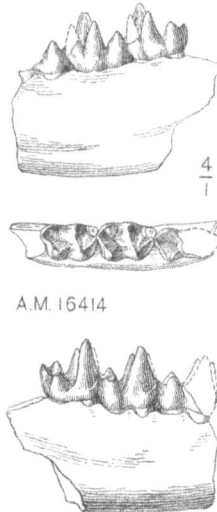

Fig. 82. *Thylacodon pusillus*, type, right lower jaw, A. M. No. 16414, internal lateral view above, crown view in middle, external lateral view below. Four times natural size.

worn and very perfectly preserved, except that the paraconids are broken off. The construction of the molars is that characteristic of the didelphids as shown by their bases, and by the height of the two trigonid cusps, which is suggestive of Leptictidae. They have not the characteristic form of the leptictids, and the hypoconulid is internal instead of median. The hypoconulid of m_1 overlaps upon the base of the trigonid of m_2, showing that the latter has emerged later from the jaw. This proves that the first tooth is a molar and not a premolar, for if it were p_4 it would be of later emergence than the tooth behind it, and could not overlap it. There are therefore four true molars, including the one represented by its alveolus. In the Leptictidae the molariform p_4 is usually distinguishable from the first true molar by being deeper set in the jaw, less completely emerged, and its heel never overlaps the base of the tooth behind it.

The jaw is compressed and of moderate depth, corresponding, as far as it goes, with that of *Marmosa*. It is considerably deeper than in *Peradectes elegans*, which is a much smaller animal with trigonid of more normal didelphid proportions.

Thylacodon is no doubt nearly allied to the smaller didelphoid genera of the Lance, but present evidence, although far from conclusive, seems to be against generic identity with any of them. It is certainly distinct from *Peradectes, Peratherium*, or any of the existing genera. Its relations to the didelphoid species of the Fort Union are wholly uncertain, as none of these have been figured or described.

III. GENERAL OBSERVATIONS AND CONCLUSIONS

A. SALIENT CHARACTERS OF THE PUERCO–TORREJON MAMMALS

GENERAL DISCUSSION

Two primary groups are included, the multituberculates and the trituberculates.

The multituberculates are a small group of genera, the affinities of which have been much disputed. Although resembling certain marsupials in dentition they represent certainly a distinct order (Multituberculata) and may even be entitled to rank as a distinct subclass, Allotheria, combining prototherian and metatherian peculiarities with others distinct from either group. They certainly are not placentals, and their relationship to the primitive placentals, or Bunotheria, is a decidedly remote one. The five genera, *Taeniolabis*, *Catopsalis*, *Ptilodus*, *Ectypodus* and *Eucosmodon*, included in the families Plagiaulacidae, Taeniolobididae and Ptilodontidae are the last survivors of a Mesozoic group. The molars are never tritubercular.

The trituberculates comprise the great bulk of the Puerco-Torrejon mammals. Their molar teeth are mostly of the typical tritubercular pattern. In the few exceptions they are so nearly related and so little altered from it that their derivation is manifest.

The tritubercular pattern is seen in a wide variety of adaptive specializations of the teeth and very evident diversity of food habits, proportions and size. *Claenodon*, with flat-crowned, frugivorous molars, *Didymictis*, with viverrine, carnivorous teeth, *Pantolambda*, with its cusps as sharply crescentic as in a ruminant, or *Ectoconus*, with rounded, pig-like cusps on the grinders—all display the unmistakable pattern, the central type of which is represented by the numerous genera with unspecialized molars very much alike and difficult to differentiate—such genera as *Chriacus*, *Oxyclaenus*, *Goniacodon*, *Triisodon*, *Mioclaenus*, *Oxyacodon*, *Carcinodon*, *Pentacodon*.

The primitive tritubercular molar type of tooth is much more definite and precise in its constructive details than the mere name " tritubercular " would convey. The upper molar consists of three main cusps, the two outer ones (paracone and metacone) of subequal size, tending to be conical, the inner one (protocone) larger and more or less clearly crescentic (convex inwardly). An external cingulum is frequently present and may develop minor marginal cusps at three points, the anterior angle (parastyle), the posterior angle (metastyle), and half way between (mesostyle). A pair of small cusps (paraconule, metaconule) may be present on or within the two wings of the protocone crescent; these sometimes become important cusps. A postero-internal cingulum is developed to a varying extent on the flank of the protocone and may develop into an important cusp (hypocone). A cusp of similar function but dissimilar origin (pseudo-hypocone) may (1) result from the twinning of the protocone, or (2) bud off from its posterior wing. A cingulum is often present on the anterior flank of the protocone, and sometimes gives rise to a cusp (protostyle).

The lower molar consists of three main cusps comprising the trigonid and a basined heel or talonid. The trigonid fits into the interspace between two upper teeth, the basined heel receives the protocone of the upper molar. Of the three trigonid cusps the anterior one (paraconid) frequently decreases and disappears very commonly in correlation to the growth of the hypocone and squaring up of the upper molar. The posterior cusps of the trigonid are the protoconid and the metaconid. The talonid has normally an outer and inner cusp (hypoconid, entoconid) and frequently an intermediate posterior cusp (hypoconulid).

The above are the essentials of the tritubercular molar. Excepting the multituberculates, all Paleocene mammals are either of this type or quite near to and obviously derivable from it.

All the trituberculate genera, as far as known, except *Thylacodon?* either have the normal primitive placental dental formula i_3^3, c_1^1, p_4^4, m_3^3, or are not far removed from it.

Normal: *Tetraclaenodon, Periptychus, Ectoconus, Pantolambda, Claenodon, Oxyclaenus, Chriacus, Hemithlaeus, Conacodon, Anisonchus, Haploconus, Mioclaenus, Ellipsodon, Loxolophus, Triisodon?, Oxyacodon, Diacodon, Onychodectes, Goniacodon, Protogonodon, Eoconodon, Dissacus, Protoselene.*

P¹ absent: *Conoryctes, Wortmania.*

$P_{\overline{1}}$ absent: *Deltatherium, Tricentes.*

M³ absent: *Didymictis.*

Incisors reduced: *Mixodectes, Conoryctes, Wortmania, Psittacotherium.*

There is positive evidence in several genera of the normal placental milk dentition preceding the incisors, canines and premolars. In the remaining genera the emergence and wear of the teeth and other indirect evidence indicate that they also were completely diphyodont (save perhaps for p_1). No evidence of replacement of the true molars has been seen, nor of a pre-lacteal series; the latter would probably not be recognizable in the fossils if it existed, but the non-replacement of the true molars may be regarded as sufficiently proven.

The skull characters of all the tritubercular genera are much alike and conform wholly to the primitive placental type. They do not show any approach to the marsupial type in the relations of the jugal, the alisphenoid, the angle of the jaw, but only in such characters as may be regarded as persistently primitive in marsupials, inheritances from a " metatherian " stage in the evolution of Mammalia. Such are the small brain, high sagittal and occipital crests, long nasals widely expanded posteriorly, lack of postorbital process, narrow postorbital constriction, posterior position of orbits with moderate expansion of lachrymal upon the face, but lachrymal foramen within orbital margin, long and narrow parietals, considerable lateral exposure of mastoid, prominent mastoid, paroccipital and postglenoid processes, lower jaw with transverse condyle, deep, flat, vertical angle projecting backward in a sharp spine, high recurved coronoid process, long and moderately deep jaw with usually heavy canines, small incisors, trenchant premolars and tritubercular molars.

The skeleton characters, while similar for the most part in all the trituberculates, show a few deep-seated distinctions which would scarcely be inferred from the underlying unity of tooth structure. These are best illustrated in the structure of the feet, and in particular of the astragalus. There are certainly two, and possibly four, fundamental groups:

a. Creodont-Condylarth group, with asymmetric trochlea, distinct neck, oval or rounded head, astragalar foramen present.

b. Taligrade group, closely related to *a* but with short neck and wide, shallow trochlea.

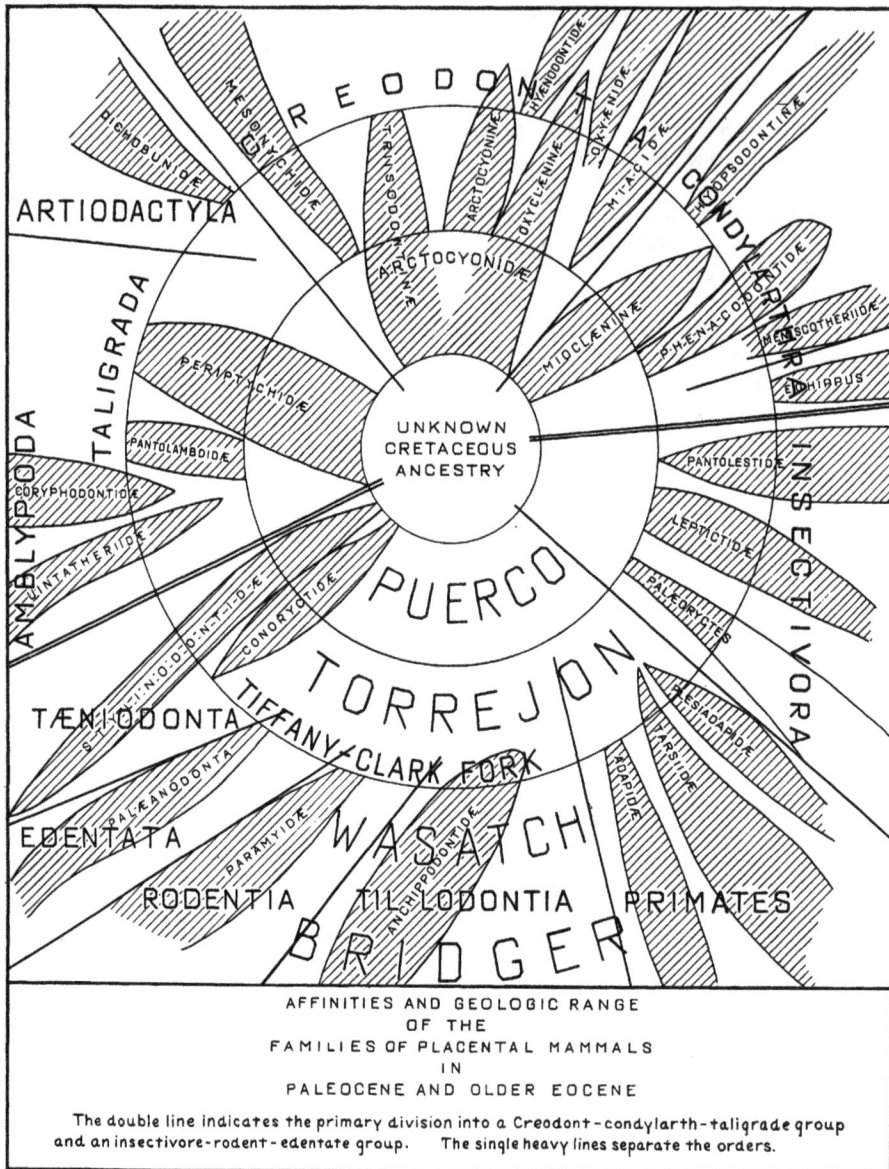

FIG. 83. Affinities and Geologic Range of the Families of Placental Mammals in the Paleocene and Older Eocene.

c. Insectivore-Rodent-Edentate group, with wide symmetric trochlea, bounded by sharp crests, distinct neck, oval to flattened head, no astragalar foramen.

d. Primate group, apparently related to *c* but with trochlea less symmetric, shallow, crests obscure, neck long and head round or oval. No astragalar foramen, but usually a considerable palmad process behind the trochlea.

Group a	Group b	Group c	Group d
Creodont-Condylarth (ancestral to Carnivora, Artiodactyla, Perissodactyla, Notoungulata)	Taligrade (ancestral to Amblypoda, Arsinoitheria, Pyrotheria, Astrapotheria, Hyracoidea)	Insectivore (ancestral to Insectivora, Rodents, Edentates)	Lemuroid (ancestral to Primates)
Astragalus	Astragalus	Astragalus	Astragalus
trochlea asymmetric, shallow or flat	trochlea asymmetric, flat	trochlea symmetric, wide and short, sharp crests	trochlea asymmetric, flat or shallow
neck distinct	neck short or none	neck distinct, rather short	neck rather long
head round-oval, strongly convex	head oval, little convex	head expanded, flat-oval	head rounded, not expanded
foramen present	foramen present	foramen absent	foramen absent

It is not altogether certain that type *d* is represented in the Paleocene; it is apparently absent from the Torrejon and Puerco. The very distinctive perissodactyl and artiodactyl types of astragalus are conspicuously absent from the Paleocene. They appear, fully developed, in the earliest Eocene strata, and while they may be ultimately derived from the Creodont-Condylarth type of astragalus the relationship is certainly a distant one. The Creodont-Condylarth type, on the other hand, leads up quite readily and directly into any of the specialized Creodonta, Carnivora and Notoungulata. Fundamentally it has the body of the astragalus wedged in between tibia and fibula, the primary reptilian position. The progressive evolution of the foot involves a shifting of this relation, the astragalus coming to lie wholly under the tibia, while the fibula is reduced to an external malleolus of the tibia, with or without a rest on the calcaneum.

The second or taligrade astragalus is clearly ancestral to various types of " subungulata "—the Amblypoda, the Proboscidea, the arsinoitheres, pyrotheres, ?astrapotheres, etc. It is probably a derivative of type *a*; at all events the smaller and more primitive taligrades approach that type. The astragalar foramen is a distinct feature of both these types, continued in various derivative phyla up to a certain grade of specialization in digitigrade or unguligrade evolution and then disappearing, retained in plantigrade types unless the plantigradism is secondary.

The third type of astragalus is still the prevalent type among insectivores and in a slightly modified form among rodents. In this form the body of the astragalus lies directly under the tibia, the fibula being more or less closely united to the tibia, the tarsus relatively short and flat, the astragalus without foramen and cut short posteriorly, the neck distinct and head wide and flat-oval.   While this type is theoretically derivable from the primitive creodont type, yet it is distinct and definite as far back as the early Paleocene; it is as old as the other type, as far as the actual record goes, for of pre-Tertiary astragali we know nothing whatsoever.   It is found in three types—leptictids, pantolestids, metacheiromyids —which have otherwise but little in common.   Perhaps it signifies merely that they had all reached an equivalent stage of specialized adaptation of similar type and are not especially related in any other way.

The fourth or lemuroid type is found in the Notharctidae; whether in any others of the so-called primates is open to question.   It is not altogether clear whether this is a distinct primary type or a partial reversion from the primitive insectivore type towards the primitive creodont type.   The latter is plausible, for reasons elsewhere to be considered (p. 317).

TRITUBERCULY AND THE CHARACTERS OF THE TEETH

All Paleocene mammals, aside from multituberculata, have teeth of the tritubercular pattern or nearly related thereto, but the animals are not so nearly related, as the tooth formula and skeleton characters show.

The tritubercular pattern must therefore be viewed not wholly as proof of near relationship but as indicating similar adaptation of teeth in several stocks long since separated, but retaining a primitive and rather limited environmental adaptation.   Among modern mammals the pattern is most nearly retained in stocks of arboreal habit and a somewhat generalized insectivorous-frugivorous diet, such as the opossums, tree-shrews and smaller lemuroids.

The Cretaceous ancestors of the Paleocene mammals must therefore have included several distinct stocks of similar habit.

The known Cretaceous mammals, aside from multituberculates, unless the Paskapoo fauna be included, are opossum-like marsupials.   No conclusive proof exists of Cretaceous placentals; nevertheless there must have been such.[197]   Their habitat or geographic distribution must be the explanation of their non-occurrence in the Lance and other dinosaur faunas.

The Jurassic mammals are associated also with dinosaurs, and represent probably a similar facies, but much more ancient.   It is therefore against the probabilities that the known Mesozoic mammals include the direct ancestry of any large part of the Paleocene fauna.

Enlarged and gibbous premolars are characteristic of a considerable number of Paleocene placentals.   It is evidently an adaptive, not an inherited, character and is associated with low-crowned crushing molars.   It also appears in some of the Lance didelphids which assuredly have no very close relations with the Paleocene periptychids, mioclaenids, etc.   It does not appear in the Jurassic-Comanchic mammal fauna.   Perhaps this is to be associated with absence of deciduous hard-shelled berries or nuts.

The mammalian dental formula and tooth succession appears to be inherited from cynodont or theriodont reptiles.   The differentiation between marsupials and placentals appears to turn on the partial suppression of the milk dentition (*vide* Matthew, 1916).

[197] They have been reported by Simpson, since the above lines were written, in the Lance fauna, from isolated teeth, and positively identified in the Cretaceous of Mongolia.—w.d.m.

The numerous teeth of some Jurassic mammals are very probably a secondary matter, as they are in *Myrmecobius*, some armadillos, *Otocyon*, *Centetes*, Sirenia and Cetacea.

The Paleocene mammals may be classified in regard to their teeth as follows:

A. Three premolars and four molars...Marsupials.
    Peradectes. Tritubercular molars, stylar cusps rudimentary.

B. Four premolars and three molars except as noted....................................Placentals.
    I. Carnivora as to skeleton (astragalus beneath tibia and fibula, feet sub-prehensile). Large sharp canines, tritubercular molars.
        a. Three premolars.
            Deltatherium. Strong external cingulum, rudimentary stylar cusps, high trigonids.
            Tricentes. Weak external cingulum, no styles, low trigonids.
        b. Four premolars.
            Chriacus. Sharp cusps, wing-hypocone.
            Protogonodon, Claenodon, Neoclaenodon. Low cusps, sub-round teeth.
            Loxolophus, Oxyclaenus. Low cusps, rather sharp, small hypocones.
        c. Four premolars, two molars, p^4, m_1 enlarged and specialized as carnassials.
            Didymictis.

    II. Condylarthra as to skeleton (feet locomotive). Small canines, more or less bunodont cheek teeth.
        a. Sexitubercular molars, p^4_4 submolariform.
            Tetraclaenodon.
        b. Tritubercular molars, p^4_4 simpler.
            Mioclaenus.
            Ellipsodon.

    III. Taligrada as to skeleton (feet sub-rectigrade). Molars with tritubercular construction.
        a. Bunodont, canines not enlarged.
            Periptychus. Premolars enlarged, ribbed, molars low-cusped, polybunar.
            Conacodon.
            Hemithlaeus.
            Haploconus. }High cusps, internal styles variously developed.
            Anisonchus.
            Ectoconus. Molars enlarged, polybunar.
        b. Selenodont, canines dicotylid.
            Pantolambda.

    IV. Insectivora as to skeleton (astragalus beneath tibia, feet sub-prehensile). Molars tritubercular, weak external cingulum.
        Palaeoryctes.
        Diacodon. }P^4_4 molariform. Cusps of cheek teeth acute, paraconid slow, protoconids
        Leptacodon. and metaconids subequal.
        Mixodectes. P^4_4 simple, incisors diprotodont.
        Pentacodon. P^4_4 simple, anterior premolars small, front teeth unknown.

    V. Taeniodonta as to skeleton (feet fossorial). Molars and premolars with low conical cusps capping an inflated massive crown, and connate roots. Diprotodont front teeth.
        a. Tritubercular, taeniodont characters very rudimentary in teeth or skeleton.
            Onychodectes.
         b. Upper molars with paracone and metacone low, separate, taeniodont characters moderately developed.
            Conoryctes.
         c. Upper molars with paracone and metacone high connate, taeniodont characters well developed.
            Wortmania.
            Psittacotherium.

    VI. Primate as to skeleton (feet scandent). Molars low-crowned, tritubercular.
        Plesiadapis.

The Paleocene mammals are by no means the oldest-known Mammalia, but they are the first of which we know the skulls and skeletons sufficiently to afford a really certain and indisputable basis for their affinities to each other and to earlier and later Mammalia. They may be grouped to advantage around the best-known types:

Ectoconus, Periptychus and *Pantolambda* are typical of the Taligrada. They are known from the skulls and skeletons herein figured and their detailed characters are elsewhere described. They present three distinct but related types of dentition. They show the placental dental formula and the placental tooth-replacement. The pattern of the molars is tritubercular; the premolars are simple but with internal cusps on the posterior upper premolars and posterior heels on the posterior lower premolars.

Ectoconus has bunodont teeth, the molars large and tending to polybuny, the premolars reduced.

Periptychus has bunodont teeth, the molars tending to polybuny, the premolars enlarged and gibbous, with a very characteristic vertical ribbing.

Pantolambda has selenodont teeth, the molars larger than the premolars and both with strongly crescentic cusps.

Haploconus, Anisonchus, Conacodon and *Hemithlaeus* are unmistakably associated with *Periptychus* and *Ectoconus* by the detailed pattern of the teeth. All show obscurely the characteristic vertical ribbing of the premolars; all have the premolars to a moderate extent enlarged and gibbous, the molars develop additional cusps but retain tritubercular symmetry as they do in *Periptychus* and *Ectoconus*.

The Mioclaenidae are likewise related to the taligrade group by their tooth characters, but have simpler tooth patterns; there is no trace of a vertical ribbing on the cusps, but the premolars are to a varying extent enlarged and gibbous. The skeleton is imperfectly known but appears to be of condylarth rather than taligrade affinities.

Tetraclaenodon (Euprotogonia) is known from incomplete skulls and skeletons. It is a primitive condylarthran, unmistakably related to *Phenacodus*. The molars are quadritubercular below, sextubercular above, the last premolar partly molariform. Feet slender with narrow, claw-like hoofs. In teeth, skull and skeleton construction it approaches the oxyclaenid creodonts.

Claenodon, Loxolophus and *Deltatherium* are known from skulls and characteristic parts of the skeleton. These are primitive Carnivora, as shown in the basicranial structure, limbs and feet. They have tritubercular tooth pattern, simple premolars, the posterior upper with internal cusp, the posterior lower with heels; as compared with taligrades and condylarths the molar cusps are mostly sharper, the premolars higher and compressed, not inflated, the canines large, sharp, often spaced. *Claenodon* is exceptional in the flattening of the molar crowns.

Chriacus, Tricentes, Oxyclaenus, Protogonodon, Neoclaenodon and *Carcinodon* are associated by tooth characters with the above.

Didymictis has the characteristic carnassident specialization of the teeth, associating it with the modern Carnivora, although in teeth as well as in skeleton it retains many characters that connect it with the Arctocyonidae.

Eoconodon, = " *Triisodon*," has teeth of much the same pattern as the Arctocyonidae, somewhat more simple and massive. *Dissacus* has both teeth and feet of mesonychid type; *Goniacodon* and *Microclaenodon* a type of teeth intermediate between *Eoconodon* and *Dissacus;* the skeleton is, however, oxyclaenid as far as known.

Diacodon has tritubercular molars; the fourth premolar is molariform. The skull and skeleton relate it to the erinaceid Insectivora, also to the Menotyphla.

Leptacodon is related to *Diacodon* by its tooth pattern, the skeleton being unknown.

Mixodectes has tritubercular molars with a prominent hypocone, reduced and peculiarly specialized premolars, and enlarged soricoid incisors. The astragalus appears to be of insectivore type. *Indrodon* is much like *Mixodectes* in teeth.

Palaeoryctes has the zalambdodont tooth pattern, with some approach to the tritubercular. Skeleton unknown.

Peradectes has the didelphid tooth formula, jaws and tooth pattern.

Onychodectes has tritubercular molars and simple premolars, with the beginnings of the taeniodont pattern in elevation of protocone and conules into a wide massive crescent, slight inflation of premolars. The skeleton is of proto-insectivoran type. In *Conoryctes*, *Wortmania*, and *Psittacotherium* the taeniodont specializations are progressively developed, in teeth, skull and skeleton.

Plesiadapis has tritubercular molars, reduced premolars and specialized front teeth. The details show but little resemblance to *Mixodectes*.

Pentacodon is somewhat like *Mixodectes* in the style of the cusps, but is very imperfectly known. It also has characters (construction of teeth, mental foramen) indicating relationship to *Palaeosinopa* and *Pantolestes*, which is not in disaccord with mixodectid affinities; in either case its position is with the primitive Insectivora.

To these must be added two groups represented each by a single specimen from the Tiffany, unknown in Puerco and Torrejon:

Palaeoryctes is a zalambdodont insectivore, referred provisionally to the Centetidae, as no distinctions of family value are shown in the teeth.

Zanycteris is a chiropteran referred provisionally to the phyllostomids for the same reason.

It is to be observed that the external stylar cusps of the upper molars are rarely prominent in any Torrejon, still more rarely in any Puerco, mammal. In certain phyla, e.g. *Tetraclaenodon*, *Phenacodus*, *Periptychus coarctatus*, *P. rhabdodon*, *Eohippus*, *Orohippus*, *Pelycodus*, *Notharctus*, etc., the styles are demonstrably an acquired character, a secondary upgrowth from the cingulum; in the series of didelphids from *Peradectes* through *Peratherium* to the modern opossum, the same appears to be the case, and it is more or less positively proven in many other cases.

Winge holds that the stylar cusps are always primitive, and consequently denies the validity of all such phyla as I have instanced—or rather of such few as have come under his notice—in spite of the concordant evidence derived from the characters of all the teeth, of the skull and skeleton, of the geologic succession, and the series of intermediate stages of intermediate geologic age. The essential basis for this belief appears to be the prominence of the stylar cusps in the modern opossums. But it is not safe to assume that because the opossums are primitive in most respects they are primitive in all respects, and it appears that the direct recorded evidence is overwhelmingly strong in favor of the stylar cusps being very frequently, or indeed generally, secondary in Tertiary mammals. It would not be safe to assert that they are always so. But it appears that the premolar succession affords an illustration of the presumptive method of evolution of the tritubercular from the simple haplodont molar, which to prove false would require very strong and positive evidence, and no such evidence has been produced.   If so, the stylar cusps are clearly secondary; the primary cusp is the connate paracone-plus-metacone—*not* the paracone alone—and the splitting apart of these two cusps is correlated with the upgrowth of the heel of the lower molars, and the growth of the protocone with the broadening and basining of this heel.   In the specialization of the carnassial teeth of Carnivora and in other instances, the process is reversed: the para- and metacone become connate as the heel sinks below the trigonid, the protocone disappears as the heel becomes narrowed and trenchant.   In Taeniodonta, Mesonychidae, etc., we may likewise trace the progressive union of paracone-plus-metacone, but the correlation is not altogether exact, nor the process always uniform.

The zalambdodont type of molar is clearly a very ancient one.   It is well shown in *Palaeoryctes*, and appears to be approached in *Dryolestes* and its allies of the Jurassic. Matthew agrees with Gregory in his interpretation of these teeth, although his figure appears misleading as to the character of the stylar cusp.[198]

The Origin of the Mammalian Molar

The subject of the tritubercular theory has been admirably discussed by Dr. W. K. Gregory.[199] My views are substantially in accord with his, and my present aim is to present, in somewhat more detail than he has done, the evidence afforded by the Paleocene fauna and its relations to earlier and later mammal faunas.   It is apparently necessary to draw attention to the weighty character of the palaeontological evidence, which seems to have been little appreciated by various anatomists and embryologists who have evolved more or less fantastic theories upon data selected from modern mammals.

When, as in the Equidae, Canidae, Camelidae, Felidae, Elephantidae and numerous other lines of animals living and extinct, we have a succession of stages from successive formations, represented not by selected teeth but by complete skulls and skeletons, and when the correspondence of adjacent stages is close, not merely in teeth but in every part of skull and skeleton, we cannot avoid the conclusion that these successive stages are genetically related, approximately if not exactly.   And if they be even approximately in the relation of ancestor and descendant, then the changes that the teeth undergo in the successive members of each series must needs represent the evolutionary history of the teeth in that series.

On the other hand, it does not seem that the comparative structure of modern types or their ontogeny gives any sure clue to their true phylogeny.   Yet this is the evidence that has been chiefly used by those who have written extensively on this subject, especially with the critics of the "tritubercular theory."   It has been shown that the paracone appears before the protocone in the ontogeny of most modern animals, and assumed as an indisputable conclusion that this is the order of appearance of the cusps in the phylogeny. But if the ontogeny of the teeth were a certain guide to their phylogeny, it would prove that the points of the cusps were older than their bases, that the crowns originated and grew together by concrescence long before the roots were formed, and various other absurdities.

---

[198] Gregory, W. K., 1916, Bull. Amer. Mus. Nat. Hist., XXXV, pl. I.

[199] Gregory, W. K., 1916, *op. cit.*, pp. 239–257.

Most of the present memoir on the Paleocene was written about 1917, but these remarks require no change in the light of subsequent discoveries.—W.D.M. This note was written before 1930. The author's remarks do not entirely apply to Dr. Gregory's most recent views.—THE EDITORS.

While there is a modicum of truth in the recapitulation theory, yet ontogenetic changes are in the main a matter of the necessities and conveniences of mechanical development and of adaptation to the environment and requirements of the growing animal or the growing structure.   No anatomist ever accepted the theory in full; always certain features are selected which support some favored views, and these are declared to be proof from recapitulation; other features which would support some totally different theory are ignored or explained as merely adaptive or accidental.

Nor is the case any better when the anatomist compares a series of modern forms and derives one from the other.   One author declares that the simpler are necessarily the more primitive and arranges his " phylogeny " on this principle—with very interesting results which bear little or no relation to the facts disclosed by the palaeontologic record.   Many writers, starting from the conclusion that the Insectivora among all modern placentals most nearly represent the primitive mammals, have assumed that their teeth are in all respects the most primitive, and that all others must be derived from them.   Again these views have a modicum of truth in them—how much truth can be found out by the simple test of examining the teeth of the oldest placental mammals that we know and seeing what they *are* like, and through study of such phyla as are actually known to us by adequate series of well-preserved stages in *geologic* (not anatomic) succession, we may discover in what manner the modern types of teeth have evolved from the primitive type or types. But unless adequately checked and tested by the facts of palaeontology, the inferences of comparative anatomy are likewise a doubtful and dangerous guide.

One would naturally suppose that the evolution of teeth was a problem on which the palaeontologist would be accorded especial authority and his evidence given the most weight.   But it does not seem to be so.   Comparative anatomists whose knowledge of palaeontology is measurable from the fact that they lump all extinct mammals from this country as " from the American Tertiary," have no apparent misgivings over forcing them into preconceived arrangements based upon such evidence as I have criticized, no matter what absurd associations and sequences these arrangements may entail for the rest of the skeleton and for the geologic succession.

Such methods are analogous to those of the sociologist who, utterly ignorant of history, and classing Napoleon and Julius Caesar together as " historical personages " in his discussion, would attempt to trace the origin and evolution of European civilization solely upon the basis of modern sociology.

The evidences on the evolution of the mammalian molar may be arranged in order of importance as follows:

A. Palaeontological evidence.
1. Evolutionary history in known phyla, as far back as they can be positively and accurately traced.
2. General characters and scope of variation in the teeth in the successive geologic epochs.
B. Structural and mechanical evidence.
1. Premolar analogies.
2. Interrelations of upper and lower teeth.
3. Adaptation and changes therein.
C. Taxonomic evidence.
D. Ontogenetic evidence.

It is obvious that any theory is valid only if it is supported by or in accord with all the evidence from all these sources.   If not, then either the inferences that support it or the inferences that conflict with it are misinterpretations of the data and must be reconsidered and harmonized.

The Evolution of the Teeth in Known Phyla

In numerous mammalian phyla we can trace the actual evolution of the molar through a series of successive stages which are in geologic succession and which are shown by the conformant evolution of all other parts of the skeleton to be more or less exactly in genetic succession.   This appears to be the most weighty and direct of evidence.   But it is not easy for anyone save the working palaeontologist who has collected, identified and studied the original material to appreciate how conclusive the evidence is in some cases, how inconclusive in others.   Even among the active workers there are none whose practical knowledge covers more than a small portion of the field.   Beyond that each is dependent upon the judgment of others and upon such evidence as they are able to put into printed form, which can never be more than a brief and imperfect summary.   It is very difficult for the anatomist to discover what and how much evidence of this kind exists, and to assign due weight to conflicting opinions.

Probably there is no other phylum of Mammalia the history of which is more exactly recorded than the equine series.   But the evidence has been very imperfectly published, almost wholly confined to brief summaries and resumés and scattered descriptions of certain portions of the data.   These apparently have not received the universal acceptance that an adequate publication would entail; the cautions and exceptions made by writers anxious not to overstate their evidence, and the objections made by others but slightly acquainted with the facts and finding the conclusions in conflict with some favorite theory, have been seized upon and exaggerated by many writers who for one reason or another are equally indisposed to accept it.   In fact, no one who had examined the skeletons, skulls and innumerable fragmentary specimens of each of the ten or more intermediate stages between *Eohippus* and *Equus* that are preserved in the American Museum of Natural History, each stage characteristic of and mostly limited to its appropriate geologic horizon, would think of questioning the evolutionary history of the molars that they illustrate.   To those who question the conclusions based upon such series as this the only reply is, if you will not accept the judgment of those who have seen the evidence, come and see it for yourself; it stands always open to examination.

Assuming the acceptance of the succession, what does it prove?

1. The hypsodont elaborate molars of the horse are derived from brachyodont bunodont teeth, six-cusped, the inner pillar from the protocone, the inner crescents from the conules, the outer crescents from the paracone and metacone, and the external pillars of secondary origin.

2. The premolars carry the evolution back to simpler stages than this.   The fourth premolar carries it back to a strictly tritubercular stage, with a pair of outer cusps and a single subcrescentic inner cusp.   The third premolar takes us back to a stage with a single outer cusp, the conjoined paracone and metacone, and a single inner cusp, the protocone, of smaller size.   The second premolar takes us back to a practically haplodont tooth, a single main cusp, paraconid + metaconid and a rudimentary protocone.   The first premolar has evolved but little and finally degenerated and almost disappeared.

Here, then, is one actual instance of the way in which complex molars or molariform teeth *have* originated. It carries the molars back to a stage when they are bunodont, six-cusped, imperfectly crested, and without stylar cusps. The premolars are carried back to a simple haplodont stage. Now undoubtedly it may be objected theoretically that the earlier history of the molars was different from that of the premolars. In some particulars the objection is undoubtedly valid, for the evolutionary stages do not precisely correspond in the premolars themselves, and there are certain differences of detail in the observed evolution of other phyla. But in the points here cited the evolutionary history accords with that of other phyla, as will be seen.

The stylar cusps of the molars are absent in *Eohippus*, Lower Eocene; small in *Orohippus*, Middle Eocene; progressively larger in *Epihippus*, Upper Eocene, *Mesohippus*, Lower Oligocene, *Miohippus*, Upper Oligocene, *Parahippus*, Lower Miocene, *Merychippus*, Upper Miocene, *Hipparion* or *Pliohippus*, Pliocene, and the Pleistocene *Equus* and *Hippidium*.

The tapirs, rhinoceroses, and other extinct Perissodactyla can be traced back, although not upon such exact or complete evidence, to types substantially the same in molar construction as *Eohippus*.

The external stylar cusps are, in Osborn's nomenclature, the parastyle, mesostyle and metastyle; in Winge's, Nos. 1, 2 and 3. They are prominent in the opossum and in many Insectivora, variously modified and often absent in other mammals. Winge regards them as primary elements of the mammalian molar. According to the Cope-Osborn theory of trituberculy, and equally according to Huxley's premolar analogy theory, they are not primary elements but secondary upgrowths from the cingulum. The palaeontological evidence is as follows:

a. DIDELPHIIDAE: In *Peratherium* (Oligocene) and *Peradectes* (Paleocene) the stylar cusps are less prominent than in the modern *Didelphis*. However, in the allied Cimolestidae of the still older Lance they are sometimes well developed (*Didelphops*); other isolated molars attributed to the same family show very little of the stylar cusps. The Jurassic and Comanchic mammals include some like *Dryolestes* which may perhaps be marsupials, but their molars are of the zalambdodont pattern; the styles are very prominent. The affinities of the Mesozoic forms are so very doubtful that little weight can be attached to their evidence; the Paleocene *Peradectes*, which is an unquestionable didelphid, gives some, but very slight, support to the view that the stylar cusps are secondary.

b. INSECTIVORA: Two distinct types of insectivore dentition are characteristic of the Paleocene—the leptictid and palaeoryctid. The former is distinctly tritubercular in pattern, with the protocone and paracone external in position, well separated, and the external styles only moderately developed at the angles of the tooth, the protocone a large embracing crescentic cusp occupying the inner half. The latter is zalambdodont in pattern, with paracone and metacone connate, high, forming a double central cusp, the styles large and prominent at the external angles of the tooth, and the protocone an internal heel or stylar cusp. Most of the Paleocene Insectivora belong to the former group. In the Leptictidae the hypocone is developed, the paraconid reduced and the fourth premolar is molariform. In *Onychodectes* the hypocone is absent, paraconid reduced, fourth premolar simple. In the Mixodectidae the external shelf is more developed, the mesostyle is distinct, hypocone present, fourth premolar simple. In *Pentacodon* the stylar cusps are absent and hypocone minute.   In the leptictids and *Onychodectes* enough is known of the skull and skeleton to show a real relationship; the astragalus of *Mixodectes* also suggests real affinity, but the remainder of the trituberculate group are only provisionally associated.   The skeleton of *Palaeoryctes* is unknown.

The relationship of these two groups to modern Insectivora is not demonstrated through any such convincing series as may be seen in some other groups.   The erinaceoids have been regarded as derived from leptictids and the earliest genera undoubtedly make a considerable approach to that family, but there is quite a gap between.   The soricoids have not been traced back to any satisfactory ancestral group, and it is still open to question whether they fall in with the erinaceoid or the zalambdodont ancestry; if the latter, they would afford some evidence for the view that the zalambdodont molar is a derivative of the tritubercular type.   The modern zalambdodonts include two groups of but distant relationships.   Little is known of the chrysochloroid ancestry; the centetoid group may be connected with *Apternodus* and *Palaeoryctes*, but the exact relations are not yet clear.

In describing the skull of *Apternodus* Matthew (1910) suggested that the zalambdodont and tritubercular types of molar might be regarded as of independent origin from primitive Reptilia, the former never having acquired or passed through a tritubercular stage; but in 1913 the characters of *Palaeoryctes* suggested that the Cenozoic zalambdodont molar might better be regarded as a reverted form, modified from a tritubercular ancestry.   The more recent discoveries of placental mammal skulls in the Cretaceous of Mongolia have been very ably discussed by Simpson, who has pointed out that the two types were already distinct in the Mongolian genera, but not so far apart as to preclude a common, more or less intermediate, ancestry.   It is probable that a much more extensive knowledge of the various Tertiary Insectivora is needed to clear up their real phyletic relations, but it is at least evident that the principal group of the Paleocene Insectivora is in a broad way structurally ancestral not only to the later Insectivora but to rodents, edentates and perhaps to some other orders, but any common ancestry must date well back into the Cretaceous.

Simpson has recently summed up the present status of the evidence as to the stylar cusps as follows:

Winge, in an early and important paper on molar evolution, long ago suggested that the external styles and cingulum of the upper molar are extremely ancient structures. (Winge, H., 1882. Om Pattedyrenes Tandskifte, etc. Vidensk. Meddel. f. d. naturh. Foren. i Kjöbenhavn, 1882, pp. 15–69.) For his extreme view that they are the *most* ancient part of the tooth there seems no real evidence and a vast body of facts now opposes it, but more and more items of evidence, of which these Cretaceous mammals are not the least, are appearing to demand a greater antiquity and importance for the part of the upper molar external to the paracone and metacone than has been commonly granted. In the most primitive living mammals and in the majority of the early Tertiary forms the upper molars usually have a strong external shelf, in some cases, which may offer a real clue to the whole process, agreeing with the Deltatheridiidae in occupying nearly half the total width. Not only is such a structure seen in ancient and primitive zalambdodonts, but also relatively little modified in many creodonts, which are the most primitive and central members of a group including carnivores and ungulates and related to the ancestry of other orders. The soricoids and bats also have a specialized molar structure which could be derived from that of the Deltatheridiidae by wider separation of paracone and metacone, their acquisition of a lambdoid shape, and the upgrowth of a hypocone. In the more primitive members of many groups of mammals the paracone and metacone are not really external and there appears to be no real evidence that in these groups they have migrated inward from a strictly buccal position.

It is not to be assumed that no new styles have arisen, or that all which occupy analogous positions are homologous. It is suggested only that the ancestral condition, the condition in the Cretaceous insectivores which gave rise to all higher mammals, was near that of the Deltatheridiidae, with a large bifid central cusp, an internal heel, and a more or less broad external cingulum or shelf. These are probably the only elements which were present in the common ancestry and which are strictly homologous (when correctly identified, of course) throughout all placentals. Within each line of descent the teeth went their own way, hypocones (not really homologous in the different orders), conules, styles, supplementary cusps arose or were lost, paracone and metacone became more distinct (most mammals) or fused (zalambdodonts and some carnivores), premolars became molarized by steps which followed the general history of the molars but, since they started from a different basis, could not be expected to recapitulate the exact history.[200]

c. CARNIVORA: The secondary development of the stylar cusps independently in several different lines of Creodonta and Fissipedia is conclusively shown. The primitive type of molar in the Paleocene is well represented by *Loxolophus*. The paracone and metacone are well separated, sub-external, but with a well-developed external cingulum rising to sharp angles at the outer corners. None of the Paleocene creodonts is far from this pattern, except for the mesonychids in which the external cingulum and styles are reduced, apparently as a secondary change.

d. CONDYLARTHRA: In this group the external styles are variously developed, varying from absence to prominence, but appear to be secondary when they can be traced through any definite phylum. Mioclaenidae apparently represent the primitive type most nearly. It would seem probable that this order, like the Artiodactyla, began with tritubercular molars which had lost whatever they formerly had of external styles and cingula.

e. TALIGRADA: The progressive development of external stylar cusps can be traced to some extent in *Periptychus*, but in *Ectoconus* the mesostyle is already prominent. The smaller periptychids have slight or no external cingula and no stylar cusps. In *Pantolambda* on the other hand the styles are strongly developed, quite obviously in relation to the strongly crescentic form of the cusps. As we know nothing of the ancestry of *Pantolambda*, we have no direct proof that these crescentic cusps are a secondary specialization, but for many reasons it is a probability almost overwhelmingly strong, and it carries with it the corollary that the development of the stylar cusps in this genus is also secondary.

Mesozoic Mammals and the Origin of the Molars

One looks naturally to the Jurassic and Cretaceous mammals as likely to afford the best evidence on the origin of the mammalian molar, but the results are singularly unsatisfactory. There is a rather imposing array of names recorded from the " Cretaceous." In the first place, however, they are all very late Cretaceous, most of them from the Lance, whose assigment to the Cretaceous or Tertiary is disputed; in the second place most of them are based upon separate teeth, some upon fragments of jaws, only two (*Eodelphis* and *Thlaeodon*) upon sufficiently complete jaws to afford any definite idea of the succession of the teeth. It is generally recognized that there are only a few genera, which may be summarized as follows:

Multituberculates (Plagiaulacidae):
   Ptilodus group. Probably several closely allied genera.
   Meniscoëssus.
Trituberculates (" Cimolestidae," close to Didelphiidae):
   Eodelphis.
   Thlaeodon.

[200] Simpson, G. G., 1928, Amer. Mus. Novitates No. 330, Oct. 30, pp. 9–10.

Diprotodont types related doubtfully to *Caenolestes:*
  Batodon, Telacodon, etc., imperfectly known.

The whole fauna is composed of plagiaulacid multituberculates and didelphoid and
?caenolestid marsupials.   There is no indisputable evidence of placentals, although Simpson
has tentatively referred certain isolated teeth to the Leptictidae.

The fauna immediately precedes the Puerco fauna, or may indeed be contemporaneous,
yet its relations are of the slightest, less than to the later Torrejon fauna.   It is probable
that it represents a different facies, but whatever the explanation of its unlikeness it is
perfectly obvious that we have not in this fauna the ancestors of the Puerco and Torrejon
faunas save for a very minor and doubtful portion.

The American Cretaceous mammals cannot therefore afford any light upon the pre-
Tertiary evolution of the placental trituberculates.   The recent discoveries of Cretaceous
mammal skulls in Mongolia is on the other hand most illuminating.   As carefully studied
and figured by Gregory and Simpson in 1926 and by Simpson in 1928, they appear as
unquestionable placental mammals with two fairly distinct types of dentition, tritubercular
and zalambdodont.   The former is clearly comparable with those of leptictid insectivores
and the earliest creodonts, and yet shows an approach toward the zalambdodont construction so marked that they were at first associated with that group, until better-preserved
molars showed that they were tritubercular.   The second group, more clearly zalambdodont
in structure, might afford, as Simpson observes, a structural ancestral type in many respects
for the zalambdodont insectivores, but it appears more probable that their real relations
are with some of the creodont-like genera provisionally referred to the Leptictidae but
perhaps really related to the pseudo-creodine creodonts.   These relationships probably
will remain more or less doubtful until skeletal material is found which will show the foot-
structure.   The point to be emphasized here is that, however interpreted, we have in these
Mongolian Cretaceous placentals a clear approach between the tritubercular and zalambdodont types of dentition, with suggestions of affinities so confusingly intermingled that a
common ancestral stock at no remote date has to be predicated.   Whether such ancestral
stock could be classed as tritubercular or as zalambdodont, or whether it should be regarded
as represented by the Pantotheria of the late Jurassic and early Cretaceous, may be a
matter of technical dispute, but it does not really leave a very wide margin of real disagree-
ment.   Simpson's conclusion as presented in his memoir on the European Mesozoic
Mammalia [201] that the Pantotheria are not far removed from the common ancestral stock
of both marsupials and placentals appears to be in accord with all the evidence at present
available.

The status of the " Jurassic mammals " may be summarized as follows:

1. Whether they be regarded as all Upper Jurassic or in part Lower Cretaceous
(Comanchic), they are in any case vastly more ancient than the earliest Tertiary mammals.
The time interval between them and the Paleocene Mammalia is at least as great as, perhaps
considerably greater than, that between the Paleocene and modern mammals.

2. While the material is fragmentary it does include lower jaws, fairly complete, of a
considerable variety of mammals, and upper jaws of a few.

3. The fauna is of respectable size, although not so large as the imposing list of published genera and families might indicate.

[201] Simpson, G. G., 1928, A Catalogue of the Mesozoic Mammalia . . ., p. 182.

CHARACTERS OF THE FEET AND LIMBS

Carpus and Tarsus in Paleocene Mammals

The characters of the carpus and tarsus in primitive Mammalia were ably discussed by Baur [202] and Cope [203] half a century ago.   Considerable evidence has since accumulated to test the conclusion reached by these authorities, in some respects confirming, in others disproving them.

Around the theory that the primitive mammalian carpus and tarsus were serially arranged, Cope built up a very important series of morphogenic concepts, which colored his classifications to no small degree.   As usual with this author, he rested the concept upon certain fundamental facts of observation.   Primarily in this case it rested upon the facts that in *Hyrax*, the most primitive living ungulate, and in *Phenacodus*, the oldest ungulate in which the foot structure was fully and exactly known, the carpus and tarsus were serial. *Phenacodus* belonged to the Condylarthra, which Cope regarded as essentially prot-Ungulata, and the serialism of *Phenacodus* was ascribed to the order.

The evolution of the serial into the various types of displaced or alternating carpus was explained upon theoretical grounds of stresses or of mechanical advantage.   It was very generally if not universally accepted and was incorporated into various textbooks.

Matthew in 1897, while privately skeptical of the soundness of the theoretical mechanics involved, did not formally challenge the entire theory, but placed on record the plain facts with regard to the carpus and tarsus of the Paleocene mammals as far as known, and showed that in *Phenacodus* the serial carpus was not primitive but secondary.   Osborn in 1898 pointed out the important bearing of these facts as bringing into question the primitive serial carpus.   Since that date there has been, without any formal discussion of the subject, a gradual abandonment of the doctrine of the primitive serialism of carpus and tarsus on the part of many workers in mammalian palaeontology.   Others no doubt still hold to the old belief.

The facts only with regard to the Paleocene mammals are presented here.   It is hardly worth while to revive the theoretical discussion, which appears rather meaningless and based more upon diagrams and upon drawings of the dorsal surface of the carpus than upon the real relations and articulations of the bones.   In dealing with the stiff and rigid locomotive foot of the later Tertiary ungulates the difference between serial and displaced or alternating arrangement of the joints is no doubt an important one, with clearly defined points of advantage and disadvantage in the mechanics of locomotion.   But in dealing with these Paleocene ungulates we have to do with foot-structure much more of the " unguiculate " type and so different from the ordinary ungulates that the question of serialism becomes quite a minor matter.

Primary Types of Feet in Paleocene Mammals

The general resemblance in teeth, skull and most skeleton characters among the Paleocene mammals is a very notable characteristic.   The teeth show no such contrasts in structure as we find even among the members of a single modern order—the Carnivora, for instance, or the Artiodactyla.   They are practically all brachydont; nearly all are tritubercular; the dentition is usually the complete formula and never very greatly reduced.

[202] Baur, G., 1885, Amer. Nat., XIX, pp. 86–8, 195–6.
[203] Cope, E. D., 1881, Amer. Nat., XV, pp. 269–73.

The skull construction and proportions show but little diversity. They are all very much alike in basicranial characters, as far as is known, nor is any fundamental diversity noted in the axial skeleton or girdles or limbs.

The feet show more diversity than one would expect from the above, and they differ in characters that in the later Tertiary mammals are clearly of a fundamental type. We are fortunate in being able to study fairly complete foot material of four very distinct and characteristic types of Puerco mammals and of eight Torrejon types, as follows:

	PUERCO	TORREJON	LOWER EOCENE RELATIVES
MULTITUBERCULATA		*Ptilodus*	
CREODONTA	*Loxolophus*	*Claenodon*	*Miacis* and *Didymictis*
		?Goniacodon	*Sinopa* and *Limnocyon*
			Chriacus
TAENIODONTA	*Onychodectes* *Wortmania*	*Psittacotherium*	
INSECTIVORA		*Palaeolestes*	*Diacodon*
TALIGRADA and AMBLYPODA	*Periptychus* *Ectoconus*	*Periptychus* *Pantolambda*	
			Coryphodon
ACREODI		*Dissacus*	*Pachyaena*
CONDYLARTHRA		*Tetraclaenodon*	*Phenacodus*
PRIMATES			*Pelycodus*
PERISSODACTYLA			*Eohippus*, etc.
ARTIODACTYLA			*Diacodexis*
RODENTIA			*Paramys*

The specialized types of feet that characterize the Lower Eocene genera of the four last-named orders are easily recognized by the astragalus and other characteristic bones. They are not found in the Paleocene either associated with teeth or as isolated bones. Among the great numbers of isolated or unnumbered minor " lots " of associated fragments no sign of any bone referable to these orders is to be found. Their absence from the fauna is thus practically demonstrated, and we can be reasonably certain that none of the various types of dentition that are " incertae sedis " belongs to any of these orders, nor do we find among them any strange types other than those known to be associated with certain dentitions. The numerous genera which have not been associated with these specific types of teeth may therefore be regarded as probably belonging to one or another of the seven known ordinal or subordinal groups of Paleocene mammals.

The four known types of feet of the Puerco are of particular interest:

1. *Loxolophus* seems to be the primitive foot of the true Carnivora. It is unguiculate, with long slender phalanges, slender metapodials, the distal ends rounded, keeled inferiorly. The astragalus has a rather long neck, round-oval head, very oblique trochlea, the body being wedged up between tibia and fibula, which are well separated, and the fibula a moderately strong shafted bone. The astragalar foramen is well developed. The calcaneum does not support the fibula. The navicular is of the usual creodont type. This is the type of foot of most of the older and less specialized creodonts. It is more or less approached among the smaller and more generalized fissipede Carnivora. Adaptively it would seem to be an arboreal foot, and the oldest and least specialized primates approach it in very marked degree. No primate, however, has the astragalar foramen, and in all the early types there are already present divergences in proportions and construction approaching the more typical primate specializations.

2. A distinct and marked type of foot is presented by *Onychodectes*. In this we have short, wide, flattened phalanges, with limited play on the facets, small ungues, short metapodials with somewhat "squared" distal facets, astragalus with short but distinct neck, flattened oval head, the trochlea short, broad and shallow, nearly symmetrical with inner and outer crests subequal. The head rests beneath the tibia, the fibular facet wholly lateral and matched with the inner malleolus. The fibula tends to coössify with the tibia. It does not articulate with the calcaneum. The astragalar foramen is not present. This is essentially the insectivore type of foot. It is essentially the same as in the Leptictidae, Erinaceidae, etc., more or less modified in most modern Insectivora. It is rather nearly related to the foot structure of the most primitive rodents (*Paramys*) and of the sciuridae among modern rodents. It is connected through *Metacheiromys* and the armadillos with the highly specialized foot structures of the Xenarthra and Pholidota. It is perhaps not remote in construction from the taligrade pes, but differs in some important points. With the oxyclaenid foot it is contrasted in adaptive features, which perhaps accentuate the apparent remoteness, as it parallels in some respects the earliest condylarth foot, though probably remote in real affinities.

3. The third type of foot from the Puerco is that of *Ectoconus*, representing the Taligrada. It is short and stout, with very short phalanges, distinct flattened hoofs, "squared" distal ends to the metapodials, flattened phalangeal facets with very restricted motion; the astragalus has a very short neck, oval head with but little convexity; trochlea short, very flat, but oblique, the inner crest little developed, the head set between tibia and fibula, the fibula well separated, and with a considerable calcanear facet. The astragalar foramen is well developed. The navicular is represented by two distinct bones. The cuboid has a broad astragalar facet. This is the type of foot from which the specialized foot of the Amblypoda is derived, and probably that of other subungulate orders, although these may be parallelisms, for their early history is not known. In spite of a considerable adaptive resemblance to the *Onychodectes* type, I suggest that it is fundamentally more nearly related to the *Loxolophus* foot.

4. A fourth type of foot is represented by *Wortmania* ("*Hemiganus*"), but we know only a metapodial and a claw. It is similar to that of *Psittacotherium* of the Torrejon noted below.

These four types seen in the Puerco are supplemented by others in the Torrejon.

Most important is the condylarth foot as exemplified in *Tetraclaenodon*. This foot has a great deal of resemblance to that of the earliest creodont, as Matthew pointed out in 1897. The phalanges are shortened up, the unguals commence to broaden out into hoofs, the astragalus begins to take on the phenacodont characters, but all is in its inception.

A second interesting new type is seen in *Claenodon*, even more unmistakably a creodont, nearly related to *Loxolophus* but more massively proportioned, and with high, sharp claws.

Pantolambda and *Periptychus* show more progressive stages of the taligrade foot, especially the former, which is clearly on its way toward the amblypod type.

Palaeolestes shows the foot of the primitive Insectivora of the leptictid family. In essentials it is like that of *Onychodectes*, differing only in the more slender proportions.

Mixodectes has an astragalus of the same insectivore type, but the feet are otherwise unknown.

Haploconus, Hemithlaeus and *Anisonchus* are known from a few fragments of the feet to be more or less intermediate in construction between the taligrade and condylarth type.

Psittacotherium has a peculiar and highly specialized foot, too imperfectly known to be placed with certainty. Along with *Wortmania*, of which only a metacarpal and claw are known, it represents a distinct foot-group, apparently a specialized derivative of the *Onychodectes* type. The metapodials are very short and stout, with the distal ends hinge-jointed but no keel, the phalanges extremely short and claws of great size, compressed and high in the fore foot but not compressed in the hind foot.

One of the most important of the known types of foot in the Torrejon is that of *Ptilodus*, typical of the Multituberculata. This is essentially the marsupial foot, with its peculiar primitive type of astragalus lacking any distinct neck or head and consisting merely of a small, flattened, rudely quadrangular bone.

Another type of foot very important to the present discussion is that of *Plesiadapis*, representing the foot structure of the Plesiadapidae. This is imperfectly known, but the parts available compare most nearly with the modern Menotyphla. The astragalus has a moderate neck and round-oval head, shallow unsymmetric trochlea, and no astragalar foramen. The metapodials are slender, with rounded heads, with obscure keels beneath, the phalanges are very long, the unguals so far as known claw-formed. It is intermediate between the primitive insectivore and primitive creodont types, but adaptively much nearer the latter.

In the Lower Eocene four new types of foot appear:

Eohippus, representing the Perissodactyla, has a foot already highly specialized, but with suggestions of condylarthran relationships.

Diacodexis, representing the Artiodactyla, has a still more highly specialized foot, the relationships of which are obscure.

Paramys, representing the Rodentia, has a type of foot with much resemblance to that of the earliest Insectivora, but with certain peculiarities of its own, notably the functionally tetradactyl manus with pentadactyl pes.

Pelycodus is clearly the lemuroid primate foot, with very long, slender phalanges, one or more of the unguals converted nails, the others small claws, very convex, rounded distal metapodial facets, highly opposable, large inner digit on manus and pes, astragalus with long neck and round-convex head, trochlea little grooved, head set between tibia and fibula, no astragalar foramen, but astragalus extended backwards.

The interrelationships of these different types may be thus represented:

Stem Groups	Puerco	Torrejon	Tiffany	Lower Eocene
Multituberculata	....	Ptilodus		
Creodonta		..... Claenodon		
	Oxyclaenus.....	....		Didymictis
				.................Sinopa
				Dipsalidictis
				Oxyaena......Limnocyon, etc.
Condylarthra				.................Hyopsodus
		......Tetraclaenodon.	....	Phenacodus
Perissodactyla				Eohippus
Mesonychia	.........Dissacus......			Pachyaena....Mesonyx
Artiodactyla				Diacodexis
Amblypoda	Ectoconus......	Periptychus		
		Pantolambda...	....	Coryphodon...Uintatherium
Edentata		....		Palaeanodon...Metacheiromys
Insectivora	Onychodectes...	....		
	Wortmania....	Psittacoth'm....	....	Calamodon...Stylinodon
	....	Palaeolestes....	....	Diacodon
	......			
Rodentia		....	....	Paramys
Menotyphla	....	....	Nothodectes	
Primates		....	....	Pelycodus...Notharctus

The Paleocene mammal fauna does not comprise the direct ancestry of the Tertiary mammalian stocks. For the most part these appear suddenly at the base of the Wasatch, the four new orders comprising nine-tenths of the later Tertiary and existing placentals. It does contain the direct or approximate ancestry of a number of groups that mostly become extinct in the Eocene or Oligocene, and of two orders that still survive—the Carnivora and Insectivora—and more doubtfully of a third, the Edentata. The foot of *Plesiadapis* gives us a clue to the derivation of the primate foot, that of *Tetraclaenodon* to the derivation of the perissodactyl foot, and less certainly that of *Dissacus* to the derivation of the artiodactyl foot. The *Paramys* foot is clearly a derivative of the insectivore foot. Again the feet of the Taligrada apparently are ancestral to the proboscidean and other subungulate types of foot.

From the condylarth foot we may probably also derive that of the Notoungulata of Tertiary South America, and perhaps the modern *Orycteropus*.

The *Loxolophus* foot appears to be the most primitive of the several types of the Puerco.

(1) The astragalus is retained in its primary position, wedged in between the distal ends of tibia and fibula. (2) The astragalar foramen (through which passes the peroneal artery) is retained. (3) The cuboid does not support the astragalus (or navicular), and the unciform supports the lunar only to a minor degree. If it be primitive in these essential features it is presumptively primitive in other features which are adaptively correlated with it— e.g. the long, slender phalanges and small, claw-like unguals.

These feet are clearly adapted to arboreal habits. The arboreal adaptations are progressively exaggerated in the primitive menotyphlan foot (cf. *Plesiadapis, Tupaia*) and primitive primate foot (cf. *Notharctus* and Lemuroidea).

The primitive insectivore foot appears to be an adaptation to terrestrial life of small animals, and the primitive subungulate foot to terrestrial life of large animals.

In the insectivore type (cf. *Palaeolestes* and the Leptictidae, and the hedgehogs, shrews, centetids, etc.) the foot becomes more digitigrade, the phalanges much shorter and less flexible. The astragalus is shifted over so as to lie under the tibia, the fibula tends to consolidate with the tibia and the joint is thus made tighter and stronger but loses its lateral mobility.

In the taligrade type (*Ectoconus*, etc.) the foot takes on the " rectigrade " adaptations and is structurally ancestral to the amblypod foot. The entire foot is much shortened and broadened, the pes remaining semi-plantigrade, the claws converted into hoofs and the proximal podial bones partly shifted over to gain support from the outer digits, while the fibula gains a footing on the calcaneum. Apparently heavy padding underneath the foot conditions the development from the first; at a later date the same factor brings about the true rectigrade type of the subungulates.

Another type of pes is represented by *Psittacotherium* and its allies, apparently a graviportal, fossorial, terrestrial adaptation, but too imperfectly known to estimate its affinities very closely.

Onychodectes-Palaeanodon-Metacheiromys present a series of stages leading from the primitive insectivore type into the terrestrial fossorial specializations of the armadillos. Thence we get the primitive gravigrades and the tardigrades as stages leading into a re-adaptation to arboreal life, the glyptodonts as graviportal specializations from the armadillos, the anteaters and *Manis* as parallel specializations for fossorial habits partly readapted to arboreal life.

The condylarth adaptations are parallel to the primary insectivoran but closer to the creodont stock. From these we have a further cursorial specialization in the perissodactyl stock, which in its turn breaks into cursorial and graviportal adaptations.

The precise affinity of the artiodactyl stock is not at all clear. We may regard this type of foot as mechanically derivable from some of the primitive insectivoran types, but there is no direct evidence of intermediate primitive stages to support this view.

The above data and conclusions carry as corollaries:

(1) That the primitive creodont type represents the primary type of the placental mammals and was of fully arboreal adaptation.

(2) That the primitive Insectivora type of foot represents an adaptation from this primitive arboreal placental to terrestrial life of a microfauna.

(3) That the primitive subungulate type is likewise a terrestrial adaptation from the primitive arboreal placentals but of larger terrestrial animals (macrofauna).

(4) That the insectivores, rodents, edentates (excluding *Orycteropus*) are derived from a common Upper Cretaceous stock.

(5) That the subungulate groups are very likely derived from the Taligrada.

(6) That the Carnivora are all derived from the primitive creodonts, and that the Condylarthra and notoungulates, and probably through them the perissodactyls, are derived from the same Cretaceous source, the perissodactyls dating perhaps well back into the Cretacic.

(7) That the common stock of the placentals should be dated well back in the Cretaceous, possibly as far back as the Comanchic.

It will be apparent that this record involves changes in adaptation and habitat which are by no means in accord with the theories of certain palaeontologists as to the " irreversibility of evolution." I do not refer here to this law as formulated by Dollo, but to the interpretations that have been placed upon Dollo's law by other less learned or less judicial expositors.

If Dollo's law be understood to mean that a special adaptive trend of habit with its consequent or concurrent progressive specializations of structure is maintained unchanged and irreversible until the phylum becomes extinct, then the " law " is not a natural law or uniformity at all; it is not even so prevalent as to be called normal if the scope of its application covers more than a brief stage in the evolution of a phylum. I deem it unnecessary to cite exceptions to any such dictum; they are so obvious and so numerous that they cannot but occur to everyone.

Dollo explains his meaning clearly and succinctly as that no phylum can so alter its adaptation and structure as to retrace its exact path or to become again precisely what it was in a previous stage of existence. And this he says is true in virtue of the " indestructibility of the past." The influence of the intermediate stage of adaptation is indelibly impressed upon its structure. Gregory has called such characters as are due not to the present adaptation of a phylum but to a previous adaptation or adaptations of a different kind " palaeotelic," as opposed to the " caenotelic " characters moulded by its present mode of life.

In this sense the law is not merely true but it is fundamental to our interpretation of the history of any phylum. It is true not only of the animal as a whole but of any individual structure of any reasonable degree of complexity. One must make this last reservation indeed because there are structures of such simple form that the end results of a readaptation cannot be practically distinguished from the primary adaptation. One might seek in vain, for instance, in the external form of the tooth of certain seals or cetaceans for any proof that either had passed from the correspondingly simple tooth of a reptile through the tritubercular complex molar type of early terrestrial Mammalia; yet the anatomy of other parts of the animal and the palaeontologic evidence demonstrate that this was the case, certainly in regard to the seal and almost certainly in regard to the cetacean.

CHARACTERS OF THE SKULL

The skull is fairly well known in the Torrejon *Claenodon, Deltatherium, Periptychus, Pantolambda, Conoryctes, Haploconus,* and in the Puerco *Loxolophus, Ectoconus* and *Onychodectes.* A large part of the skull is known in the Torrejon *Dissacus* and *Psittacotherium* and in the Puerco *Wortmania* and *Hemithlaeus.* In all these skulls there are many points in common:

1. The brain-case is always small, the brain of very primitive type with the cerebral lobes little developed, lissencephalic so far as known, comparatively narrow and not overlapping the olfactory lobes or cerebellum. Casts of the brain cavity have been made in *Pantolambda* and *Periptychus;* in the other genera the general proportions appear to be much the same as in these two Taligrada. In no case is there any reason to believe that the brain was of higher type than those described by Cope and Marsh in *Phenacodus, Coryphodon* and *Uintatherium.*

2. The basicranial and basifacial axes are parallel.

3. The sagittal and occipital crests are nearly always well developed.

4. The postorbital process is usually weak, sometimes absent, never united with the jugal.

5. The nasal bones are long and expanded toward the posterior ends as in marsupials.

6. The lachrymal bone has a considerable expansion on the face but the foramen is marginal or within the orbital rim.

7. The premaxilla extends well back between the anterior part of the nasal and the maxilla.

8. The jugal is of moderate development, sometimes much reduced (*Onychodectes*). It never extends back to the glenoid cavity.

9. The alisphenoid is relatively small, as in placentals generally, not so large as it is in marsupials.

10. The glenoid articulation is transversely extended. A postglenoid foramen present.

11. No otic bulla. Tympanic unknown, presumably ring-like. Alisphenoid does not form any false bulla. Petrosal prominence high, rounded or pyriform, the fenestra rotunda on its posterior flank. ?Cochlea of two or three coils.

12. The palate is completely ossified and margined at the back by a slight transverse crest. It is not extended much behind the molars.

13. The post-nareal gutter is long, the post-palatine crests at its sides are long and high, sometimes overarching, and made up chiefly of the palatines with pterygoids and alisphenoid plates behind.

These are the characters one must conclude of the primitive placental skull. In many respects they are approached by the more primitive living placentals. In some features they are more primitive than any living mammal. In all respects they define a common ancestral type from which all the diverse mammalian groups are structurally derivable, although it does not appear that they are directly ancestral to the Tertiary and modern Mammalia.

They are distinguished from primitive marsupials by the small alisphenoid and jugal, non-inflected angle, etc., as well as by the dentition, all typically placental. They show one peculiarity—the long backwardly expanded nasals retained by the most primitive marsupials, lost by all modern placentals—but on the whole they show no significant approaches to marsupials, as should be the case if the latter really represented a " metatherian " stage of evolution more primitive than the placentals. Truth is that whatever may be said as respects the soft parts or the broader relations of the two groups, the marsupials are quite as specialized in skull and skeleton as the more primitive placentals and have departed quite as far from the common ancestral group. *Didelphys* and its allies are more primitive on the whole than any *one* living insectivore. But the placentals are not descended from marsupial ancestry, any more than the marsupials from placental ancestry. The view set forth by Simpson that both are derived from a group of Jurassic Metatheria of which the pantotheres are nearly related representatives appears to be fully in accord with the evidence of the Paleocene mammals. This ancestral group however was as much pro-placental as pro-marsupial in its osteologic characters.

The Mongolian Cretaceous placentals or pro-placentals are not far from representing the more immediate ancestors of some Paleocene mammals, but probably only of a minor part of the fauna. Their relations to the taligrades are not so apparent.

The relationship of the Paleocene mammals to those of the Eocene are more clearly and exactly shown. The Eocene condylarths, amblypods and taeniodonts are specialized descendants, unprogressive in brain. The Eocene creodonts are more progressive descendants. The Eocene Glires, Perissodactyla, Artiodactyla and Primates are less closely related, more progressive, but derivable from stocks related to the known Paleocene mammals.

Characters of the Vertebrae and Ribs

The vertebrae are not adequately known in any but *Ectoconus*, *Pantolambda* and *Tetraclaenodon*, but fragmentary data are available for *Deltatherium*, *Claenodon*, *Onychodectes*, *Psittacotherium*, *Periptychus*, *Dissacus*, *Loxolophus*. These all have much in common, particularly the following:

1. The cervicals are short in taligrades and taeniodonts, of moderate length in the remaining genera. The vertebrarterial foramen is present on C_{2-6}, absent on C_7. The axis has a moderately high spine somewhat hatchet-shaped, but not approaching the peculiar form of didelphids. The odontoid process is peg-like, moderately long.

2. The dorsals are of relatively small size.

3. The lumbars are large, six or seven in number, with long centra, transverse processes flat, moderately wide, the zygapophyses strongly convex and concave, not reflected. Anapophyses are present on the anterior lumbars and posterior dorsals.

4. The sacrals are usually three in number (four in *Ectoconus*), the first bearing the pelvic attachment, the others free or nearly so.

5. The caudals are exceptionally large, the median and distal caudals elongate, the proportions of the tail following those of the Eocene creodonts, but of even larger size.

6. The ribs are comparatively short, small, not much flattened nor strongly curved.

The most noticeable feature in the vertebral column is the length and massiveness of the tail. This character of the tail is most like that in the Eocene creodonts. Among modern mammals some of the long-tailed Carnivora and Primates—e.g. *Potos*, *Nasua*, *Galago* or Cebidae—make the nearest approach, but in none of these are the caudal arches and processes carried so far out on the tail. It is not unlike the tail of *Didelphis*, but larger and longer throughout. It is quite unlike the tail of the edentates and Pholidota, with their uniformly decreasing vertebrae, all of similar type and proportions. Apparently it is adapted to arboreal life, but whether actually prehensile or not we cannot determine. For the rest, the proportions are most nearly those of the more primitive and flexible-bodied Carnivora. In the taligrades the neck is shortened, lumbar region and caudals somewhat reduced, but the proportions not greatly changed in other respects.

Ecologic Adaptation of the Paleocene Mammals

The general characters of vertebrae, limbs and feet are suggestive throughout of animals either of arboreal habits or derived from arboreal ancestry. The neck is flexible and of medium length. The lumbar region is long and large. The ribs are short, not flattened. The tail is very long and large and in the lengthening of median and distal caudals finds analogues in the thoroughly arboreal raccoons and arboreal Primates. The limbs are of medium length, flexible and most like those of climbing or tree-dwelling animals, but lack the extreme elongation of the primate limbs. The feet are five-toed, mobile, the inner digit more or less opposable as well as divergent, the phalanges of moderate length in the more primitive types, with small sharp claws.

This is the general character of the Paleocene mammals as seen in the majority of the fauna, and modified to a limited extent in three diverse directions in the larger and more specialized members of the taligrades, condylarths and taeniodonts, which nevertheless retain a great deal of the generalized characters cited, are connected with the generalized primitive stock by intermediate stages, and carry their specializations further along the same lines in the Eocene. It is hardly open to reasonable question that such types as the smaller Paleocene creodonts and insectivores (including *Onychodectes*) represent the ancestral adaptive type of the Paleocene placentals. As far as the evidence goes, they may fairly be considered as representing the primitive type of the placentals as a whole.

In 1904 Matthew [204] cited the above characters of primitive placental mammals as evidence that the placentals were primarily arboreal in habit, as Dollo and Bensley had previously interpreted the very similar evidence in regard to primitive marsupials. It is in reality more cogent evidence than with the marsupials, because it is not a hypothetical but an actual primitive ancestral type of mammal which displays these features, and it is possible to trace most of the existing placental specialized types by actual record backward through the Tertiary to or toward this same primitive type. The Tertiary record of the marsupials is only partially known. The arboreal ancestry of the placentals has not been so generally accepted however, partly because the evidence in its behalf has never been set forth, save in the very brief summary given in the little article in 1904, partly perhaps because other theories as to the primitive adaptations and specializations of various groups of placental mammals already held the field.

In 1919 Gidley [205] published an article attacking this view and denying that the primitive mammals excepting the Primates had the pollex or hallux in any degree opposable or were of arboreal habit. He states—quite incorrectly—that " this latter belief was apparently based, principally at least, on the fact that the first digit, in both the fore and hind feet of the early Eocene mammals of generalized type, is so frequently found in a divergent position, and on the condition found in the earliest Primates." It was in fact based upon the entire characters of the skeleton, which are those of animals that can climb well or live partly or chiefly in the trees; and the nearest approaches to the Paleocene primitive type are found in the most arboreal of modern Carnivora, rodents and Insectivora. This created a general probability, the validity of which was tested by the specific feature of the partly opposable pollex and hallux; but it would remain as a general probability even

---

[204] Matthew, 1904, Amer. Nat., XXXVIII, pp. 811–18.
[205] Gidley, J. W., 1919, Proc. Wash. Acad. Sci., IX, pp. 273–80.

though it were true that the pollex and hallux of the Paleocene mammals were in no degree opposable.

So far as the matter of opposability is concerned, that is a question of fact, and in the detailed descriptions of various genera in the systematic part of this memoir I have shown that the facets of the trapezium and first metacarpal and, to a less extent, those of the ento-cuneiform and first metatarsal are so formed as to permit of a partial opposition of the first digit to the rest of the manus or pes. The other digits, although more or less divergent, and allowing of a large amount of flexibility and range of movement, are not specifically opposable in this manner. This primitive opposability *explains* the fact that in all mammals, even in those whose inner digit is not now opposable, the first digit has a peculiar and unusual set of muscles to move it, whose characters are explained not by a digit primarily divergent, but by a digit primarily opposable.

Gidley's comparison with the foot of *Sphenodon* is unconvincing, because this is a reptile of extremely remote relationships to Paleocene mammals, and its divergent digits have no more bearing on the case than the divergent digits of birds. A much more profitable comparison might be made with the theriodont reptiles of the Permian, the feet of which have been very admirably described and discussed by Broom, Williston, Watson and Gregory in various articles. These have an important bearing on the osteological development of the mammalian pes, and show that the suggestion made by Matthew of certain peculiar relations of the first digit as a result of arboreal adaptation in Mesozoic mammals is untenable. These relations, whatever their primary cause, were already forecast in the Permian pro-Mammalia, but on the question of arboreal habitat in late Mesozoic Mammalia the Karroo reptiles have no evidential weight. There is a vast interval of time between the Permian and the Paleocene and sound geological and faunal reasons for the belief that the arid climate to which the Karroo fauna was adapted, widespread in the Permian, gave place to one in which forests prevailed widely over large portions of the continents during most of the later Mesozoic. Corresponding changes in the adaptation of various groups of land vertebrates are to be expected, and the little that is known of Mesozoic mammals is all suggestive of their being forest dwellers, with the probable exception of the Mongolian Cretaceous mammals.

During the Cretaceous the terrestrial field, both forest and such open country as may have existed, appears to have been pretty closely held by the dinosaurs, and the opportunities for mammals to expand were thereby limited. Toward the end of the Cretaceous a progressive extinction of the dinosaurs took place. It is often stated, but quite against the evidence, that the extinction of the dinosaurs took place suddenly at the end of the Cretaceous. The number and variety of species appear to have been progressively reduced throughout the late Cretaceous, leaving at the end (Lance, Denver, etc.) only a few highly specialized types, mostly gigantic. Their disappearance at the end of the Cretaceous, to whatever cause it be assigned, left the field of terrestrial adaptation open for mammalian occupation, and the mammals proceeded to expand first into terrestrial forest dwellers and then more slowly into terrestrial open country, plains and desert adaptations. The first phase of this is seen in the Paleocene faunas, among which the larger taligrades, condylarths and taeniodonts and the larger creodonts (*Claenodon, Dissacus*) show characters that must be interpreted as adaptive to terrestrial habitat but not yet any of the adaptations to open plains that progressively appear and become dominant through the Eocene, Oligocene and Miocene.

The larger taligrades, with their hoofed toes, short rounded feet, short and heavy limbs, short thick neck and long heavy fleshy tail, were obviously terrestrial animals. The periptychids, like the pigs and peccaries, probably subsisted on fruit, nuts and miscellaneous food chiefly vegetarian. *Pantolambda* was probably a more strictly browsing type. The taeniodonts with their digging feet may have been adapted to feed upon roots and tubers in addition to miscellaneous surface food. *Claenodon*, with its curious analogy to the bears, may be supposed to have had a similarly omnivorous-frugivorous diet. *Dissacus* is a more puzzling type, the teeth unlike any modern adaptation, but suggestive of some hard-shelled food, combined with the predaceous habits of the Canidae; perhaps turtles, crustacea and fresh-water mollusca formed a part of the diet. *Dissacus* shows only the rudiments of the cursorial specializations of the later Mesonychidae, which are more nearly analogous to the wolves in proportions.

Tetraclaenodon would seem to have been omnivorous like the Periptychidae but with the beginnings of cursorial specialization in the skeleton, carried a little further in *Phenacodus* of the Eocene, but not to an extent that would suggest either being a dweller in open country.

The smaller Creodonta were all like *Loxolophus* in proportions and adaptive characters, so far as the skeleton is known. They find their nearest living analogues in the modern Procyonidae, especially *Nasua*, *Potos* and *Bassariscus*. The smaller Insectivora (*Prodiacodon*, *Plesiadapis*, *Onychodectes*, etc.) find their nearest analogues in Tupaiidae. The little that is known of the skeleton in Mixodectidae and others agrees in proportions. The smaller Periptychidae and the Mioclaenidae may well have been arboreal; the little that is known of their skeleton conforms to the characters and proportions of *Loxolophus* but is not sufficient to be decisive. A considerable part of the skeleton of *Nothodectes* (= *Plesiadapis*) is known, and is intermediate between Tupaiidae and the Eocene lemuroids in proportions and adaptive features.

In sum, we find the larger and more specialized Paleocene mammals to be of terrestrial adaptation but not far removed from a generalized primitive type which has the characters of most arboreal mammals (flexible neck of moderate length, small dorsals, large mobile lumbars, long heavy tail with median and distal caudals elongate, moderately slender and very mobile limbs, feet five-toed with slender well-separated digits, ball-headed distal metapodial joints, phalanges medium or long, convex dorsad, ungues sharp small claws, first digit partly opposable, fifth divergent but not opposable, etc., etc.) and agrees most nearly in proportions and skeletal characters with the most arboreal members of modern Carnivora, rodents, Insectivora and marsupials, differing from arboreal Primates chiefly in that these have carried certain arboreal specializations still further, and among Primates finding its nearest analogue in the oldest and the most primitive members of the order.

The smaller and more generalized Paleocene mammals all are of this type and are remarkably similar in general proportions of skeleton and in dentition, although closer study of their osteology shows that they are not so nearly related as they might seem, and include apparently a number of distinct groups of similar dentition and skeleton proportions and presumably of similar habits and environment.

B. THE PALEOCENE REPTILIA

GENERAL DISCUSSION

Although the reptiles are not included in the present revision, for the sake of completeness it has seemed better to cite the original descriptions and published revisions.   Very little has been done with the Paleocene reptiles except for the Chelonia, of which a number of species based upon excellent material has been described by Hay [206] and Gilmore.[207]

Turtles and crocodiles are the most abundant fossil vertebrates in the Paleocene, as also in most Eocene formations.   No crocodiles, however, have been described from the Puerco or Torrejon.[208]   Cope indicates three species but does not name or distinguish them in his list of the Puerco fauna in 1888.[209]   Gilmore (loc. cit.) lists *Crocodilus stavelianus* Cope as from the Puerco, but Cope states in his description [210] that it was associated with dinosaur, teeth, etc., from the " Laramie," i.e. Ojo Alamo or Kirtland shales probably. There are certainly two genera in the Puerco—a small form allied to *Allognathosuchus* of the Wasatch and a larger species probably referable to *Crocodylus* or some nearly related genus.

Champsosaurs are represented only by rare and fragmentary specimens, chiefly weathered vertebrae.   Two species have been described by Cope, one reduced to synonymy by Brown in his revision of the group,[211] but neither exactly determinable.

Lizards are likewise rare and fragmentary.   An imperfect skull has been named and described by Gilmore but its exact affinities are uncertain.   Another specimen has been provisionally referred to the Oligocene genus *Peltosaurus* by the same author.

The Chelonia, as revised by Gilmore in 1919, include five families: Pleurosternidae, Baënidae, Dermatemydidae, Plastomenidae and Trionychidae.   All these are Mesozoic families, of which the Pleurosternidae, a very primitive group, make their last appearance in the Paleocene and the Baënidae, Dermatemydidae and Plastomenidae disappear with the Eocene except for a few species of dermatemydids which have lingered down to the present day in Central America.   The Trionychidae have continued abundant throughout the Tertiary and to modern times.   The two important Tertiary and modern families of Emydidae and Testudinidae have not yet appeared in the Paleocene.   It would appear therefore that the Chelonia of the Paleocene are more nearly related to the Cretaceous turtles than to those of the Eocene.   The geologic range of the families of turtles is very like that of the orders of mammals except that the Cretaceous ancestors of the turtles are known, those of the mammals inferred.   But in both we find the Paleocene groups partly disappearing at the end of that epoch, partly surviving to the end of the Eocene, or still represented by a few lingering remnants in tropical regions, while the principal Tertiary and modern groups appear suddenly in our record at the beginning of the Eocene and flourish down to the present day.   The Paleocene turtles, like the Paleocene mammals, are the

[206] Hay, O. P., 1908, Carn. Inst. Wash. Publ. No. 75; 1910, Proc. U. S. Nat. Mus., XXXVIII.

[207] Gilmore, C. W., 1919, U. S. Geol. Surv. Prof. Paper 119.

[208] Recently crocodiles have been described from the Puerco and from the Torrejon by Simpson and by Mook. *See* Simpson, G. G., 1930, Amer. Mus. Novitates No. 445; Mook, C. C., 1930, Amer. Mus. Novitates No. 447.—THE EDITORS.

[209] Cope, E. D., 1888, Trans. Amer. Phil. Soc., XVI, N.S., p. 301.

[210] Cope, E. D., 1885, Amer. Nat., XIX, p. 986.

[211] Brown, B., 1905, Mem. Amer. Mus. Nat. Hist., IX, pt. 1.

culmination of the Cretaceous rather than the beginning of the Tertiary fauna. As these are the only two groups of Paleocene vertebrates in which the material is abundant and well studied, the agreement is significant. It could and should be checked by a competent and thorough revision of the Cretaceous and Tertiary Crocodilia, which would likewise afford adequate evidence, when a sufficient number of good skulls has been collected, described and critically compared. There is no prospect that the fossil record of birds, lizards or amphibians will ever be adequate for this purpose. The more or less provisional identifications and classifications of fragmentary material may look like evidence from the outside, but it is apt to be of very little real weight in this field.

The real status of the Paleocene fauna has been obscured by the obvious and well-known facts that dinosaurs are not found in it and that mammals appear for the first time in force. This to elementary science is a decisive turning point and has conditioned the general placing of the dividing line between Cretaceous and Tertiary at this point. The palaeobotanists, familiar with the close resemblance between the Paleocene floras and those of the latest dinosaur beds, have insisted that the latter should be placed in the Tertiary, while Schuchert and others have argued from the palaeozoologic evidence that the logical division lay at the base of the Wasatch. The evidence herein discussed undoubtedly supports Schuchert's view, but it may be doubted whether logic is likely to prevail over custom in this matter.

SYSTEMATIC REVISION

1. ORDER RHYNCHOCEPHALIA

Suborder Choristodera

CHAMPSOSAURIDAE Cope, 1884

Cope, E. D., 1884, Rept. U. S. Geol. Surv. Terr., III., p. 104

Champsosaurus Cope, 1876

Cope, E. D., 1876, Paleont. Bull. No. 23, p. 10, and Proc. Acad. Nat. Sci. Phila., XXVIII, p. 348; 1884, Amer. Nat., XVIII, p. 815; Dollo, L., 1884, Bull. Mus. Roy. Hist. Nat. Belg., III, pp. 151–86, pls. viii, ix; Lemoine, V., 1885, Comptes Rend. Acad. Sci. Paris, C, pp. 753–5; Dollo, L., 1885, Bull. Soc. Géol. de France (3), XIV, pp. 95–6; Cope, E. D., 1885, Tertiary Vertebrata, p. 104; Dollo, L., 1891, Mem. Soc. Belge de Géol. . . ., V, pp. 147–99, pls. vi–viii; Dollo, L., 1893, Bull. Soc. Belge de Géol. . . ., VII, p. 79; Brown, Barnum, 1905, Mem. Amer. Mus. Nat. Hist., IX, pp. 1–26, pls. I–V.

Type: Champsosaurus profundus Cope, from Judith River beds (Cretaceous) of Montana.

Champsosaurus australis Cope

Champsosaurus australis Cope, 1881, Amer. Nat., XV, p. 670; 1885, Tertiary Vertebrata, p. 107, pl. xxiiib, figs. 1–4; Brown, B. 1905, Mem. Amer. Mus. Nat. Hist. IX, p. 6.

Type: A. M. Cope Coll. No. 1626, " portions of a dozen vertebrae . . . appear to belong to one animal, and are unworn . . . the neural arches are lost." [212]

Horizon and Locality: Middle Paleocene, Torrejon formation, near Canyon Largo, San Juan basin, New Mexico.

[212] Cope, 1881, *loc. cit.*

Author's Diagnosis: Cervical vertebrae distinguished by the superior transverse extent as compared with the longitudinal and vertical. The dimensions are about those of the *C. laticollis.* There is a similar median inferior low keel. The outline of the articular face for the neural arch is pyriform, the wide portion concave, with its external edge decurved. The decurvature is sometimes sufficient to resemble part of a rib-facet. Articular faces of centra nearly plane. Sides of centra very little concave, pierced by a foramen below the base of each diapophysis. Non-articular surfaces of centrum marked with a delicate thread-like sculpture. [Measurements follow.] The cervical vertebra is wider and more transverse than in either of the four known American species.

In 1885 (Tertiary Vertebrata, p. 107) Cope had recognized the supposed cervicals as dorsal vertebrae and modified the description in accordance. The essential distinctions of the species are given in a key (p. 106), as "small; dorsal centra with semicircular faces, much wider than deep; anterior dorsal keeled below."

In 1905 Brown distinguishes the species as follows:

Vertebrae distinguished by superior transverse as compared with longitudinal and vertical extent. The outline of the articular face for the neurocentrum is pyriform with strongly decurved edge. Articular faces of centra nearly plane.

Champsosaurus puercensis Cope

Champsosaurus puercensis Cope, 1882, Proc. Amer. Phil. Soc., XX, Feb. 20, p. 195; 1885, Tertiary Vertebrata, p. 107, pl. xxiii *b*, figs. 5–10; (= *C. saponensis* Cope) Brown, Barnum, 1905, Mem. Amer. Mus. Nat. Hist., IX, p. 6.

Type: [Presumably in the Cope Collection, American Museum, but not located.—The Editors.]

Horizon and Locality: ?Paleocene, " near the Puerco river, west of the Nacimiento mountain, New Mexico, in the typical locality of the Puerco formation." Probably Torrejon if Paleocene.

Author's Diagnosis: An animal of larger size than any of those heretofore referred to *Champsosaurus,* excepting the *C. vaccinsulensis.* In all of the vertebrae the neural arch is more or less coössified with the centrum, and the animal had probably reached its full size. . . .

The articular faces of the dorsal centra are a little wider than deep, and the depth about equals the length of the body. They are not nearly so depressed as those of *C. australis,* and their outline is different. This is wider above and narrows below; in both *C. australis* and *C. saponensis* the inferior outline is part of a circle. None of the dorsals preserved are keeled below. There is a fossa below the diapophysis which has a subvertical posterior boundary. The general surface (somewhat worn) does not display wrinkles near the articular faces. An anterior dorsal has a short compressed diapophysis with a narrow figure 8 articular surface, and its superior border is in line with the roof of the neural canal. The anterior caudals have subround articular faces; the posterior are more oval and the bodies compressed. With greater compression, the length increases. [Measurements follow.]

Brown in 1905 makes this species a synonym of *Champsosaurus saponensis* of even date: " *C. puercensis,* invalid, founded on fragments and badly weathered vertebrae of an old individual which is probably referable to *C. saponensis.*"

Champsosaurus saponensis Cope

Champsosaurus saponensis Cope, 1882, Paleont. Bull. No. 34, Feb. 20, and Proc. Amer. Phil. Soc., XX, p. 196; 1885, Tertiary Vertebrata, p. 109, pl. xxiii*b*, figs. 11–22; Brown, Barnum, 1905, Mem. Amer. Mus. Nat. Hist., IX, p. 6.

Type: A. M. Cope Coll. No. 1627, " six cervical and several dorsal vertebrae, one only of the latter with well preserved centrum, parts of ribs, and various other bones, whose reference is not yet certain."

Horizon and Locality: Paleocene, " Puerco beds " (= Nacimiento group), San Juan basin, New Mexico.

Author's Diagnosis: The description deals almost wholly with the generic or family characters as illustrated in the type.   In 1885 (Tertiary Vertebrata, p. 106) Cope states the essential specific distinctions of *C. saponensis* in a key, as follows:

Medium; length, width and depth of dorsal centra equal; faces subround; not keeled below; axis not keeled below.

Brown in 1905 adds to the characters given by Cope: The transverse bar of the interclavicle is deeply excavated along its median line on the dorsal surfaces, and a process of the clavicle extends up to the excavation on the anterior dorsal surface.

2. ORDER LACERTILIA

Two specimens from the Paleocene of the San Juan basin have been referred to the Lacertilia (Sauria).

Machaerosaurus Gilmore, 1928

Gilmore, C. W., 1928, Mem. Nat. Acad. Sci., XXII, Third Memoir, p. 155

Type: Machaerosaurus torrejonensis, A. M. No. 5184, fragmentary skull and jaws.

Horizon and Locality: Torrejon formation, east fork of Torrejon Arroyo.   Collected by Sinclair, 1913.

Type Description: The frontal, parietal and upper portion of the maxillary bones have their surface sculptured by a distinctive roughening which may be described as *punctate-vermiculate.*   I have failed to find among living lizards any sculpturing closely resembling it, and among fossil forms none with which it can be confused.   This ornamentation of the cranial bones in conjunction with tooth characteristics appear sufficient to definitely distinguish the genus and species.   The bones of the left side are so little disturbed from their normal relationships as to show the subcircular shape and large size of the orbit, which has a greatest vertical diameter of 7 millimeters.   The frontal participates in the formation of the upper boundary of the orbital rim.   The prefrontal is large and sends backward and upward a slender pointed process which underlaps the forward outer border of the frontal, but is terminated in advance of the center of the orbit.

The vermiculate sculpturing on the maxillary vanishes toward the alveolar border leaving a narrow, smooth band which is perforated at intervals by a row of foramina.   The incomplete left maxillary shows evidence of having carried eleven teeth; eight of these occupy a space 5 millimeters in length.   The teeth of both upper and lower jaws have simple, compressed, sharply pointed, slightly recurved, dagger-like crowns with sharp cutting edges both front and back.   In cross section the bases of the crowns would be lenticular fore and aft.

All of the elements of the mandible, so far as they can be observed, are distinct.   The dentary appears to extend behind the coronoid on the external face of the jaw.   The coronoid is moderately produced forward on the outside of the dentary and less so backward.   It has a decided upward prolongation.   The surangular is incomplete but appears to be slender, is in about the same vertical plane as the dentary, and is elongated posteriorly.   The posterior angle of the ramus is obscure.   The skull is slightly larger than the cranium of a large *Gerrhonotus.*

Beyond determining its undoubted lacertilian affinities, I have been unable to arrive at any conclusion as to the family or other relationships of this form.

Peltosaurus sp.

Gilmore, C. W., 1928, *op. cit. supra,* p. 137

Referred: A. M. No. 5187, an upper jaw from the Torrejon formation; reference provisional.

3. ORDER SERPENTES

CROTALIDAE Gray, 1825

Gray, J. E., 1825, Ann. Philos., London, XXVI, (N.S., X), p. 204

Helagras Cope, 1883

Cope, E. D., 1883, Proc. Amer. Phil. Soc., XX, Feb. 14, and Paleont. Bull. No. 36, p. 545

Type: Helagras prisciformis of the Paleocene of New Mexico.

Author's Diagnosis: The generic characters are drawn from vertebrae only. These display a modified form of the zygosphen articulation, as follows: The roof of the zygantrum is deeply notched on each side of the median line so as to expose the superior lateral angles of the zygosphen. This separate median portion of the roof of the zygantrum forms a wedge-shaped body which may be called the *episphen.* It is surmounted by a tuberosity, which constitutes the entire neural spine. The latter is thus entirely different in form from that of other serpents. Articular extremities of centrum round, the ball looking somewhat upwards. Costal articulation 8-shaped, the surfaces convex and continuous. Hypapophyses none on the two vertebrae preserved. Zygapophyses prominent. Free diapophyses none.

Helagras prisciformis Cope

Helagras prisciformis Cope, 1883, Proc. Amer. Phil. Soc., XX, Feb. 14, p. 545, and Paleont. Bull. No. 36; 1885, Tertiary Vertebrata, p. 731, pl. xxiv*g*, fig. 2.

Type: A. M. No. 1628, two vertebrae.

Horizon and Locality: Paleocene of San Juan basin, New Mexico.

Author's Description: A section of the vertebra at the middle is pentagonal, the inferior side slightly convex downwards. The lateral angle is the section of the angular ridge which connects the zygapophyses. The episphen has a shallow rounded groove on its inferoposterior side, which is bounded by a projecting angle on each side at its middle. The episphen does not project so far posteriorly as the postzygapophyses, and the degree of its prominence differs in different parts of the vertebral column. In one of the two vertebrae in my possession its prominence is small. The tuberosity on its summit is a truncate oval with the long diameter anteroposterior, and equaling two-fifths the length of the arch above. It is elevated above the rest of the median line, which is roof-like, with obtuse angle. The tubercular articular facet is entirely below the prezygapophysial surface, but the free part of the prezygapophysis extends well in front of it. It is distinguished from the capitular surface by a very slight constriction. A slight ridge extends from the capitular articulation to the edge of the ball of the centrum. Below this, the surface is slightly concave, and the middle line is gently convex. The latter terminates in an obtuse angled mark just in front of the edge of the ball. This edge is also slightly free from the ball. The capitular costal surfaces do not project inferiorly quite to the line of the inferior surface of the centrum. [Measurements follow.]

This snake was about the size of the black snake, *Bascanium constrictor.* It is an interesting species for two reasons. First, it is the oldest serpent known from North America. Second, in the imperfection of the zygantrum we observe an approximation to the ordinary reptilian type of vertebra, from which the ophidian type was no doubt derived. In the former there is no zygosphen or zygantrum.

This important if somewhat inadequate type does not appear to have been critically reëxamined since Cope described it. It should be in the American Museum collections but cannot now be located.

4. ORDER CHELONIA (= TESTUDINES)

The following list of the turtles described from the Puerco and Torrejon is from Gilmore's revision of 1919.

	Puerco	Torre-jon		Puerco	Torre-jon
Fam. Pleurosternidae			Fam. Dermatemydidae—continued		
Compsemys parva Hay	×	×	Hoplochelys bicarinata Hay	×	
" vafer Hay	×		" laqueata Gilmore	×	
" puercensis Gilmore	×		" saliens Hay		×
" torrejonensis Gilmore		×	" paludosa Hay		×
Fam. Baënidae			" elongata Gilmore		×
Baëna escavada Hay		×	Fam. Trionychidae		
" sp.	×	×	Conchochelys admirabilis Hay	×	
Fam. Plastomenidae			Aspideretes sagatus Hay	×	
Plastomenus acupictus Hay		×	" puercensis Hay	×	?
" torrejonensis Gilmore		×	" reesidei Gilmore	×	
" sp.	×	×	" vegetus Gilmore	×	
Fam. Dermatemydidae			" quadratus Gilmore	×	
Adocus hesperius Gilmore	×		" perplexus Gilmore	×	
" substrictus (Hay)		×	" singularis Hay		×
" onerosus Gilmore		×	" sp.		×
" annexus (Hay)		×	Platypeltis antiqua Hay		×
Hoplochelys crassa (Cope)	×		Amyda eloisae Gilmore		×

PLEUROSTERNIDAE Cope, 1868

Cope, E. D., 1868, Proc. Acad. Nat. Sci. Phila., XX, p. 282

Compsemys Leidy, 1856

Leidy, J., 1856, Proc. Acad. Nat. Sci. Phila., VIII, p. 312

Genotype: Compsemys victa Leidy, type from " Laramie formation," Long Lake, North Dakota.   The range of the genus extends from the Judith River to the Fort Union.

Diagnosis: Gilmore, 1919, defines the genus as follows, referring to it four species out of the Torrejon and Puerco:

A genus of Pleurosternidae.   External surfaces of the shell ornamented with small, close-set enameled tubercles; first peripherals meet on mid-line and exclude the nuchal from reaching the anterior margin of the carapace; vertebrals broader than long; first suprapygal absent; inguinal buttresses uniting exclusively with the fifth costals; mesoplastrals joining broad at mid-line, often narrowing toward outer extremities; costo-marginal sulcus entirely below costo-peripheral sutures; gulars reduced; humerals meeting narrow at midline; entoplastron wider than long; posterior lobe of plastron deeply notched; inframarginals only partially on plastron.[213]

Compsemys parva Hay

Compsemys parva Hay, 1910, Proc. U. S. Nat. Mus., XXXVIII, p. 308, pl. x, figs. 1–3.

Type: U. S. Nat. Mus. No. 6548, portions of plastron.

Topotype: U. S. Nat. Mus. No. 8528, most of carapace and plastron.

[213] Gilmore, C. W., 1919, U. S. Geol. Surv. Prof. Paper 119, p. 12.

Horizon and Locality: Topotype fifty feet above base of Puerco formation, 4 miles E. of Kimbetoh, San Juan basin, New Mexico. Reeside, 1916. A second specimen, U. S. Nat. Mus. No. 8598, from the Torrejon formation is referred by Gilmore to the same species, the only vertebrate known upon adequate evidence to pass through without change of species.

Diagnosis (Gilmore, 1919, p. 16): Small size; costo-marginal sulcus crossing peripherals of posterior half of carapace, on their upper fourths, whereas in all other species it crosses them midway or below; posterior lobe of plastron with a median V-shaped notch; eighth neural relatively large.

Compsemys vafer Hay

Compsemys vafer Hay, 1910, Proc. U. S. Nat. Mus., XXXVIII, p. 311, pl. x, figs. 4, 5; pl. xi, figs. 1, 2.

Type: U. S. Nat. Mus. No. 6551, fragments of carapace and some few of plastron.

Topotypes: U. S. Nat. Mus. Nos. 8529, 8530, 8600, the first including the greater part of plastron and a considerable part of carapace.

Horizon and Locality of type and topotypes: Puerco formation, San Juan basin, New Mexico.

Diagnosis (Gilmore, 1919, p. 18): Medium size; costo-marginal sulci crossing posterior peripherals at middle of their height or below; sulcus between third and fourth vertebrals crossing fifth neural; posterior lobe of plastron with a wide median notch and prominently developed xiphiplastral processes.

Compsemys puercensis Gilmore

Compsemys puercensis Gilmore, 1919, U. S. Geol. Surv. Prof. Paper 119, pp. 9, 10, 19, pl. iii, figs. 1, 2.

Type: U. S. Nat. Mus. No. 8544, consists of posterior two-thirds of plastron, first six neurals with portions of attached costals, and various fragments.

Horizon and Locality: Puerco formation, 4 miles N.W. of Kimbetoh, San Juan basin, New Mexico.

Author's Diagnosis: Medium size; sulcus between third and fourth vertebrals crosses fifth neural, as in *C. vafer;* posterior lobe of plastron with wide median V-shaped notch, and without prominent xiphiplastral processes.

Compsemys torrejonensis Gilmore

Compsemys torrejonensis Gilmore, 1919, U. S. Geol. Surv. Prof. Paper 119, pp. 10, 21, pl. iv, figs. 1, 2.

Type: U. S. Nat. Mus. No. 8549, carapace and plastron nearly complete.

Horizon and Locality: Torrejon formation 8 miles N.E. of Kimbetoh, San Juan basin, New Mexico.

Author's Diagnosis: Typically largest known species; costo-marginal sulcus crossing all peripherals at middle of their height or below; sulcus between vertebrals 3 and 4 crossing sixth neural; posterior lobe of plastron with a deep U-shaped median notch; xiphiplastral processes prominently developed; eighth costal plate relatively small.

BAËNIDAE Cope, 1882

Cope, E. D., 1882, Proc. Amer. Phil. Soc., XX, p. 143

Baëna Leidy, 1871

Leidy, J., 1871, U. S. Geol. Surv. of Wyoming . . . 2d (4th) Ann. Rept., p. 367

Type: Baëna arenosa from the Middle Eocene Bridger formation of Wyoming.

Revised Diagnosis, Hay, 1908:[214] No interhumeral scute. . . . No preneural so far as known. Plastron projecting little, if at all, beyond front of carapace. Skull with choanae well in front. . . . Shell firmly joined to the carapace by sutural union with the lateral peripherals and by broad and high axillary and inguinal buttresses. Hinder border of the carapace scallopt, and with an extensive excavation over the tail. Nuchal bone in contact with the first neural; no preneural; no supramarginal scutes; anterior lobe of plastron not extended in front of the carapace. Mesoplastra large, with the outer ends expanded. Posterior plastral lobe slightly emarginated. Intergulars, gulars and inframarginals present. Skull broad, with the temporal region extensively rooft, the squamosals in contact with the parietals. Jugal forming a part of the rim of the orbit. Triturating surface of the maxilla furnisht with a prominent longitudinal ridge. Choanae opening on a line joining the fronts of the orbits.

Baëna has a geological range from the Judith River to the Uinta, but is characteristically a Mesozoic survival in the Eocene, and like all such it does not last later than Eocene. The Paleocene species is based upon a fine shell.

Baëna escavada Hay

Baëna escavada Hay, 1908, Carn. Inst. Wash. Publ. No. 75, p. 65, pl. xi, figs. 1, 2, text figs. 42, 43.

Type: A. M. No. 1203, nearly complete shell.

Horizon and Locality: Middle Paleocene, Torrejon formation, head of Escavada Canyon, San Juan basin, New Mexico, A. M. Exped., 1896.

Author's Description (p. 66): All traces of the sutures between the bony elements of the shell are obliterated thru coössification. The sutures between the epidermal scutes are distinct, but narrow and only moderately impresst. The total length of the carapace [figured] is 381 mm. In the midline the length is about 12 mm. less, on account of the slight emargination in front and the excavation in the rear. The breadth was close to 300 mm. The carapace is broad, ovate, obtusely pointed in front, truncate behind. On each side of the anterior emargination the border was gently repand. The hinder border, on each side of the excavation for the tail, was scallopt. On the hindermost part of the carapace there is a moderate rounded keel. From this there may be traced forward faint indications of the two parallel grooves so distinct in *B. arenosa*. Except these grooves, the ornamentation seen in the latter species is absent.

The most distinctive character of this species is seen in its narrow and spatulate anterior plastral lobe [figured]. The plastron as a whole is relatively small. The total length is 295 mm. The anterior lobe has a length of 83 mm. and a breadth, at its base, of 96 mm. From the base the lateral borders run forward and inward to within 30 mm. of the front, at the ends of the sutures between the gular and humeral scutes. Here the width is 61 mm. Beyond these points the lobe expands again to 64 mm.; then curves forward and inward to the ends of the intergular sutures. The extreme end of the lobe is truncated and falls about 30 mm. short of the anterior border of the carapace.

The posterior lobe has a width of 114 mm. at the base, and a length of 75 mm., failing to reach the excavation in the carapace by about 47 mm. It narrows, as it passes backward, more rapidly than does that of *B. arenosa*. The hinder border is slightly concave.

The bridge has a width of 135 mm. Its length, to outer margin of the shell, is 105 mm.

The vertebral scutes are relatively narrow, the second having a length of 88 mm. and a width of 65 mm.;

214 Hay, O. P., 1908, Carn. Inst. Wash. Publ. No. 75, pp. 58, 59.

the third a length of 85 mm. and a width of about 70 mm.; the fourth a length of 80 mm. and a width of 70 mm.; the fifth a length of 48 mm. and a width of 80 mm.   In the case of the first vertebral the same lack of symmetry is to be seen as has been observed in so many other specimens belonging to the genus *Baëna*.

The costo-marginal sulci are distant from the edge of the carapace about 22 mm.   Those subdividing this anterior marginal region are too obscure for certain determination.   There are indications of one which crosst this region about 15 mm. to the left of the midline.   There was, therefore, probably a nuchal scute 30 mm. long from side to side.   Beyond this the marginals increase in length and breadth.   Over the bridge they rise on the sides of the carapace 45 mm.

On the plastron are distinct gulars and intergulars.   The humero-pectoral sulcus crosses the midline on the line joining the axillary notches.   Laterally the sulcus is suddenly turned forward and outward. The pectoral scutes meet along the midline for a distance of 53 mm., and extend laterally about 72 mm. The abdominal scutes occupy 52 mm. of the midline; the femorals 67 mm.; the anals 50 mm.   As in other species of *Baëna*, the suture between the femorals and the anals runs outward, then turns backward for some distance, then again outward.

There are 4 large inframarginals.

Baëna sp.

Two specimens from the Puerco are referred to this genus by Gilmore in 1919.

PLASTOMENIDAE Hay, 1902

Hay, O. P., 1902, Bull. U. S. Geol. Surv. No. 179, p. 452; 1908, Carn. Inst. Wash. Publ. No. 75, p. 466

Author's Diagnosis: Trionychoidea with skull like that of the Trionychidae.   Neck unknown, but probably like that of the Trionychidae.   No peripheral bones.   Epiplastra separated from the hyoplastra by the large, crescentic entoplastron.   Hyoplastra, hypoplastra, and xiphiplastra closely united, as in the Emydidae.   Feet unknown.

One genus, which apparently ranges from Judith River Cretaceous to Bridger Eocene. The Cretaceous and Paleocene species are, however, of more or less doubtful reference to the genus because the specimens are so fragmentary.

Plastomenus Cope

Cope, 1873, Proc. Acad. Nat. Sci., p. 278; 1877, Rept. U. S. Geol. Surv. (Wheeler Survey), IV, p. 47; Hay, 1908, Carn. Inst. Wash. Publ. No. 75, p. 466

Type: Plastomenus thomasi Cope, of the Bridger formation.
Diagnosis: According to Hay (1908, p. 466), " the generic characters of *Plastomenus* cannot be separated from those assigned above to the family."   The reader is referred to the diagnosis for the family, above.

Plastomenus acupictus Hay

Plastomenus acupictus Hay, 1907, Bull. Amer. Mus. Nat. Hist., XXIII, p. 852, pl. LIV, figs. 1–3, text fig. 8; 1908, Carn. Inst. Wash. Publ. No. 75, p. 470, text fig. 629.

Type: A. M. Cope Coll. No. 1025, fragments of costal and plastral bones.
Horizon and Locality: Paleocene, probably Torrejon, of New Mexico.
Author's Description: That the species belongs to *Plastomenus* is shown by the fact that the hypoplastron and xiphiplastron were both suturally joined to their fellow bones as far as represented by the specimen.   [Measurements of the costals follow.]

The free border of the costals is beveled off on the upper side.   A fragment of the nuchal shows that its free border was similarly beveled.

On the hinder costals are seen six or seven welts which run backward and somewhat outward. The whole upper surface of the shell, except the beveled border, is ornamented with small pits and narrow intervening ridges. There are five of the pits in as many millimetres. [Measurements of plastral bones follow.] The sutural border of the hyoplastron in this [bridge] region is concave . . . the free border of the xiphiplastron . . . is thin and acute.

The sculpture of the xiphiplastron resembles that of the carapace, but there are no welts and the pits are somewhat smaller.

Plastomenus torrejonensis Gilmore

Plastomenus torrejonensis Gilmore, 1919, U. S. Geol. Surv. Prof. Paper 119, p. 55, pl. xix, fig. 2, text fig. 25.

Type: U. S. Nat. Mus. No. 8543, about half of carapace and fragment of plastron.

Horizon and Locality: Torrejon formation (upper), 8 miles E.N.E. of Kimbetoh, San Juan County, New Mexico.

Author's Diagnosis: Disk of carapace elongated oval; nuchal not suturally united to first costals; longitudinal welts crossing costals; some of neurals keeled; six or seven pits in a line 10 millimetres long.

. . . Distinguished from *Platypeltis trepida* Hay by its narrower shell and the nonsutural union of the nuchal with the first costals. It resembles *Platypeltis antiqua* in the proportions of the carapace and the lack of the sutural articulation of the nuchal with the first costal, but its wider neurals and the lack of longitudinal welts crossing the costals appear to indicate specific distinctness.

Plastomenus torrejonensis is distinguished from *P. acupictus* . . . by the larger pits forming the sculpture, the smaller number of longitudinal welts crossing the costals, and the very much smaller seventh costal.

DERMATEMYDAE Gray, 1870

Dermatemydidae emend. Hay, 1908

Gray, J. E., 1870, Proc. Zool. Soc. London, p. 711; Hay, O. P., Carn. Inst. Wash. Publ. No. 75, p. 224

Adocus Cope, 1868

Cope, E. D., 1868, Proc. Acad. Nat. Sci. Phila., XX, p. 235

Genotype: Emys beata Leidy from Upper Cretaceous of Monmouth Co., N. J.

Synonym: Alamosemys Hay, 1908, *auct.* Gilmore, 1919; type: *A. substricta* Hay, from Torrejon formation of New Mexico.

The range of the genus is from the Upper Cretaceous Kirtland formation, New Mexico, and the greensands of New Jersey, through the Puerco and Torrejon. One described species from the Puerco, three from Torrejon, distinguished by Gilmore (1919, p. 26):

Puerco species:
  Costal scutes higher than wide; posterior marginals reduced in height but overlapping costal bones.
  Adocus hesperius

Torrejon species:
  Entoplastron rhombic:
    Anterior lobe rounded with slight median notch, the length of lobe 43 per cent of the width at
      the base. . . . .*Adocus annexus*
    Anterior lobe truncated; the length of lobe 33 per cent of the width at the base.
      Adocus onerosus Gilmore
  Entoplastron broad behind, pointed in front:
    Anterior lobe rounded; the length of lobe 47 per cent of the width at the base. .*Adocus substrictus*

Adocus hesperius Gilmore

Adocus hesperius Gilmore, 1919, U. S. Geol. Surv. Prof. Paper 119, pp. 9, 26, 33, pl. ix.

Type: U. S. Nat. Mus. No. 8596, consists of the carapace lacking the anterior border and the peripherals and distal parts of the costals of the left side.

Horizon and Locality: Puerco formation, 4 miles N.W. of Kimbetoh, San Juan basin, New Mexico.

Author's Diagnosis: Seventh and eighth neurals suppressed, sixth reduced; vertebral scutes longer than wide; fifth vertebral unusually large; eighth pair of costals meeting wide on the midline; costal scutes higher than long; posterior marginals reduced in height; sculpture made up of rounded pits; three pits in 5-millimetre line.

Adocus onerosus Gilmore

Adocus onerosus Gilmore, 1919, U. S. Geol. Surv. Prof. Paper 119, pp. 10, 26, 35, pl. x, figs. 1, 2.

Type: U. S. Nat. Mus. No. 8594, a considerable part of a poorly preserved carapace, which lacks most of the peripherals except those of the median anterior border, and the plastron, which lacks the posterior ends of the xiphiplastral bones.

Horizon and Locality: Torrejon formation 8 miles N.E. of Kimbetoh, San Juan basin, New Mexico.

Author's Diagnosis: Anterior lobe of plastron truncated; length 33 per cent of width at base; epiplastrals meeting for a relatively short distance on midline; entoplastron rhombic; overlapped in front by intergulars; anterior border of carapace with pronounced median excavation for reception of neck. Sculpture having obscure pattern.

Adocus annexus (Hay)

Alamosemys annexa Hay, 1910, Proc. U. S. Nat. Mus., XXXVIII, p. 318, fig. 19.

Type: U. S. Nat. Mus. No. 6539, a well preserved plastron.

Horizon and Locality: Torrejon formation, Ignacio quadrangle, La Plata Co., Colorado (northern part of San Juan basin).

Gilmore refers to this species a plastron with damaged carapace from the typical Torrejon beds eight miles northeast of Kimbetoh, and gives the following revised specific diagnosis: [215]

Anterior lobe rounded, with slight median notch; length 43 per cent of width at base; posterior lobe rounded; seventh and eighth neurals suppressed; vertebrals wider than long; eighth pair of costals meeting for a considerable distance on midline; entoplastron rhombic overlapped by both intergular and pectoral scutes.

Adocus substrictus (Hay)

Alamosemys substricta Hay, 1908, Carn. Inst. Wash. Publ. No. 75, p. 260, pl. xxxix, figs. 1, 2, text figs. 323, 324.

Type: A. M. No. 1204, a complete shell, considerably obscured by a hard investing matrix.

Horizon and Locality: Middle Paleocene, Torrejon formation, Alamosa Creek, San Juan basin, New Mexico, A. M. Exped., 1896.

Author's Description: Hay's description cited above is an extended one; the following is a condensation:

[215] 1919, U. S. Geol. Surv. Prof. Paper 119, p. 40.

Carapace elongated, moderately elevated, decidedly constricted at inguinal notches, peripherals behind these flaring. As in *Adocus* a large part of the shell (60 per cent) lies behind sutures separating 3rd and 4th costals and suture separating hypoplastrals from hyoplastrals. Nearly 50 per cent of shell is behind inquinal notches. Carapace border emarginate at nuchal, feebly serrate posteriorly. Seven neurals, the two last reduced and crowded. No keel. Supra-pygal single. Costals 6–8 meet in midline. Peripherals unusually high; free peripherals thin at borders. Carapace ornamented with narrow, low, longitudinal wrinkles.

Plastron flat, anterior lobe projecting to equal anterior border of carapace; posterior lobe 115 mm. short of posterior border. Slight anterior emargination; no posterior notch.

Anterior lobe decreases in width gradually at first, then rounds rapidly to the notch in front. . . . Entoplastral bone is short and broad and with a nearly straight posterior border. . . . Gular and intergular scutes seem to be present. . . . The humero-pectoral sulcus crosses the midline at the hinder end of the entoplastron. . . . On the bridges of the plastron are at least four inframarginals. . . . Surface of the bones of the plastron appear for the most part to have been smooth, but there are some indications of the same kind of wrinkling, as is seen on the upper surface.

Hoplochelys Hay, 1905

Hay, O. P., 1905, Amer. Geol., XXXV, p. 339; 1908, Carn. Inst. Wash. Publ. No. 75, p. 263

Type: Chelydra crassa Cope, 1888, from the Puerco formation of New Mexico.

Author's Diagnosis: Shell thick and solid. Peripherals united to the plastral bones by means of digitations and dentated sutures. Carapace furnisht above with three carinae. Plastron with anterior lobe immovable and with the posterior lobe narrow.

The characters of this genus are wholly derived from the shell. The plastral structures are not well known in any of the species. The genus appears to be related to *Staurotypus*, now living in Central America. The latter possesses only 10 peripheral bones on each side. It appears not improbable that *H. saliens* and *H. paludosa*, having the peripherals sutured to the costals, really belong to a distinct genus.

Revised Diagnosis (Gilmore) : External surfaces of the shell finely shagreened. Carapace flattened on top and having three dorsal carinae. Eight neurals. Eleven peripherals, which unite with the plastral bones by means of digitations and dentated sutures; with the costals by gomphosis, and in young by simple apposition with pits for distal prolongations of costals, in adults by close sutures. Plastron cruciform, with anterior and posterior lobes immovable, and with the posterior narrow end pointed. A row of inframarginal scutes on each bridge. Pectoral and femoral scutes meeting and crowding the abdominals from mutual contact at the midline. Intergulars and gulars wanting or consolidated with the humerals. . . .

With the exception of *H. paludosa* Hay, based on a single peripheral, the other Torrejon species of the genus *Hoplochelys* may at once be distinguished from the Puerco representatives by their relatively wider vertebrals and neurals. . . . There was a full complement of eight neural bones, a feature in which this genus differs from all other members of the family Dermatemydidae, except *Baptemys*, thereby requiring a slight modification in the definition of that family.[216]

The genus is known only from the American Paleocene, one species from the Fort Union, three each from Torrejon and Puerco.

Hoplochelys crassa (Cope)

Chelydra crassa Cope, 1888, Trans. Amer. Phil. Soc., XVI, N.S., p. 306.
Hoplochelys crassa Hay, 1905, Amer. Geol., XXXV, p. 339; 1908, Carn. Inst. Wash. Publ. No. 75, pp. 263–5, text fig. 325, pl. 38, figs. 4–9; Gilmore, 1919, U. S. Geol. Surv. Prof. Paper 119, p. 41, pl. XIII, text fig. 16.

Type: A. M. No. 6091, fragments of carapace and plastron.

Horizon and Locality: Puerco formation (Puerco or Torrejon). ?Chaco Canyon, San Juan basin, New Mexico.

[216] U. S. Geol. Surv. Prof. Paper 119, p. 40.

Author's Description: The bones of both carapace and plastron are relatively much thicker than the corresponding parts of the snapping tortoise, equaling in this respect the largest species of Emys. The bridge of the plastron is not so slender as in *C. serpentina.* The vertebral bones have a median keel-like angle, which becomes at the anterior part of each vertebral scutum a prominent rib. This results from the abrupt depression of the surface on each side immediately posterior to the transverse dermal suture. In the larger specimen [type] this suture is deeply notched anteriorly, and its anterior border is so prominent posteriorly as to give an imbricate appearance, the anterior vertebral scute rolling over the posterior by an obtuse border. The marginals of the bridge are very massive, and the pit for the process of the costal is at one side of the middle, and is nearly round. It is flat in the *C. serpentina.* The pits for the plastral fingers are three, on the inner inferior edge of each marginal, and are directed obliquely. The external face of the marginals is distinguished by a rabbet, the inferior margin of which projects as a ridge beyond the external face. Inferior face convex. No other except the fine mutual sutures on the marginals of the bridge. The free marginals, of which I have two, and one of them the anal, have no gomphosis nor suture with the costals or pygal, being held in place by the integuments and by the mutual marginal sutures. The dermal scuta are well marked, the marginals having their bounding suture below that of the marginal bones. Surface of the shell everywhere smooth. [Measurements follow.]

Hay distinguishes the species from *H. saliens* " in having the bones of the plastron less firmly sutured with the peripherals and in having the latter bones, at least those of the bridge region, without close suture with the costals. The bridge is much narrower." [217]

Gilmore in 1919 refers to *H. crassa* three specimens from the Puerco, U. S. Nat. Mus. Nos. 8525, 8641 and 8643, points out that the characters relied upon by Hay as distinctive of the species are really characters of immaturity, and re-defines the species upon other characters shown in the National Museum specimens as follows:

Carapace oval, pointed behind; breadth of carapace 68 per cent of length; lateral carina well defined, sharp; carina crossing bridge peripherals well defined, sharp; excavation for neck narrow.

Hoplochelys bicarinata Hay

Hoplochelys bicarinata Hay, 1910, Proc. U. S. Nat. Mus., XXXVIII, pp. 321–4, pl. xii, text figs. 20–3; Gilmore, 1919, U. S. Geol. Surv. Prof. Paper 119, p. 43, pl. xiv, text fig. 17.

Type: U. S. Nat. Mus. No. 6549, parts of carapace and plastron.

Horizon and Locality: Puerco formation, near Ojo Alamo, San Juan basin, New Mexico.

Type Diagnosis: Differs from *H. crassa* (Cope) in having the lateral keels of the carapace broader and more obtuse. *H. crassa* also evidently had the abdominal scutes pushed away from the midline. The width of these at the inguinal notch was about 13 mm.; whereas, in *H. bicarinata,* a larger individual, these scutes are only 5 mm. wide.

From *H. caelata* the present species differs in not having the bones sculptured with oblique ridges. The outer faces of the hinder peripherals are not flat, as they are in *H. caelata,* but more or less concave, with the free borders somewhat upturned. In *H. caelata* the hypoplastron did not enter the eighth peripheral. The hinder end of the seventh is thin, as is also the whole of the eighth. In *H. bicarinata* the anterior end of the eighth is much thickened and receives a process from the hypoplastron. The hypoplastron of *H. crassa* (Cope) does not pass behind the seventh peripheral, resembling in this respect *H. caelata.*

Gilmore (1919, *loc. cit.*) refers to this species a nearly complete carapace and plastron, U. S. Nat. Mus. No. 8524, from the Puerco formation, 4 miles east of Kimbetoh, and re-defines the species as follows:

Carapace relatively short, broadly rounded behind; posterior end descending abruptly to peripherals; breadth of carapace 69 per cent of length; lateral carina well defined, obtuse; carina crossing bridge peripherals well defined, sharp; excavation for neck wide; first neural hexagonal, wider end directed posteriorly.

[217] Hay, 1908, *op. cit.*, p. 265.

Hoplochelys laqueata Gilmore

Hoplochelys laqueata Gilmore, 1919, U. S. Geol. Surv. Prof. Paper 119, p. 47, pl. xv, figs. 1, 2, text figs. 19, 20.

Type: U. S. Nat. Mus. No. 8527.

Horizon and Locality: Puerco formation, 4 miles E. of Kimbetoh, San Juan County, New Mexico. Locality and horizon same as No. 8524, referred to *H. bicarinata.*

Author's Diagnosis: Carapace elongated oval, pointed behind; posterior end 45° slope to peripherals; breadth of carapace 78 per cent of length; lateral carina not well defined, broadly rounded; carina crossing bridge peripherals faintly indicated; excavation for neck wide; humeral scute reaching posterior border of entoplastron.

Hoplochelys saliens Hay

Hoplochelys saliens Hay, 1908, Carn. Inst. Wash. Publ. No. 75, p. 265, text figs. 326, 327.

Type: A. M. No. 1200, crushed carapace and plastron.

Horizon and Locality: Torrejon formation, head of Escavada Canyon, Rio Arriba County, New Mexico.

Type description: In form the carapace [figured] was oval, rounded in front and behind, and apparently of some considerable elevation. The total length was close to 325 mm.; the width, close to 200 mm. There was a median keel of moderate height. On each side of this, at a distance of 50 mm. in front, 30 mm. behind, there was a more prominent lateral keel. Along none of these keels was the bone suddenly sunken just behind the transverse sulci, as it is in *H. crassa.* In addition to these keels, there is on each side, on the bridge peripherals, a narrow ridge which passes in front and behind into the free edges of the peripherals. [An extended detail description follows.]

Gilmore [218] refers to this species two better-preserved specimens, U. S. Nat. Mus. Nos. 8605 and 8646, from the Torrejon formation, 8 miles E.N.E. of Kimbetoh, and redefines the species:

Largest species of the genus; lateral carina broad, external sides dropping perpendicularly to general surface level, not sunken where crossed by sulci; vertebrals and neurals relatively broad; excavation for neck broad.

Hoplochelys elongata Gilmore

Hoplochelys elongata Gilmore, 1919, U. S. Geol. Surv. Prof. Paper 119, p. 50, pl. xvi, figs. 1, 2, text figs. 21, 22.

Type: U. S. Nat. Mus. No. 8553, greater part of carapace and plastron.

Horizon and Locality: Torrejon formation, 8 miles E.N.E. of Kimbetoh, San Juan County, New Mexico.

Author's Diagnosis: Carapace elliptical; shell narrow, elongated, width being 62 per cent of the length; lateral carinae broadly rounded; carinae crossing bridge peripherals well defined, sharp; excavation for neck narrow; vertebrals and neurals relatively broad; posterior end of carapace descending abruptly to peripheral border.

Hoplochelys paludosa Hay

Hoplochelys paludosa Hay, 1908, Carn. Inst. Wash. Publ. No. 75, p. 266, fig. 328.

Type: A. M. No. 6079, peripheral bone.

Horizon and Locality: Torrejon formation, Escavada Canyon, San Juan basin, New Mexico.

[218] 1919, U. S. Geol. Surv. Prof. Paper 119, p. 49.

Type Description: Differs from the latter species [*H. saliens*] in lacking the longitudinal ridge along the convexity of the outer surface of the peripherals, and in having the suture with the costals much higher above the costo-marginal sulci. . . .

From *H. crassa* the present species differs in having the supposed seventh peripheral strongly sutured to the contiguous costal bone, in having no lateral carina, and in having the longitudinal sulcus at the middle of the height, instead of near the costal border.

TRIONYCHIDAE Bonaparte, 1838

Bonaparte, C. L., 1838, Arch. Naturges., IV, I, p. 136

Of the four genera represented in the Paleocene one is peculiar to it, known only from a skull. The other three range from Cretaceous to Recent.

Conchochelys Hay

Hay, O. P., 1905, Bull. Amer. Mus. Nat. Hist., XXI, p. 335, figs. 1–3; 1908, Carn. Inst. Wash. Publ. No. 75, p. 483, pl. LXXXVIII, figs. 1–3

Type: Conchochelys admirabilis Hay, the only known species.

Conchochelys admirabilis Hay

Conchochelys admirabilis Hay, *loc. cit. supra.*

Type: A. M. No. 6090, a skull.

Horizon and Locality: Puerco formation, Coal Creek Canyon, San Juan County, New Mexico.

Author's Diagnosis (p. 338): Skull broad. Masticatory surfaces broad and concave. Pterygoid region narrow. Basioccipital region short. Choanae behind the orbits. Posterior squamosal process apparently much shortened.

A detailed description of the species is given.

Aspideretes Hay

Hay, O. P., 1903, Proc. Amer. Phil. Soc., XLII, p. 274

Type: Trionyx gangeticus Cuvier.

Generic Diagnosis (Hay, 1908, pp. 483, 485): Skull usually not so broad; pterygoid region of the palate broad. Carapace and plastron ornamented with ridges and pits. . . . Carapace with 8 pairs of costal plates; one or more of the posterior pairs in contact on the midline. A preneural plate between the nuchal and the true first neural.

Aspideretes reesidei Gilmore

Aspideretes reesidei Gilmore, 1919, U. S. Geol. Surv. Prof. Paper 119, p. 56, pl. XX, text fig. 26.

Type: U. S. Nat. Mus. No. 8537, almost complete carapace.

Horizon and Locality: Puerco formation, 4 miles E. of Kimbetoh, San Juan County, New Mexico.

Author's Diagnosis: Disk of carapace slightly longer than wide; nuchal relatively narrow transversely, joined by its whole length to preneural and first costals, and extending forward beyond the general outline of the disk; free ends of first costals looking strongly forward; notch on posterior median border of disk. Outer two-thirds of costals crossed by longitudinal ridges; five ridges or four valleys in a line 10 millimeters long.

Aspideretes vegetus Gilmore

Aspideretes vegetus Gilmore, 1919, U. S. Geol. Surv. Prof. Paper 119, p. 57, pl. xviii, fig. 2, pl. xxi, text figs. 27, 28.

Type: U. S. Nat. Mus. No. 8539, considerable portions of the carapace and plastron.

Horizon and Locality: Puerco formation, 4 miles E. of Kimbetoh, San Juan County, New Mexico.

Author's Diagnosis: Disk of carapace oval in outline; entire upper surface of nuchal sculptured; joined by whole length to preneural and first costals; notched on median posterior border for preneural; costals crossed by numerous longitudinal welts; sculpture consisting of shallow rounded pits, usually four in a 10-millimeter line; plastron with twisted ridges.

Aspideretes quadratus Gilmore

Aspideretes quadratus Gilmore, 1919, U. S. Geol. Surv. Prof. Paper 119, p. 59, pl. xxii, text fig. 29.

Type: U. S. Nat. Mus. No. 8545, carapace, lacking nuchal and distal portions of costals of left side.

Horizon and Locality: Puerco formation, 4 miles N.W. of Kimbetoh, San Juan County, New Mexico.

Author's Diagnosis: Disk of carapace broader than long; posterior border truncated and broadly concave transversely; lateral border turning in abruptly toward the center on middle of second costal; from three to five pits in a 10-millimeter line.

Aspideretes perplexus Gilmore

Aspideretes perplexus Gilmore, 1919, U. S. Geol. Surv. Prof. Paper 119, p. 60, pl. xix, fig. 1, text fig. 30.

Type: U. S. Nat. Mus. No. 8532, carapace lacking the nuchal, seventh neural and parts of first costals.

Horizon and Locality: Puerco formation, 4 miles E. of Kimbetoh, San Juan County, New Mexico.

Author's Diagnosis: Carapace with seven pairs of costals; seventh pair considerably reduced; notch on median posterior end; sculpture of medium coarseness, five or six pits in a 10-millimeter line.

Gilmore refers this species with doubt to *Aspideretes,* " as it could with almost equal propriety be assigned to *Plastomenus,* and its true generic assignment must await the discovery of the plastral bones." The reduction of the costals is approached in other species of *Aspideretes* but not in *Plastomenus.*

Aspideretes sagatus Hay

Aspideretes sagatus Hay, 1908, Carn. Inst. Wash. Publ. No. 75, p. 497, pl. 93, figs. 1–3, text fig. 652; Gilmore, 1919, U. S. Geol. Surv. Prof. Paper 119, p. 61, pl. xxiii, fig. 1, text fig. 31.

Type: A. M. No. 1201, a complete carapace.

Horizon and Locality: " Puerco formation " (?Torrejon) of New Mexico. A. M. Exped., 1892.

Author's Diagnosis (p. 486): Disk of carapace longer than wide; nuchal joined by its whole length to preneural and first costals; the outer end hardly reaching the rib of the first costal; 5 to 7 pits in 25 mm. line.

Gilmore (1919, *op. cit.*) refers to *A. sagatus* two specimens in the National Museum collections, Nos. 8554, 8652, obtained from the upper Torrejon formation about 8 miles E.N.E. of Kimbetoh, New Mexico.

Aspideretes singularis Hay

Aspideretes singularis Hay, 1907, Bull. Amer. Mus. Nat. Hist., XXIII, p. 853, pl. LIV, fig. 4, text figs. 9–17.

Type: A. M. No. 1028, carapace, plastron and nearly complete skeleton.

Horizon and Locality: Torrejon formation, Chaco Canyon, San Juan County, New Mexico.

Author's Description: No other trionychoid skull so old is known that is accompanied by the shell. The specimen is therefore interesting because it shows how little change has taken place in the group since Basal Eocene times.

The carapace [figured] is elliptical and appears to have been rather convex. . . . The carapace is composed of the nuchal, a preneural, seven neurals, and eight pairs of costals. . . . The outer ends of the nuchal overlap strongly the projecting ends of the ribs of the costals of the first pair. . . .

The costals of the eighth pair join on the midline and extend laterally each 60 mm. The free borders of the carapace are beveled off and are smooth. The ends of the ribs extend beyond the borders of the disk about 50 mm. and vary in width from 25 mm. to 33 mm.

The sculpture [figured] consists of pits with prominent intervening ridges. On the neurals and the proximal ends of the costals the pits are nearly circular and there are from six to eight in a line 20 mm. long. Toward the distal ends of the costals the pits increase in size. On the distal fourth of the costals the pits are arranged in longitudinal rows, with broad ridges between the rows. In some places there are only three or four rows in a 20 mm. line. On the rear of the carapace the pits are of size larger than the average.

. . . The xiphiplastron is of triangular form, with the proximal processes interdigitated with the processes of the hypoplastron. Each is about 110 mm. long and 75 mm. wide. With the exception of the anterior and posterior processes the whole lower surface of each xiphiplastron is sculptured.

. . . The skull resembles closely that of *Platypeltis ferox*, but the profile descends more abruptly from the middle of the orbits.

Aspideretes puercensis Hay

Aspideretes puercensis Hay, 1908, Carn. Inst. Wash. Publ. No. 75, p. 499, pl. 94, figs. 1–3, pl. 104, figs. 2, 3, text figs. 654, 655.

Type: A. M. No. 1202, most of carapace and plastron.

Horizon and Locality: " Puerco " beds of New Mexico (Torrejon).

Author's Diagnosis (p. 486): Disk longer than wide; nuchal loosely joined to preneural and costals; fontanels behind it.

Platypeltis Fitzinger, 1836

Fitzinger, L., 1836, Ann. Wien Mus. Zool. I, pp. 120, 127

Type: Testudo ferox Schneider.

Platypeltis antiqua Hay

Platypeltis antiqua Hay, 1907, Bull. Amer. Mus. Nat. Hist., XXIII, p. 859, fig. 18.

Type: A. M. No. 1036, parts of carapace, limb bones, etc.

Horizon and Locality: ?Torrejon formation, San Juan County, New Mexico.

Author's Diagnosis: This species differs from *P. serialis*, of the Wasatch and Bridger, in having no welts on the upper surface and in having smaller pits.   From *P. trionychoides* of the Bridger, it differs in having coarser pits, as well as in other respects.

Amyda Oken, 1816

Oken, L., 1816, Lehrbuch Zool., II, p. 348

Type: Trionyx euphraticus Daudin.

The known geological range of this modern genus extends back to the Puerco in the western United States, but certain Upper Cretaceous species of the Atlantic coast have been referred to it.

Amyda eloisae Gilmore

Amyda eloisae Gilmore, 1919, U. S. Geol. Surv. Prof. Paper 119, p. 63, pl. xxiv.

Type: U. S. Nat. Mus. No. 8540, a nearly complete carapace.

Horizon and Locality: Puerco formation, 2¾ miles N. by E. of Kimbetoh, San Juan County, New Mexico.

Author's Diagnosis: Disk of carapace slightly broader than long; sculpture four to five pits in 10-millimeter line, irregular in character; greatest width slightly posterior to middle; nuchal more than half the width of the carapace.

5. ORDER CROCODILIA

Two crocodiles from the Paleocene of New Mexico have been described since Matthew's final work on this monograph.   These are listed below, with the references for the publications in which they were described, and expanded in the Addendum (p. 366).—The Editors.

Allognathosuchus mooki Simpson, 1930.   From the Puerco formation.   *See* Simpson, G. G., 1930, Amer. Mus. Novitates No. 445.

Leidyosuchus multidentatus Mook, 1930.   From   the   Torrejon   formation.   *See* Mook, C. C., 1930, Amer. Mus. Novitates No. 447.

C. CORRELATION OF THE PUERCO AND TORREJON FAUNAS

Earlier and Later Correlation Work

The Wasatch fauna in the overlying Lower Eocene of the San Juan basin was described fully by Cope in 1877 [219] and extended and revised by Granger, Sinclair and Matthew in 1914–18.[220]   It is well known as equivalent to the northern Wasatch of Wyoming and to the Suessonian of the typical Tertiary succession of western Europe.

---

[219] Cope, E. D., 1877, Rept. U. S. Geog. Surv. W. of 100th Mer., IV.

[220] Sinclair, W. J., and Granger, W., 1914, Bull. Amer. Mus. Nat. Hist., XXXIII, pp. 297–316. Granger, W., 1917, *ibid.*, XXXVII, pp. 821–30.   Matthew, W. D., and Granger, W., 1915, *ibid.*, XXXIV, pp. 1–103, 311–61, 429–83; 1918, *ibid.*, XXXVIII, pp. 565–657.

The Puerco formation was named and described by Cope in the same report, but fossils were not found in it until later. David Baldwin in 1881–5 secured the large fauna which Cope described in various papers; all Baldwin's collections came from the northern and western side of the basin, while Cope's type locality for the Puerco was near Cuba over to the east and south. Granger and Sinclair conclude, as a result of their studies of 1913, that the Puerco formation as shown at the type locality is chiefly Torrejon, although it may include a little of the lower formation at the base. Cope, however, in 1881–8, extended the scope of the formation to include the whole of the true Puerco as subsequently limited, as well as Wortman's " Torrejon." Cope's contributions deal almost wholly with the faunas and he makes very little note of the stratigraphy.

The Puerco and Torrejon faunas were placed by Cope at the base of the Eocene succession. He emphasized, however, their separation from the Wasatch and stratigraphic and faunal association with the " Laramie Group " and adopted at one time the suggestion that both should be included in a separate Post-Cretaceous period.[221] This radical proposal has met with little support and was not maintained by Cope in his later writings. He correlated the fauna with the Cernaysian of France and regarded the Fort Union as a provisional equivalent in the absence of known fossil mammals from that group.

With the progressive recognition of the distinctions of the Puerco and Torrejon faunas, by Cope, Osborn, Wortman and Matthew, it became evident that the Cernaysian was more nearly related to the Torrejon, the Puerco being decidedly older, with no known equivalent elsewhere. The recognition of the Tiffany as a distinct faunal division by Granger and Matthew made it further evident that this rather than the Torrejon was the nearest equivalent of the Cernaysian, a conclusion confirmed by Teilhard's revision of the Cernaysian fauna. Schlosser, however, maintained in 1920 that the Cernaysian was between Torrejon and Puerco in age.

In the meantime a considerable fauna has been described from different localities and horizons in the Fort Union by Douglass,[222] Gidley,[223] and recently by Simpson.[224] While considerable parts of Gidley's collection still remain undescribed it is sufficiently known to approximate its faunal relations, and Simpson has recently published a series of advance notes on the correlation of these various Fort Union and European faunas which bring their status down to the present date.[225]

The beds underlying the Puerco were referred by Cope and Wortman to the Laramie group—a term then generally understood as covering all the formations overlying the Fox Hills and beneath the Fort Union. In 1910 Brown examined these beds and secured a fine skull and jaws of *Kritosaurus* from shales underlying the heavy conglomerates beneath the Puerco. To these beds, previously called Laramie, he gave the name of Ojo Alamo, as the limitation of the term Laramie by the U. S. Geological Survey in 1907–9 made it undesirable to apply it to this series. Bauer in 1916 made a detailed study of part of this field and split up the " Laramie " into two formations, Fruitland and Kirtland, and re-

---

[221] Cope, E. D., 1885, Tertiary Vertebrata, p. 43, table.
[222] Douglass, Earl, 1908, Ann. Carn. Mus., V, pp. 11–26.
[223] Gidley, J. W., 1909, Proc. U. S. Nat. Mus., XXXVI, pp. 611–26; 1915, *ibid.*, XLVIII, pp. 395–402; 1919, Bull. Amer. Mus. Nat. Hist., XLI, pp. 541–55; 1923, Proc. U. S. Nat. Mus., LXIII, pp. 1–38.
[224] Simpson, G. G., 1928, Amer. Mus. Novitates No. 297; 1929, Ann. Carn. Mus., XIX, pp. 115–22; 1929, Amer. Mus. Novitates No. 345.
[225] Simpson, G. G., 1929, Amer. Mus. Novitates No. 354.

stricted Brown's Ojo Alamo to the upper conglomerate beds, thus excluding the typical part of Brown's formation.   Osborn in 1917 described the finely preserved skull of *Pentaceratops* from the Fruitland beds and Gilmore in 1919 described the Chelonia of the San Juan Cretaceous and Paleocene.   Reeside in 1924 added a fourth formational name, the McDermott, intercalated between Kirtland and Bauer's restricted Ojo Alamo.   This new formation included the shales from which Brown had obtained his typical Ojo Alamo fauna. The vertebrate fauna appears to be practically identical through all four divisions, but while the Kirtland and Fruitland floras were regarded by Knowlton as of Montana age, that of the McDermott was identified as of intermediate character between Cretaceous and Tertiary, and that of Bauer's Ojo Alamo seemed to him (on very slight evidence) Tertiary rather than Cretaceous.

These four " formations " are probably not very distinct nor as clearly separable as they appear to be on the Survey maps, nor is it by any means always easy to distinguish the base of the Puerco.   No two authors are in agreement as to this line, nor is the usage uniform on other formational lines.

The accompanying table, modified from Reeside, 1924, serves to show the usage of formational names by various writers and the correlation here adopted.

Cope 1874–7	Wortman 1892–6	Gardner 1907	Brown 1910	Granger 1913–16	Bauer 1916	Reeside 1924	This paper S.W. side	This paper N. side
Wasatch	Wasatch	Wasatch		Wasatch { Largo / Almagre }	Wasatch	Wasatch	Largo / Almagre	Tiffany
Puerco	Torrejon / Puerco	Naeimiento Group { Torrejon / Puerco }	Puerco	Torrejon / Puerco	Torrejon & Puerco	Torrejon & Puerco	Torrejon / Puerco	Torrejon
—?—	Laramie	Laramie	Ojo Alamo	Ojo Alamo	Ojo Alamo s.s.	Ojo Alamo s.s.	Clays Ojo Alamo s.s. / McDermott / Kirtland / Fruitland (Ojo Alamo)	Animas { McDermott / Kirtland / Fruitland }
Laramie / —?—	—?—	—?—	—?—	—?—	Kirtland / Fruitland / Pictured Cliffs	McDermott / Kirtland / Fruitland / Pictured Cliffs	Pictured Cliffs	Pictured Cliffs

Although a considerable vertebrate fauna has been described by Brown, Hay and Gilmore from the San Juan Cretaceous, very little of it has much weight in exact correlation. The chelonian and crocodilian genera run through a wide geologic range without clearly progressive change; the species have not the definite distinctness of the species of Paleocene mammals and identity or distinctness of species of these genera has little value in correlation. The genera of dinosaurs have great value if complete skulls or skeletons or thoroughly characteristic parts are found; but only two complete skulls are recorded, *Kritosaurus* from the McDermott, *Pentaceratops* from the Fruitland. These two specimens probably outweigh in correlation the rest of the list of vertebrate fauna. *Pentaceratops*, as pointed out by Osborn, suggests an age later than Belly River, somewhat earlier than Lance. Other unpublished skulls confirm this allocation. This conforms exactly to the invertebrate evidence. *Kritosaurus* is comparable with Belly River or Edmonton.

FORMATIONS OF THE SAN JUAN BASIN

The geological succession in the San Juan basin as now known extends from the upper Cretaceous to the Lower Eocene. The series of formations is mostly fresh-water, the marine members being limited to the lower part of the Cretaceous series. They are nearly all flat-lying save on the northern and western margins, with the usual irregularities of fresh-water formations, so that the stratigraphy has been difficult to work out in detail. The upper and lower limits are fairly well fixed, the upper by the characteristic Wasatch fauna, the lower by the marine faunas of upper Pierre age. Above the latest marine Cretaceous (Pictured Cliffs sandstone) lies a series of fresh-water dinosaur-bearing formations, succeeded by the Paleocene mammal beds and these in turn by the Wasatch. Owing to the irregular character of this fresh-water sequence the succession and correlation of the formations has been difficult to work out in detail, and many points are still far from clear.

1. *Pictured Cliffs.* This has yielded a considerable littoral marine fauna which is regarded by Reeside as of Montana age, somewhat older than the Fox Hills, and presumably equivalent to the upper part of the Pierre.

2. The *Fruitland* has given a small vertebrate fauna of which the most important specimens are dinosaurs, the crocodiles and turtles having no great value in correlation. Among the dinosaurs the fine skull of *Pentaceratops* is by far the most significant evidence and, as pointed out by Osborn, it suggests an age decidedly later than Judith River and somewhat earlier than Lance; it may be provisionally correlated with the Edmonton of Alberta. Other undescribed skulls support this view. A number of fragmentary specimens of dinosaurs have been compared by Gilmore to Judith River forms, but they could just as well be of Edmonton age as far as the evidence goes.

The Fruitland has also yielded a considerable invertebrate fauna, which, according to Stanton,[226] favors assignment " to an epoch considerably later than Mesaverde and Judith River and possibly somewhat earlier than Lance," agreeing quite exactly with the vertebrates. The flora is fairly large and well-determined and is of Montana age according to Knowlton.

3. The *Kirtland* shales contain a vertebrate fauna which appears to be of the same age as in the Fruitland although no very complete specimens of dinosaurs have been described. The flora as far as known appears to be identical with that of the Fruitland.

[226] Stanton, T. W., 1916, U. S. Geol. Surv. Prof. Paper 98-R, p. 310.

4. The *McDermott* contains also the same dinosaur fauna as far as known, but only one complete specimen has been described, the type of *Kritosaurus*. This is comparable with various incomplete specimens in the Fruitland, Kirtland and Ojo Alamo, and with fine skulls and skeletons found in the Belly River and Edmonton. So this evidence indicates continuance of the Fruitland-Kirtland vertebrate fauna. It is not any more advanced than the Fruitland. The flora as listed by Knowlton is a mixture of Cretaceous and Tertiary species. Knowlton remarks: [227] " Two explanations are possible to account for this apparent confusion. The material, especially that indicating Tertiary affinity, may not have been correctly identified; or the beds may be so high in the Cretaceous section that the plants foreshadowed a Tertiary facies, but this is, of course, unsubstantiated conjecture." Bearing in mind Knowlton's fundamental view that a tremendous unconformity and time gap everywhere separate Cretaceous from Tertiary formations, this may be interpreted as signifying, to those who find no reason to believe in the " Great Unconformity," that the older flora was at this time in process of giving place to the newer one.

5. The *Ojo Alamo* vertebrates include some dinosaur material, mostly rather fragmentary but agreeing so far as it goes with the Kirtland-Fruitland-McDermott fauna, except for one remarkable specimen, the scapula and ischium of a sauropod dinosaur *Alamosaurus*. This is the only North American record of a sauropod above the Lower Cretaceous, although the group is known to have survived in South America until the end of the Cretaceous.[228] Whether it represents an otherwise unrecorded survival throughout the Cretaceous in North America, or a re-invasion from South America made possible by the diastrophic uplift at or towards the end of the Cretaceous is not determinable until we know more of its exact affinities to South American sauropoda. In the present state of knowledge it throws no light on the correlation.

The few fragmentary plant remains from the Ojo Alamo are compared by Knowlton with Tertiary species but regarded as indecisive.

6. The *Puerco* is marked by the complete disappearance of dinosaurs and the appearance of the Puerco mammalian fauna described in this memoir. This fauna is not known anywhere else; the only comparison that can be made is with the Torrejon fauna in the overlying beds. As shown on preceding pages, the evidence seems to be conclusive for a very considerable time gap between Puerco and Torrejon, although no stratigraphic break can be detected. A small flora from the Puerco is stated by Knowlton to agree closely with the Animas flora which in turn agrees with the Denver-Raton.

From the above data it would appear that a dinosaur fauna approximately equivalent to the Edmonton is found through the Fruitland-Kirtland-McDermott-Ojo Alamo series, lasting without any real evidence of change from the bottom to the top, and that it was then suddenly replaced by the Paleocene mammalian fauna. The flora, however, seems to have changed from the Cretaceous (Montana) to the Tertiary (Denver-Raton) flora at an earlier point, the McDermott being transitional and the Puerco and Ojo Alamo wholly of Tertiary character. This would signify that the change in flora was followed, but not immediately, by an even more complete change in fauna, except for the Chelonia, etc.,

---

[227] Knowlton, F. H., 1924, U. S. Geol. Surv. Prof. Paper 134, p. 77.

[228] See Von Huene's recent studies of the Argentinian sauropoda. Von Huene, F., 1929, Anales del Museo de La Plata, III (2).

which persist through with but little change.   This is not unexpected.   The Lance and Denver equally show a Cretaceous fauna with a Tertiary flora, while the Edmonton of Canada with a Cretaceous fauna seems, according to Knowlton's identifications, to have had a Cretaceous flora in its lower and a Tertiary in its upper beds; i.e. the flora was transitional as in the McDermott.

7. The *Torrejon* mammal fauna is comparable with that of the middle and lower part of the Fort Union, as shown by identity of a number of species and more or less close equivalence of others.

8. The *Tiffany* fauna, although found in the lower member of the Wasatch formation and separated by a distinct erosional unconformity from the Torrejon formation beneath it, is nevertheless a continuation of the Torrejon fauna, and contains none of the new invading elements which make up the bulk of the true Wasatch fauna found in the beds above it.   Yet its time relations with the Wasatch, as judged from the nearness in evolutionary stage of phyla that pass through, are nearer than with the Torrejon below it.

The Tiffany fauna has been found only north of the San Juan River.   South of the river the fossiliferous Wasatch has beneath it a great thickness of barren sandstone, which may represent this stage; the fossil-bearing strata are divided by Granger into two horizons about equivalent to the middle and upper divisions of the Wasatch in the Bighorn basin; the lower division and the underlying Clark Fork are not fully accounted for, as the Tiffany appears to be somewhat older than Clark Fork; however—the fauna is small and from a limited zone, so it is quite likely that future exploration will fill some of the gaps.

CORRELATION OF CRETACEOUS AND TERTIARY FORMATIONS OF THE SAN JUAN BASIN

Wasatch {	Largo	= Upper Wasatch of Bighorn Basin	= Ageian
	Almagre	= Middle Wasatch of Bighorn Basin	= Sparnacian
	Tiffany [229]	= Upper Fort Union = Clark Fork	= Cernaysian

Torrejon   Middle and Lower Fort Union   ?Thanetian

Puerco   } = Animas }? = Lance [229]   ?Montian
Ojo Alamo *s.s.*   }   }   Laramie [230]
McDermott   }   Fox Hills = Edmonton   ?Danian
Kirtland   }
Fruitland   }
Pictured Cliffs   = Upper Pierre   = Belly River   Maestrichtian

[229] Recently Jepsen has maintained that the basal Fort Union is equal to the Puerco and later than the Lance, while the Tiffany is earlier than the Clark Fork.—THE EDITORS.

[230] Some recent studies by Dobbin, Reeside, Ward and others have led them to the conclusion that the Laramie and Lance are practically equivalent in time.   Whether this be fully authenticated or not, they appear to have shown that the two are not far apart in time, and that the Great Unconformity is a myth.—W.D.M.

COMPARATIVE LIST OF GENERA OF FOSSIL MAMMALS

	San Juan Basin			Montana and Bighorn Basin (Fort Union)	Lebo	Sentinel Butte	Clark Fork	Cernaysian
	Puerco	Torrejon	Tiffany					
Creodonta								
Didymictis........		×						
Dissacus..........		×	×	Dissacus..........		×		Dissacus
Microclaenodon....		×						
Claenodon.........		×	×					Arctocyon
Neoclaenodon......		×		Neoclaenodon.......	×			"Arctocyonides" (Schlosser)
Protogonodon......	×							
Loxolophus........	×							
Carcinodon........	×							
Thryptacodon......			×	Thryptacodon.......		×		Arctocyonides
Chriacus..........		×						
Deltatherium......		×						
Triisodon.........		×						
Eoconodon........	×							
Goniacodon.......		×						
Condylarthra								
Mioclaenus........		×						
Ellipsodon........	×	×		Ellipsodon.........	×			
Protoselene.......		×						
Oxyacodon........	×							
Mixoclaenus......		×		Coriphagus.........	×			
Tetraclaenodon.....		×		Tetraclaenodon......	×			
Phenacodus.......			×	Phenacodus........			×	
Taligrada								
Periptychus.......	×	×	×					
Ectoconus.........	×							
Conacodon........	×							
Anisonchus.......	×	×		Anisonchus........	×			
Hemithlaeus.......	×							
Haploconus.......		×						
				Titanoides.........	×			
Pantolambda......		×	×	Pantolambda.......	×			

FAUNAL CORRELATIONS
Montana and Wyoming

In the foregoing comparative list the scanty character of the faunas is painfully evident. The " Lebo " fauna as designated by Simpson is the Fort Union fauna obtained by Douglass and Gidley mostly from the Silberling quarry in Sweetgrass County, Montana. This quarry is near the top of the Lebo member of the Fort Union, but the Torrejon fauna is also recognized near its base. The Sentinel Butte fauna is that recently described by Simpson from the Eagle coal mine, Bear Creek, Montana.

COMPARATIVE LIST OF GENERA OF FOSSIL MAMMALS—*Continued*

	San Juan Basin			Montana and Bighorn Basin (Fort Union)	Lebo	Sentinel Butte	Clark Fork	Cernaysian
	Puerco	Torrejon	Tiffany					
Insectivora								
Palaeoryctes.......			×	*Picrodus*...........	×			*Tricuspiodon*
				Megapterna........	×			
				Protentomodon......		×		*Pleuraspidotherium* *Orthaspidotherium*
Prodiacodon.......		×						
				Myrmecoboides......	×			
Acmeodon.........		×						
Leptacodon........			×	*Leptacodon*........		×		*Adapisorex*
Mixodectes........		×						
				Planetetherium......		×		
Pentacodon........		×		*Pentacodon*........		×		
Plesiadapis........			×	*Plesiadapis*........			×	*Plesiadapis*
Carpodaptes.......			×	*Carpolestes*........		×		*Chiromyoides*
Labidolemur.......			×	*Labidolemur*........		×		
Primates								
				Paromomys........	×			
				Elphidotarsius......	×			
				Palaechthon........	×			
Taeniodonta								
Onychodectes.......	×							
Conoryctes........		×						
Wortmania........	×							
Psittacotherium....		×		*Psittacotherium*.....	×	×		
Marsupialia								
Thylacodon........	×							
Multituberculata								
Taeniolabis.......	×							
Catopsalis........		×						
Ptilodus..........		×		*Ptilodus*...........	×			
Eucosmodon.......	×							*Neoplagiaulax*
Ectypodus.........			×					

The equivalence of the Lebo fauna to the Torrejon is fairly clear. All its genera are Torrejon genera or closely related thereto, and the species are in several instances identical or closely comparable with those from the San Juan basin.

The relations of the Sentinel Butte or Bear Creek fauna to the Tiffany are not so close, but Simpson has made out a strong case for its equivalence. It is found in beds which by Thom and Dobbin are associated with the Wasatch rather than with the Fort Union—a parallel situation to the relationships of the Tiffany.

The Puerco fauna stands unique. It has no known equivalent elsewhere, unless it be the Lance, which has nothing whatsoever in common with it. The possibility has been suggested by Matthew (1921) of its being equivalent in age to the Lance though wholly different in facies. This rests upon the facts (a) that it overlies the Ojo Alamo-Kirtland-Fruitland series without any greater structural break than may be seen within this series, which although divided by the U. S. Geological Survey geologists into four formations, contains but a single fauna from bottom to top, a fauna approximately equivalent to the Edmonton and older than the Lance; (b) that the Fort Union, which similarly overlies the Lance, contains unquestioned Torrejon and Tiffany faunas but no Puerco mammals have been found even near the base; [231] (c) that the Lance mammals contain nothing that can be regarded as more primitive stages of Puerco phyla. *Meniscoëssus* was once regarded as an ancestor of *Taeniolabis*, but Simpson has disproved this; the ptilodonts show no progressive specialization through the Paleocene; the leptictids, if represented at all in the Lance, are represented by species not known to be any more primitive than the Paleocene forms.

Simpson has dissented strongly from the view that the difference between Puerco and Lance is a matter chiefly of facies but his argument is not so convincing as it might be. He apparently does not understand just what was meant by Matthew's remark that " the Paleocene mammals belong to Cretaceous orders and are the culmination of a Cretaceous mammal fauna." The conclusion seems explicit enough in connection with its context. The main *ordinal* differentiation of the Paleocene mammals must be referred back to the Cretaceous period; the later stages of their culmination and detail specialization precedent to their extinction took place during the Paleocene and Eocene. This is in accord with the usual " pattern of evolution " seen in all groups, and unnecessary to discuss here. The Tertiary mammals do their chief branching out in the Eocene and the later Tertiary faunas are the culmination of this process. The Paleocene faunas excepting the Carnivora had attained a certain fixity of type which finds its correspondence in the fixity of type seen in the later Tertiary mammalian species and genera. It was not intended to express any judgment as to the geographic range of these Cretaceous ancestors of the Paleocene faunas. Simpson regards it as " highly probable that the Eocene orders also were already differentiated in the latest Cretaceous." But the degree of divergent specialization of these orders during the Eocene, if projected back into the equally long Paleocene, would suggest that the Eocene orders were differentiated *not* during the Cretaceous but during the Paleocene from unknown groups of the Paleocene orders. Such discussions are not of much practical importance. The essential point is that Simpson explains the difference between the Lance and Puerco faunas as due almost entirely to migration. It is quite true that the character of the sediments suggests no definite difference in facies; it can hardly be said that the floras are almost identical, although they are both nearly related to the Animas flora, and also to the Fort Union. It certainly is not true that such differences of facies among land faunas are otherwise unknown among vertebrates. They are quite familiar to students of the ecology and distribution of modern faunas, as well as in innumerable cases among fossil faunas, which although of the same age, are of chiefly or wholly different content. It is obscured in the minds of many paleontologists by the fact that sedimentation

[231] Sinclair and Jepsen have recently reported the finding of a Puerco fauna in the lower Fort Union of the Bighorn basin. If this is really Puerco (which is doubtful at present writing) and overlies Lance (not Fox Hills) it would settle the matter, which otherwise remains an open question.—W.D.M.

and the preservation of fossil vertebrates in the sediments are very *apt to be* limited to certain facies, the same facies occurring in many localities over a wide region, just as certain types of modern mammals are found in numerous " stations " over a wide area, but in other stations under different environment quite other types may be found over the same or overlapping areas.   A change from marshy to dry-land forest conditions might not be reflected in any obvious or recognized change in the sediments.   It would be reflected in the flora and fauna.   But the flora of the Puerco is imperfectly known.   The late Cretaceous and Paleocene include a widespread group of formations that no doubt were each of them laid down in a region of varied environment from some localities whence have been obtained floras, from others limited faunas, but nowhere associated floras and faunas.   The assumption that the Paleocene mammals must necessarily have been absent during the Lance from the whole of the " large area " through which it is found goes far beyond what the facts really warrant, nor is there any evidence that it was present.   It may have been, in suitable stations, and the lack of close agreement between the major part of the European and American Paleocene faunas suggests at least separate centers of dispersal, Eurasiatic and North American, for the groups peculiar to each, just as the almost complete identity of the Lower Eocene fauna of these two regions suggests a single common center (?northern Asia) of dispersal for these mammals.

The Lance and Hell Creek beds are known to have been deposited in low-lying deltas close to sea-level (alternation with marine intercalation in the Cannonball member; logs bored by marine worms associated with dinosaurs.)   The Animas and Ojo Alamo overlie a great thickness of fluviatile beds, the elevation above sea-level in the San Juan basin beginning at an earlier time in the Cretaceous than in central Montana, and the area may have been considerably above sea-level at the time of the Puerco deposition.   The difference in flora between Animas and Lance may reflect this difference in facies.   And as a further item, of no great weight but pointing in this same direction, may be noted that in the Monument Creek formation, Upper Cretaceous, dinosaurs are associated with a mammal tibia referred by Gidley to the Creodonta, and have been recently reported from the mammaliferous Fort Union of the Bighorn.

It is not at all necessary to assume that the Paleocene mammal fauna, if present in North America during the time of the latest dinosaur faunas, ranged out into the low-lying swamp regions close to sea-level.   They may have been rather recent invaders from the unknown northern half of North America, spreading down as the uplands and mountains extended, and slow to reach the wide swamps.   Compare ranges of most Quaternary immigrants into the United States from the northern regions.   That the appearance of the Paleocene faunas in the record was due to a large but unknown degree to great migration movements is probable, but it is not possible at present to determine how much of the difference between the late Cretaceous and Tertiary faunas is due to differences in facies. The same may be said of the floras, in spite of the positive dicta of certain palaeobotanists.

Simpson (1929) [232] is disposed to correlate the Puerco with the Tullock member of the Lance as defined by Thom and Dobbin (1924) [233] equivalent according to their correlation to the Ludlow and to the marine Cannonball member of the Lance.   This, however, is quite as arbitrary a correlation as with the dinosaur-bearing beds.   Simpson further con-

[232] Amer. Mus. Novitates No. 345, p. 6.
[233] Bull. Geol. Soc. Amer., 35, p. 491.

cludes that if the Tullock proves to be the equivalent of the Puerco " its continued inclusion in the Lance would be quite improper." This is no more so than the continued inclusion of the Tiffany in the Wasatch.

A more recent study by Dobbin and Reeside (1929, U. S. Geol. Surv. Prof. Paper No. 158-B) has brought these authors to the conclusion that the actual relations of the Lance to the underlying Fox Hills are essentially the same as those of the Laramie, and they believe that the Lance is practically contemporaneous with the Laramie as well as with the Arapahoe and a large part of the Denver formation. So far as the argument has weight of corresponding relationship and essentially continuous sequence of sedimentation in each area, without evidence of any real break or hiatus of importance, it applies equally to the Cretaceous-Puerco succession in the San Juan basin; and certainly the contrast in fauna is equally great between Lance and Laramie or Lance and Cannonball as between the Lance and Puerco.

In view of these recent contributions to the problem of Cretaceous-Paleocene correlation, it would seem advisable to avoid dogmatic or overly precise correlations of the Puerco with other formations until a common fauna or flora is available to support them; at present the only such evidenced correlation is with the Animas of Colorado.

Paleocene of Europe

The Cernaysian fauna of France compares best with the Tiffany and Sentinel Butte faunas in America. The faunas are all scanty, probably somewhat different in facies, and the common elements are not the most abundant elements of the fauna. *Arctocyon* is close to the Torrejon and Tiffany *Claenodon*. *Plesiadapis* is common to Tiffany and Cernaysian, the species quite nearly related. *Arctocyonides* of the Cernaysian is close to *Thryptacodon* and *Neoclaenodon*. *Adapisorex* is related to *Prodiacodon*. But the abundant Cernaysian *Pleuraspidotherium* has no relative in the American Paleocene, unless it be *Meniscotherium*, which is essentially a Wasatch genus, although first appearing in the Clark Fork beds. Nor is there any sign in the European Paleocene of periptychids or phenacodonts, the common forms of the American Paleocene.

Perhaps these discrepancies are partly accident, but the relationship with the American Paleocene is not close, in contrast with the almost complete agreement between European and American Lower Eocene. In so far as they are really due to geographic differences in the faunas and not to facies or to the accidents of collecting, they would suggest that the regions in which the European and American Paleocene faunas originated and from which they invaded their Paleocene habitats were not a single center of dispersal, but two or more separated centers. The pleuraspidotheres appear earlier in Europe than the related meniscotheres in America; *per contra*, the phenacodonts much earlier in America than in Europe.

D. ORIGIN AND RELATIONSHIPS OF THE TERTIARY MAMMALIA

In the preceding sections I have summarized the characters and adaptation of the Paleocene mammals.   It is the oldest mammalian fauna of which I have any competent knowledge.   It is the most primitive, as measured by the general directions of diverse specializations, that I see in the succession of Tertiary faunas.   In a very general way it may be regarded as representing the common ancestral group from which the various orders of placental mammals are derived.   But it is quite clear that it is directly ancestral to only a minor part of the Tertiary phyla, a part that disappears progressively through the Eocene, leaving only a few survivors in the Oligocene and practically none in the later Tertiary.

The principal Tertiary stocks appear suddenly at the beginning of the true Eocene both in America and western Europe, not directly derivable from anything in the Paleocene, although somewhat distantly related.   The natural inference has been a great faunal invasion at that time, presumably from the intervening land region of northern Asia. It might be expected therefore that a knowledge of the Paleocene fauna of Asia would reveal the more direct ancestors of the Tertiary mammals.   The recently discovered Gashato fauna of Mongolia does not however conform to this expectation.   It does not, with one very doubtful exception (*Eurymylus*), contain anything that could be regarded as ancestral to the perissodactyls, artiodactyls, rodents and Primates of the Eocene.   It is a very interesting and important fauna and throws valuable light on the ancestry of notoungulates and uintatheres, but it clearly is not the fauna from which the Eocene mammals were derived.   This is explained by the environmental facies which it represents.   The Eocene faunas, like the Paleocene, have been shown to be a forest facies.   The Gashato apparently was an arid or semi-desert facies, judging from the character of the beds and the geological history of the region.   It is a reasonable interpretation of the evidence that the mammals primarily forest-dwelling had their chief center of dispersal in the forests of northern Eurasia at the beginning of the Tertiary, the Gashato representing a specialized facies adapted to arid regions, while the Puerco-Torrejon and the Cernaysian represent a post-Cretacic dispersal wave from that center or from secondary centers in northern Europe or North America.

These facts are shown in detail in the body of this memoir and are broadly summarized in the accompanying table (Geologic Range of Families, etc.).   They may be variously interpreted.

1. We may note without seriously considering them the interpretation placed upon such data by special creationists or the advocates of discontinuous evolution—"maximutationists," if one may coin a word—who assert that the sudden appearance in the record of new groups widely different from those previously known is to be explained, in some or all cases, as a new creation, whether by an " act of God " (i.e. by a departure from the known laws that have guided the history of life on the earth) or by a mutation analogous to the major abnormalities or " sports " that appear from time to time in experimental breeding and are shown to breed true.   The earlier and cruder speculations of the Mendelian experimentalists made much of this aspect in attempting to interpret the evolution of life on the earth, and a valiant defender of a substantially similar viewpoint has recently appeared in Dr. Austin Clark.

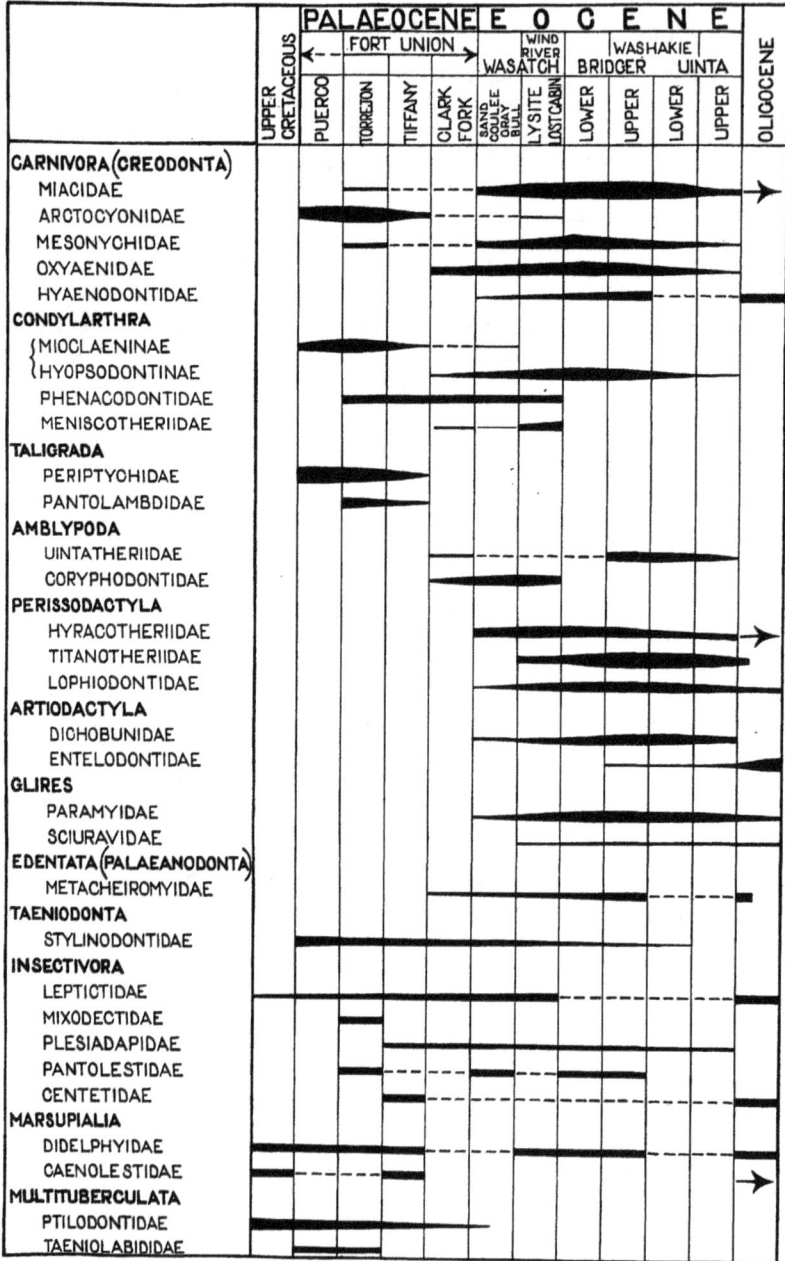

FIG. 84. Geologic Range and Relative Abundance of Paleocene and Eocene Mammals.

All such interpretations fail to understand the real nature of the geological succession and of the fossil record.   No practical paleontologist fails to be convinced of the continuity of the evolutionary process down to the stage of interspecific differences at least, and finds other interpretations of the major gaps in the record more probable than sudden creation of the new types.

2. Assuming such continuity, these sudden faunal changes may be attributed to one or more of three factors:

a. *Lapse of time*, unrecorded locally, whether or not it be recorded elsewhere.

b. *Change of facies*, the older formation representing a different environment than its successor and containing consequently a different faunal facies.

c. *Migration*, introducing suddenly (in a geologic sense) more or less important new elements to the fauna, these reacting on the stocks already there and resulting in extinction or modifications of these autochthonic elements.

The first may sometimes be indicated by geologic breaks and unconformities between the older and the later formations, but more often by change in those phyla that pass through from one to the other.

The second is indicated by change in the ecology and adaptation of the fauna or flora, changes in lithology or other geologic evidence.

The third is indicated by the sudden appearance of new groups which take the place of older ones and cause their extinction.

Time Lapses (*Disconformities*).   The time gap between the Ojo Alamo Cretaceous and the Puerco at the base of the Paleocene is difficult to estimate, as few if any phyla pass through among the mammals.   The reptiles which do pass through (Champsosauridae, Crocodilidae, Baënidae, Trionychidae, Dermatemydidae) show so little progressive change during the late Cretaceous on one hand or during the Tertiary on the other, that their negative evidence has no serious weight.   The geologic evidence does not show any wide time gap; there is said to be an erosional but no angular unconformity between the two, and a casual re-examination does not suggest that the break is so clearly marked that much weight should be placed upon it.

Between the Paleocene and Eocene the faunal break can be estimated in terms of change of several phyla that pass through: Phenacodontidae, Leptictidae, *Didymictis*, *Dissacus*, Arctocyoninae, *Chriacus*, *Thryptacodon*, Stylinodontidae, *Eucosmodon* and in the Clark Fork *Coryphodon*, *Oxyaena*, Limnocyoninae, Bathyopsinae.   In all these we find a wide gap between Torrejon and Wasatch representatives, but between Tiffany and Wasatch it is comparatively small, and even less between Clark Fork and Wasatch.   The data indicate a long lapse of time between Middle Paleocene and the beginning of the Eocene, but a short or very short interval between Upper Paleocene and Eocene.   This conforms with the stratigraphic data which place the Tiffany at the base of the Wasatch formation with an unconformity of erosion between it and the Torrejon and no break between it and the Eocene.   In the Bighorn basin the relations between the Clark Fork and the beds above and below it are apparently rather close, no very great break being indicated.

Changes of Facies.   There is no well-marked change of facies indicated between the Paleocene and Eocene in the San Juan basin, nor is there any between the Ojo Alamo and the Puerco, so far as the stratigraphy goes.   But the very great change between Cretaceous and Tertiary faunas has been generally interpreted as associated with an environmental change of some kind.   The continuance through of crocodiles, turtles and champsosaurs unchanged does not support this interpretation very well; nor does the abundance of fossil logs in the Puerco as well as in the Ojo Alamo suggest any great environmental change to account for the disappearance of dinosaurs and appearance of mammals.   In fact, a better argument might be made for facial change between the Paleocene and Eocene; there is more change in the character of the sediments, and likewise the disappearance of champsosaurs, change in the genera of crocodiles and appearance of the emydid turtles, replacement of *Tetraclaenodon* by the more cursorial *Eohippus*, appearance of notharctid primates, etc., might be attributed to an environmental change extending their range into the area, thus supplementing and locally modifying the effects of broader migration movements.

Migrations.   The appearance at the beginning of the Eocene in such distant regions as central Europe and the Rocky Mountain basins of nearly related species of the same genera *Eohippus, Coryphodon, Palaeonictis, Pachyaena, Paramys* can hardly be explained otherwise than by immediate migration from an intervening centre of dispersal where their immediate ancestors had been living at the end of the Paleocene.   That this migration was caused and accompanied by diastrophic and climatic changes in this intervening region, extending very probably to or towards the outlying areas invaded, is a simple and probable explanation of it, but at present there is no direct evidence as to the location or geologic history of such areas of dispersal..   The Mongolian later Eocene faunas have thus far been insufficiently analyzed to clear up these problems of migration, although some suggestions have been made in interpreting them.   Like the Paleocene of Mongolia they evidently indicate relatively arid conditions in the early Tertiary and represent apparently precocious specializations, partly paralleling, partly ancestral to the mid-Tertiary faunas of the western plains and elsewhere.

The sudden appearance of the Paleocene fauna in the record is not explicable as due to a wide lapse of time, nor is there any clear evidence that it is due to a change of facies. These factors may indeed have played a part, but we have no reason to believe that it was a principal or important part.   In default of evidence to the contrary, we must attribute it to a great migration movement, although the positive evidence is not at hand of the simultaneous appearance of identical new types in widely separated regions as at the beginning of the Eocene.   The Lower Paleocene fauna is unknown except in a limited area in the San Juan basin.   The Paleocene mammals of Europe (Cernaysian, Orsmael, etc.) appear to correlate with Upper and doubtfully Middle Paleocene, and their relationship to the American Paleocene is not so clearly shown as with the Sparnacian faunas of the two regions.   Nevertheless, so far as the evidence goes, it points to Paleocene affinities that may have been equally close at the beginning of the epoch.

In order to explain these relationships we must assume then:

A. A late Cretaceous dispersal center of the archaic placental mammals probably in northern Asia.   The migration movement that introduced the Paleocene fauna into America and Europe was presumably conditioned by the great diastrophism towards and at the end of the Cretaceous period.

B. A Paleocene dispersal center of the Tertiary placental mammals in northern Asia or in northern Holarctica.   The migration movement that introduced them into the western part of the United States and western Europe at the beginning of the Eocene was conditioned presumably by diastrophism that may have been intense in the north, but is

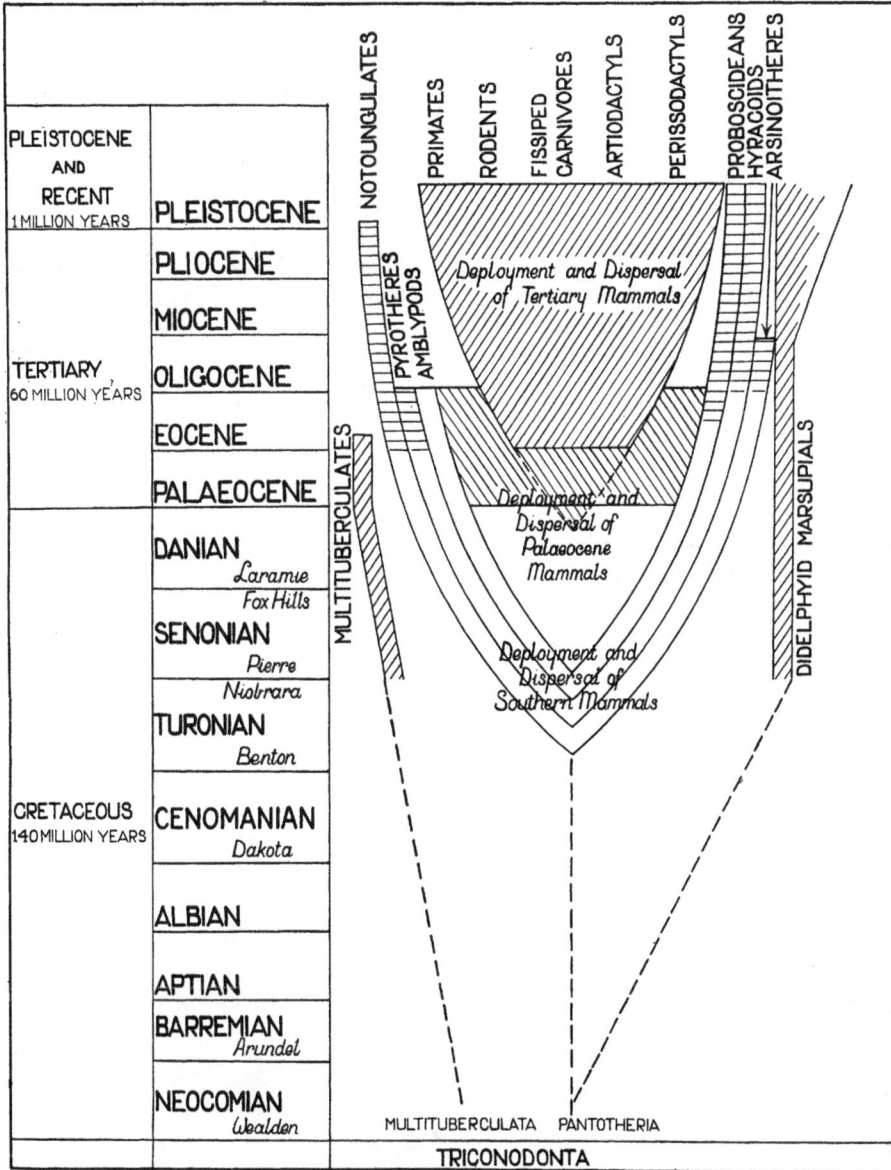

Fig. 85. Deployment and Dispersal of the Mammalia, from the Upper Cretaceous through the Tertiary and Quaternary Periods.

not so indicated in the geological succession of the Rocky Mountain basins or of western Europe. So far as the evidence goes it indicates that diastrophism was more intense in the northern part of the United States and Canada than farther to the south. There is a marked and significant parallelism between the history of the Paleocene groups of mammals during the Eocene, after the invasion of the Wasatch Tertiary fauna, and the history of the neo-tropical groups of mammals during the Pleistocene after the invasion of the nearctic fauna. In both cases, certain groups that come directly into competition with the invaders are promptly eliminated, and this includes most of the generalized adaptable and smaller members. The invading hyaenodont-oxyaenid-miacid carnivores quickly displace the arctocyonids, leaving only a single peculiarly specialized survivor in *Anacodon*, while the aberrantly specialized mesonychids hold along to the end of the Eocene. The taligrades promptly give way to coryphodonts and uintatheres; the phenacodonts give way to perissodactyls more slowly but are all gone by the end of Wasatch time. Mioclaenids give way to *Hyopsodus;* plesiadapids are replaced by Primates, save for a few highly specialized survivors. Taeniodonts peculiarly specialized hold along to the end of the Eocene; again we see no competing types, for the tillodonts, although also having diprotodont front teeth, are very different in cheek teeth and in claws.

Compare with this record the prompt disappearance in South America, at the beginning of the Pleistocene invasion, of the marsupial carnivores, of the pseudo-horses, of most of the notoungulates, but the survival of two large and specialized types, *Toxodon* and *Macrauchenia*, which were not in competition with any of the invaders, and the uniquely specialized edentates to the end of the Pleistocene.

IV. ADDENDUM

By The Editors

It has been the purpose of the editors to keep the manuscript of this memoir in its original form, so that it would represent the work of the author, free from subsequent changes that might be affected in any way by bias. However, since Dr. Matthew at the time of his death had not completed his work on the manuscript, some minor inconsistencies had not been corrected; as far as possible these have been rectified by the editors.

There are also a few omissions of Puerco and Torrejon species in the manuscript. It is certain that Dr. Matthew would have included them in this great work had he lived to complete it, and these species are therefore added below.

Oxyclaeninae

Paradoxodon Scott, 1892

Scott, W. B., 1892, Proc. Acad. Nat. Sci. Phila., XLIV, p. 322

Type: Chriacus rütimeyeranus Cope, from the Puerco formation, New Mexico.

Author's Diagnosis: This curious form, the systematic position of which is altogether uncertain, is known only from inferior molars, though the alveoli of the premolars indicate that these teeth were extremely compressed and recall in their proportions those of the primitive artiodactyls; the molars also suggest relationship to that group. The latter increase in size from the first to the third and the trigonid rises very little above the talon. In $m_{\overline{1}}$ the proto- and metaconids are of about the same size and on the same transverse line; they are both compressed and the protoconid shows a tendency to become crescentic; the small paraconid is placed immediately in front of the metaconid from which it is separated by a slight notch, while a low ridge connects it with the protoconid; the hypoconid is also somewhat crescentic, the entoconid lower and more conical. In $m_{\overline{3}}$ the trigonid is curiously asymmetrical, which is caused by the backward inclination of the metaconid, the larger size of the paraconid than in $m_{\overline{1}}$ and the greater prominence of the ridge connecting this element with the protoconid; the hypoconid is somewhat crescentic and the entoconid reduced, the valley opening inward in advance of that cusp; the hypoconulid is much enlarged and carried on a distinct fang. It would require but relatively little alteration to convert this tooth into one of true selenodont pattern.

Paradoxodon rütimeyeranus (Cope)

Chriacus rütimeyeranus Cope, 1888, Trans. Amer. Phil. Soc., XVI, N.S., p. 340; (*Paradoxodon*) Scott, 1892, *loc. cit.*; (*Mioclaenus*) Matthew, 1899, Bull. Amer. Mus. Nat. Hist., XII, p. 28.

Type: A. M. Cope Coll. No. 3125, lower jaw, m_{1-3}, l., ?" *Chriacus* " *rütimeyeranus*.

Horizon and Locality: Puerco formation, lowest horizon, San Juan basin, New Mexico.

Author's Diagnosis: This is the largest species referred to the genus, and is the only one as yet represented by a single individual. I possess part of the left mandibular ramus which exhibits the true molars, and the roots of the first and second premolars.

The last molar is as large as the second, and its anterior cusps are not opposite to each other as in other species of Chriacus, but the external is in advance of the internal. The fifth cusp is not elevated as in other species, but is represented by an internal angle of an interior ledge, which is quite wide. The anterior cusps of the m.i are opposite, those of m.ii are injured. The heel has the external cusp of crescentic section, and the internal elevated border. No cingula. Enamel obsoletely coarsely rugose. The heel of the m.iii is well developed, as are also the adjacent internal and external marginal cusps. [Measurements follow.]

Tetraclaenodon minor (Matthew)

Euprotogonia minor Matthew, 1897, Bull. Amer. Mus. Nat. Hist., IX, pp. 310–11, fig. 13.

Protogonia zuniensis Cope, 1888 (name only, in part), Trans. Amer. Phil. Soc., XVI, N.S., p. 305.

Tetraclaenodon minor Hay, 1902, Bull. U. S. Geol. Surv. No. 179, p. 603.

Type: A. M. No. 3896, occiput and maxilla, with m^{1-3}, l.; lower teeth in matrix.

Paratypes: A. M. No. 3897, upper and lower jaws; No. 3904, p^4, m^3, r., m^{1-3}, l.

Horizon and Locality: Torrejon formation, Middle Eocene of New Mexico.

Author's Description: Upper true molars sextubercular, differing from those of *E. puercensis* in the smaller size, the series measuring .0191 [m.] as against .0222 to .0255. Last upper and lower molar disproportionately small, the upper (.0048 × .0062) more rounded than in *E. puercensis*, and not appressed against the second molar. (The shape of m^3 in the common species is quite variable, so that this is not a very good distinction.) The paraconid is well developed in the three specimens referred to *E. zuniensis*. Fourth permanent premolar unknown, the last lower milk premolar (No. 3897) is like that of the larger species except in size. The trigon is large, the paraconid strong, not quite equalling the proto- and meta-conid, and placed far forward. The tooth is in this respect not unlike the *permanent* fourth premolar of *Phenacodus*. The proto- and metaconid are well separated, and all three cusps strongly trihedral, with a crest curving forward and inward from proto- to paraconid. The heel is like that of the first molar but considerably narrower. Dimensions of the tooth in this species: Length, .0073, width of trigon, .0039, of heel, .0042. The corresponding measurements in *E. puercensis* (No. 3833) are, .0100, .0057, .0063.

PALAEORYCTIDAE Simpson, 1931

Palaeoryctes (see p. 212) was regarded by Dr. Matthew as belonging to the Centetidae-Tenrecidae. It is now placed in a separate family.

Palaeoryctes Matthew, 1913

Type: Palaeoryctes puercensis Matthew, 1913, from the Torrejon formation of New Mexico. See the discussion below.

Palaeoryctes puercensis Matthew

Palaeoryctes puercensis Matthew, 1913, Bull. Amer. Mus. Nat. Hist., XXXII, pp. 309–14, figs. 2, 3.

Type: A. M. No. 15923, under-surface of a skull with lower jaws, with both upper and lower dentitions well preserved.

Horizon and Locality: Torrejon formation of Middle Paleocene age. From the San Juan basin, New Mexico.

Author's Description: Dentition $\frac{?.1.3.3}{?.1.3.3}$ upper molars sharply triangular, very wide transversely, with a high sub-crescentic outer cusp slightly twinned at the tip, strongly compressed crescentic inner cusp, and broad external shelf raised into crests at postero-external and antero-external angles. Last upper molar unreduced, transverse, without metastyle. Lower molars with very high trigonid, protoconid over-topping metaconid, paraconid antero-internal, smaller than metaconid, heel small, deeply basined, hypoconid and entoconid prominent. Premolars progressively reduced in size and complexity, the second very small and the first absent. Upper premolars non-molariform, P^2 very small, simple, somewhat compressed with minute posterior basal cusp. P^3 a high compressed cusp, with imperfectly separated posterior and no internal cusp. P^4 with high subtrigonal central cusp, large compressed postero-external, and large, well separated compressed-triangular internal cusp, set somewhat anteriorly; a minute rudiment of the antero-external cusp is also present.   A small part of the alveolar border of c^1 is preserved, indicating that there was no spacing behind it.   Upper incisors unknown.   Lower premolars simple, p_2 minute, all with high, simple, moderately compressed principal cusp, and trenchant heel.   No internal or anterior cusp on p_4. Lower canine with single oval root, the crown not preserved, but apparently the tooth was as large as p_2 and had a small heel-cusp.   In front of the canine is an alveolus for a large procumbent or semi-procumbent tooth presumably i_2; i_3 if present must have been very small; there are no indications in regard to the development of i_1.   The muzzle is comparatively short, the middle portion of the skull of moderate length and the basicranial region broad and flat.   There are no palatal vacuities, and the posterior border of the palate is a little behind m_3.   The presence or absence of zygomatic arches cannot be determined.   The palatal and basicranial axes are approximately parallel.

The otic region is well preserved. . . .   Its construction is as follows:

The auditory prominence is a round eminence rising considerably above (properly below) the level of the basioccipital and separated from it and from the basisphenoid by a well defined suture.   Its anterior end is prolonged alongside the basioccipital and basisphenoid in a short ridge sharply crested, but the crest is not extended to take any considerable part in the bulla.   On the postero-external face of the prominence is the fenestra ovalis, well marked, but not lying in a depression.   The mesotympanic fossa lies anterior and external to the prominence, and on one side a considerable part of the bulla has been preserved; it roofs over the antero-internal portion of the mesotympanic fossa extending posteriorly alongside the ridge of the auditory prominence just within it and quite distinct from either petrosal or alisphenoid.   This is apparently a true tympanic ring, and constitutes apparently the whole of the osseous bulla, the petrosal and alisphenoid taking no part in it, although the ridges described above represent in form and position the petrosal and alisphenoid portion of the bulla as developed in other insectivora.

The alisphenoid crest is continuous with the post-glenoid crest of the squamosal and with the pterygoid crest, and the foramen ovale lies close in front of it.   The posterior lacerate foramen lies behind and somewhat external to the auditory prominence, while the carotid canal apparently occupies the deep furrow internal to the prominence and does not perforate the basisphenoid.   The alisphenoid canal was apparently not present, a deep groove, incompletely bridged for a short distance, occupying its place.

The following new genera and species have been described recently by Simpson:

MULTITUBERCULATA

PTILODONTIDAE

Kimbetohia Simpson, 1936

Simpson, G. G., 1936, Amer. Mus. Novitates No. 849, p. 2

Type: Kimbetohia campi Simpson.

Author's Diagnosis: A genus of very small, earliest Paleocene multituberculates.   Four upper premolars.   Antepenultimate premolar with three cusps, one external and two internal, wider than long.   Penultimate premolar nearly as large, rounded quadrate, with four equal cusps.   Last premolar with well-developed inner cusp row, second row nearly complete but with several fewer cusps, and outer, third, row incipient (one cusp in type).

Kimbetohia campi Simpson

Kimbetohia campi, Simpson, 1936, Amer. Mus. Novitates No. 849, p. 3, fig. 1.

Type: Mus. Paleo. Univ. Calif. No. 31305, skull fragment with last three premolars and part of first of the right side.   Collected by Camp and Vander Hoof, 1928.

Horizon and Locality: Lower fossil level, Puerco formation, Bitonitsoseh Arroyo, San Juan basin, New Mexico.

Author's Diagnosis: Only species surely referable to the genus as defined above.   Last premolar with cusp formula probably 1:4:7 (outer:middle:inner). . . .

Measurements: Antepenultimate P   { Length 1.9 / Width 2.2 }

Penultimate P   { Length 1.7 / Width 1.8 }

Last P   { Length 4.1 / Width 2.8 }

CREODONTA

ARCTOCYONIDAE

Chriacus antiquus Simpson

Chriacus antiquus Simpson, 1936, Amer. Mus. Novitates No. 849, pp. 4, 5, fig. 3.

Type: A. M. No. 27714, right upper jaw with P^3 and M$^{1-3}$. Found by A. C. Sawtelle, American Museum Expedition, 1929.

Horizon and Locality: Upper level, Puerco formation, Barrel Spring Arroyo, San Juan basin, New Mexico.

Author's Diagnosis: Length M$^{1-3}$ about 19.5 mm.   P^3 with distinct parastyle and small protocone.   Hypocone strong on M^1 and somewhat projecting internally, present on M^2 but less internal, absent on M^3 except as a cingulum.   Cingula not complete around inner faces of protocones.   No distinct protostyles, but anterior cingula present.   External cingula evenly developed, with weak median embayment on M^1, not on M^2.   M^3 obliquely triangular, with strong parastylar projection.   Enamel nearly smooth.

Protogonodon grangeri Simpson

Protogonodon grangeri Simpson, 1936, Amer. Mus. Novitates No. 849, pp. 5–7, fig. 4.

Type: A. M. No. 27713, lower jaws, cemented together, with roots of right C–P$_3$, crowns of right P$_4$–M$_3$, roots of left P$_{2-3}$, and crowns of left P$_4$–M$_3$ (that of M$_1$ obscure on this side).   Found by G. G. Simpson, American Museum Expedition, 1929.

Horizon and Locality: Upper level, Puerco formation, Barrel Spring Arroyo, San Juan basin, New Mexico.

Author's Diagnosis: Length M$_{1-3}$ 29.5 mm.   Molar proportions about as in *P. pentacus*, except that M$_3$ is small, long and narrow, hypoconulid projecting and tending to form a third lobe.   P$_4$ low, protoconid little if any higher than that of M$_1$, antero-internal basal cusp (paraconid?) distinct, metaconid dependent on protoconid but likewise fully distinct, heel broad and well developed.

CONDYLARTHRA

HYOPSODONTIDAE

Mioclaeninae

Tiznatzinia Simpson, 1936

Simpson, G. G., 1936, Amer. Mus. Novitates No. 849, pp. 7, 8

Type: Tiznatzinia vanderhoofi, new species.
Horizon and Locality: Both fossil levels of the Puerco formation, New Mexico.

Author's Diagnosis: A small condylarth most nearly similar to *Oxyacodon* and *Ellipsodon*. P_4 little enlarged, slightly higher than M_1 and not wider, with distinct but single-cusped heel, and very small incipient metaconid. Molars low-crowned, cusps relatively low and blunt, trigonids little elevated above talonids. Paraconids more or less reduced, but always present, closely approximated and usually almost directly anterior to metaconids, partly confluent with the latter. Trigonids with shallow basins, nearly or quite closed. Deep talonid basins open through a narrow notch on the inner side, hypoconids large, entoconids and hypoconulids smaller and poorly differentiated, frequently with other minute cuspules on the talonid crescent. M_3 (referred species) reduced, hypoconulids not strongly projecting.

Tiznatzinia vanderhoofi Simpson

Tiznatzinia vanderhoofi Simpson, 1936, Amer. Mus. Novitates No. 849, pp. 8, 9, fig. 5.

Type: Mus. Paleo. Univ. Calif. No. 31264, part of left lower jaw with P_4–M_2. Collected by Camp and Vander Hoof, 1930.

Horizon and Locality: Upper fossil level, Puerco formation, Barrel Spring Arroyo, San Juan basin, New Mexico.

Author's Diagnosis: Size small (see measurements). P_4 compressed. Trigonids of M_{1-2} subquadrate, paraconid wholly internal and larger and higher than in *T. priscus*. M_{1-2} longer than wide. Trigonid slightly narrower than talonid on M_1, wide on M_2. Mandible shallow.

Measurements: P_4 $\begin{cases} \text{Length} \ 3.5 \\ \text{Width} \ \ 2.3 \end{cases}$

M_1 $\begin{cases} \text{Length} \ 3.2 \\ \text{Width} \ \ 2.9 \end{cases}$

M_2 $\begin{cases} \text{Length} \ 3.5 \\ \text{Width} \ \ 2.9 \end{cases}$

Carsioptychus Simpson, 1936

Simpson, G. G., 1936, Amer. Jour. Sci. (5), XXXII, p. 234

Type: Periptychus coarctatus Cope, 1883 (= *Carsioptychus coarctatus*).

Author's Diagnosis: Principal cusps of premolars pitched obliquely backward, the anterior basal cusps and deuteroconids of lower premolars wider than long, the basal portion of the crown extended inwardly, supplementary cusps not developed on molars. (Matthew. *See* p. 122, footnote 77.)

Carsioptychus matthewi (Simpson)

Plagioptychus matthewi Simpson, 1936, Amer. Mus. Novitates No. 849, pp. 9–11, fig. 6.
Carsioptychus matthewi Simpson, 1936, Amer. Jour. Sci. (5), XXXII, p. 234.

Type: A. M. No. 27712, right lower jaw with P_2–M_3. Found by G. G. Simpson.

Horizon and Locality: Upper level, Puerco formation, Barrel Spring Arroyo, San Juan basin, New Mexico.

Diagnosis: Over 10 per cent larger than average specimens of *P. coarctatus*. Lower premolars large and heavy. P_{3-4} with small but distinct antero-internal basal cusp and well-developed heels. M_{1-3} with well-defined median external basal cuspule and M_{1-2}

with smaller median internal basal cuspule. M₃ elongate, with hypoconulid lobe partly differentiated.   M₃ at least (M₁₋₂ being too worn for determination) with incipient central seventh cusp.

CROCODILIA

Allognathosuchus Mook, 1921

Mook, C. C., 1921, Bull. Amer. Mus. Nat. Hist., XLIV, pp. 105–16, pl. xv

Type: Crocodilus polyodon Cope.

Allognathosuchus mooki Simpson

Allognathosuchus mooki Simpson, 1930, Amer. Mus. Novitates No. 445, p. 3.

Type: A. M. No. 6780, skull, jaws, and most of skeleton.   Collected by Simpson, 1929.

Horizon and Locality: Upper fossil level ("*Taeniolabis* zone"), Puerco formation, Barrel Spring Arroyo, San Juan basin, New Mexico.

Author's Diagnosis: The earliest known species of *Allognathosuchus*.   Size relatively large, skull somewhat more elongate, less distinctly triangular than in *A. heterodon*.   Five premaxillary, fourteen maxillary, and twenty mandibular teeth.   Third and fourth maxillary and fourth and thirteenth mandibular teeth much enlarged, conical.   Third maxillary teeth larger than in *A. heterodon;* posterior upper teeth somewhat less enlarged and bulbous. Lower jaw very stout, dental border depressed and flattened, symphysis deeper than in *A. polyodon,* surangular rising less abruptly posterior to teeth, alveoli separated by stouter, more complete walls.

Leidyosuchus Lambe, 1907

Lambe, L. M., 1907, Trans. Roy. Soc. Canada (3), I, Sec. IV, pp. 219–44, pls. i–v

Type: Leidyosuchus canadensis Lambe.

Leidyosuchus multidentatus Mook

Leidyosuchus multidentatus Mook, 1930, Amer. Mus. Novitates No. 447, p. 1.

Type: A. M. No. 5179, complete lower jaws, 12 vertebrae, left ilium, tibia, several isolated teeth.

Horizon and Locality: Torrejon beds, Paleocene, Torrejon Arroyo, New Mexico. Collected by Granger, 1912.

Author's Diagnosis: Mandible very long in proportion to its breadth, the breadth constituting only 29 per cent. of the length, compared with 54 per cent. in *L. sternbergii.* Undulation of alveolar border slight.   Number of teeth greater than in other species, being 28 in each ramus.   Splenials just reach symphysis, which extends back to the level of the fifth teeth.

TIFFANY FAUNAL LISTS

While not discussing the Tiffany formation in this work, Dr. Matthew included a faunal list with those of the Puerco and Torrejon. This list is set forth below, in comparison with the list from a recent revision by Simpson.[234]

MATTHEW	SIMPSON
Multituberculata:	Multituberculata:
Plagiaulacidae:	Ptilodontidae:
Ectypodus musculus Matthew and Granger, 1921	*Ectypodus musculus*
Marsupialia:	Marsupialia:
Didelphiidae:	Didelphiidae:
Peradectes elegans Matthew and Granger, 1921	*Peradectes elegans*
Creodonta:	Carnivora:
Oxyclaenidae:	Arctocyonidae:
Chriacus sp.	*Chriacus* sp.
Thryptacodon australis (Matthew, 1915)	*Thryptacodon australis*
Mesonychidae:	Mesonychidae:
Dissacus sp.	*?Dissacus* sp.
Arctocyonidae:	
Claenodon sp.	
Condylarthra:	Condylarthra:
Phenacodontidae:	Phenacodontidae:
Phenacodus sp.	*Phenacodus grangeri*
?Tetraclaenodon	*Phenacodus matthewi*
	Phenacodus gidleyi
	Phenacodus sp.
Taligrada:	Amblypoda:
Periptychidae:	Periptychidae:
Periptychus sp.	*Periptychus superstes*
Insectivora:	Insectivora:
Leptictidae:	Leptictidae:
Leptacodon tener Matthew and Granger, 1921	*Leptacodon tener*
Xenacodon mutilatus Matthew and Granger, 1921	*Xenacodon mutilatus*
Chiroptera:	?Chiroptera:
?Phyllostomatidae:	?Phyllostomatidae:
Zanycteris paleocenus Matthew, 1917	*Zanycteris paleocena*
Primates:	Primates:
?Tarsiidae:	Plesiadapidae:
Ignacius frugivorus Matthew and Granger, 1921	*Plesiadapis gidleyi*
Navajovius kohlhaasae Matthew and Granger, 1921	Apatemyidae:
Menotyphla:	*Labidolemur soricoides*
Plesiadapidae:	Carpolestidae:
Plesiadapis gidleyi (Matthew, 1917)	*Carpodaptes aulacodon*
Labidolemur soricoides Matthew and Granger, 1921	Anaptomorphidae:
Carpodaptes aulacodon Matthew and Granger, 1921	*Navajovius kohlhaasae*
	Family uncertain:
	Phenacolemur frugivorus

[234] Simpson, G. G., 1935, Amer. Mus. Novitates No. 795.

V. CHRONOLOGIC LIST OF PUBLICATIONS DEALING WITH PALEOCENE FAUNAS OF SAN JUAN BASIN AND VICINITY

Mar. 25,[235] 1881. COPE, E. D. Mammalia of the Lower Eocene Beds. Amer. Nat., XV, 337–8.
Periptychus, with *P. carinidens*, and *Deltatherium*, with *D. fundaminis*, described.

July 27, 1881. COPE, E. D. The Temporary Dentition of a New Creodont. Amer. Nat., XV, 667–9.
Triisodon quivirensis described.

COPE, E. D. A Laramie Saurian in the Eocene. Amer. Nat., XV, 669–70.
Champsosaurus, recorded from Puerco, *C. australis* described.

Sept. 22, 1881. COPE, E. D. Mammalia of the Lower Eocene. Amer. Nat., XV, 829–31. Date of separata; October number of Naturalist appeared Sept. 23.
Conoryctes comma, "*Catathlaeus*" *rhabdodon*, *Mioclaenus turgidus*, "*M.*" *angustus*, "*M.*" *sectorius* and "*M.*" *mandibularis* described.

Sept. 30, 1881. COPE, E. D. On Some Mammalia of the Lowest Eocene Beds of New Mexico. Paleont. Bull. No. 33, Proc. Amer. Phil. Soc., XIX, 484–95.
"*Mesonyx*" *navajovius*, *Anisonchus*, "*Mioclaenus*" *subtrigonus*, "*Phenacodus*" *puercensis* and *zuniensis*, *Protogonia subquadrata* described, and tentatively referred to the Puerco formation described by Cope in 1877 from unfossiliferous beds beneath the Wasatch in northwestern New Mexico. A number of descriptions first published in the American Naturalist are repeated in this paper without alteration.

Oct. 28, 1881. COPE, E. D. Eocene Plagiaulacidae. Amer. Nat., XV, 921–2. Separata dated Nov. 12.
Ptilodus described with *P. mediaevus*.

Nov. 29, 1881. COPE, E. D. Notes on *Creodonta*. Amer. Nat., XV, 1018–20. Date of separata; November Naturalist dated Dec. 3.
Genus *Dissacus* established, *Lipodectes* described with two species, *L. penetrans* (= *Deltatherium fundaminis*) and *L. pelvidens* (= *Chriacus pelvidens*).

Jan. 25, 1882. COPE, E. D. A New Genus of Tillodonta. Amer. Nat., XVI, 156–7. Separata dated Jan. 30.
Psittacotherium described with *P. multifragum*.

Feb. 20, 1882. COPE, E. D. Contributions to the History of the Vertebrata of the Lower Eocene of Wyoming and New Mexico, made during 1881. Part II, The Fauna of the Catathlaeus Beds or Lowest Eocene of New Mexico. Paleont. Bull. No. 34, Proc. Amer. Phil. Soc., XX, 191–7.
Psittacotherium aspasiae, *Sarcothraustes* with *S. antiquus*, and *Champsosaurus puercensis* and *saponensis* described.

Apr. 24, 1882. COPE, E. D. A Second Genus of Eocene Plagiaulacidae. Amer. Nat., XVI, 416–17. Separata dated Apr. 25.
Catopsalis described with *C. foliatus*.

COPE, E. D. Two New Genera of the Puerco Eocene. Amer. Nat., XVI, 417–18. Separata dated Apr. 25.
Haploconus described with *H. lineatus*, and *Pantolambda* with *P. bathmodon*.

May 20, 1882. COPE, E. D. The Ancestry and Habits of *Thylacoleo*. Amer. Nat., XVI, 520.
Thylacoleo regarded as descended from *Ctenacodon* and *Plagiaulax*, *Ptilodus* and *Catopsalis* being approximately intermediate types. Key to genera of Plagiaulacidae, recognizing that premolars of *Ptilodus* and *Catopsalis* are functionally reduced to one.

---

[235] For dates of publication of the numbers of American Naturalist for 1881 see Amer. Nat., 1882, XVI, p. 34, except where an earlier publication date appears on the separata. For 1882 dates of publication see Amer. Nat., XVII, p. 60.

June 22, 1882. COPE, E. D. A New Genus of Taeniodonta. Amer. Nat., XVI, 604–5.
Taeniolabis described, with *T. sulcatus*.

July 28, 1882. COPE, E. D. New Marsupials from the Puerco Eocene. Amer. Nat., XVI, 684–6.
Polymastodon with *P. taöensis*, and *Catopsalis pollux* (= lower dentition of *Polymastodon*) described, also *Ptilodus trovessartianus*, "*Haploconus*" *entoconus* and "*H.*" *gillianus*.

Sept. 28, 1882. COPE, E. D. A New Form of Taeniodonta. Amer. Nat., XVI, 831–2. Separata dated Oct. 5.
Hemiganus (= *Psittacotherium*) described, with *H. vultuosus*.

Jan. 31,[236] 1883. COPE, E. D. New Mammalia from the Puerco Eocene. Amer. Nat., XVII, 191.
Abstract of article in Proc. Amer. Phil. Soc., XX, 545–63. No tenable descriptions except for *Helagras prisciformis*.

Mar. 15, 1883. COPE, E. D. The Ancestor of *Coryphodon*. Amer. Nat., XVII, 406–7.
Diagnosis of skeletal characters of *Pantolambda* and of new sub-order Taligrada and new family Pantolambdidae.

Apr. 17, 1883. COPE, E. D. First Addition to the Fauna of the Puerco Eocene. Paleont. Bull. No. 36, Proc. Amer. Phil. Soc., XX, 545–63.[237]

July 3, 1883. COPE, E. D. On Some Fossils of the Puerco Formation. Read before Proc. Acad. Nat. Sci. Phila., July 3, 1883, 168–70; see paper for August 15.
Descriptions of *Periptychus coarctatus*, *Pantolambda cavirictus* and of *Zetodon* with *Z. gracilis*.

July 16, 1883. COPE, E. D. The Puerco Fauna in France. Amer. Nat., XVII, 869–70.
Remarks upon the Cernaysian fauna and comparisons with the Puerco-Torrejon.

Aug. 15, 1883. COPE, E. D. Some New Mammalia of the Puerco formation. Amer. Nat., XVII, 968.
Abstract of paper read before Acad. Nat. Sci. Phila.
"*Periptychus*" *ditrigonus* (*Ectoconus ditrigonus*) referred to *Conoryctes*. Tenable description of *Zetodon* with *Z. gracilis*.

Dec. 29, 1883. COPE, E. D. On New Lemuroids from the Puerco Formation. Amer. Nat., XVIII, 59–61.
Key to families and genera of Eocene Lemuroidea. *Indrodon* described with *I. malaris*.

Jan. 2, 1884. COPE, E. D. Second Addition to the Knowledge of the Fauna of the Puerco Epoch. Paleont. Bull. No. 37, Proc. Amer. Phil. Soc., XXI, 309–24 (date of publication stated on last page of separate).

Feb. 17, 1884. COPE, E. D. The Creodonta. Amer. Nat., XVIII, March, April and May numbers,
Mar. 16 255–67 (Feb. 17), 344–53 (Mar. 16), 478–85 (Apr. 19).
Apr. 19 First published illustrations of Paleocene Creodonta.

June 17, 1884. COPE, E. D. The Tertiary Marsupialia. Amer. Nat., XVIII, July number, 686–97.
First illustrations of Paleocene Multituberculata.

July 17, 1884. COPE, E. D. The Choristodera. Amer. Nat., XVIII, 815–17.
Affinities of *Champsosaurus*.

July 17, 1884. COPE, E. D. The Condylarthra. Amer. Nat., XVIII, August and September numbers,
Aug. 15 790–805 (July 17), 892–906 (Aug. 15).
First illustrations of Periptychidae and other Paleocene types.

Oct. 20, 1884. COPE, E. D. The Amblypoda. Amer. Nat., XVIII, November and December numbers,
Nov. 19 1110–21 (Oct. 20), 1192–1202 (Nov. 19), and XIX, January (1885) number, 40–55
Dec. 30 (Dec. 30).
First published illustrations of *Pantolambda*.

[236] For dates of publication of the numbers of the American Naturalist for 1883, see Amer. Nat., 1884, XVIII, p. 41.

[237] These pages are dated February 14 (to p. 554) and March 16 (for the remainder), as reprinted in the Proceedings. Cope in 1888 (Trans. Amer. Phil. Soc., XVI, N. S., 300) gives the date of the paper as "January 1883," but this was the date of reading, not of publication.

Feb.,[238] 1885.   COPE, E. D.   The Vertebrata of the Tertiary Formations of the West.   Book I.   Report
U. S. Geol. Surv. Terrs., F. V. Hayden in charge, III, xi–xxxiv, 1–1009, pls. i–lxxv a.
Descriptions of Paleocene fauna revised to about July or August, 1883, and numerous
illustrations.   Many of these illustrations had been copied and used in the series of
"monographs" in the American Naturalist for 1884 as cited above.   Most of the
descriptions are reprinted with little or no alteration from advance notices in the
American Naturalist for 1881–3 or Palaeontological Bulletins 33–6.

Mar. 21, 1885.   COPE, E. D.   The Oldest Tertiary Mammalia.   Amer. Nat., XIX, April, 385–7.
Recognizes the "Lower Puerco."   *Loxolophus* described with *L. adapinus*, also
Polymastodon latimolis, "*Chriacus*" *hyattianus* and "*Sarcothraustes*" *coryphaeus*.

Apr. 20, 1885.   COPE, E. D.   The Lemuroidea and the Insectivora of the Eocene Period of North America.
Amer. Nat., XIX, May, 457–71.
Tricentes doubtfully referred to Mixodectidae.   Arctocyonidae provisionally referred
to Insectivora.   No other new points.

Apr. 20, 1885.   COPE, E. D.   The Mammalian Genus *Hemiganus*.   Amer. Nat., XIX, May, 492–3.
"*H.*" *otariidens* described.

Apr. 20, 1885.   COPE, E. D.   Marsupials from the Lower Eocene of New Mexico.   Amer. Nat., XIX,
May, 493–4.
"*Neoplagiaulax*" *americanus*, *Ptilodus trovessartianus* and "*Polymastodon*" *attenuatum*
described, also *Polymastodon taoensis* and *Taeniolabis scalper*.

Sept. 22, 1885.   COPE, E. D.   The Relations of the Puerco and Laramie Deposits.   Amer. Nat., XIX,
985–6.
Puerco and "Laramie" distinct formations; both may be included in post-Cretaceous,
which is characterized.

Apr. 24, 1886.   COPE, E. D.   The Plagiaulacidae of the Puerco Epoch.   Amer. Nat., XX, 451.
"*Neoplagiaulax*" *molestus* described.

May 1, 1887.   COPE, E. D.   The Mesozoic and Cenozoic Realms of the Interior of North America.
Amer. Nat., XXI, 445–62.
Post-Cretacic system redefined, with Laramie and Puerco included.

May 1, 1887.   COPE, E. D.   Some New Taeniodonta of the Puerco.   Amer. Nat., XXI, 469.
Psittacotherium megalodus described.

June, 1887.   COPE, E. D.   The Marsupial Genus *Chirox*.   Amer. Nat., XXI, 566–7, one figure.
Palate of "*Chirox*" *plicatus* described.   Family Chirogidae proposed.

1887–90.   SCHLOSSER, MAX.   Die Affen, Lemuren, Chiropteren u.s.w. des europ. Tertiärs.   Beit.
Pal. Oest.-Ung., VI, VII, VIII.

Jan. 1, 1888.   COPE, E. D.   The Mechanical Causes of the Origin of the Dentition of the Rodentia.
Amer. Nat., XXII, 3–13, figs. 1–9.
Taeniodonta considered as possible ancestral stage.   Parallelism in multituberculate
jaw.

Feb. 1, 1888.   COPE, E. D.   The Vertebrate Fauna of the Puerco Epoch.   Amer. Nat., XXII, 161–3.
Abstract of conclusions in Amer. Phil. Soc. paper (infra).   No tenable descriptions.

Mar. 1, 1888.   OSBORN, H. F.   A Review of Mr. Lydekker's Arrangement of the Mesozoic Mammalia.
Amer. Nat., XXII, 232–6.
Critical remarks on affinities of Multituberculata.

Mar. 1, 1888.   COPE, E. D.   The Multituberculata Monotremes.   Amer. Nat., XXII, 259.
Poulton's discovery of teeth in *Ornithorhynchus* "renders it extremely probable that
the Multituberculata are monotremes."

Aug. 1, 1888.   COPE, E. D.   Synopsis of the Vertebrate Fauna of the Puerco Series.   Trans. Amer.
Phil. Soc., XVI, N.S., 298–361, pls. iv, v.

1892.   SCOTT, W. B.   A Revision of the North American *Creodonta*. . . . Proc. Acad. Nat. Sci.
Phila., 291–323.

1892.   SCOTT, W. B.   The Genera of American *Creodonta*.   Princeton Coll. Bull. IV, 76–81.

[238] Date of publication as cited by Cope in 1888, Trans. Amer. Phil. Soc., XVI, N. S., 300.

1893. EARLE, CHARLES. The Structure and Affinities of the Puerco Ungulates. Science, XXII, 49–51.

1893. OSBORN, H. F. The Collection of Fossil Mammals in the American Museum of Natural History. Science, XXI, 261.

1895. OSBORN, H. F. Vertebrate Paleontology in the American Museum. Science, N. S., II, 178.

1895. SCOTT, W. B. On the *Creodonta*. Report Brit. Ass. Adv. Sci., LXV, 719–20.

1896. WORTMAN, J. L. *Psittacotherium*, a Member of a New and Primitive Sub-Order of the Edentata. Bull. Amer. Mus. Nat. Hist., VIII, Art. XVI, 259–62.

1897. COPE, E. D. The Position of the Periptychidae. Amer. Nat., XXXI, 335–6.

1897. MATTHEW, W. D. A Revision of the Puerco Fauna. Bull. Amer. Mus. Nat. Hist., IX, 259–323, 20 text figures.

1897. WORTMAN, J. L. The Ganodonta and their Relationship to the Edentata. Bull. Amer. Mus. Nat. Hist., IX, 59–110, 36 text figures.

1897. OSBORN, H. F. The Ganodonta or Primitive Edentata with Enameled Teeth. Science, N. S., V, 611–12.

1898. MATTHEW, W. D. (Remarks on *Mixodectes*.) Trans. N. Y. Acad. Sci., XVI, 369–70.

1898. MATTHEW, W. D. On Some New Characters of *Claenodon* and *Oxyaena*. Science, N. S., VIII, 880.

1899. MATTHEW, W. D. A Provisional Classification of the Fresh-water Tertiary of the West. Bull. Amer. Mus. Nat. Hist., XII, 19–75.

1900. SCHLOSSER, MAX. Review of Matthew 1897 (A Revision of the Puerco Fauna). Neues Jahrb., i, Ref., 299–308.

June 19, 1909. GIDLEY, J. W. Notes on the Fossil Mammalian Genus *Ptilodus*, with Descriptions of New Species. Proc. U. S. Nat. Mus., XXXVI, 611–26.
Definition of the genus and review of the New Mexican species in connection with description of skull and partial skeleton from Fort Union of Montana and discussion of the affinities of the Multituberculata.

1910. GARDNER, J. H. The Puerco and Torrejon Formations of the Nacimiento Group. Jour. Geol., XVIII, 702–41, 1 pl., 9 text figures.
Stratigraphy of Puerco and Torrejon (limits erroneously taken).

1914. GRANGER, WALTER. On the Names of Lower Eocene Faunal Horizons of Wyoming and New Mexico. Bull. Amer. Mus. Nat. Hist., XXXIII, 201–7.
Stratigraphy and faunal divisions of the Wasatch.

1914. SINCLAIR, W. J., AND GRANGER, WALTER. Paleocene Deposits of the San Juan Basin, New Mexico. Bull. Amer. Mus. Nat. Hist., XXXIII, 297–316.
Stratigraphy of the Puerco and Torrejon.

1916. BAUER, C. M., GILMORE, C. W., STANTON, T. W., AND KNOWLTON, F. H. Contributions to the Geology and Paleontology of San Juan County, New Mexico. U. S. Geol. Surv. Prof. Paper 98, 271–353.

1917. GRANGER, WALTER. Notes on Paleocene and Lower Eocene Mammal Horizons of Northern New Mexico and Southern Colorado. Bull. Amer. Mus. Nat. Hist., XXXVII, 821–30.
Torrejon beds of Angel Peak south of San Juan River, and of Animas Valley north of the river. Tiffany beds at base of Wasatch north of the San Juan where it crosses the Colorado line.

1917. MATTHEW, W. D. A Paleocene Bat. Bull. Amer. Mus. Nat. Hist., XXXVII, 569–71, 1 text figure.
Preliminary notice of *Zanycteris*.

1917. MATTHEW, W. D. The Dentition of *Nothodectes*. Bull. Amer. Mus. Nat. Hist., XXXVII, 831–9, pls. xcix–cii.
Description of complete palate and jaws.

Dec. 8, 1919. GIDLEY, J. W. New Species of Claenodonts from the Fort Union (Basal Eocene) of Montana. Bull. Amer. Mus. Nat. Hist., XLI, 541–55.
Figures skull, femur and pes of *Claenodon* from New Mexico and gives a revised

definition of the genus in connection with description of the new genus *Neoclaenodon* from Fort Union of Montana.

1919. GILMORE, C. W. Reptilian Faunas of the Torrejon, Puerco, and Underlying Upper Cretaceous Formations of San Juan County, New Mexico. U. S. Geol. Surv. Prof. Paper 119, 1–68, 71, pls. i–xxvi, 33 text figures. (Issued Apr. 20, 1920.)
Many new species and specimens described, principally Chelonia.

1921. MATTHEW, W. D. Fossil Vertebrates and the Cretaceous-Tertiary Problem. Amer. Jour. Sci. (5), II, 220–7.
Correlation discussed; Puerco possibly as old as Lance.

1921. MATTHEW, W. D., AND GRANGER, WALTER. New Genera of Paleocene Mammals. Amer. Mus. Novitates, 13, 1–7.
Diagnoses without figures of thirteen new genera.

1924. REESIDE, J. B., JR. Upper Cretaceous and Tertiary Formations of the Western Part of the San Juan Basin, Colorado and New Mexico. U. S. Geol. Surv. Prof. Paper 134, 1–70, pls. i–iv, 5 text figures.
Geological succession and correlations.

1924. KNOWLTON, F. H. Flora of the Animas Formation. U. S. Geol. Surv. Prof. Paper 134, 71–115, pls. v–xix.
Correlations of floras, Cretaceous and Paleocene.

1925. OSSENKOPP, G. J. Uebersicht unserer derzeitigen Kenntniss von den fossilen niederen Primaten. Ergeb. Anat. und Entwicklungsgesch., XXVI, 463–507, 3 text figures.
Critical review of affinities; Mixodectidae of doubtful position.

[Note: Doctor Matthew's bibliography ends at this point. The following later references have been added by the editors.]

Nov. 1, 1928. SIMPSON, G. G., AND ELFTMAN, H. O. Hind Limb Musculature and Habits of a Paleocene Multituberculate. Amer. Mus. Novitates, 333, 1–19.

1928. GILMORE, C. W. The Fossil Lizards of North America. Mem. Nat. Acad. Sci., XXII, No. 3, 1–201.

Feb. 21, 1929. GRANGER, W., AND SIMPSON, G. G. A Revision of the Tertiary Multituberculata. Bull. Amer. Mus. Nat. Hist., LVI, 601–76.

Dec. 19, 1930. SIMPSON, G. G. *Allognathosuchus mooki*, a New Crocodile from the Puerco Formation. Amer. Mus. Novitates, 445, 1–16.

Dec. 20, 1930. MOOK, C. C. A New Species of Crocodilian from the Torrejon Beds. Amer. Mus. Novitates, 447, 1–11.

1930. GILMORE, C. W. Fossil Hunting in New Mexico. Smithsonian Inst. Expl. and Field Work in 1929, 17–22.

1930. RUSSELL, L. S. Upper Cretaceous Dinosaur Faunas of North America. Proc. Amer. Phil. Soc., LXIX, 133–59.

Mar. 31, 1931. REYNOLDS, T. E. New Insectivores from the Lower Paleocene (abstract). Bull. Geol. Soc. Amer., XLII, 368.

Aug. 19, 1932. DANE, C. H. Notes on the Puerco and Torrejon Formations, San Juan Basin, New Mexico. Washington Acad. Sci. Jour., XXII, 406–11.

Sept., 1932. KEYES, C. R. Nacimientan Series of New Mexico. Pan-Amer. Geologist, LVIII, 132–5.

Apr., 1934. GREGORY, W. K. A Half Century of Trituberculy. The Cope-Osborn theory of dental evolution. Proc. Amer. Phil. Soc., LXXIII, 169–317.

Aug., 1934. JEPSEN, G. L. A Revision of the American Apatemyidae. Proc. Amer. Phil. Soc., LXXIV, 287–305.

Apr. 20, 1935. SIMPSON, G. G. The Tiffany Fauna, Upper Paleocene. I.—Multituberculata, Marsupialia, Insectivora, and ?Chiroptera. Amer. Mus. Novitates, 795, 1–19.

Aug. 16, 1935. SIMPSON, G. G. The Tiffany Fauna, Upper Paleocene. II.—Structure and Relationships of *Plesiadapis*. Amer. Mus. Novitates, 816, 1–30.

Aug. 16, 1935. SIMPSON, G. G. The Tiffany Fauna, Upper Paleocene. III.—Primates, Carnivora, Condylarthra, and Amblypoda. Amer. Mus. Novitates, 817, 1–28.

EXPLANATION OF THE PLATES

PLATE I

PALEOCENE ARCTOCYONIDAE FROM THE TORREJON FORMATION
SAN JUAN BASIN, NEW MEXICO

Fig. 1. *Neoclaenodon procyonoides*, type specimen. Upper and lower jaw fragments. A. M. No. 16554.
   (a) Right side, external view; (b) left side, external view; (c) left upper cheek teeth, crown view; (d) right upper cheek teeth, crown view.
Fig. 2. *Claenodon corrugatus*. Palate. A. M. No. 2456.
   (a) Side view; (b) crown view.
Fig. 3.. *Claenodon corrugatus*. Mandible. A. M. No. 2456.
   (a) Crown view; (b) side view.
   All three-fourths natural size.

PLATE I

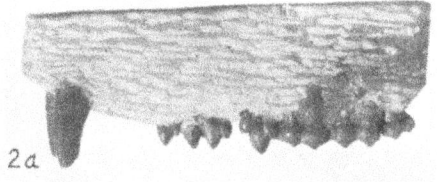

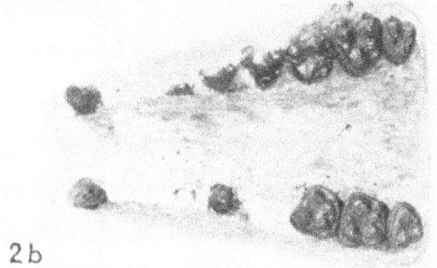

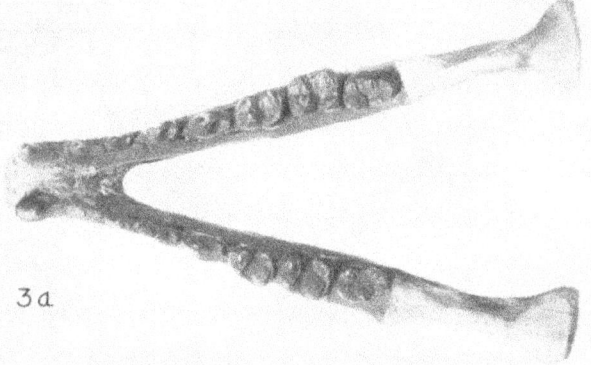

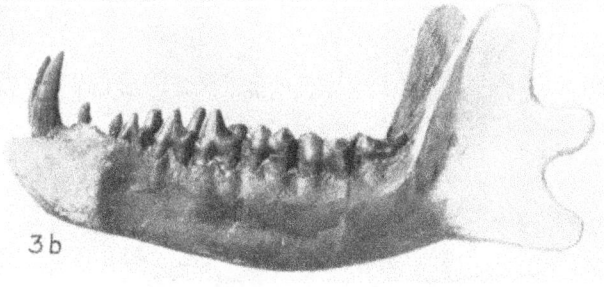

PLATE II

PALEOCENE ARCTOCYONIDAE FROM THE TORREJON FORMATION
SAN JUAN BASIN, NEW MEXICO

Fig. 1. *Claenodon ferox*. Lower jaws. A. M. No. 2459.
(*a*) Crown view; (*b*) side view.
Three-fourths natural size.

PLATE II

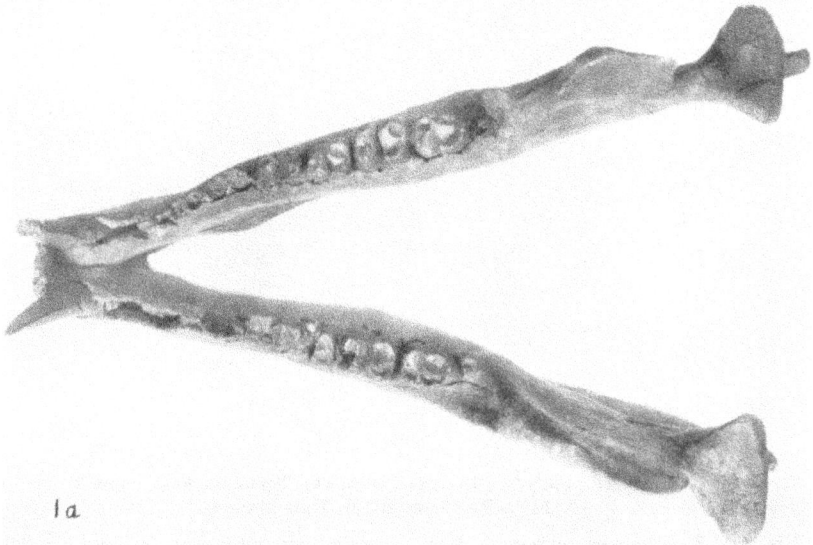

1a

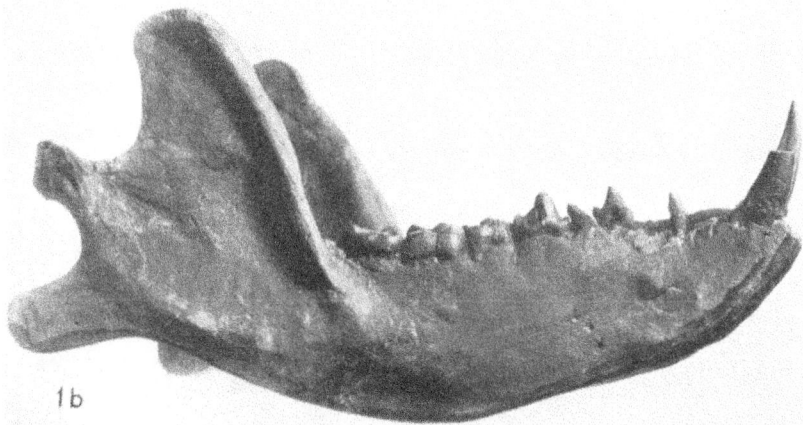

1b

PLATE III

PALEOCENE CREODONTA FROM THE TORREJON FORMATION
SAN JUAN BASIN, NEW MEXICO

Figs. 1, 2, 3. *Claenodon corrugatus.* Humerus, radius and ulna, anterior views. A. M. No. 16543.
Three-fourths natural size.

Fig. 4. *Tricentes subtrigonus.* Left ramus, external view. A. M. No. 2402. One and one-half times
natural size.

PLATE III

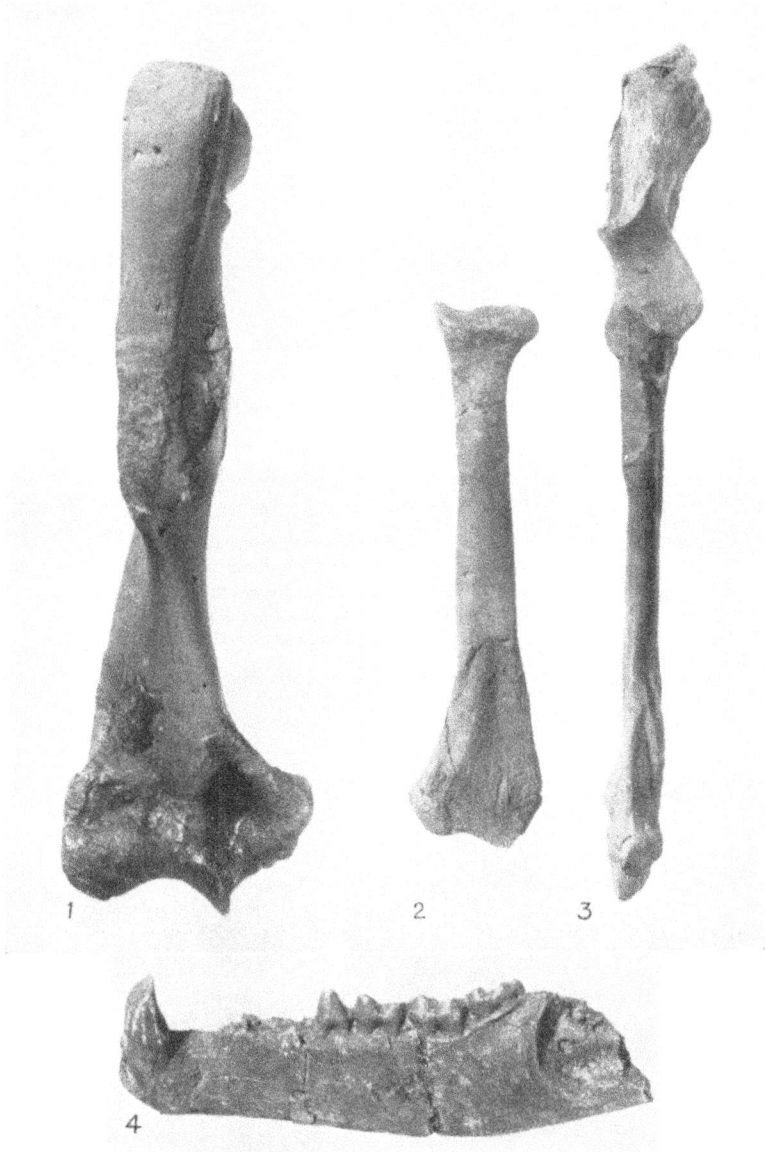

PLATE IV

PALEOCENE ARCTOCYONIDAE FROM THE TORREJON FORMATION
SAN JUAN BASIN, NEW MEXICO

Claenodon corrugatus. A. M. No. 16543.

Fig. 1. Right femur, external view.
Fig. 2. Left femur, anterior view.
Fig. 3. Left tibia, external view.
Fig. 4. Left fibula, external view.
All three-fourths natural size.

PLATE IV

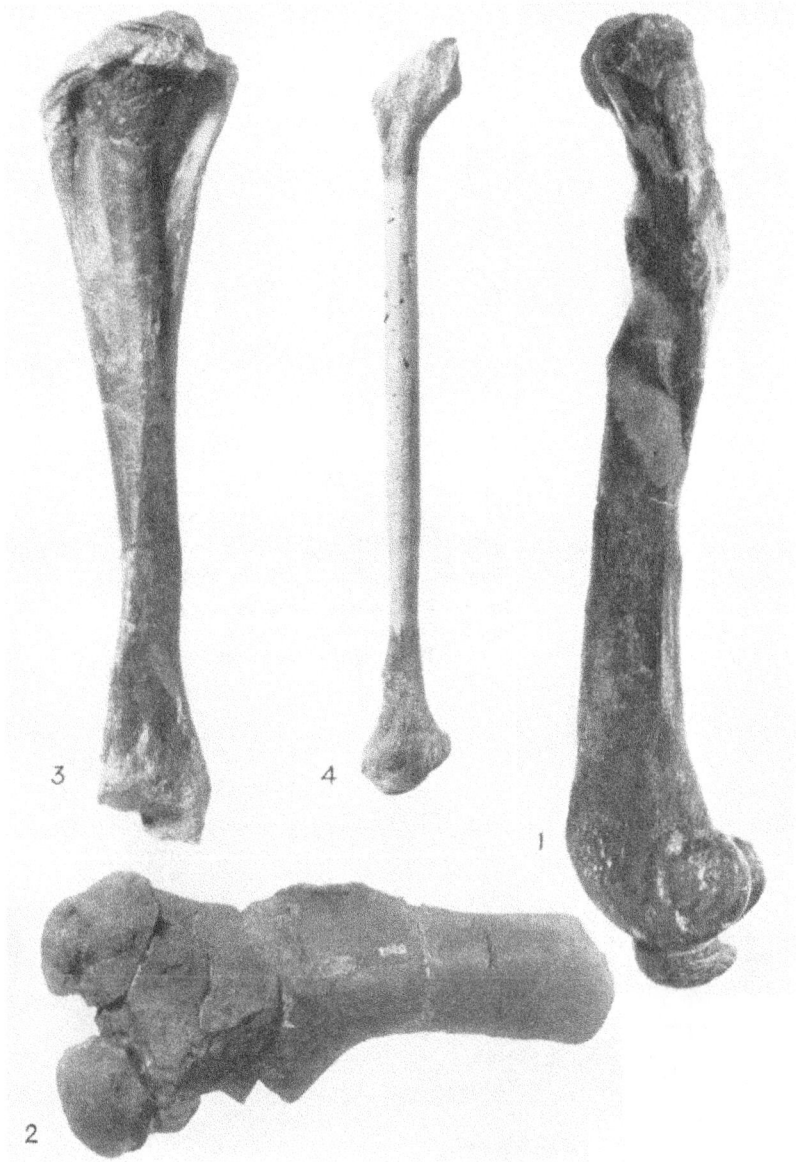

PLATE V

PALEOCENE ARCTOCYONIDAE FROM THE TORREJON FORMATION
SAN JUAN BASIN, NEW MEXICO

Claenodon corrugatus.   A. M. No. 16543.

Fig. 1.   Right manus, dorsal view.
Fig. 2.   Left pes, dorsal view.
          Three-fourths natural size.

PLATE V

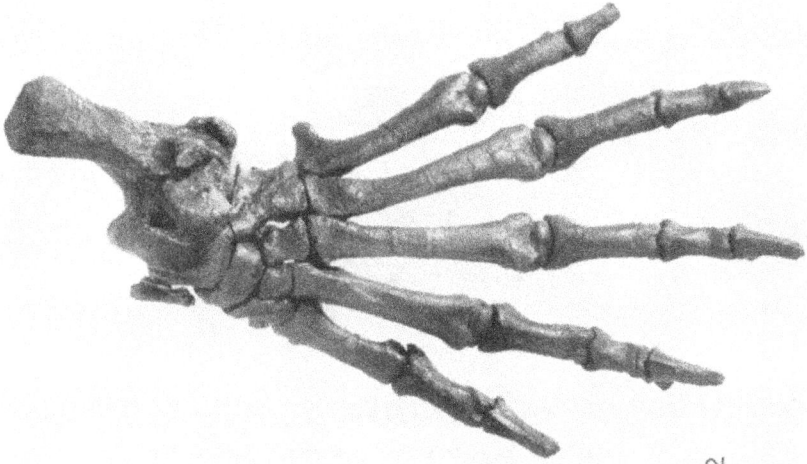

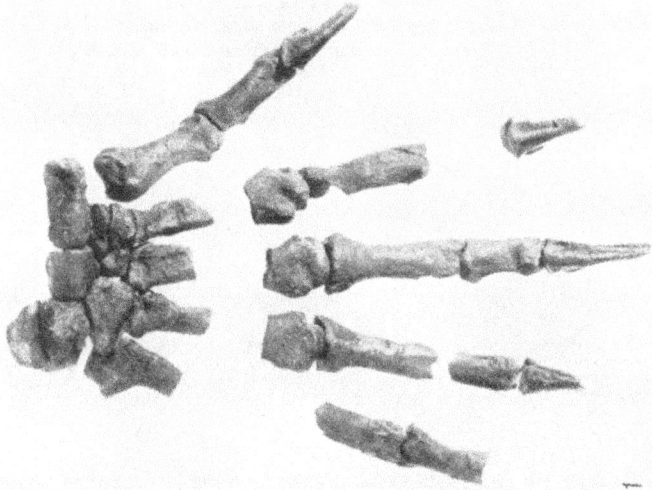

PLATE VI

PALEOCENE ARCTOCYONIDAE FROM THE TORREJON FORMATION
SAN JUAN BASIN, NEW MEXICO

Claenodon corrugatus.   A. M. No. 16543.

Fig. 1.   Right pes, dorsal view.
Fig. 2.   Left tibia, anterior view.
Fig. 3.   Left fibula, anterior view.
          All three-fourths natural size.

PLATE VI

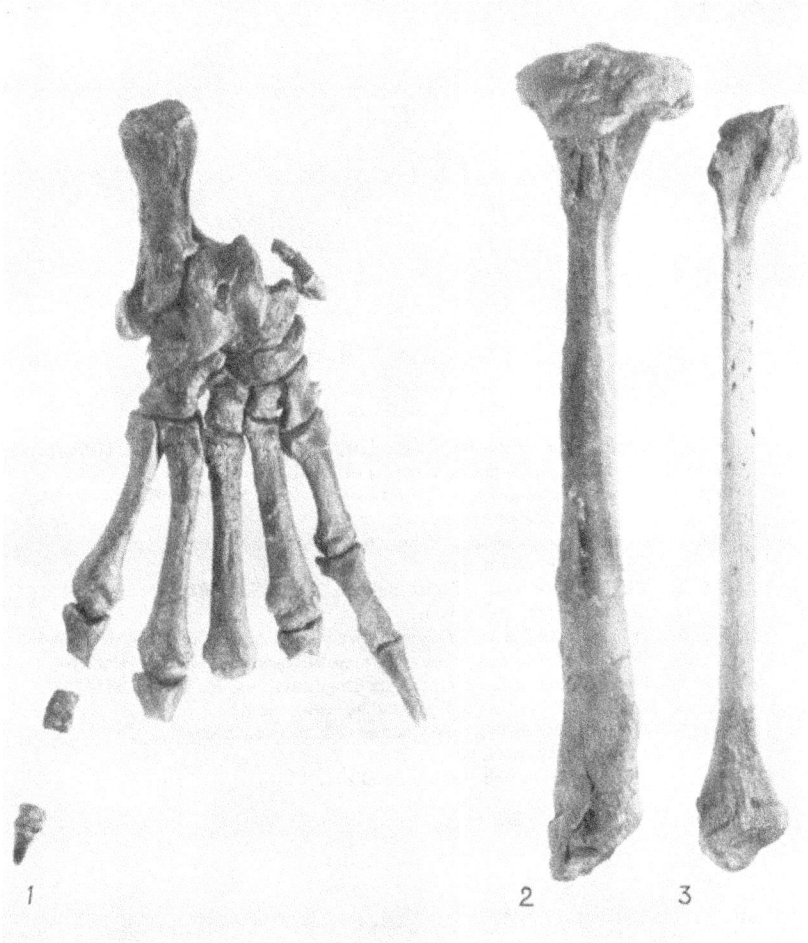

1   2   3

PLATE VII

PALEOCENE CREODONTA FROM THE TORREJON FORMATION
SAN JUAN BASIN, NEW MEXICO

Fig. 1. *Mixoclaenus encinensis.* Lower jaw, right ramus. A. M. No. 17074, paratype.
    (a) Internal view; (b) crown view; (c) external view.
Fig. 2. *Mixoclaenus encinensis.* Lower teeth. A. M. No. 16601, type.
    (a) Side view; (b) crown view.
Fig. 3. *Mixoclaenus encinensis.* Upper teeth. A. M. No. 16601, type.
    (a) Side view; (b) crown view.
Fig. 4. *Mioclaenus acolytus.* Upper molars. A. M. No. 16641.
    (a) Crown view; (b) crown view.
Fig. 5. *Tricentes subtrigonus.* Palate, crown view. A. M. Cope Coll. No. 2402.
Fig. 6. *Tricentes subtrigonus.* Lower jaw, crown view. A. M. Cope Coll. No. 2402.
Fig. 7. *Tricentes subtrigonus.* Upper jaw fragments. A. M. No. 16563.
    (a) Left side, crown view; (b) right side, crown view.
Fig. 8. *Tricentes subtrigonus.* Left ramus. A. M. No. 16581.
    (a) Side view; (b) crown view.
       All one and one-half times natural size.

PLATE VII

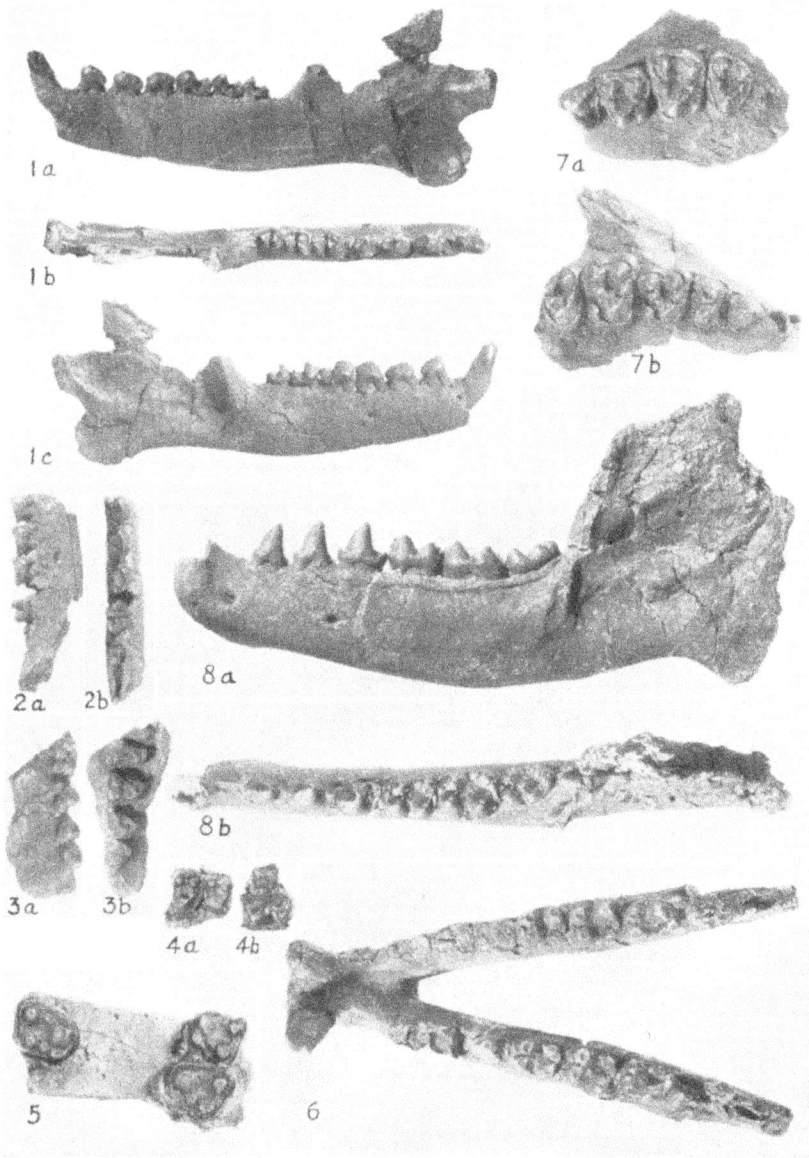

1 a

1 b

1 c

2 a   2 b

3 a   3 b

4 a   4 b

5

6

7 a

7 b

8 a

8 b

PLATE VIII

Chriacus FROM THE TORREJON PALEOCENE
SAN JUAN BASIN, NEW MEXICO

Fig. 1. *Chriacus pelvidens*. Left ramus. A. M. Cope Coll. No. 2387.
   (*a*) Crown view; (*b*) side view.
Fig. 2. *Chriacus pelvidens*. Right ramus. A. M. Cope Coll. No. 2379.
   (*a*) Crown view; (*b*) side view.
Fig. 3. *Chriacus* sp. Right humerus, posterior view. A. M. No. 16591.
Fig. 4. *Chriacus* (or *Deltatherium?*). Upper canine.
Figs. 5–9. *Chriacus schlosserianus*. A. M. No. 3117, type specimen.
   (5*a*) Humerus, distal end, anterior view.
   (5*b*) Humerus, distal end, posterior view.
   (6) Ulna, proximal end, anterior view.
   (7) Palate, crown view.
   (8) Lower jaw fragment, crown view.
   (9) Right ramus, crown view.
        All one and one-half times natural size.

PLATE VIII

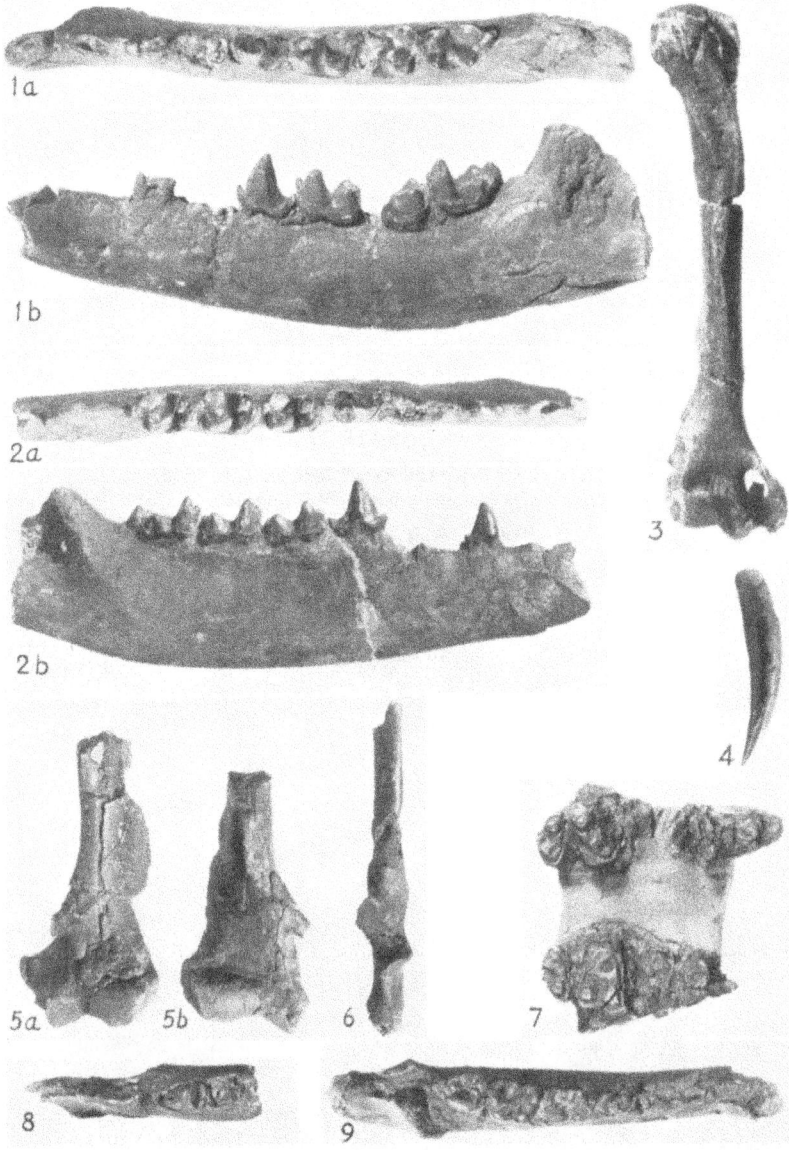

PLATE IX

Deltatherium FROM THE TORREJON PALEOCENE
SAN JUAN BASIN, NEW MEXICO

Fig. 1.   Skull.   A. M. No. 16610.
(*a*) Dorsal view; (*b*) palatal view.
Fig. 2.   Canine.   A. M. No. 16610.
Fig. 3.   Humerus.   A. M. No. 16610.
(*a*) Anterior view; (*b*) internal view.
All three-fourths natural size.

PLATE IX

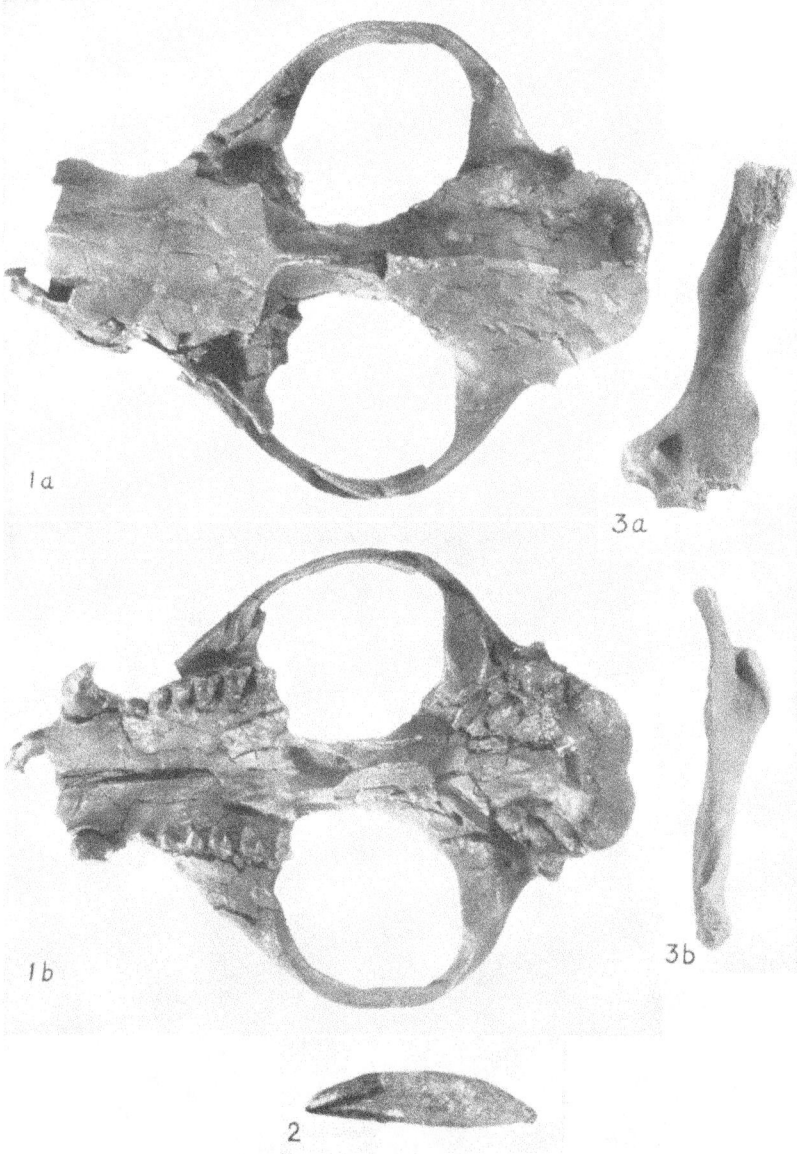

1a

1b

2

3a

3b

PLATE X

Deltatherium FROM THE TORREJON PALEOCENE
SAN JUAN BASIN, NEW MEXICO

Fig. 1. *Deltatherium* sp. Skull and canine. A. M. No. 16610.
(a) Skull, side view. Three-fourths natural size. (b) Skull, palatal view. One and one-half times natural size. (c) Canine. One and one-half times natural size. .

PLATE X

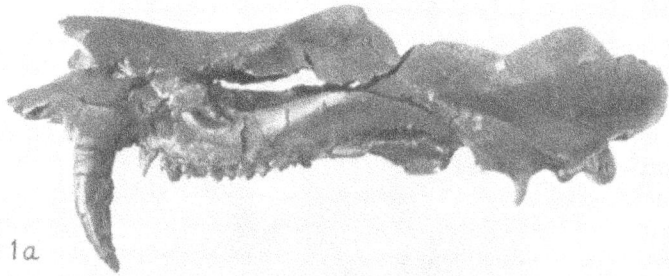

1a

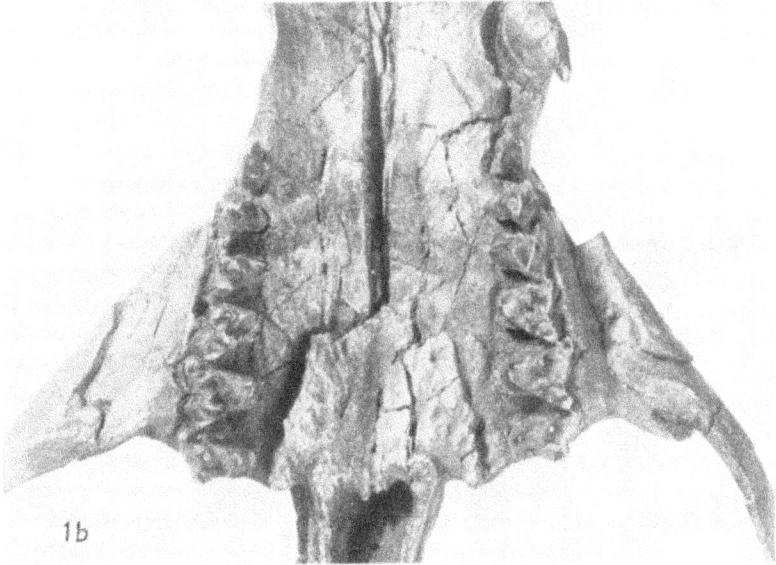

1b

1c

PLATE XI

PALEOCENE CREODONTA FROM THE PUERCO
SAN JUAN BASIN, NEW MEXICO

Fig. 1. *Triisodon quivirensis.* Type jaws. A. M. Cope Coll. No. 3352.
(a) Crown view; (b) left ramus, internal view; (c) left ramus, external view.
Fig. 2. *Protogonodon pentacus.* Right ramus, p^2-m^3, crown view. A. M. Cope Coll. No. 3192, type.
Fig. 3. *Protogonodon pentacus.* Right ramus. A. M. Cope Coll. No. 3196.
(a) Side view; (b) crown view.
Fig. 4. *Protogonodon pentacus.* Left ramus, crown view. A. M. Cope Coll. No. 3196.
All three-fourths natural size.

PLATE XI

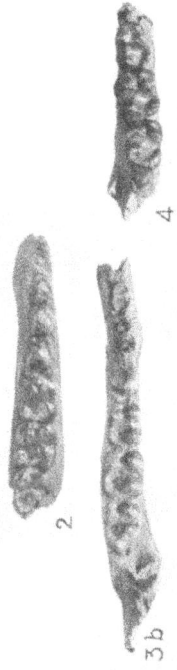

PLATE XII

Eoconodon FROM THE PUERCO PALEOCENE
SAN JUAN BASIN, NEW MEXICO

Fig. 1.   *Eoconodon heilprinianus.*   A. M. No. 16329.
(*a*) Skull and jaws, side view; (*b*) jaws, crown view.
Three-fourths natural size.

PLATE XII

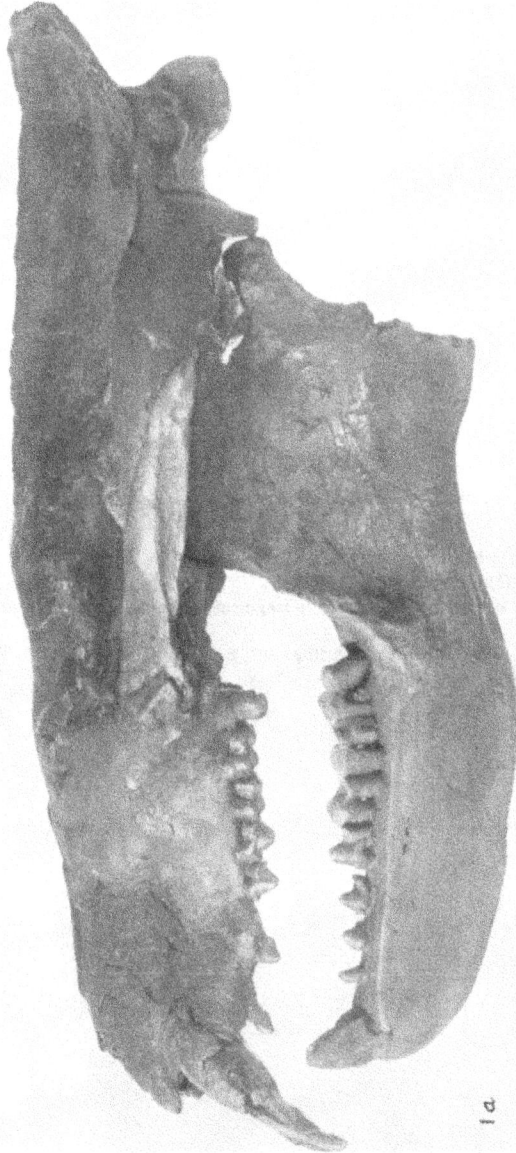

PLATE XIII

TORREJON CREODONTA FROM THE PUERCO PALEOCENE
SAN JUAN BASIN, NEW MEXICO

Fig. 1. *Eoconodon heilprinianus.* Skull. A. M. No. 16329.
  (*a*) Palatal view; (*b*) occipital view.
Fig. 2. *Protogonodon stenognathus.* Maxillary fragment, p^2–m^2 and root of p^2, crown view. A. M. No. 761.
Fig. 3. *Protogonodon pentacus.* Maxillary fragment, m^{1-3}, crown view. A. M. No. 954.
  All three-fourths natural size.

PLATE XIII

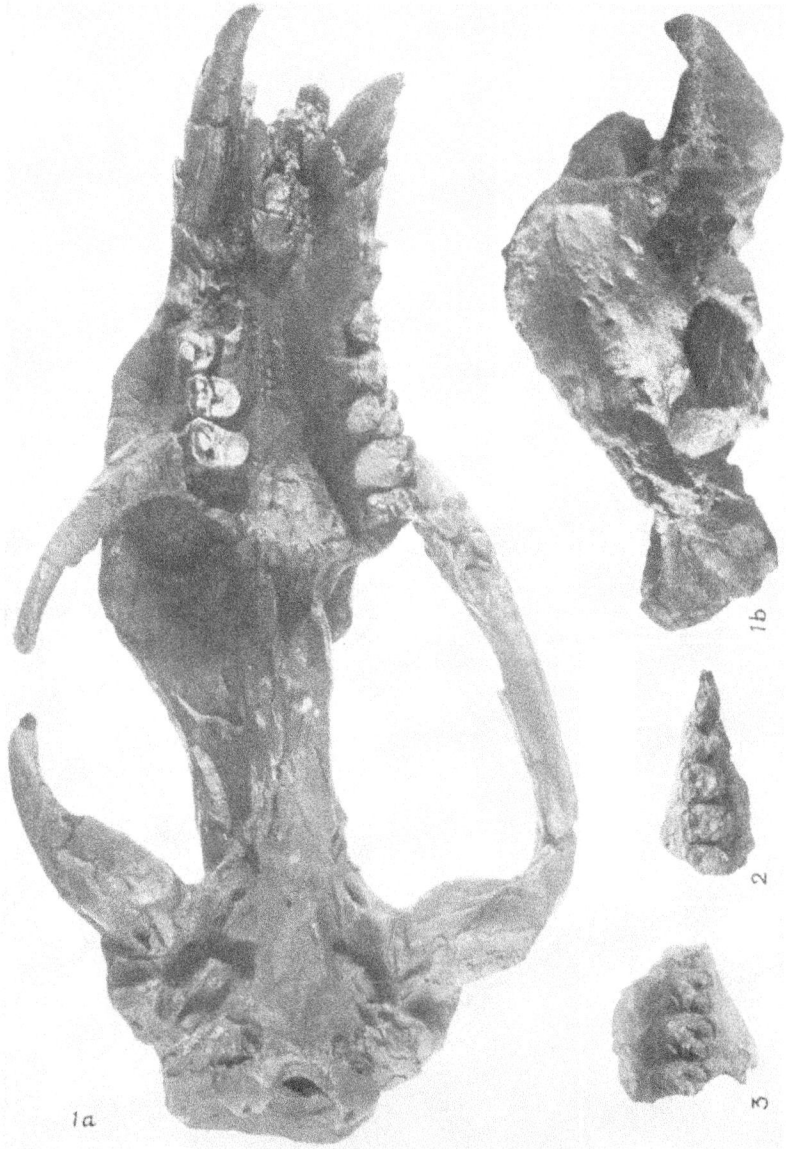

1a

1b

2

3

PLATE XIV

CREODONTA FROM THE PUERCO PALEOCENE
SAN JUAN BASIN, NEW MEXICO

Fig. 1. *Eoconodon heilprinianus.* Skull, dorsal view. A. M. No. 16329.
Fig. 2. *Protogonodon pentacus.* Right ramus, side view. A. M. Cope Coll. No. 3192, type.
Fig. 3. *Protogonodon pentacus.* Lower jaw, side view. A. M. Cope Coll. No. 3196.
About three-fourths natural size.

PLATE XIV

PLATE XV

Creodonta from the Torrejon Paleocene
San Juan Basin, New Mexico

Fig. 1. *Dissacus saurognathus.* Right manus, dorsal view, × ¾. A. M. No. 777.
Fig. 2. *Dissacus saurognathus.* Left manus, dorsal view, × ¾. A. M. No. 777.
Fig. 3. *Didymictis haydenianus.* Left ramus, × 1½. A. M. No. 15992.
   (a) .Crown view; (b) lateral view.
Fig. 4. *Didymictis haydenianus.* Lower jaw fragments, × 1½. A. M. No. 16536.
   (a) Right ramus, side view; (b) right ramus, crown view; (c) left ramus, side view; (d) left ramus, crown view.
Fig. 5. *Didymictis haydenianus.* Left ramus, crown view, × 1½. A. M. No. 16540.

PLATE XV

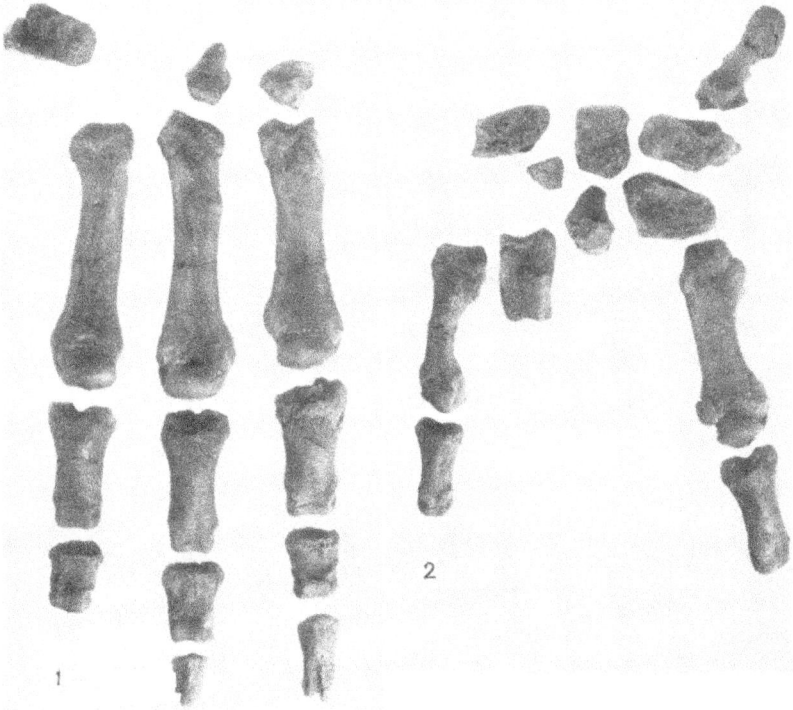

PLATE XVI

CREODONTA FROM THE TORREJON PALEOCENE
SAN JUAN BASIN, NEW MEXICO

Fig. 1. *Dissacus saurognathus.* Part of left pes, × ¾. A. M. No. 777.
(*a*) Dorsal view; (*b*) proximal view; (*c*) distal view.
Fig. 2. *Didymictis haydenianus.* Left ramus, × 1½. A. M. No. 15992.
(*a*) Crown view; (*b*) side view.
Fig. 3. *Didymictis haydenianus.* Lower jaw fragment, × 1½, side view. A. M. No. 16540.

PLATE XVI

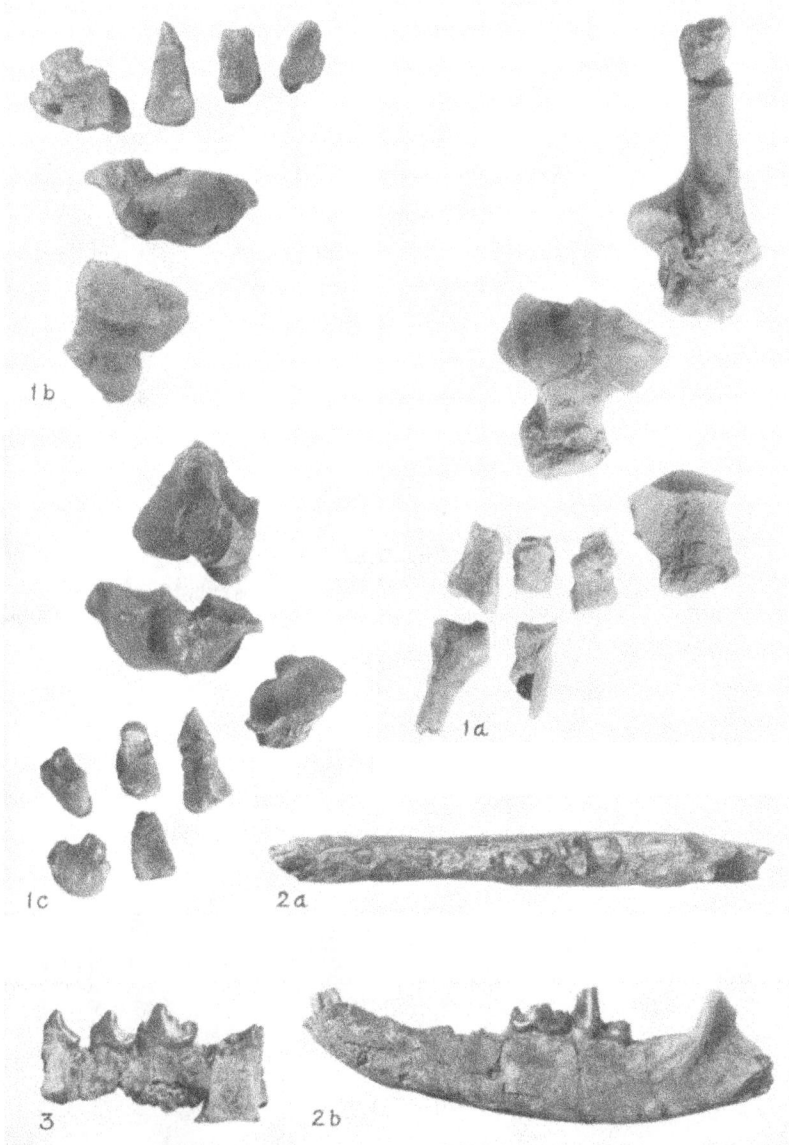

1b

1a

1c

2a

3

2b

PLATE XVII

Periptychus FROM THE TORREJON PALEOCENE
SAN JUAN BASIN, NEW MEXICO

Fig. 1. *Periptychus rhabdodon.* Skull and jaws. A. M. Cope Coll. No. 3665.
(a) Skull and mandible, side view; (b) left upper teeth, crown view; (c) mandible, crown view. Three-fourths natural size.

PLATE XVII

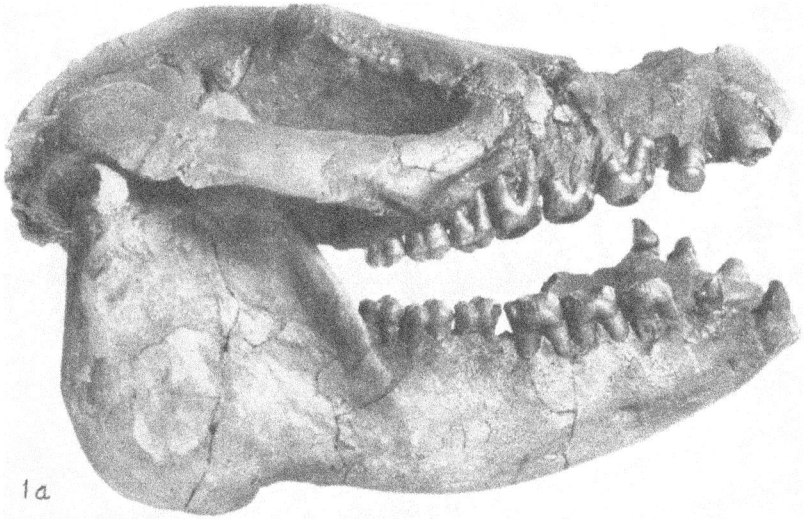

1a

1b

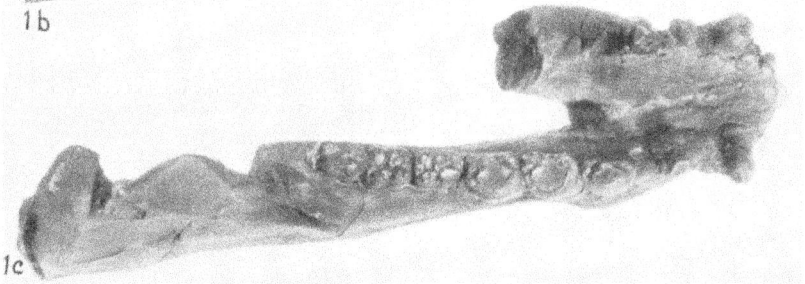

1c

PLATE XVIII

Periptychus FROM THE TORREJON PALEOCENE
SAN JUAN BASIN, NEW MEXICO

Fig. 1. *Periptychus carinidens.* Left ramus. A. M. No. 16695.
   (a) Crown view; (b) external view; (c) internal view.
Fig. 2. *Periptychus rhabdodon.* Left ramus. A. M. Cope Coll. No. 3636.
   (a) External view; (b) crown view.
      All three-fourths natural size.

PLATE XVIII

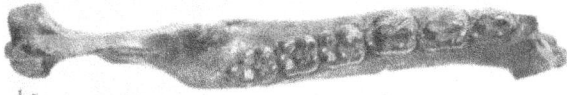

1a

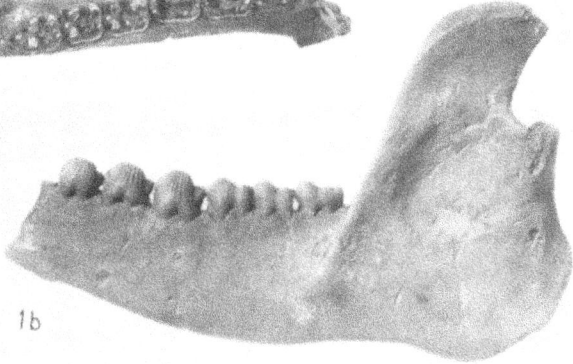

1b

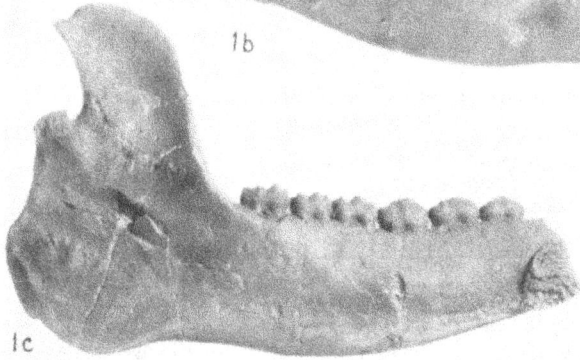

1c

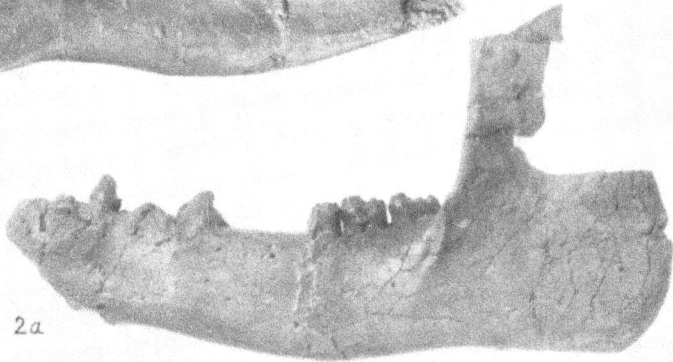

2a

2b

PLATE XIX

Periptychus FROM THE PUERCO AND TORREJON PALEOCENE
SAN JUAN BASIN, NEW MEXICO

Fig. 1. *Periptychus rhabdodon.* Right ramus. A. M. Cope Coll. No. 3636.
    (a) Crown view; (b) external view.
Fig. 2. *Periptychus rhabdodon.* Right maxilla. A. M. No. 15936.
    (a) External view; (b) crown view.
Fig. 3. *Periptychus rhabdodon.* Left maxilla. A. M. No. 15936.
    (a) Crown view; (b) internal view.
Fig. 4. *Periptychus coarctatus.* Part of right ramus, p₃–m₃. A. M. Cope Coll. No. 3772.
    (a) Crown view; (b) external view.
Fig. 5. *Periptychus rhabdodon.* Right ramus, juvenile. A. M. Cope Coll. No. 3720.
    (a) Crown view; (b) external view.
        Fig. 4, Puerco; others, Torrejon.
        All three-fourths natural size.

PLATE XIX

1a

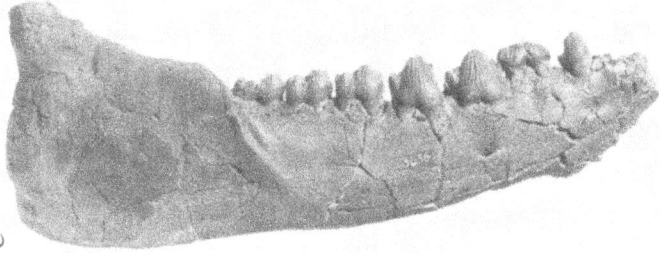

1b

2a

4a

2b

4b

3a

5a

3b

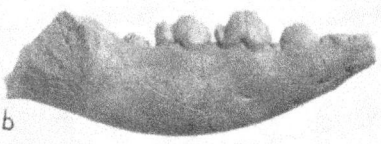

5b

PLATE XX

Periptychus FROM THE TORREJON PALEOCENE
SAN JUAN BASIN, NEW MEXICO

Periptychus rhabdodon.   Fore limb and foot.

Fig. 1.   Humerus.   A. M. Cope Coll. No. 3636.
   (*a*) External view; (*b*) anterior view; (*c*) distal view.
Fig. 2.   Ulna and radius, anterior view.   A. M. No. 17075.
Fig. 3.   Manus, dorsal view.   A. M. No. 17075.
   All three-fourths natural size.

PLATE XX

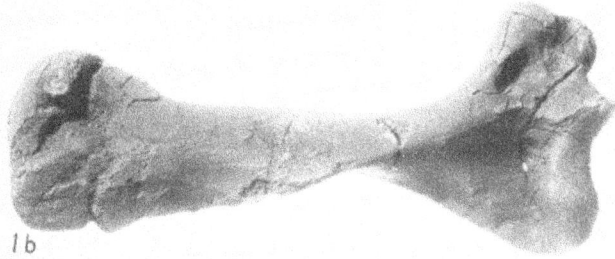
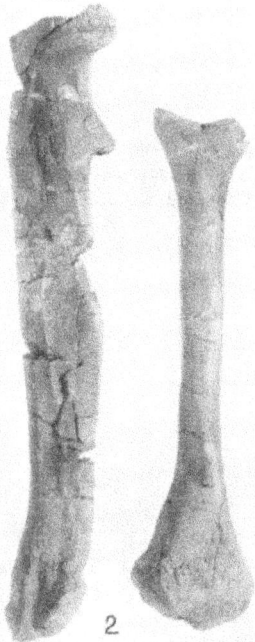
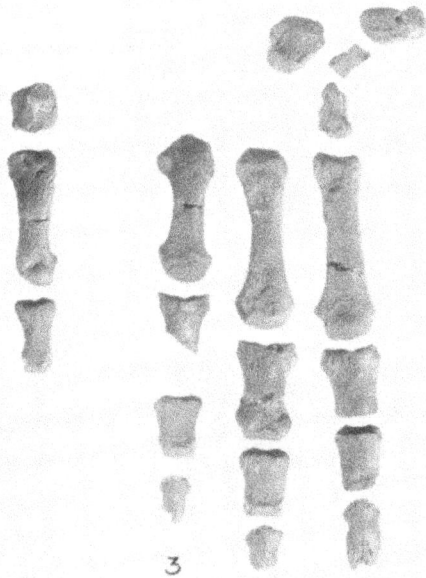

PLATE XXI

Periptychus FROM THE TORREJON PALEOCENE
SAN JUAN BASIN, NEW MEXICO

Periptychus rhabdodon. Hind limb.

Fig. 1. Femur. A. M. No. 837.
   (*a*) External view; (*b*) anterior view.
Fig. 2. Tibia and fibula. A. M. No. 837.
   (*a*) Anterior view; (*b*) external view.
Fig. 3. Humerus, anterior view. A. M. No. 17075.
Fig. 4. Ulna, external view. A. M. Cope Coll. No. 3636.
   All three-fourths natural size.

PLATE XXI

1a

1b

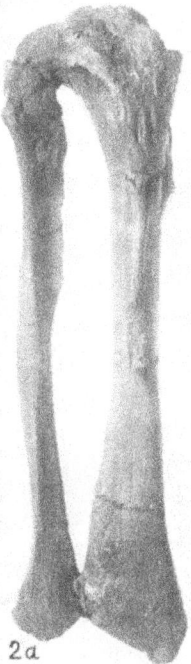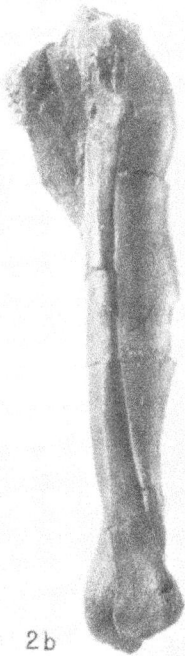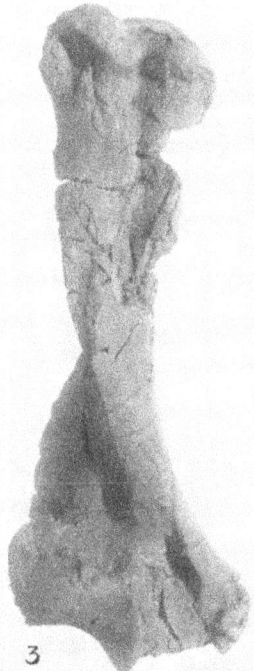

2a   2b   3   4

PLATE XXII

Periptychus FROM THE TORREJON PALEOCENE
SAN JUAN BASIN, NEW MEXICO

Periptychus rhabdodon.   Hind foot.

Fig. 1.   Pes, dorsal views.   A. M. Cope Coll. No. 3636.
    (*a*) Articulated; (*b*) unarticulated.
Fig. 2.   Tarsus, dorsal view.   A. M. No. 17075.
Fig. 3.   Pes, proximal view.   A. M. Cope Coll. No. 3636.
Fig. 4.   Astragalus.   A. M. No. 837.
    (*a*) Proximal view; (*b*) dorsal view; (*c*) external view; (*d*) plantar view; (*e*) internal view.
    All three-fourths natural size.

PLATE XXII

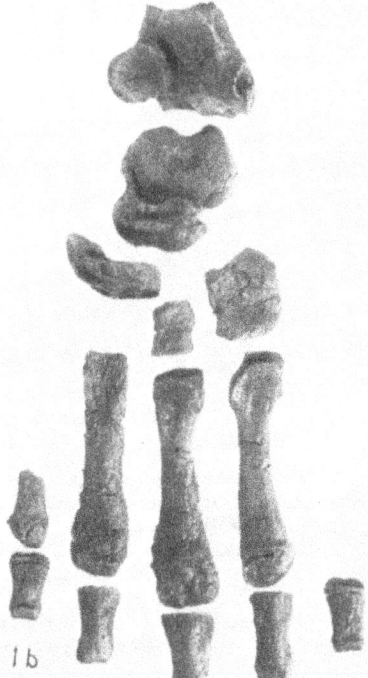

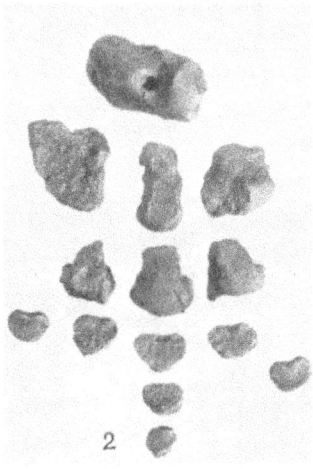

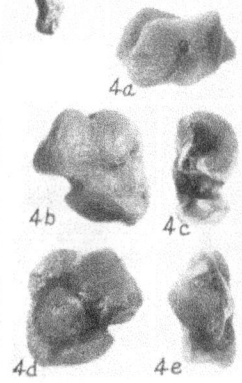

PLATE XXIII

Ectoconus FROM THE PUERCO PALEOCENE
SAN JUAN BASIN, NEW MEXICO

Fig. 1.   *Ectoconus ditrigonus.*   Upper jaws, p⁴-m³ right and p²-m³ left, crown view.   A. M. No. 16496.
Fig. 2.   *Ectoconus majusculus.*   Premaxilla, with canine.   A. M. No. 16500 (complete skeleton).
    (*a*) Side view; (*b*) crown view.
Fig. 3.   *Ectoconus ditrigonus.*   Palate, crown view,   A. M. No. 880.
Fig. 4.   *Ectoconus ditrigonus.*   Right ramus.   A. M. No. 16495.
    (*a*) External view; (*b*) crown view; (*c*) internal view.
Fig. 5.   *Ectoconus ditrigonus.*   Left ramus, crown view.   A. M. No. 16495.
Figs. 6–9.   *Ectoconus ditrigonus.*   Upper and lower jaw fragments, crown views.   A. M. Cope Coll. No. 3800.
Fig. 10.   *Ectoconus majusculus.*   Seven loose front teeth, premolars and incisors.   A. M. No. 16500 (complete skeleton).
    All three-fourths natural size.

PLATE XXIII

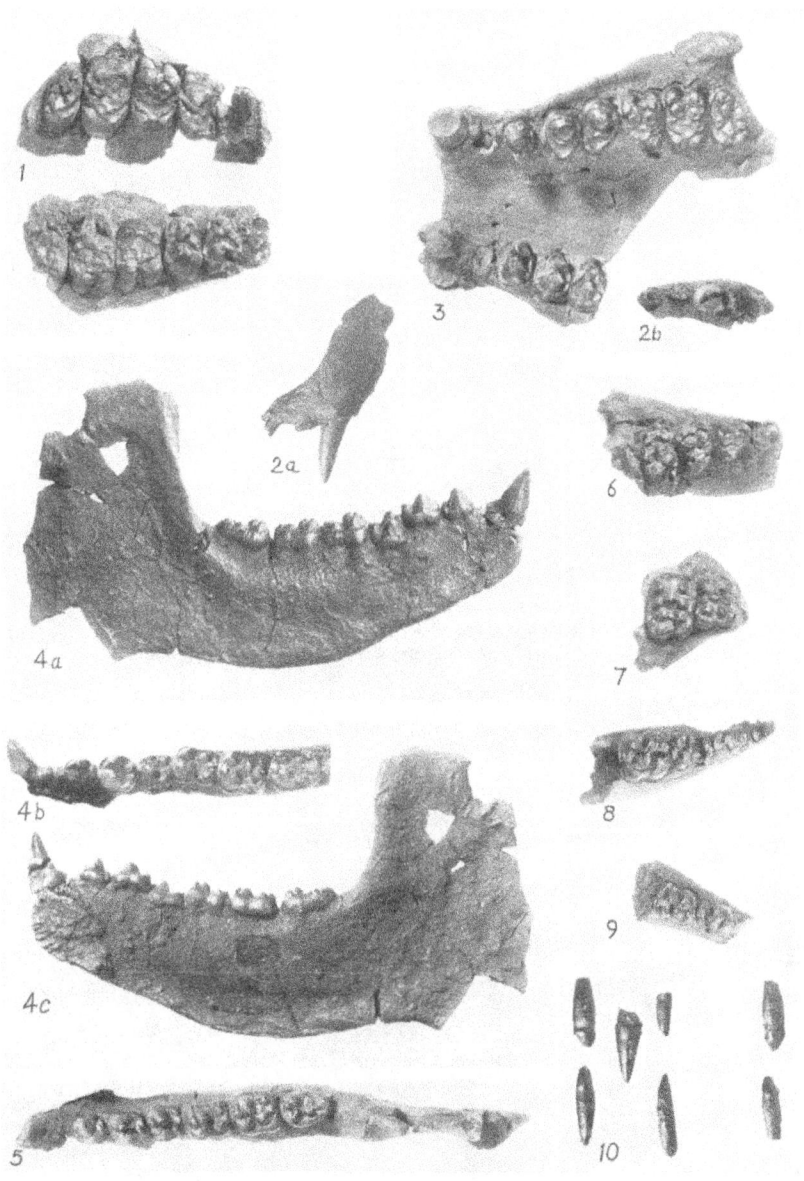

PLATE XXIV

Ectoconus FROM THE PUERCO PALEOCENE
SAN JUAN BASIN, NEW MEXICO

Fig. 1. *Ectoconus majusculus.* Skull. A. M. No. 16500 (complete skeleton).
   (a) Side view; (b) top view; (c) palatal view.
   All three-fourths natural size.

PLATE XXIV

1a

1b

1c

PLATE XXV

Ectoconus FROM THE PUERCO PALEOCENE
SAN JUAN BASIN, NEW MEXICO

Ectoconus majusculus.   Cervical vertebrae.   A. M. No. 16500 (complete skeleton).

Fig. 1.   Axis.
(a) Left side; (b) dorsal view; (c) ventral view; (d) anterior view.
Fig. 2.   Atlas.
(a) Ventral view; (b) posterior view; (c) anterior view.
Fig. 3.   Sixth cervical.
(a) Left side; (b) dorsal view; (c) posterior view; (d) anterior view.
Fig. 4.   Fourth cervical.
(a) Posterior view; (b) ventral view.
Fig. 5.   Seventh cervical.
(a) Posterior view; (b) dorsal view; (c) ventral view.
Fig. 6.   Third cervical, posterior view.
All three-fourths natural size.

PLATE XXV

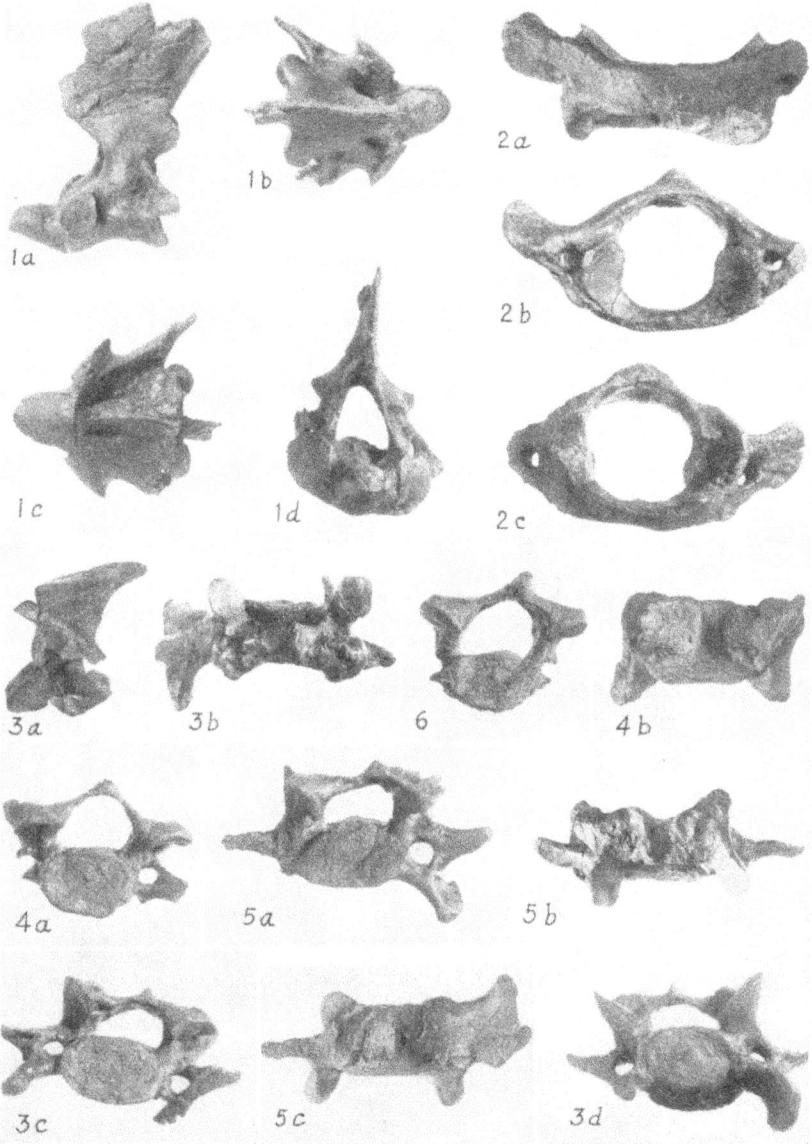

PLATE XXVI

Ectoconus FROM THE PUERCO PALEOCENE
SAN JUAN BASIN, NEW MEXICO

Ectoconus majusculus. Dorsal and lumbar vertebrae. A. M. No. 16500 (complete skeleton).

Fig. 1. First or second presacral.
  (*a*) Dorsal view; (*b*) ventral view.
Fig. 2. First dorsal.
  (*a*) Anterior view; (*b*) posterior view.
Fig. 3. Second dorsal.
  (*a*) Anterior view; (*b*) posterior view.
  All three-fourths natural size.

PLATE XXVI

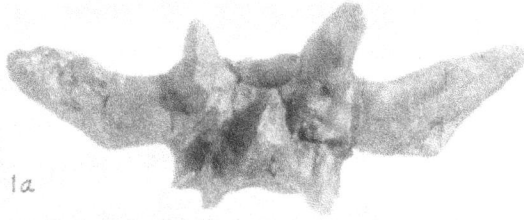

1a

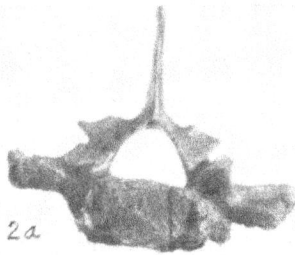

2a

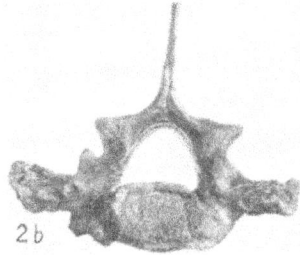

2b

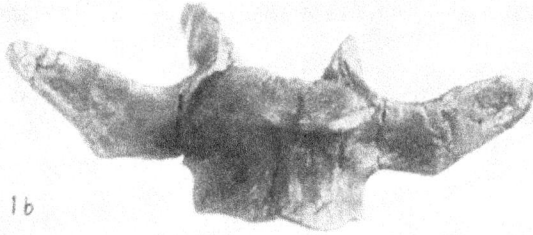

1b

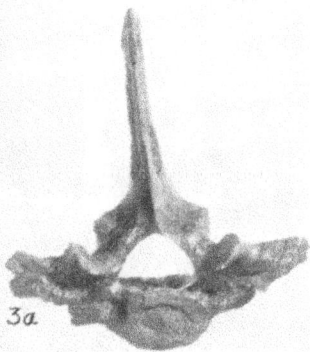

3a

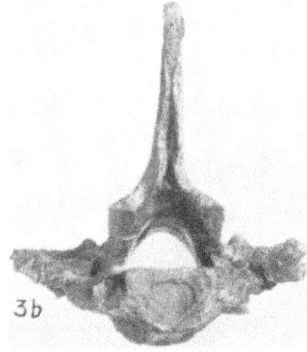

3b

PLATE XXVII

Ectoconus FROM THE PUERCO PALEOCENE
SAN JUAN BASIN, NEW MEXICO

Ectoconus majusculus.   Sacrum and lumbar vertebrae.   A. M. No. 16500 (complete skeleton).

Fig. 1.   Sacrum.
    (*a*) Ventral view; (*b*) dorsal view.
Figs. 2–4.   Third, sixth, and first or second presacrals, anterior views.
Figs. 5, 6.   Fourth and fifth presacrals, anterior views.
      All three-fourths natural size.

PLATE XXVII

PLATE XXVIII

Ectoconus FROM THE PUERCO PALEOCENE
SAN JUAN BASIN, NEW MEXICO

Ectoconus majusculus.   A. M. No. 16500 (complete skeleton).

Fig. 1.   Humerus.
    (*a*) Posterior view; (*b*) anterior view.
Fig. 2.   Scapula, external view.
    All three-fourths natural size.

PLATE XXVIII

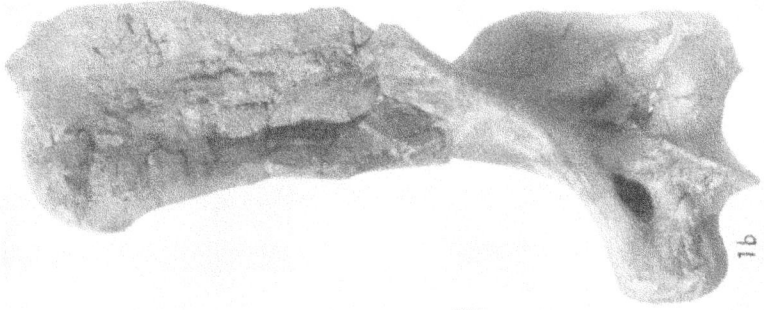

PLATE XXIX

Ectoconus FROM THE PUERCO PALEOCENE
SAN JUAN BASIN, NEW MEXICO

Ectoconus majusculus.   Fore limb.   A. M. No. 16500 (complete skeleton).

Fig. 1. Ulna.
    (*a*) Posterior view; (*b*) anterior view.
Fig. 2. Radius.
    (*a*) Posterior view; (*b*) anterior view.
Fig. 3. Clavicles.
Fig. 4. Scapula, proximal view.
    All three-fourths natural size.

PLATE XXIX

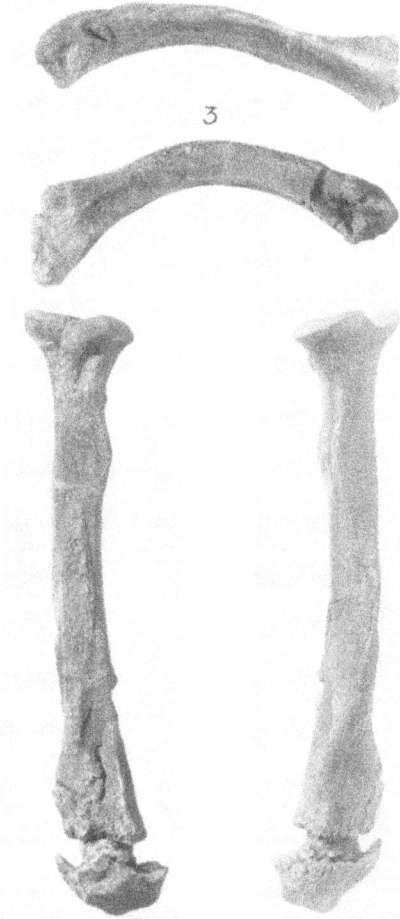

PLATE XXX

Ectoconus FROM THE PUERCO PALEOCENE
SAN JUAN BASIN, NEW MEXICO

Fig. 1.   *Ectoconus majusculus.*   Femur.   A. M. No. 16500 (complete skeleton).
   (*a*) Anterior view; (*b*) posterior view.
      About three-fourths natural size.

PLATE XXX

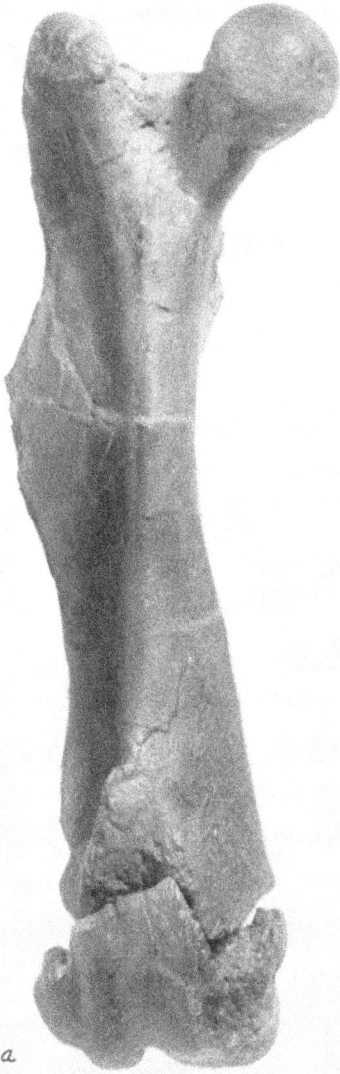

1a   1b

PLATE XXXI

Ectoconus FROM THE PUERCO PALEOCENE
SAN JUAN BASIN, NEW MEXICO

Ectoconus majusculus.   A. M. No. 16500 (complete skeleton).

Fig. 1.   Tibia.
    (*a*) Internal view; (*b*) anterior view; (*c*) external view.
Fig. 2.   Fibula.
    (*a*) External view; (*b*) anterior view.
    Three-fourths natural size.

PLATE XXXI

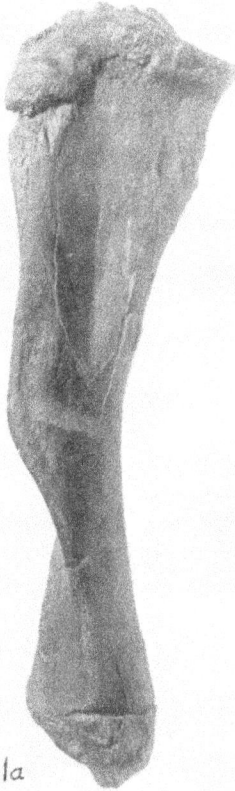

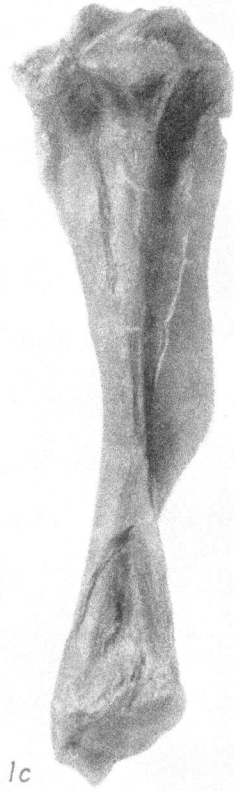

1a

1b

1c

2a

2b

PLATE XXXII

Ectoconus from the Puerco Paleocene
San Juan Basin, New Mexico

Ectoconus majusculus.   Right manus and right pes, articulated, dorsal views.   A. M. No. 16500 (complete skeleton).

Fig. 1.   Right manus.
Fig. 2.   Right pes.
        Three-fourths natural size.

PLATE XXXII

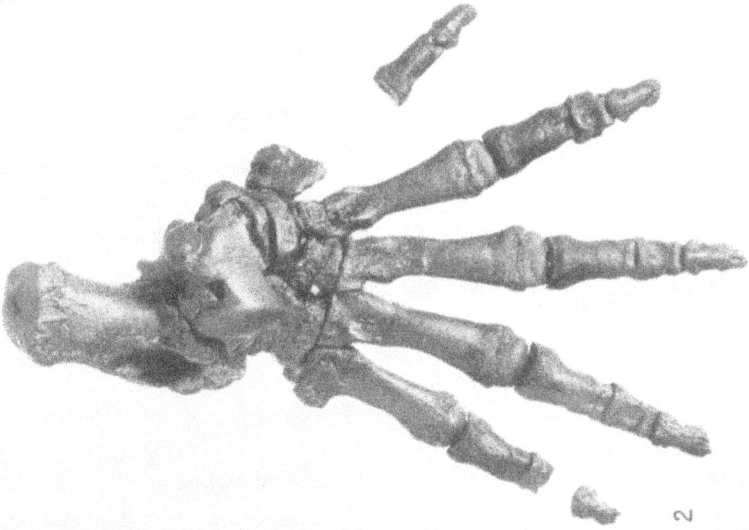

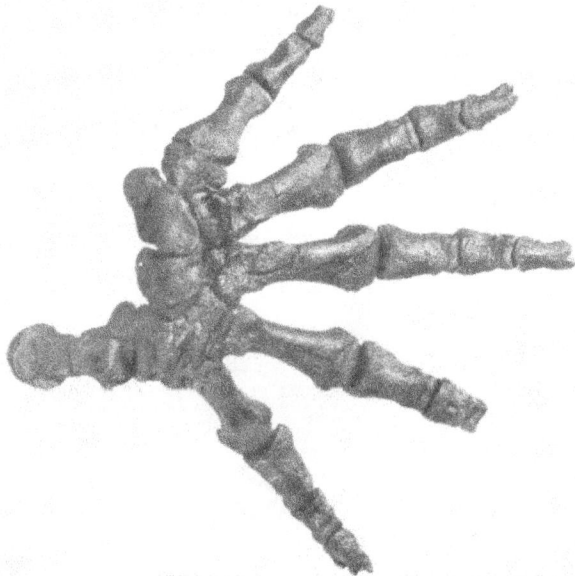

.

.

PLATE XXXIII

Conacodon entoconus FROM THE PUERCO PALEOCENE
SAN JUAN BASIN, NEW MEXICO

Fig. 1. Palate, p²–m¹ left and p²–m² right, crown view. A. M. Cope Coll. No. 3467.

Fig. 2. Right upper jaw, c and p²–m³. A. M. No. 16420.
(*a*) Side view; (*b*) crown view.

Fig. 3. Left ramus of lower jaw. A. M. No. 16424.
(*a*) Crown view; (*b*) side view.

Fig. 4. Left ramus of lower jaw. A. M. No. 16425.
(*a*) Side view; (*b*) crown view.

Fig. 5. Left ramus of lower jaw, crown view. A. M. No. 3476.
All one and one-half times natural size.

PLATE XXXIII

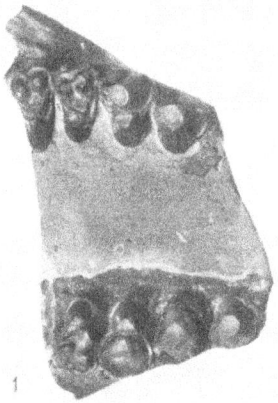

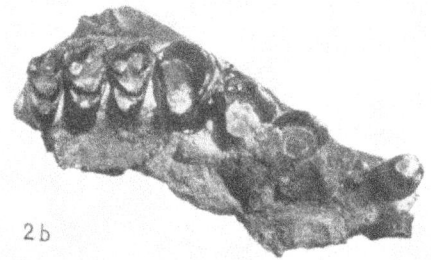

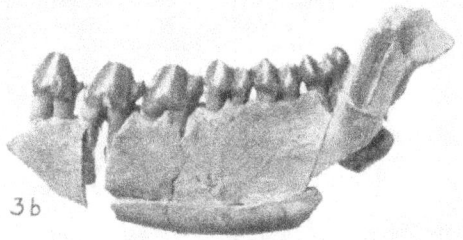

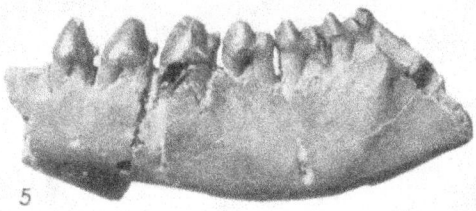

PLATE XXXIV

Paleocene Periptychidae from the Puerco and Torrejon Formations
San Juan Basin, New Mexico

Figs. 1–3. *Conacodon cophater.* Lower jaws, p₄–m₃ right and left, p₂–m₃ right. A. M. No. 16481.
  Fig. 1. Right ramus. (*a*) Internal view; (*b*) external view; (*c*) crown view.
  Fig. 2. Left ramus. (*a*) External view; (*b*) crown view.
  Fig. 3. Right ramus. (*a*) External view; (*b*) crown view.
Fig. 4. *Conacodon cophater.* Palate, crown view. A. M. Cope Coll. No. 3488.
Figs. 5–7. *Conacodon cophater.* Upper and lower jaw fragments. A. M. No. 16435.
  Fig. 5. Right maxilla, crown view.
  Fig. 6. Left maxilla, crown view.
  Fig. 7. Left ramus. (*a*) External view; (*b*) crown view.
Fig. 8. *Oxyacodon apiculatus.* Femur. A. M. No. 16368.
  (*a*) Anterior view; (*b*) posterior view.
Fig. 9. *Oxyacodon apiculatus.* Left ramus. A. M. No. 16369.
  (*a*) Crown view; (*b*) internal view; (*c*) external view.
Figs. 10, 11. *Oxyacodon apiculatus.* A. M. No. 16368.
  Fig. 10. Tibia.
  Fig. 11. Lower jaw. (*a*) Internal view; (*b*) external view.
Figs. 12, 13. *Anisonchus.* A. M. No. 16674.
  Fig. 12. Fragment of left maxilla, p⁴–m³. (*a*) Crown view; (*b*) external view.
  Fig. 13. Left ramus of lower jaw. (*a*) External view; (*b*) internal view.
    Figs. 1–11, Puerco; Figs. 12, 13, Torrejon.
    All one and one-half times natural size.

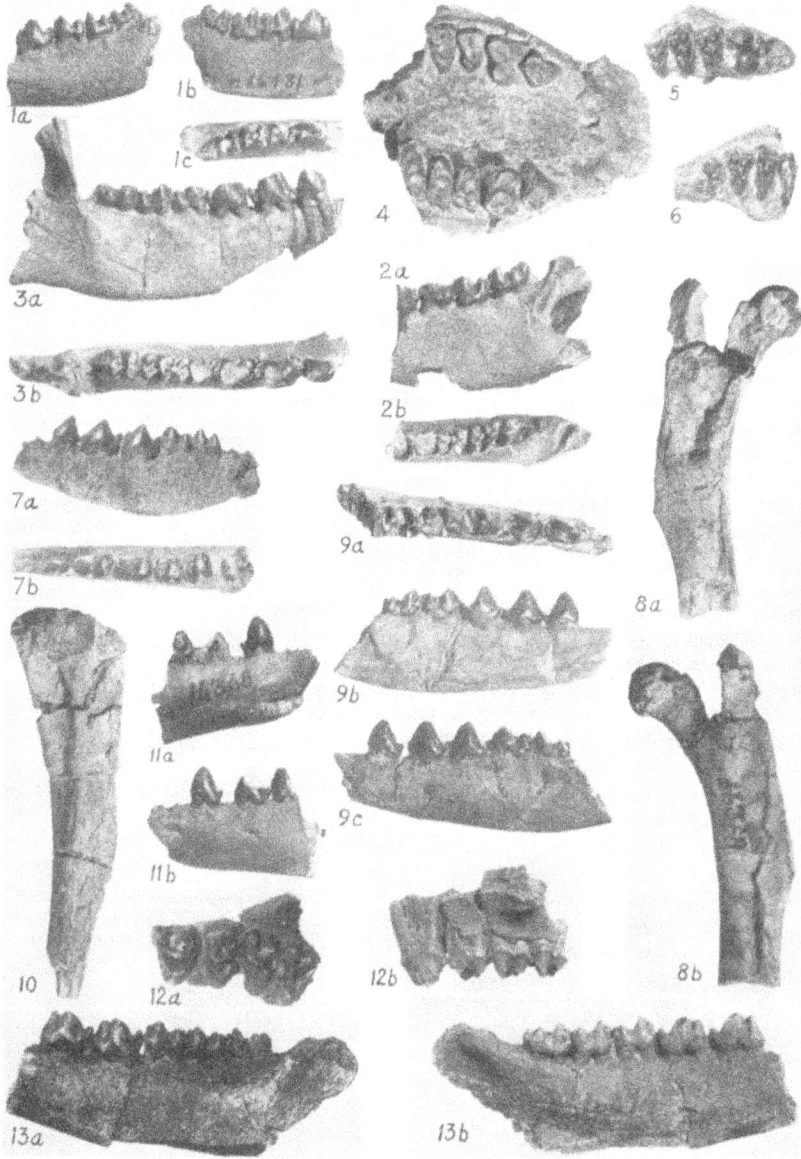

PLATE XXXIV

PLATE XXXV

Hemithlaeus kowalevskianus FROM THE PUERCO PALEOCENE
SAN JUAN BASIN, NEW MEXICO

Fig. 1. Palate, p^2–m^3 right and m$^{1-3}$ left, crown view. A. M. Cope Coll. No. 3576, type.
Fig. 2. Front of skull. A. M. No. 16439.
    (*a*) Dorsal view; (*b*) left maxilla, external view; (*c*) left and right maxillae, crown view.
Fig. 3. Right maxilla. A. M. No. 16441.
    (*a*) External view; (*b*) crown view.
Fig. 4. Lower jaws. A. M. No. 16441.
    (*a*) Right ramus, crown view; (*b*) left ramus, crown view; (*c*) right ramus, internal view; (*d*) left ramus,
    internal view.
Fig. 5. Left ramus. A. M. No. 16447.
    (*a*) Crown view; (*b*) external view; (*c*) internal view.
    All one and one-half times natural size.

PLATE XXXV

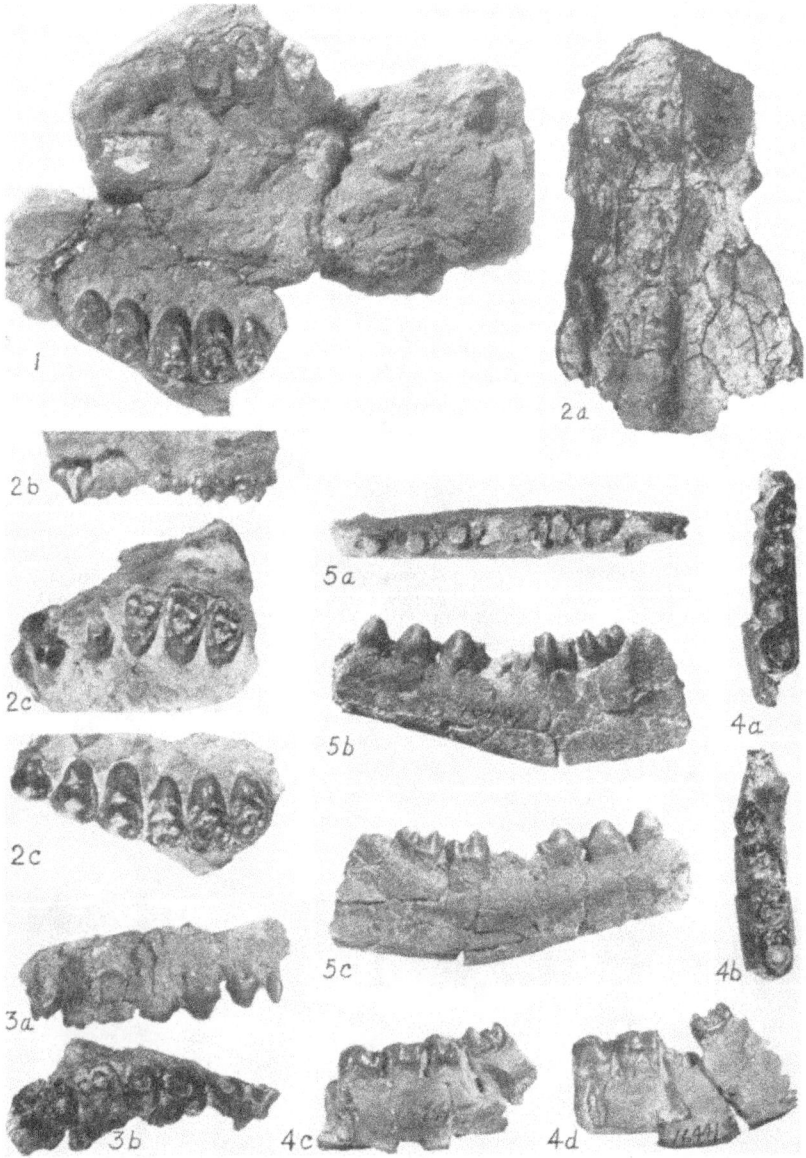

PLATE XXXVI

PALEOCENE PERIPTYCHIDAE FROM THE PUERCO AND TORREJON FORMATIONS
SAN JUAN BASIN, NEW MEXICO

Fig. 1.   *Anisonchus gillianus*.   Palate.   A. M. Cope Coll. No. 3601.
   (*a*) Side view; (*b*) crown view.
Fig. 2.   *Anisonchus sectorius*.   Maxilla, crown view.   A. M. Cope Coll. No. 3533.
Fig. 3.   *Anisonchus sectorius*.   Ramus, crown view.   A. M. Cope Coll. No. 3533.
Fig. 4.   *Anisonchus gillianus*.   Right ramus.   A. M. Cope Coll. No. 3601.
   (*a*) Crown view; (*b*) external view.
Figs. 5, 6.   *Anisonchus sectorius*.   Broken palate and right ramus.   A. M. No. 16667.
   (5*a*) Side view; (5*b*) crown view; (6*a*) external view; (6*b*) crown view.
      Figs. 1 and 4, Puerco; others, Torrejon.
      All one and one-half times natural size.

PLATE XXXVI

1a

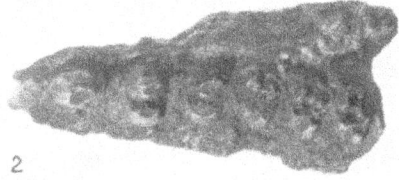

2

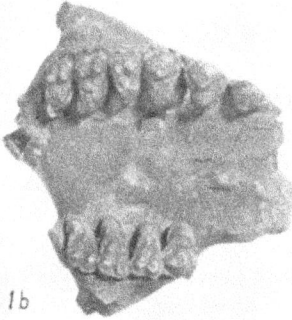

1b

3

4a

4b

5a

5b

6a

6b

PLATE XXXVII

PALEOCENE PERIPTYCHIDAE FROM THE PUERCO AND TORREJON FORMATIONS
SAN JUAN BASIN, NEW MEXICO

Figs. 1–5. *Anisonchus gillianus*. Upper and lower jaws. A. M. No. 16462.
   Fig. 1. Left maxilla. (*a*) Side view; (*b*) crown view.
   Fig. 2. Right maxilla. (*a*) Side view; (*b*) crown view.
   Fig. 3. Right maxilla. (*a*) Crown view; (*b*) side view.
   Fig. 4. Left ramus, external view.
   Fig. 5. Right ramus, external view.
Figs. 6, 7. *Anisonchus gillianus*. Lower jaws. A. M. No. 16468.
   Fig. 6. Right ramus. (*a*) Crown view; (*b*) internal view.
   Fig. 7. Right ramus. (*a*) External view; (*b*) crown view.
Fig. 8. *Anisonchus sectorius*. Left ramus. A. M. No. 15941.
   (*a*) Internal view; (*b*) external view.
Fig. 9. *Anisonchus sectorius*. Femur, anterior view. A. M. No. 16667. [See also Pl. XXXVI, figs. 5, 6.]
Fig. 10. *Anisonchus sectorius*. Left ramus, crown view. A. M. No. 16674. [See also Pl. XXXIV, fig. 13.]
   Figs. 1–7, Puerco; 8–10, Torrejon.
   All one and one-half times natural size.

PLATE XXXVII

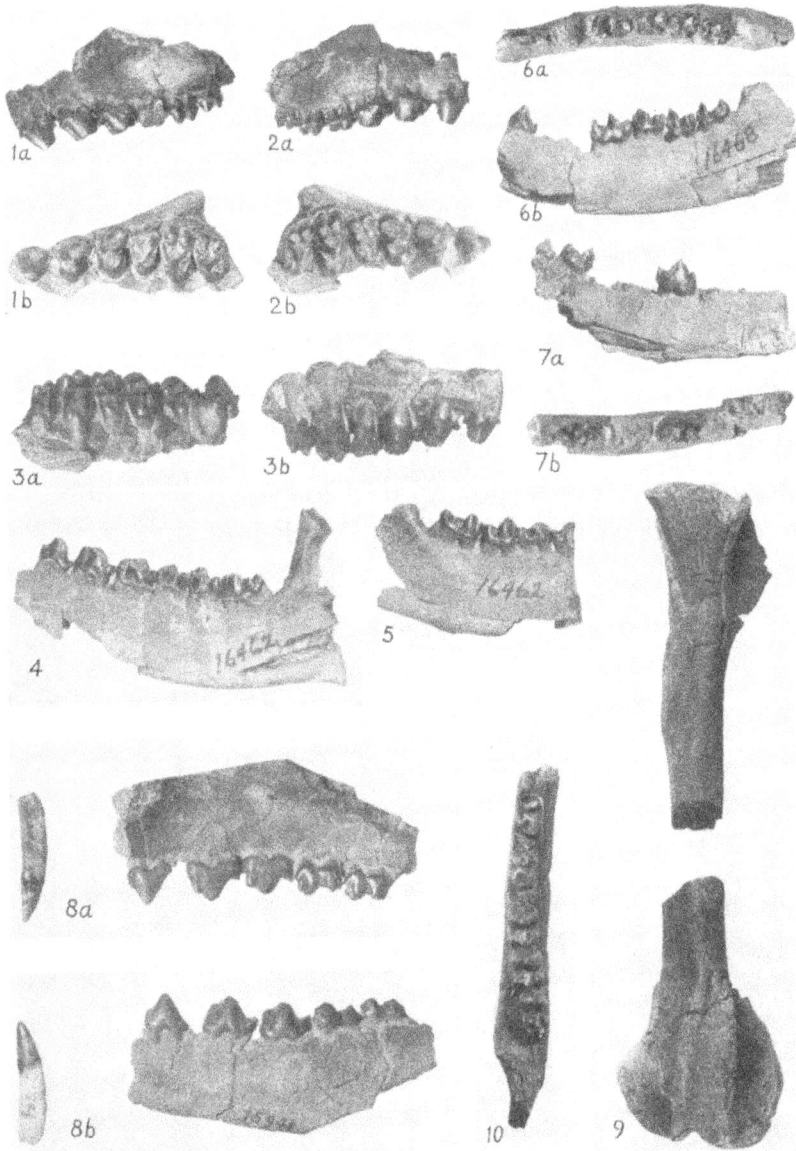

PLATE XXXVIII

Haploconus FROM THE TORREJON PALEOCENE
SAN JUAN BASIN, NEW MEXICO

Figs. 1–3. *Haploconus angustus.* A. M. No. 16680.
   Fig. 1. Skull.
      (a) Palatal view; (b) side view; (c) dorsal view.
   Fig. 2. Left ramus.
      (a) External view; (b) crown view.
   Fig. 3. Left ramus.
      (a) Crown view; (b) external view.
Fig. 4. *Haploconus lineatus,* type. Palate, crown view. A. M. Cope Coll. No. 3425.
   All one and one-half times natural size.

PLATE XXXVIII

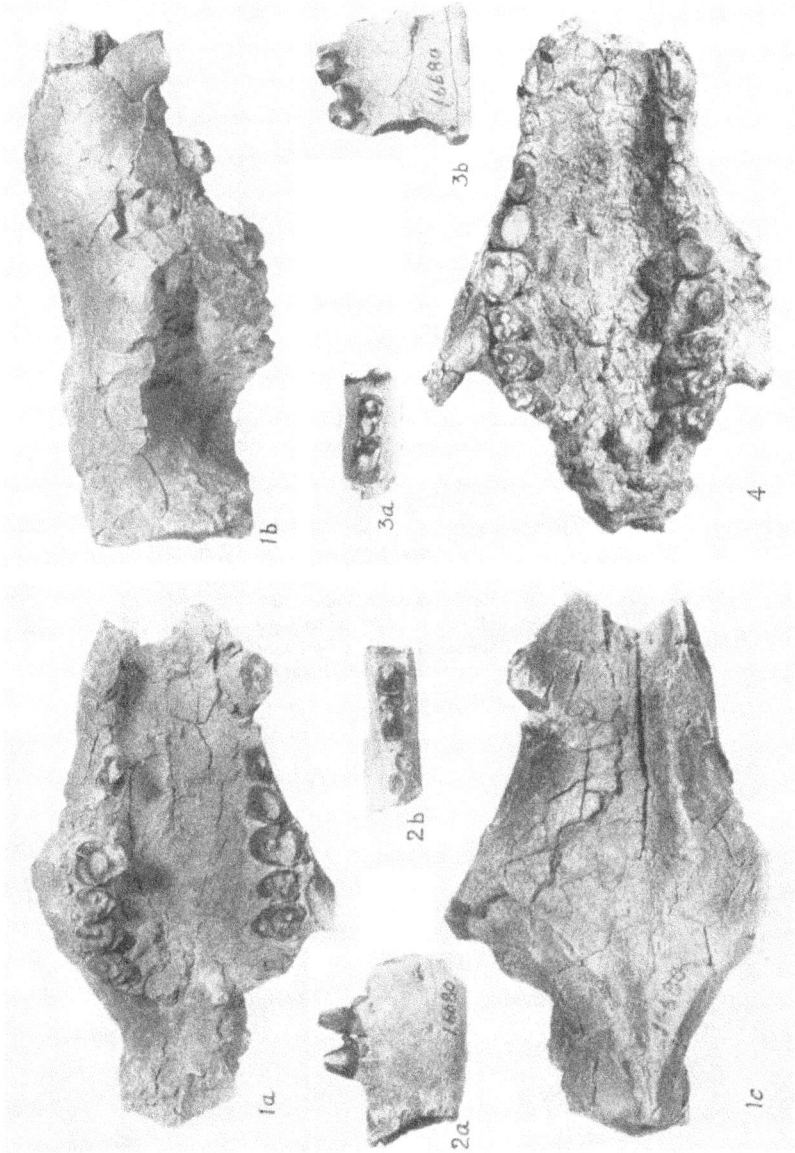

PLATE XXXIX

Haploconus angustus FROM THE TORREJON PALEOCENE
SAN JUAN BASIN, NEW MEXICO

Fig. 1.　Lower jaw.　A. M. No. 16688.
　　(*a*) External view; (*b*) crown view.
Fig. 2.　Right ramus, crown view.　A. M. No. 16688.
Fig. 3.　Right ramus, crown view.　A. M. No. 895.
Fig. 4.　Left ramus.　A. M. Cope Coll. No. 3454.
　　(*a*) External view; (*b*) crown view.
Fig. 5.　Left ramus, posterior half.　A. M. No. 16685.
　　(*a*) External view; (*b*) crown view; (*c*) internal view.
　　All one and one-half times natural size.

PLATE XXXIX

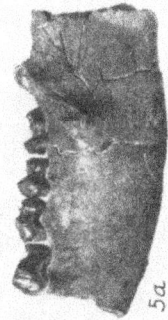

PLATE XL

Pantolambda bathmodon FROM THE TORREJON PALEOCENE
SAN JUAN BASIN, NEW MEXICO

Fig. 1.   Skull.   A. M. No. 16666.
   (a) Top view; (b) side view.   Palatal view on Plate XLII.
Fig. 2.   Skull (somewhat distorted), top view.   A. M. Cope Coll. No. 3957.
   All three-fourths natural size.

PLATE XL

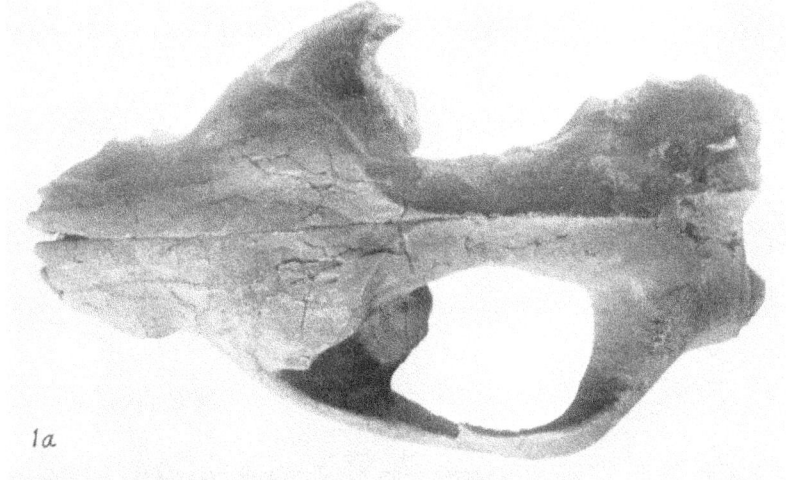

1a

1b

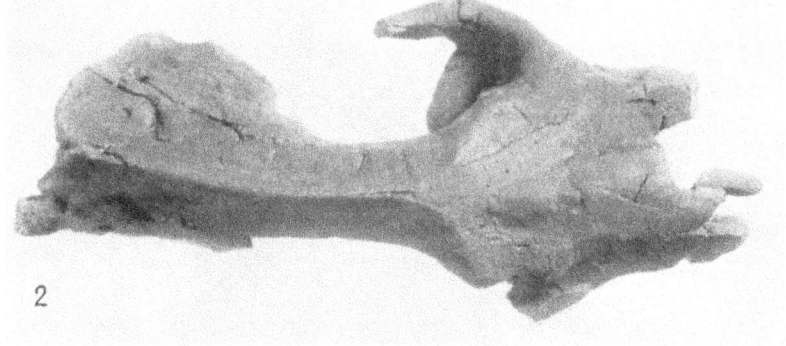

2

PLATE XLI

Pantolambda FROM THE TORREJON PALEOCENE
SAN JUAN BASIN, NEW MEXICO

Pantolambda bathmodon.   Skull and jaws and sternal segments.   A. M. No. 16663 (complete skeleton).

Fig. 1.   Skull.
    (a) Right side; (b) left side.
Fig. 2.   Sternal segments.
        Three-fourths natural size.

PLATE XLI

1a

1b

2

PLATE XLII

Pantolambda bathmodon FROM THE TORREJON PALEOCENE
SAN JUAN BASIN, NEW MEXICO

Fig. 1.   Skull, palatal view.   A. M. No. 16666.
Fig. 2.   Upper molars, crown view.   A. M. No. 15934.
Fig. 3.   Right ramus, crown view.   A. M. No. 2552.
Fig. 4.   Right ramus, crown view.   A. M. No. 2552.
Fig. 5.   Left ramus, crown view.   A. M. No. 2552.
Fig. 6.   Skull, top view.   A. M. No. 16663 (complete skeleton).
          All three-fourths natural size.

PLATE XLII

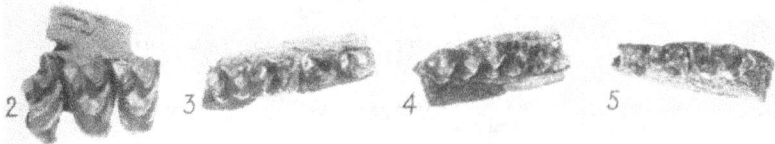

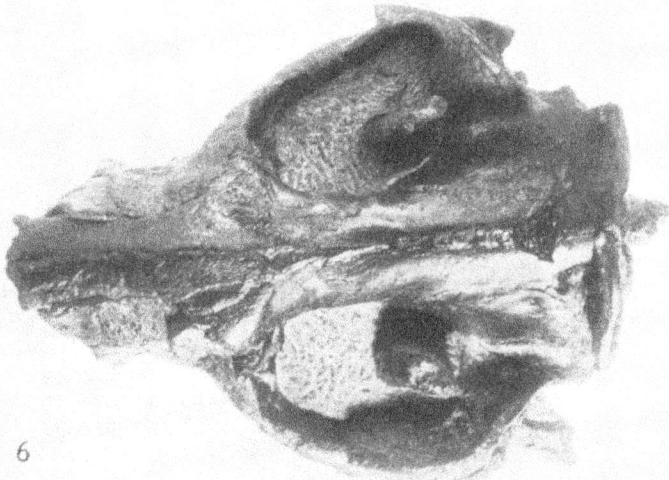

PLATE XLIII

Pantolambda FROM THE TORREJON PALEOCENE
SAN JUAN BASIN, NEW MEXICO

Pantolambda bathmodon.   Skeleton and sternal segments.   A. M. No. 16663 (complete skeleton).
Three-fourths natural size.

PLATE XLIII

PLATE XLIV

Pantolambda FROM THE TORREJON PALEOCENE
SAN JUAN BASIN, NEW MEXICO

Pantolambda bathmodon. Vertebrae, inferior views. A. M. No. 16663 (complete skeleton).

Fig. 1. Presacrals 1–5 and sacrals 5–7.
Fig. 2. Caudals.
Fig. 3. Posterior dorsals.
      All three-fourths natural size.

PLATE XLIV

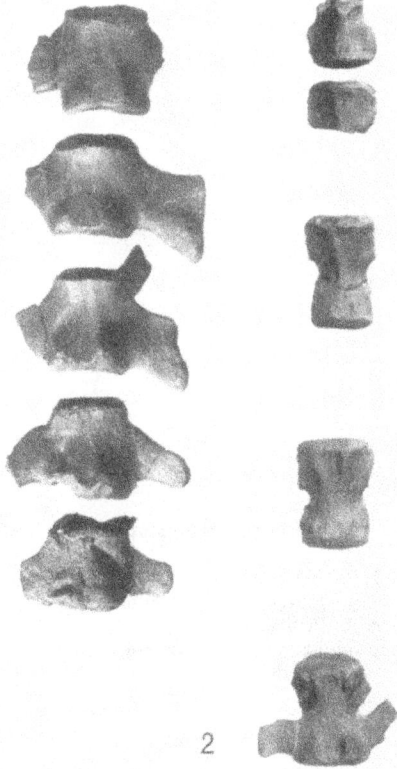

PLATE XLV

Pantolambda FROM THE TORREJON PALEOCENE
SAN JUAN BASIN, NEW MEXICO

Pantolambda bathmodon. Fore limb. A. M. No. 16663 (complete skeleton).

Fig. 1. Humerus.
    (*a*) Anterior view; (*b*) proximal view; (*c*) internal view; (*d*) posterior view; (*e*) distal view; (*f*) external view.
Fig. 2. Ulna.
    (*a*) External view; (*b*) anterior view.
Fig. 3. Radius.
    (*a*) External view; (*b*) anterior view.
    All three-fourths natural size.

PLATE XLV

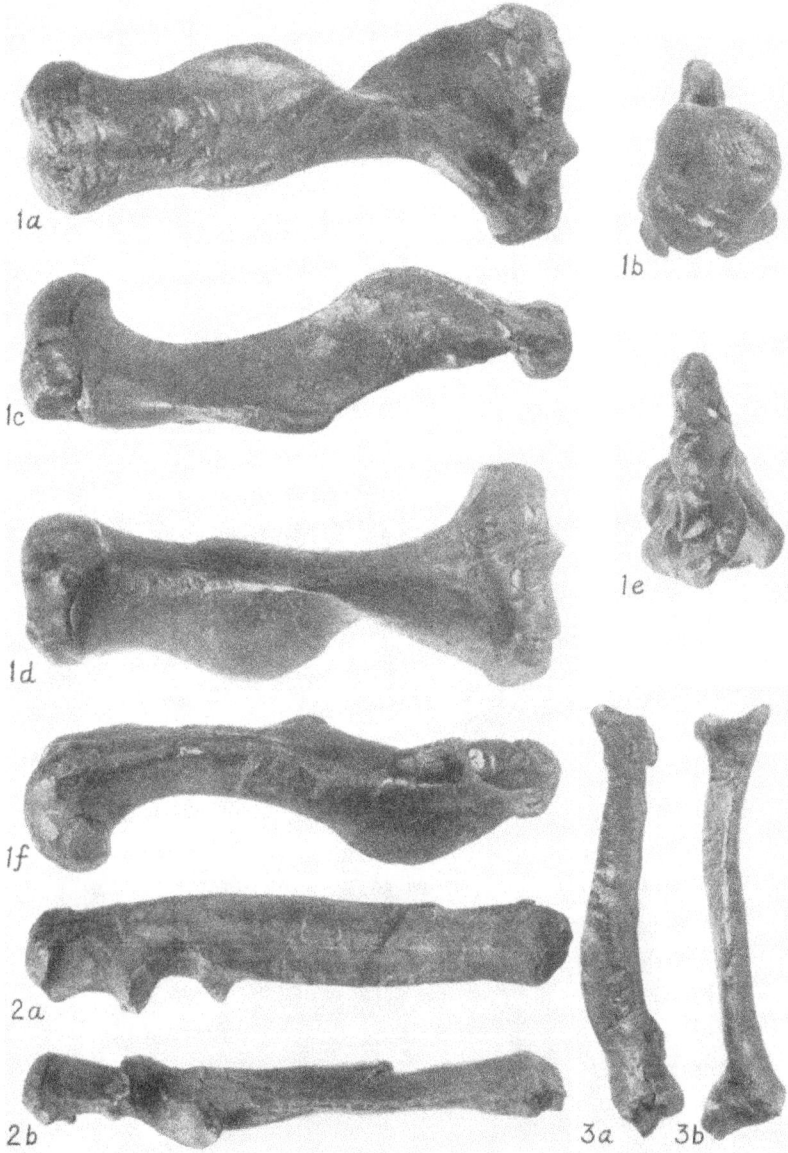

1a

1b

1c

1d

1e

1f

2a

2b

3a   3b

PLATE XLVI

Pantolambda FROM THE TORREJON PALEOCENE
SAN JUAN BASIN, NEW MEXICO

Fig. 1. *Pantolambda bathmodon.*   Right manus.   A. M. No. 16663 (complete skeleton).
   (*a*) Internal view; (*b*) external view; (*c*) dorsal view; (*d*) palmar view; (*e*) distal view; (*f*) proximal view.
   All three-fourths natural size.

PLATE XLVI

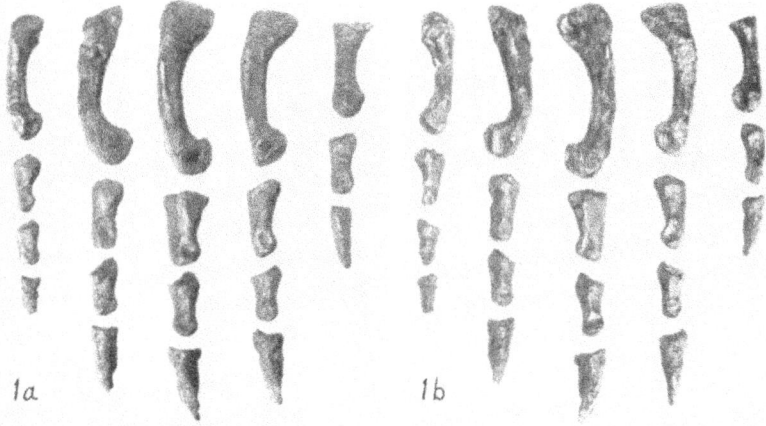

1a   1b

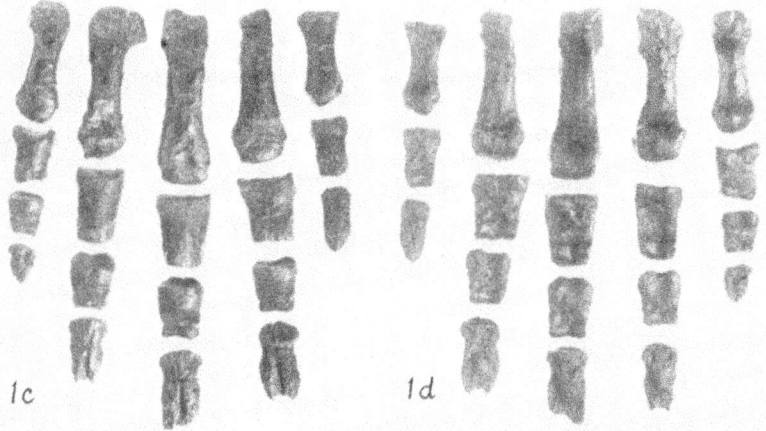

1c   1d

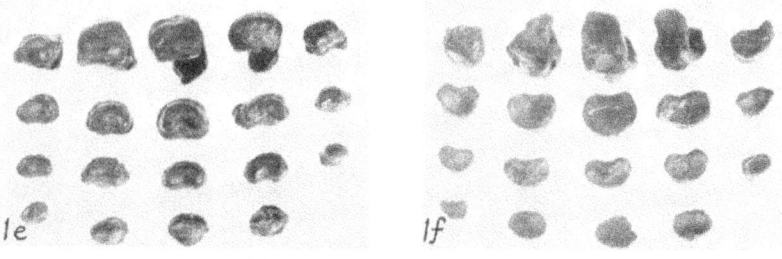

1e   1f

.

PLATE XLVII

Pantolambda FROM THE TORREJON PALEOCENE
SAN JUAN BASIN, NEW MEXICO

Fig. 1. *Pantolambda bathmodon*. Pelvis. A. M. No. 16663 (complete skeleton).
(*a*) Top view; (*b*) left side view.
Three-fourths natural size.

PLATE XLVII

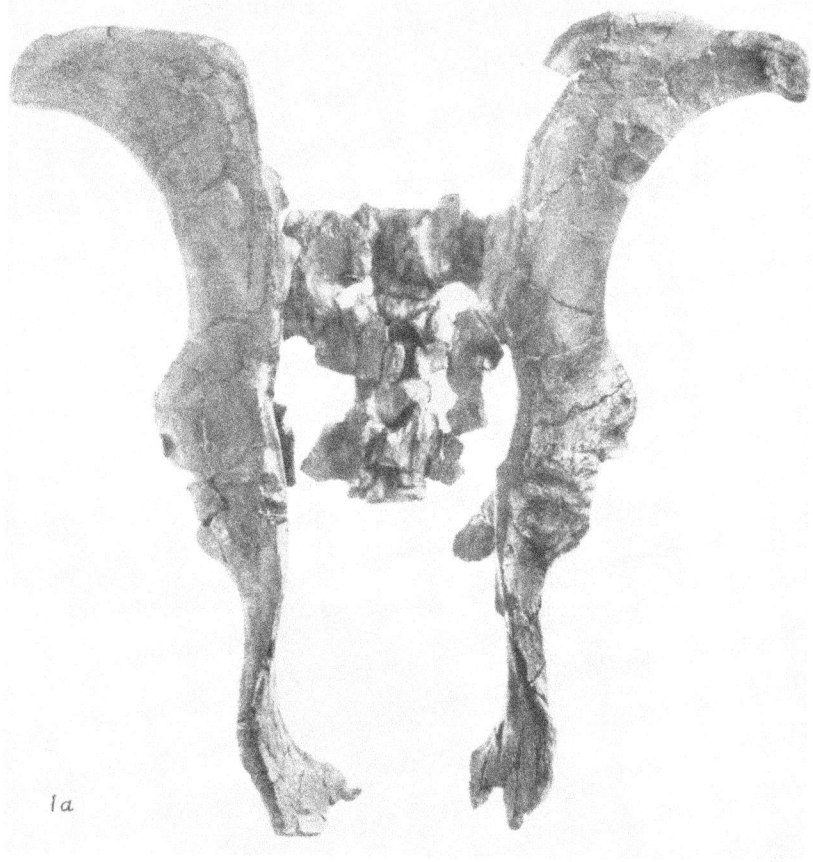

1a

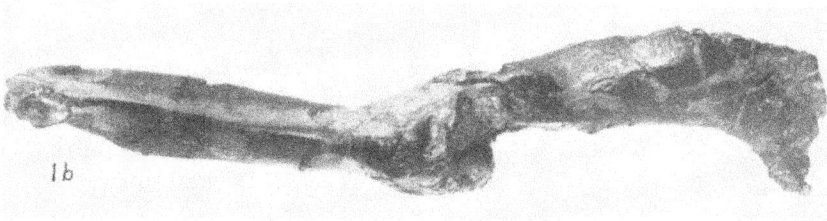

1b

PLATE XLVIII

Pantolambda FROM THE TORREJON PALEOCENE
SAN JUAN BASIN, NEW MEXICO

Fig. 1. *Pantolambda bathmodon.* Femur. A. M. No. 16663 (complete skeleton).
   (a) Anterior view; (b) external view; (c) proximal view; (d) distal view; (e) posterior view; (f) internal
      view.
   Three-fourths natural size.

PLATE XLVIII

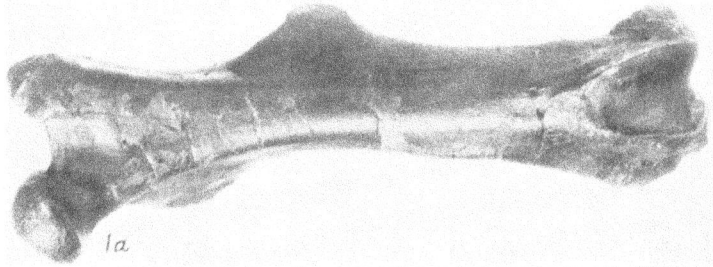

1a

1b

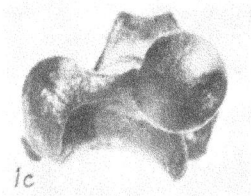

1c

1d

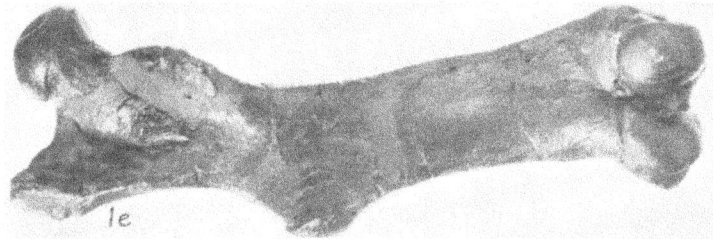

1e

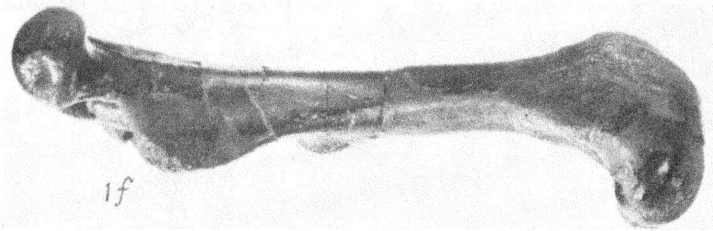

1f

.

PLATE XLIX

Pantolambda FROM THE TORREJON PALEOCENE
SAN JUAN BASIN, NEW MEXICO

Pantolambda bathmodon.　Left pes, dorsal view.　A. M. No. 16663 (complete skeleton).

Fig. 1.　Articulated.
Fig. 2.　Unarticulated.
　　　　Three-fourths natural size.

PLATE XLIX

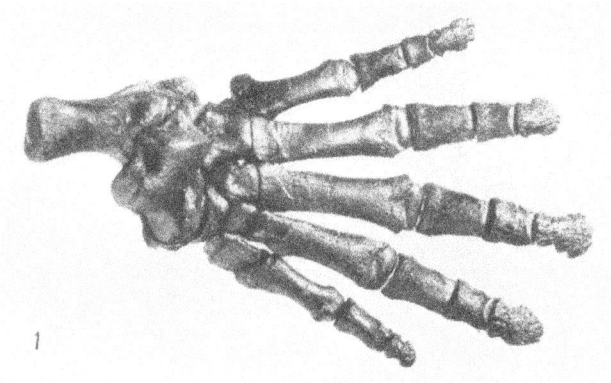

1

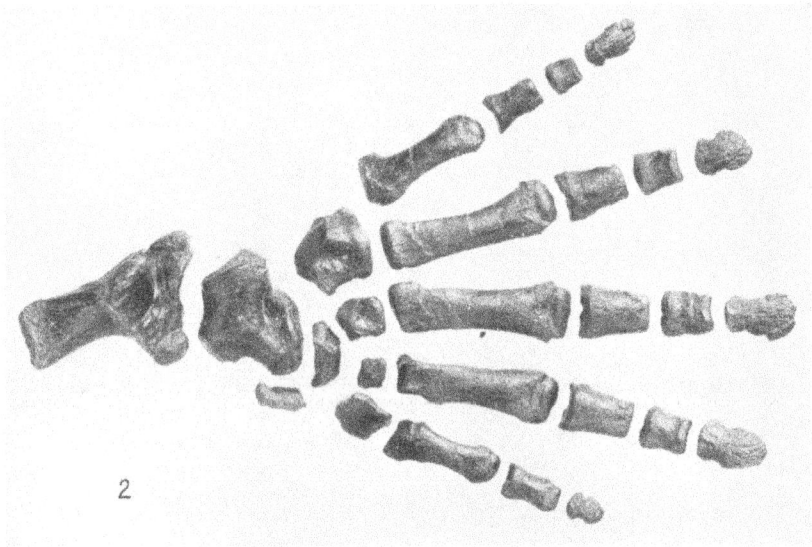

2

PLATE L

Pantolambda bathmodon FROM THE TORREJON PALEOCENE
SAN JUAN BASIN, NEW MEXICO

Fig. 1.   Tibia.   A. M. No. 16663 (complete skeleton).
  (a) Internal view; (b) anterior view.
Fig. 2.   Fibula.   A. M. No. 16663.
  (a) Internal view; (b) anterior view.
Fig. 3.   Patella.   A. M. No. 16663.
  (a) Posterior view; (b) anterior view.
Fig. 4.   Skull and jaws, left side view.   Three-eighths natural size.   A. M. No. 25.
Fig. 5.   Skull, top view.   A. M. Cope Coll. No. 3957.
  All three-fourths natural size except Fig. 4.

PLATE L

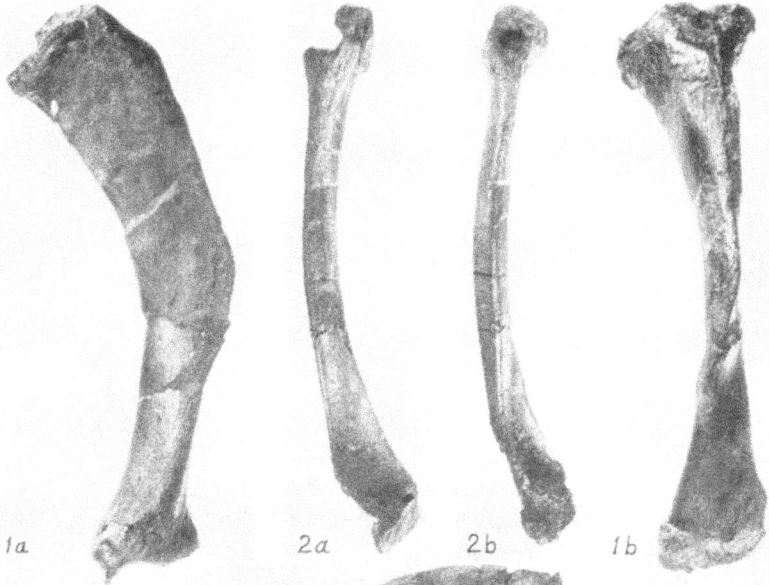

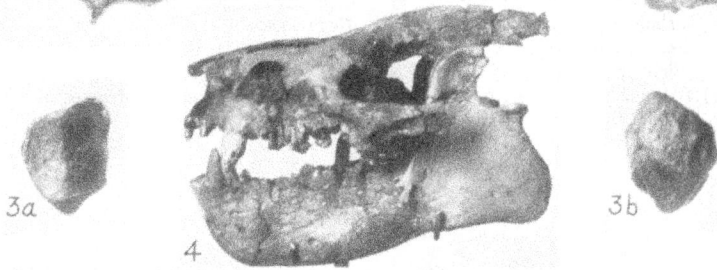

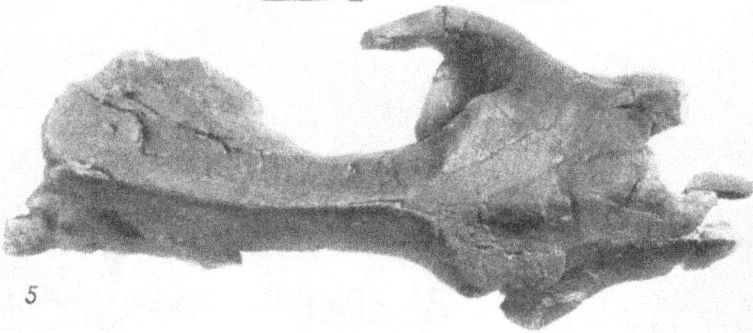

PLATE LI

Pantolambda FROM THE TORREJON PALEOCENE
SAN JUAN BASIN, NEW MEXICO

Pantolambda cavirictus.   Skull and jaws.

Fig. 1.   Palate, crown view.   A. M. No. 963.   Not to scale.
Fig. 2.   Skull, palatal view.   Three-eighths natural size.   A. M. No. 963.
Fig. 3.   Right ramus.   A. M. Cope Coll. No. 3961, type.
    (*a*) External view, three-eighths natural size; (*b*) external view, three-fourths natural size; (*c*) crown view, three-fourths natural size.

PLATE LI

PLATE LII

Pantolambda FROM THE TORREJON PALEOCENE
SAN JUAN BASIN, NEW MEXICO

Pantolambda cavirictus. Pes and jaws.

Fig. 1. Pes, dorsal view. Three-fourths natural size. A. M. Cope Coll. No. 3963.

Fig. 2. Jaws, crown view. Three-eighths natural size. A. M. Cope Coll. No. 3691, type specimen.

PLATE LII

PLATE LIII

Tetraclaenodon SKULL AND JAWS FROM TORREJON FORMATION
SAN JUAN BASIN, NEW MEXICO

Fig. 1.   Upper jaw fragments, m^{1-3}.   A. M. No. 16719.
    (a) Side view; (b) crown view.
Fig. 2.   Fragments of left ramus.   A. M. No. 16725.
    (a) Crown view; (b) external view.
Fig. 3.   Fragments of right ramus, crown view.   A. M. No. 16725.
Fig. 4.   Skull.   A. M. No. 15924.
    (a) Side view; (b) palatal view.
Fig. 5.   Left ramus, external view.   A. M. No. 15924.
Fig. 6.   Right ramus, external view.   A. M. No. 15924.
Fig. 7.   *Tetraclaenodon puercensis*.   Left ramus, crown view.   A. M. No. 2537.
Fig. 8.   Front of skull.   A. M. No. 16653.
    (a) Side view; (b) palatal view.
    All three-fourths natural size.

PLATE LIII

1a

1b

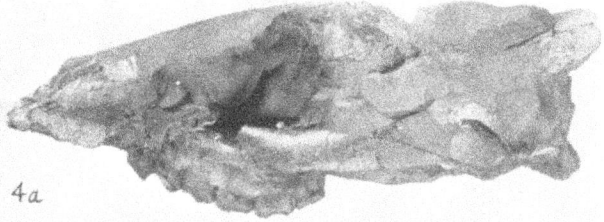

4a

2a

3

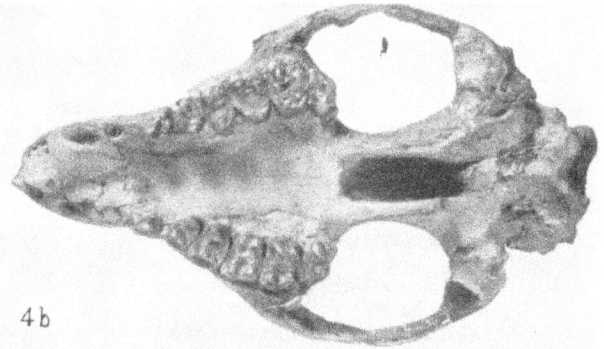

4b

2b

7

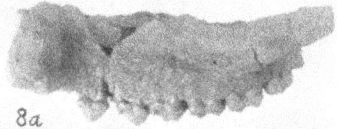

8a

5

6

8b

PLATE LIV

Tetraclaenodon FROM THE TORREJON FORMATION
SAN JUAN BASIN, NEW MEXICO

Fig. 1. Tibia. A. M. No. 15924.
    (*a*) Internal view; (*b*) external view; (*c*) anterior view.
Fig. 2. Radius. A. M. No. 15924.
    (*a*) Anterior view; (*b*) external view.
Fig. 3. Ulna. A. M. No. 15924.
    (*a*) External view; (*b*) anterior view.
Fig. 4. Patella. A. M. No. 15924.
    (*a*) External view; (*b*) anterior view.
Fig. 5. Lower jaw. P_3–m_3 and roots of c and p_{1-2}, crown views. A. M. No. 15924.
Fig. 6. Lower jaw. P_3–m_3, crown view. A. M. No. 15924.
Fig. 7. *Tetraclaenodon puercensis.* Left ramus, p_2–m_3, side view. A. M. No. 2537.
Fig. 8. *Tetraclaenodon puercensis.* Palate. A. M. No. 2537.
    (*a*) Crown view; (*b*) side view.
Fig. 9. Maxillary fragment. P^2–m^3, crown view. A. M. No. 16655.
Fig. 10. Upper teeth in broken palate. A. M. No. 16647.
    All three-fourths natural size.

PLATE LIV

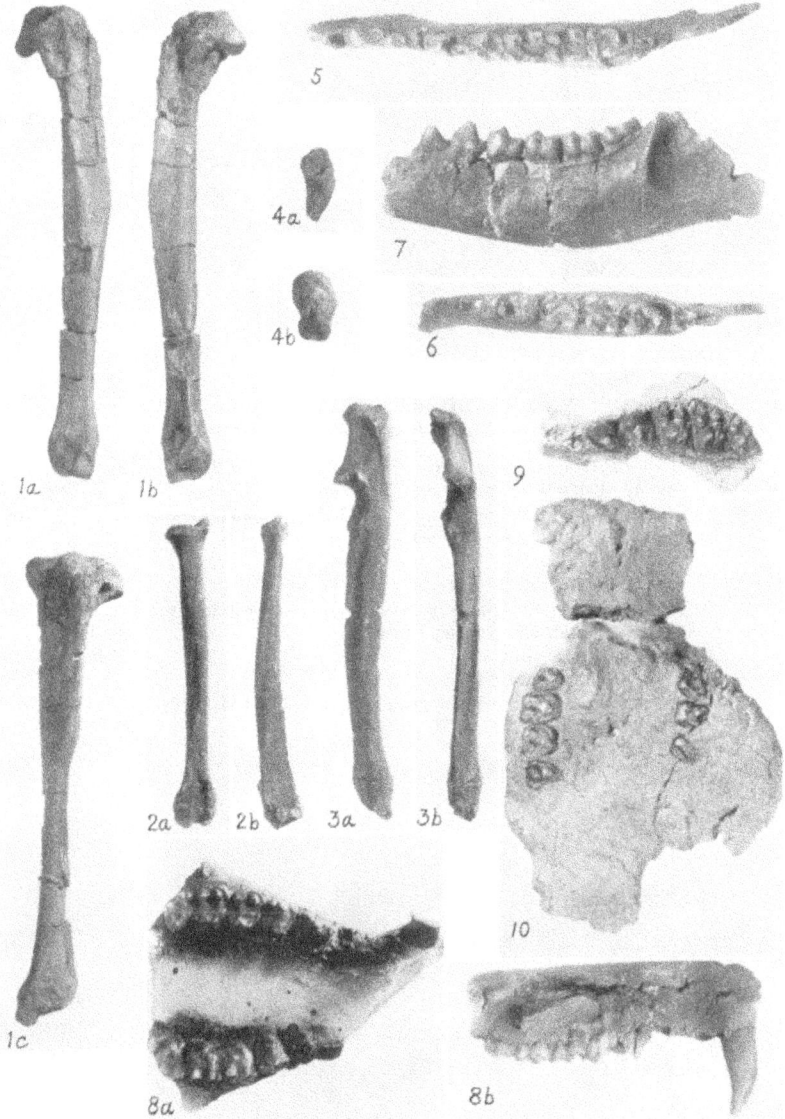

PLATE LV

PALEOCENE MAMMALS FROM THE TORREJON FORMATION
SAN JUAN BASIN, NEW MEXICO

Figs. 1–8. *Prodiacodon puercensis.* Paratype, A. M. No. 16748.
  Fig. 1. Palate. (*a*) Right side; (*b*) crown view.
  Fig. 2. Lower jaw, right ramus. (*a*) External view; (*b*) crown view; (*c*) internal view.
  Fig. 3. Femur. Anterior view.
  Fig. 4. Limb bone.
  Fig. 5. Astragalus (*a*) and calcaneum (*b*), dorsal views.
  Fig. 6. Astragalus (*a*) and calcaneum (*b*), dorsal and internal views. Also tarsals.
  Fig. 7. Tarsals and phalanges (*a–c*).
  Fig. 8. Part of pelvis.
  All two and one-fourth times natural size.
Figs. 9, 10. *Protoselene.* Humerus and femur, anterior views, × 1½. A. M. No. 16614.

PLATE LV

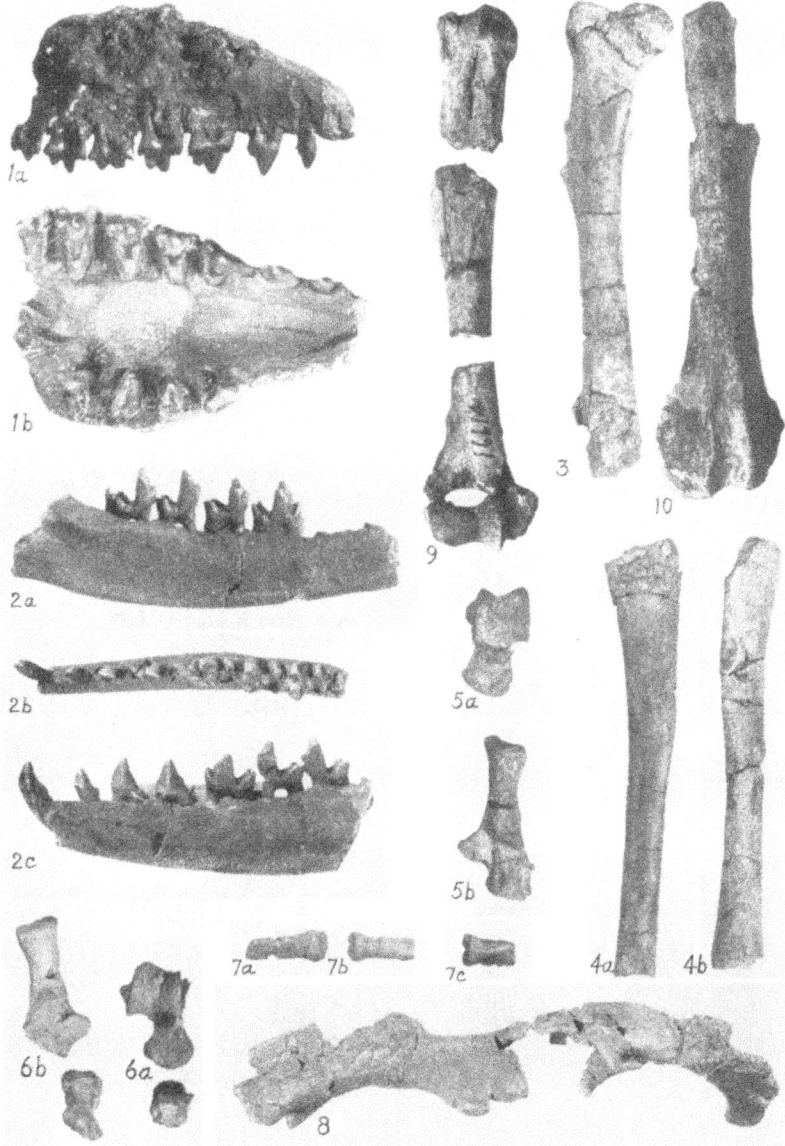

PLATE LVI

PALEOCENE INSECTIVORA FROM THE PUERCO AND TORREJON FORMATIONS
SAN JUAN BASIN, NEW MEXICO

Fig. 1. *Acmeodon secans*. Portion of left ramus, × 2¼. A. M. No. 16599.
(a) External view; (b) crown view; (c) internal view.
Fig. 2. *Acmeodon secans*. Portion of right ramus, × 2¼. Paratype, A. M. No. 16600.
(a) External view; (b) crown view.
Fig. 3. *Pentacodon*. Left ramus, × 1½. A. M. No. 16592.
(a) Internal view; (b) crown view.
Fig. 4. *Pentacodon*. Portion of left ramus, × 1½. A. M. No. 17038.
(a) Internal view; (b) crown view; (c) external view.
Fig. 5. *Pentacodon*. Part of left ramus, × 1½. A. M. Cope Coll. No. 3385.
(a) Internal view; (b) crown view.
Fig. 6. *Prodiacodon puercensis*. Part of pelvis, × 1½. Paratype, A. M. No. 16748.
All from Torrejon except Figs. 5, 6.

PLATE LVI

1a

2a

1b

2b

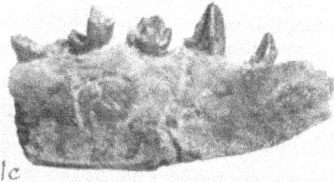
1c

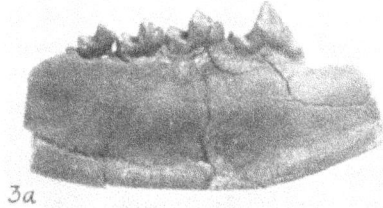
3a

4a

3b

4b

5a

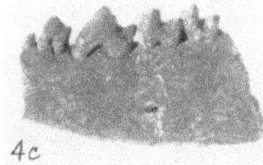
4c

6

5b

PLATE LVII

PALEOCENE INSECTIVORA FROM THE TORREJON FORMATION
SAN JUAN BASIN, NEW MEXICO

Fig. 1. *Mixodectes ?crassiusculus*. Lower jaw fragment. A. M. No. 16018.
   (*a*) Side view; (*b*) crown view.
Fig. 2. *Mixodectes ?crassiusculus*. Right ramus. A. M. No. 16018.
   (*a*) External view; (*b*) crown view.
Fig. 3. *Mixodectes*. Palate. A. M. No. 16593.
Fig. 4. *Indrodon*. Lower jaw fragment, m_{1-3}. A. M. No. 16584.
   (*a*) External view; (*b*) crown view.
Fig. 5. *Indrodon*. Left ramus. A. M. No. 17064.
   (*a*) External view; (*b*) crown view.
Fig. 6. *Mixodectes*. Lower jaw fragment, side view. A. M. No. 16595.
Fig. 7. *Palaeolestes*. Femur, anterior view. A. M. No. 16011.
   All one and one-half times natural size.

PLATE LVII

1a

1b

4a

4b

3

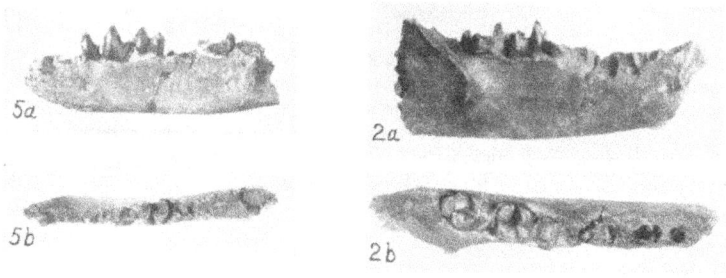

5a

5b

2a

2b

6

7

PLATE LVIII

PALEOCENE CONORYCTIDAE FROM THE SAN JUAN BASIN, NEW MEXICO

Fig. 1. *Conoryctes comma.* Mandible. A. M. Cope Coll. No. 3396.
(*a*) Crown view; (*b*) side view.
Fig. 2. *Conoryctes comma.* Palate. A. M. Cope Coll. No. 3396.
(*a*) Side view; (*b*) crown view.
Fig. 3. *Onychodectes tisonensis.* Skull. A. M. No. 785.
(*a*) Dorsal view; (*b*) side view.
Fig. 4. *Onychodectes tisonensis.* Mandible. A. M. No. 785.
(*a*) External view; (*b*) internal view; (*c*) crown view.
Figs. 1 and 2, Torrejon; 3 and 4, Puerco.
All three-fourths natural size.

PLATE LVIII

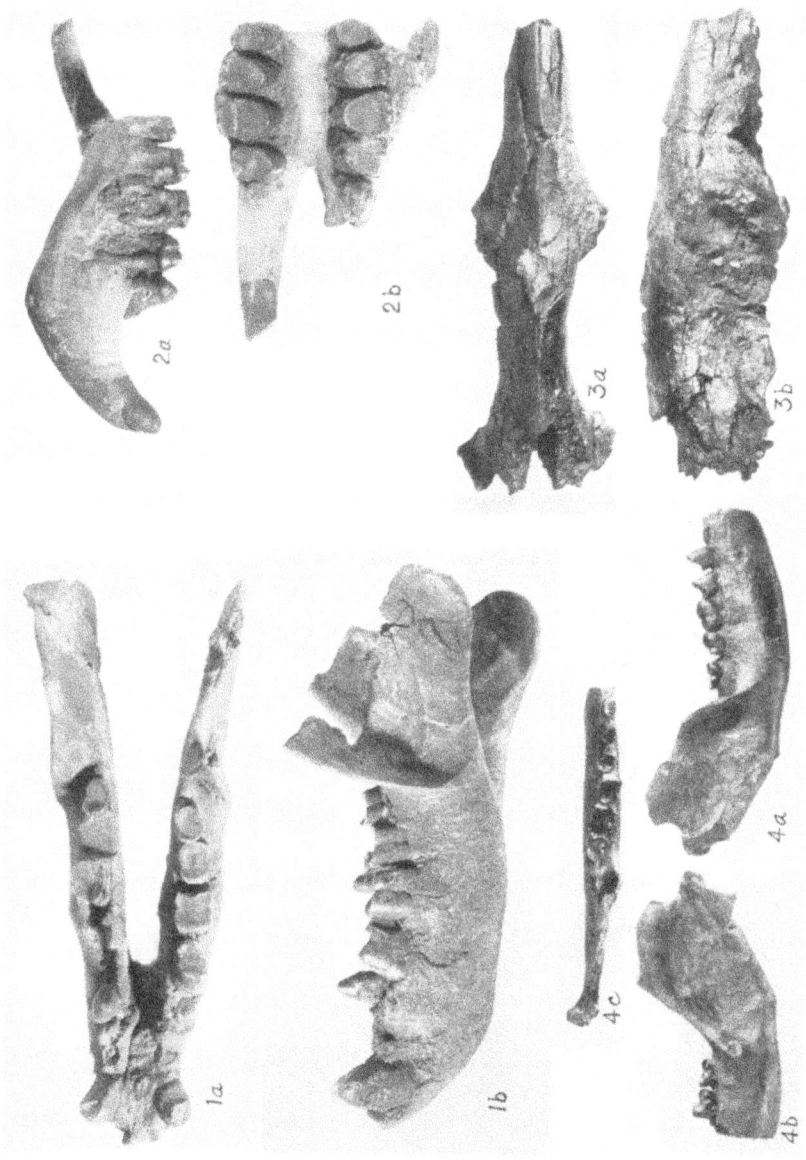

.

PLATE LIX

PALEOCENE STYLINODONTIDAE FROM THE TORREJON FORMATION
SAN JUAN BASIN, NEW MEXICO

Psittacotherium multifragum.   Teeth, skull and mandible.

Fig. 1.   Upper cheek teeth, crown view, × ¾.   A. M. Cope Coll. No. 2453.
Fig. 2.   Skull, top view, × 1½.   A. M. No. 754.
Fig. 3.   Lower jaw, × 1½.   A. M. No. 754.
    (a) Crown view; (b) side view.

PLATE LIX

1

2

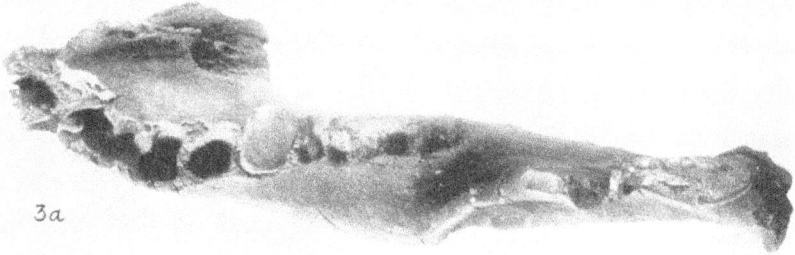

3a

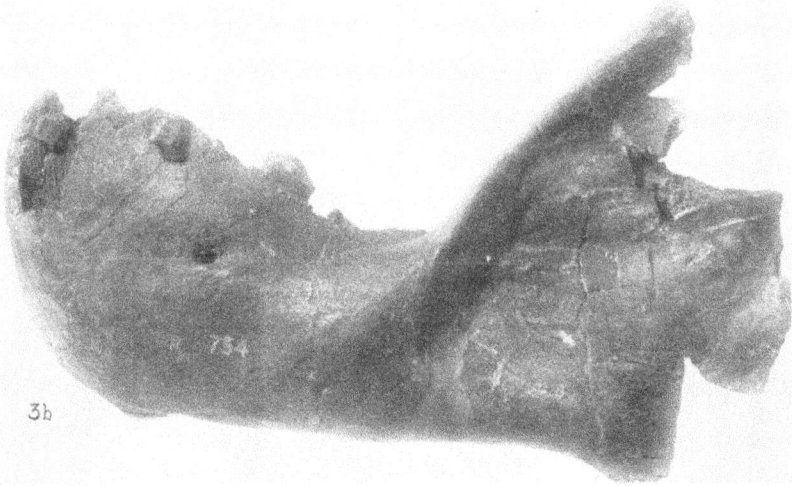

3b

PLATE LX

Paleocene Stylinodontidae from the Torrejon Formation
San Juan Basin, New Mexico

Fig. 1. *Psittacotherium multifragum.* Lower jaws. About one and one-half times natural size. A. M. No. 2453.
(*a*) Crown view; (*b*) side view.

PLATE LX

1a

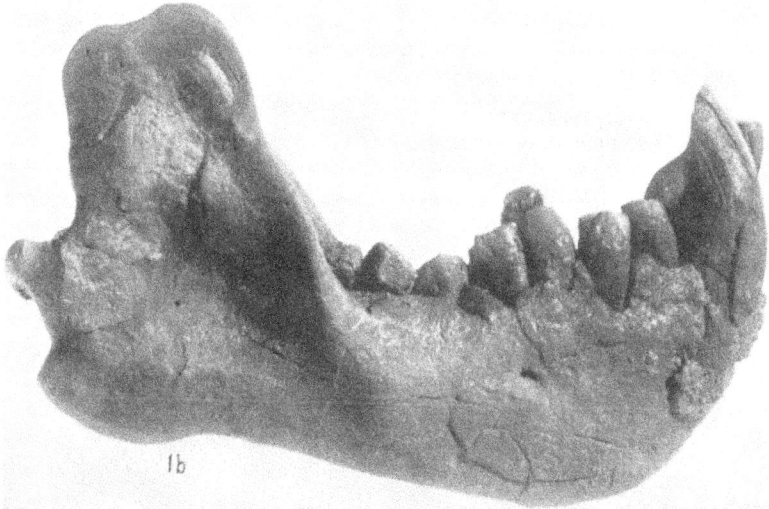

1b

PLATE LXI

PALEOCENE STYLINODONTIDAE FROM THE PUERCO AND TORREJON FORMATIONS
SAN JUAN BASIN, NEW MEXICO

Fig. 1. *Wortmania otariidens.* Tibia, anterior view. A. M. Cope Coll. No. 3394.
Fig. 2. *Psittacotherium multifragum.* Foot bones. A. M. Cope Coll. No. 2453.
  (*a*) Side view; (*b*) dorsal view.
Fig. 3. *Psittacotherium multifragum.* Ulna, anterior view. A. M. Cope Coll. No. 2453.
    Fig. 1, Puerco; Figs. 2 and 3, Torrejon.
    All three-fourths natural size.

PLATE LXI

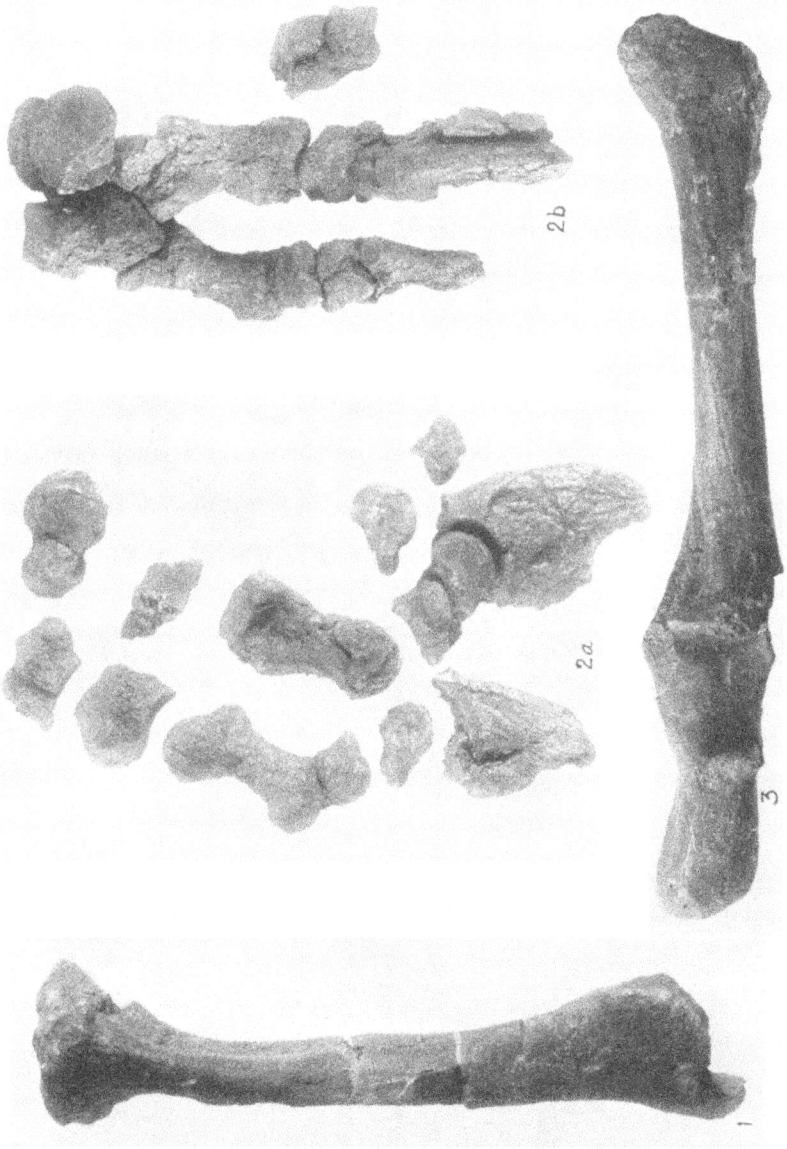

PLATE LXII

PALEOCENE STYLINODONTIDAE FROM THE PUERCO AND TORREJON FORMATIONS
SAN JUAN BASIN, NEW MEXICO

Fig. 1. *Wortmania otariidens.* Skull. Type, A. M. Cope Coll. No. 3394.
   (a) Side view; (b) top view.
Fig. 2. *Wortmania otariidens.* Ulna, proximal half. Type, A. M. Cope Coll. No. 3394.
Fig. 3. *Wortmania otariidens.* Part of palate. Type, A. M. Cope Coll. No. 3394.
   (a) Crown view; (b) side view.
Fig. 4. *Wortmania otariidens.* Radius, anterior view. Type, A. M. Cope Coll. No. 3394.
Fig. 5. *Psittacotherium multifragum.* Top of skull, side view. A. M. No. 754.
      Fig. 5, Torrejon; others, Puerco.
      All three-fourths natural size.

PLATE LXII

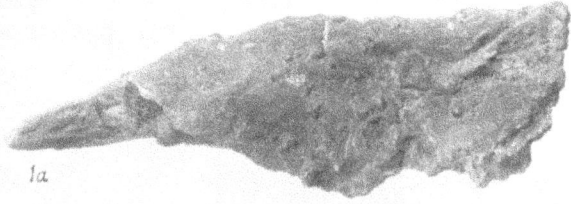

1a

2

1b

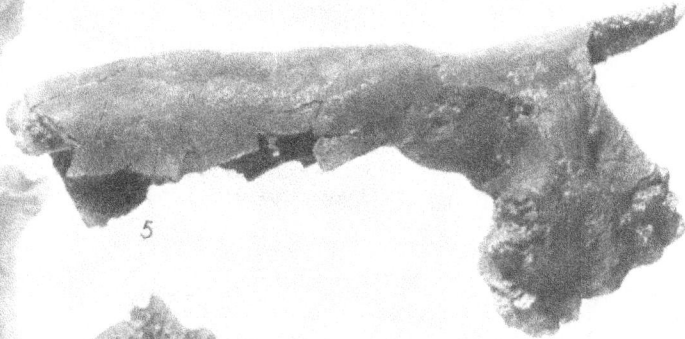

5

4

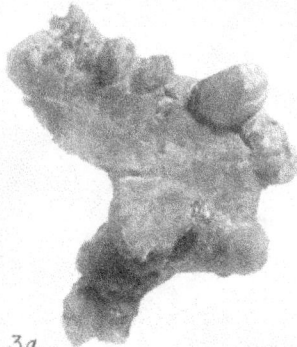

3a

3b

PLATE LXIII

PALEOCENE STYLINODONT FROM THE PUERCO FORMATION
SAN JUAN BASIN, NEW MEXICO

Wortmania otariidens.   A. M. Cope Coll. No. 3394.

Fig. 1.   Femur, anterior view.
Fig. 2.   Ungual phalanx, side view.
Fig. 3.   Phalanx, side view.
Fig. 4.   Mandible.
   (*a*) Crown view; (*b*) side view.
       All three-fourths natural size.

PLATE LXIII

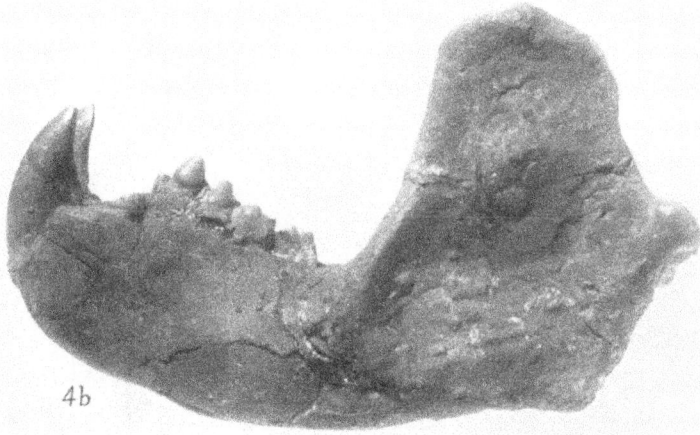

PLATE LXIV

RESTORATION OF *Psittacotherium* BY WILLIAM DILLER MATTHEW.   One-half natural size.

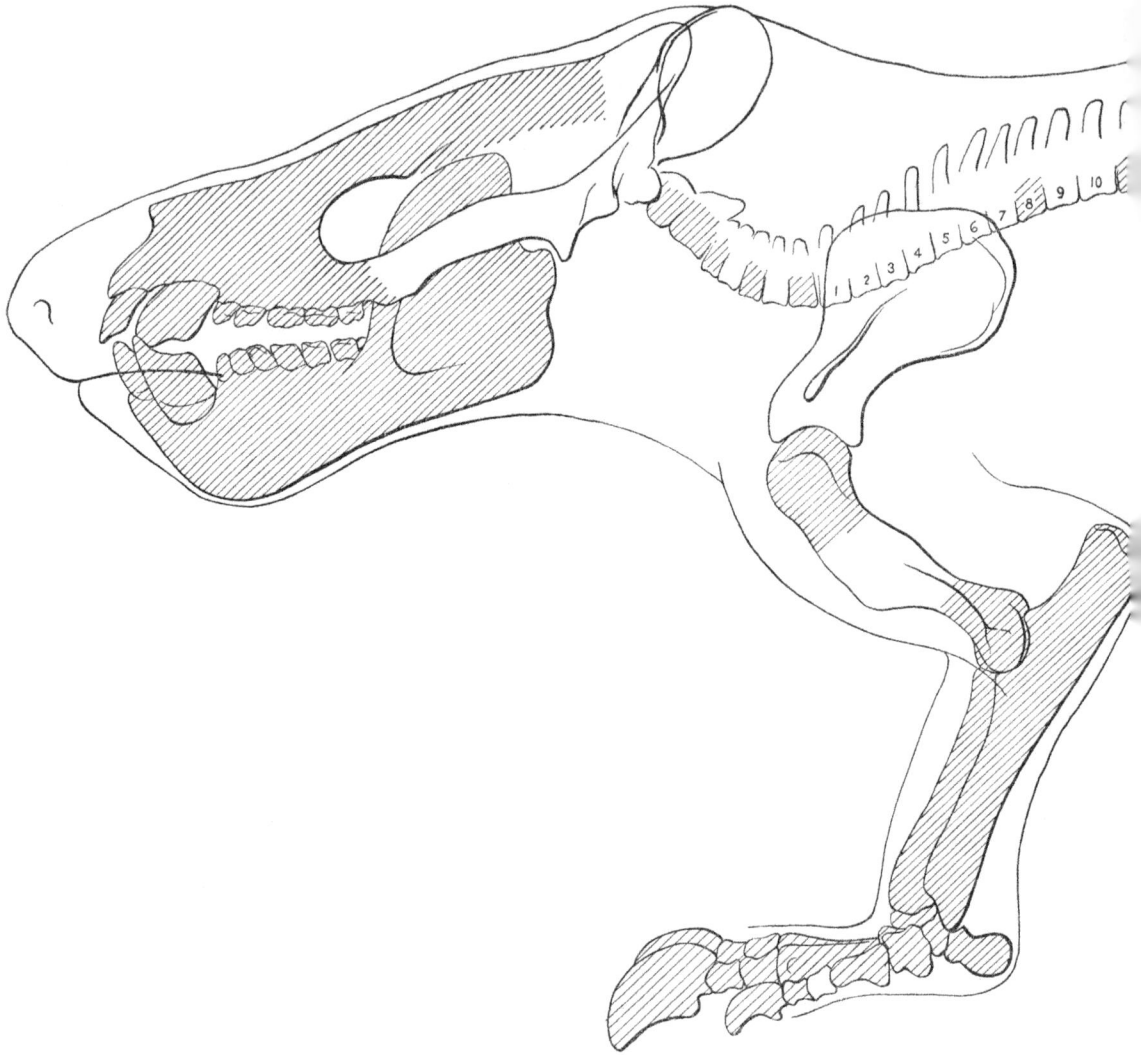

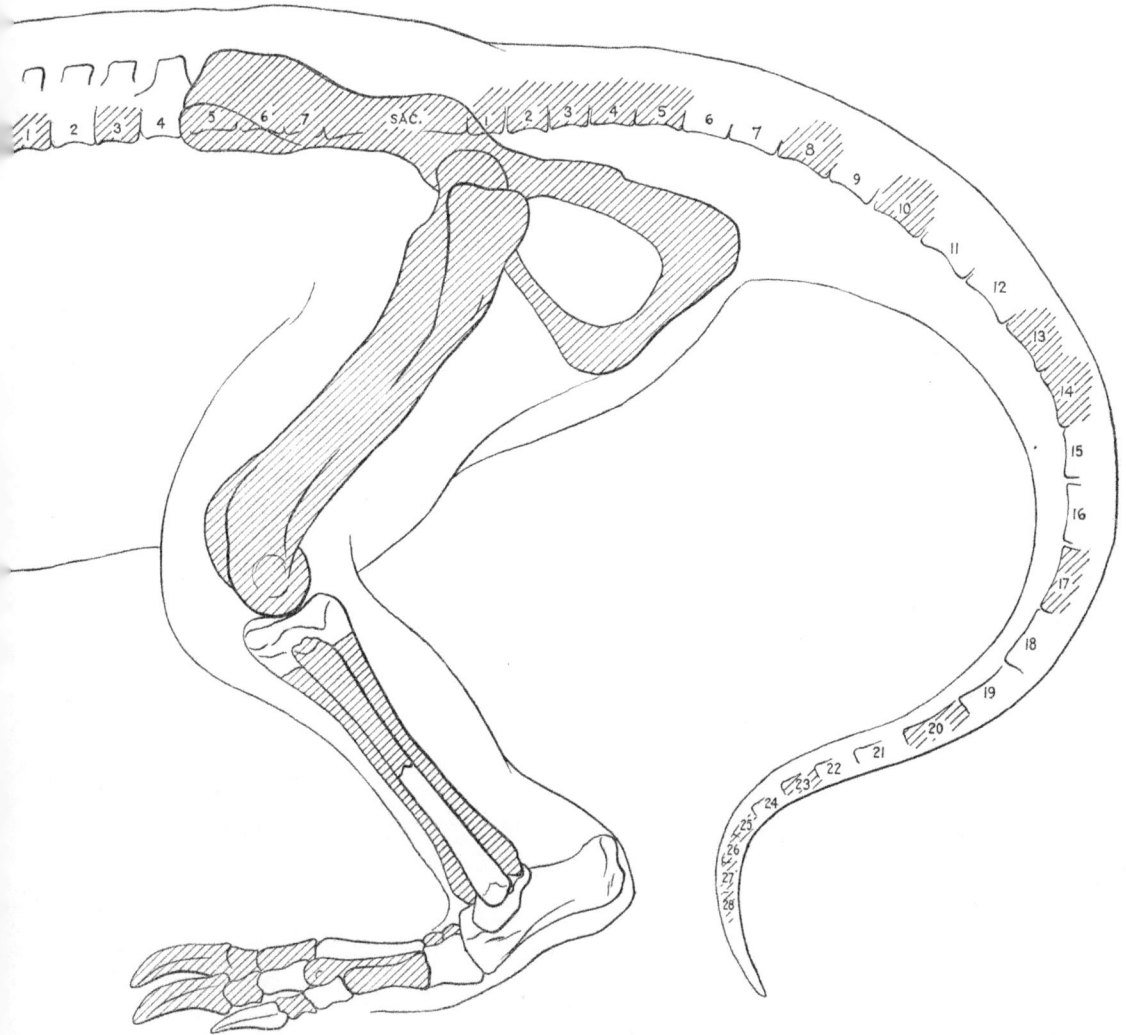

PLATE LXIV

.

PLATE LXV

MAP OF THE SAN JUAN BASIN, NEW MEXICO. Scale, one inch equals twelve miles.

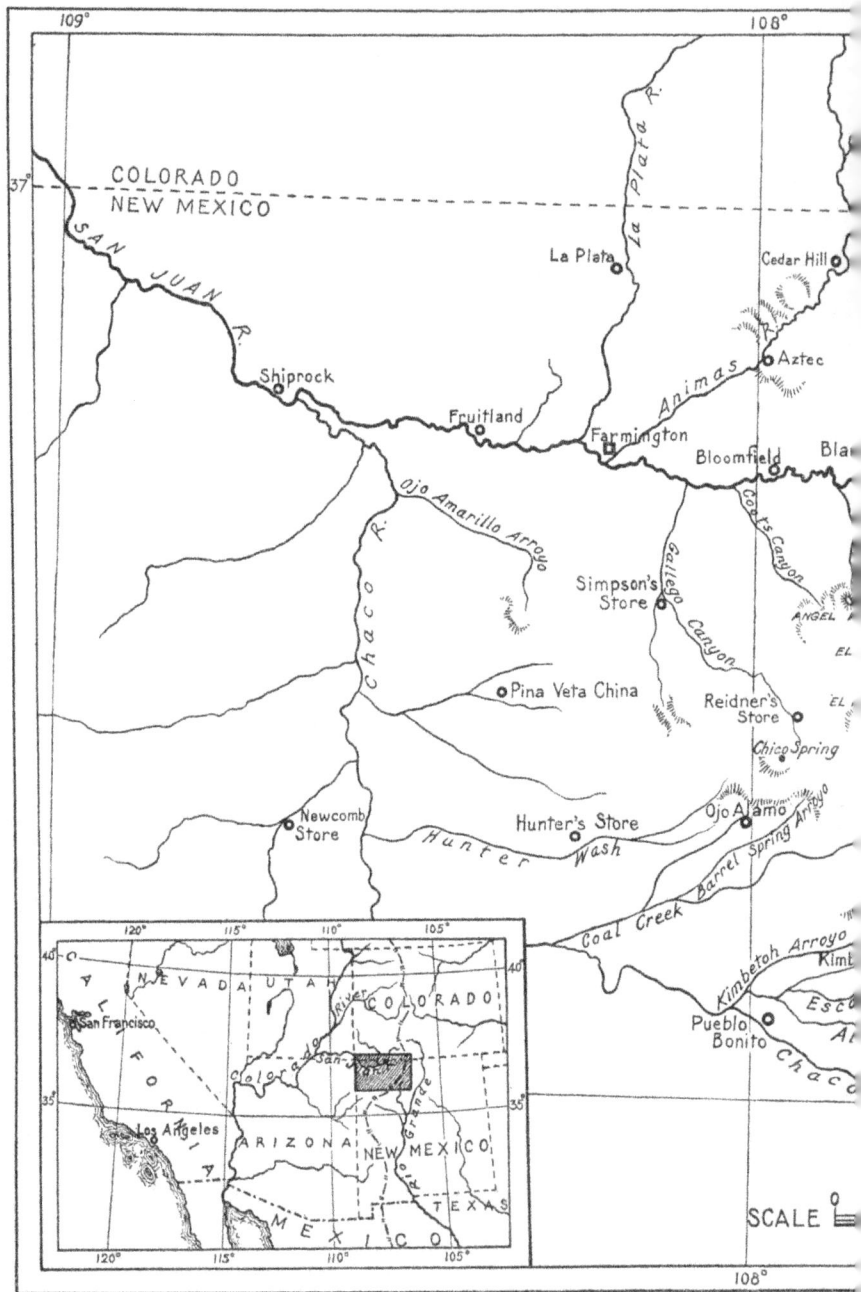

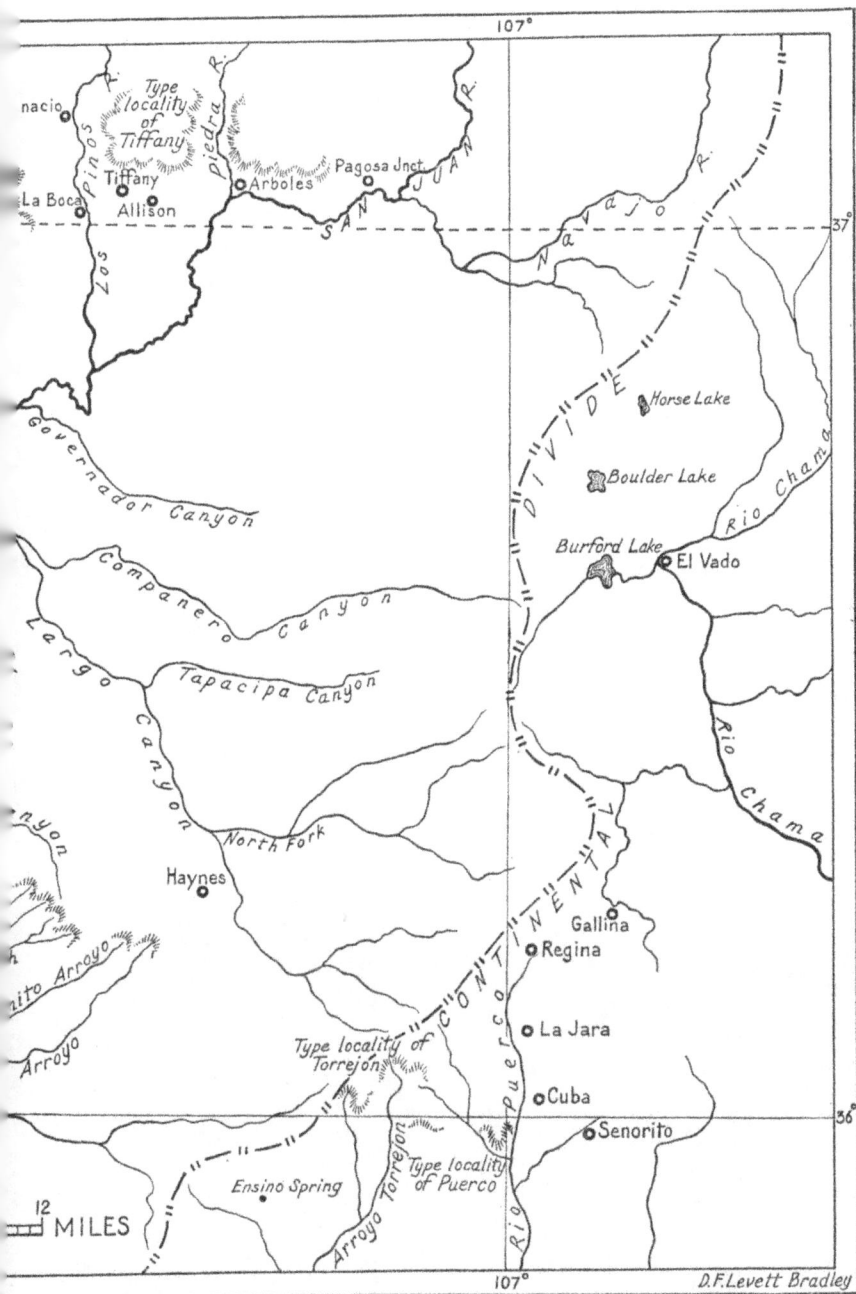

107°

Type
locality
of
Tiffany

nacio

Los Pinos R.

Piedra R.

La Boca

Tiffany

Allison

Arboles

Pagosa Jnct.

SAN JUAN R.

Navajo R.

37°

DIVIDE

Horse Lake

Boulder Lake

Rio Chama

Governador Canyon

Burford Lake

El Vado

Compañero Canyon

Largo Canyon

Tapacipa Canyon

Rio Chama

North Fork

Haynes

CONTINENTAL

Gallina

Regina

nito Arroyo

Rio Puerco

La Jara

Arroyo

Type locality of
Torrejón

Cuba

36°

Senorito

Arroyo Torrejon

Type locality
of Puerco

Rio Puerco

Ensino Spring

12
MILES

107°

D.F.Levett Bradley

INDEX

www.ingramcontent.com/pod-product-compliance
Lightning Source LLC
Chambersburg PA
CBHW081340190326
41458CB00018B/6058